PARALLELOGRAM

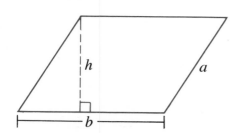

Perimeter: $P = 2a + 2b$
Area: $A = bh$

CIRCLE

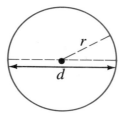

Circumference: $C = \pi d$
$C = 2\pi r$
Area: $A = \pi r^2$

RECTANGULAR SOLID

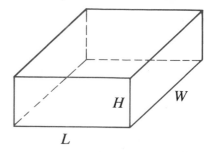

Volume: $V = LWH$
Surface Area: $A = 2HW + 2LW + 2LH$

CUBE

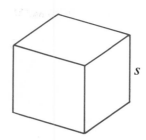

Volume: $V = s^3$
Surface Area: $A = 6s^2$

CONE

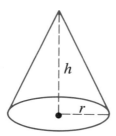

Volume: $V = \frac{1}{3}\pi r^2 h$
Surface Area: $A = \pi r \sqrt{r^2 + h^2}$

RIGHT CIRCULAR CYLINDER

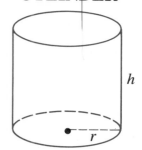

Volume: $V = \pi r^2 h$
Surface Area:
$A = 2\pi rh + 2\pi r^2$

SPHERE

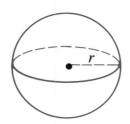

Volume: $V = \frac{4}{3}\pi r^3$
Surface Area: $A = 4\pi r^2$

OTHER FORMULAS

Distance: $d = rt$ (r = rate, t = time)

Percent: $p = br$ (p = percentage, b = base, r = rate)

Temperature: $F = \frac{9}{5}C + 32$ $C = \frac{5}{9}(F - 32)$

Simple Interest: $I = Prt$
(P = principal, r = rate, t = time in years)

Intermediate Algebra

K. Elayn Martin-Gay

University of New Orleans

Prentice Hall, Englewood Cliffs, NJ 07632

Library of Congress Cataloging-in-Publication Data

Martin-Gay, K. Elayn
    Intermediate algebra / K. Elayn Martin-Gay.
        p.   cm.
    Includes index.
    ISBN 0-13-468372-2
    1. Algebra.   I. Title.
QA152.2.M368   1993
512.9—dc20   92-26908
                                               CIP

Executive Editor: Priscilla McGeehon
Editor-in-Chief: Tim Bozik
Development Editor: Steve Deitmer
Production Editors: Ed Thomas / Virginia Huebner / Barbara Grasso Mack
Marketing Manager: Paul Banks
Copy Editor: Bill Thomas
Designer: Judith A. Matz-Coniglio
Design Director: Florence Dara Silverman
Cover Designer: A Good Thing, Inc.
Cover Artist: Mendola, Ltd. / Cliff Spohn
Photo Editor: Lori Morris-Nantz
Prepress Buyer: Paula Massenaro
Manufacturing Buyer: Lori Bulwin
Supplements Editor: Mary Hornby
Editorial Assistant: Marisol L. Torres
Photo Research: Anita Dickhuth

Photograph credits:

Richard Hutchings / Info Edit, p. 0 ● Chervenky / The Image Works, p. 28 ● M. Richards / Photo Edit, p. 86 ● Hewlett Packard, p. 146 ● James D. Wilson / Woodfin Camp & Associates, p. 248 ● Charles Gupton / TSW, p. 302 ● Stacy Pick / Stock Boston, p. 346 ● Myrleen Ferguson / Photo Edit, p. 412 ● Alan Oddie / Photo Edit, p. 466 ● Timothy Eagan / Woodfin Camp & Associates, p. 514

© 1993 by Prentice-Hall, Inc.
A Simon & Schuster Company
Englewood Cliffs, New Jersey 07632

Printed in the United States of America
10   9   8   7   6   5   4   3

ISBN 0-13-468372-2

Prentice-Hall International (UK) Limited, *London*
Prentice-Hall of Australia Pty. Limited, *Sydney*
Prentice-Hall Canada Inc., *Toronto*
Prentice-Hall Hispanoamericana, S.A., *Mexico*
Prentice-Hall of India Private Limited, *New Delhi*
Prentice-Hall of Japan, Inc., *Tokyo*
Simon & Schuster Asia Pte. Ltd., *Singapore*
Editora Prentice-Hall do Brasil, Ltda., *Rio de Janeiro*

To my husband, Clayton, and our sons,
Eric and Bryan

Contents

Appendices *548*

Index *613*

Preface

Why This Book Was Written

This book was written to bridge the gap between a beginning algebra and a college algebra course. It contains, of course, the classic topics required of an intermediate algebra course, but it also emphasizes functions, graphing (linear and nonlinear), and geometric concepts. A conscious effort was made to distinguish the content and level of this book from the beginning algebra volume which precedes it, while still meeting the needs of those students for whom this is the first algebra course taken after high school. In those cases where material is similar in content to that of a beginning algebra course, the example and exercise material reflects a greater level of difficulty and complexity.

How This Book Was Written

Throughout the writing and developing of this book, I had the help of many people. Five instructors who teach courses similar to this one were involved in the actual writing of the text, contributing their ideas for helpful examples, interesting applications, and useful exercises.

Once the first draft was complete, Prentice Hall held a two-day reviewer's conference with four reviewers, the author, and editors from Prentice Hall. We spent many hours going over the manuscript with a fine-tooth comb, refining the project's focus and enhancing its pedagogical value.

Finally, a full-time development editor worked with me to make the writing style as clear as possible while still retaining the mathematical integrity of the content.

Key Content Features

In order to distinguish this course from the preceding beginning algebra course, graphing and functions are introduced in Chapter 3. This provides a solid foundation of these very important concepts for students going on to college algebra and other mathematics courses. Too often, because graphing and functions are covered late in the semester, they are not given the proper emphasis due to lack of time.

In addition, I have included as much coverage of geometry as possible. Since many students have not taken geometry in high school, this may be their first (and only) exposure to geometry. I have tried to include those geometry concepts which are most important to a student's understanding of algebra, and to include geometric examples, exercises, and applications as much as possible. A review of geometric figures and of angles, lines, and special triangles is covered in the appendix material.

Applications are emphasized by devoting single sections to them (such as Section 6.6 on applications of rational expressions) as well as exercises throughout the book.

Key Pedagogical Features

Exercise Sets

Each exercise set is divided into two parts. Both parts contain graded problems. The first part is carefully keyed to worked examples in the text. Once a student has gained confidence in a skill, the second part contains exercises not keyed to examples. There are ample exercises throughout this book, including end-of-chapter reviews, tests, and cumulative reviews. In addition, each exercise set contains one or more of the following features:

Mental Mathematics.   These problems are found at the beginning of an exercise set. They are mental warmups that reinforce concepts found in the accompanying section and increase students' confidence before they tackle an exercise set. By relying on their own mental skills, students learn not only confidence in themselves, but also number sense and estimation.

Skill Review.   At the end of each section after Chapter 1, these problems are keyed to sections and review concepts covered earlier in the text.

Writing in Mathematics.   These writing exercises can be used to check a student's comprehension of an algebraic concept. They are located at the end of many exercise sets, where appropriate. Guidelines produced by the National Council of Teachers of Mathematics and other professional groups recommend incorporating writing in mathematics courses to reinforce concepts.

Applications.   This book contains a wealth of practical applications found in worked-out examples and exercise sets.

A Look Ahead.   These are examples and problems similar to those found in college algebra books. "A Look Ahead" is presented as a natural extension of the material and contains an example followed by advanced exercises. I strongly suggest that any student who plans to take a college algebra course work these problems.

Calculator Boxes.   Calculator Boxes are placed appropriately throughout the text to instruct students on proper use of the calculator. These boxes, entirely optional, contain examples and exercises to reinforce the material introduced.

Critical Thinking.   Each chapter opens with a critical thinking problem. The student does not need to work this problem immediately. Rather, the skills needed to solve each critical thinking problem are developed throughout the chapter. At the close of the chapter, the student is asked to apply the skills he or she has learned to answer the problem. The critical thinking problems, based on real-life situations relevant to students, require thinking "beyond the numbers." Many times, a critical thinking problem may have more than one answer. These problems are excellent for cooperative learning situations. You can assign the problem to a group of students who can then present their solution to the rest of the class for discussion.

Helpful Hint Boxes.   These boxes contain practical advice on problem-solving. Helpful Hints appear in the context of material in the chapter, and give students extra help in understanding and working problems. They are set off in a box for easy referral.

Chapter Glossary and Summary. Found at the end of each chapter, the chapter glossary contains a list of definitions of new terms introduced in the chapter, and the summary contains a list of important rules, properties, or steps introduced in the chapter.

Chapter Review and Test. The end of each chapter contains a review of topics introduced in the chapter. These review problems are keyed to sections. The chapter test is not keyed to sections.

Cumulative Review. Each chapter after the first contains a cumulative review. Each problem contained in the cumulative review is actually an earlier worked example in the text which is referenced in the back of the book along with the answer. Students who need to see a complete worked-out solution with explanation can do so by turning to the appropriate example in the text.

Supplements

For the Instructor

The following supplements are available for instructors who adopt the text:

Annotated Instructor's Edition has answers to all exercises displayed on the same page with the exercises.

Instructor's Manual with Tests, Syllabus and Instructor's Disk contains 9 tests per chapter (5 are free-response, 4 are multiple choice), suggested syllabi and homework assignments, and an ASCII disk which allows customization of syllabi.

Instructor Solutions Manual with even-numbered solutions.

IPS Testing (IBM) generates test questions and drill worksheets from algorithms keyed to the learning objectives in the book. Available free upon adoption in 3.5″ and 5.25″ IBM formats.

PHTestmanager testing (IBM and Macintosh). A bank of test items designed specifically for the book, in both free-response and multiple choice form; fully editable for flexible use.

Test Item File contains a hard copy of test questions on PHTestmanager.

For the Student

The following supplements are available for students:

Student Solutions Manual with odd-numbered solutions and solutions to all chapter tests and cumulative tests.

Math Master Tutor software (IBM and Mac) provides text-specific tutorial, exercises graduated in diffuculty which are generated new each time, fully worked-out examples, and a timed quiz.

Videotapes with class lectures by the author, closely keyed to the book itself.

Acknowledgments

Writing this book has been a humbling experience, an effort requiring the help of many more people than I originally imagined. I will attempt to thank them here.

First, I would like to thank my husband, Clayton. Without his constant encouragement, this project would not have become a reality. I would also like to thank my children, Eric and Bryan, for eating my burnt bacon. Writing a book while raising two small children is an experience that requires an infinite amount of patience and a good sense of humor.

I would like to thank my extended family for their invaluable help. Their contributions are too numerous to list. They are Peter, Karen, Michael, Christopher, Matthew, and Jessica Callac; Stuart, Earline, Melissa, and Mandy Martin; Mark Martin; Barbara and Leo Miller; and Jewett Gay.

I would like to thank the following excellent writers for their work. Creating the first draft manuscript would not have been possible without them. Each of them—Cynthia Miller, Lea Campbell, Ned Schillow, Myrna Mitchell, Cathy Pace—provided invaluable help.

I would like to thank the following reviewers for their suggestions:

Robert Baer, *Miami University-Hamilton*
Jack Barone, *City University of New York, Bernard M. Baruch College*
Sandra Belew, *Mesa College*
James Blackburn, *Tulsa Junior College*
B.P. Bockstege, *Broward Community College*
Dee Ann Christianson, *The University of the Pacific*
Juan Gatica, *University of Iowa*
Anita Harkness, *Louisiana State University*
Peter Herron, *Suffolk Community College*
William Magliaro, *Bucks County Community College*
Kathryn T. McClellan, *Tarrant County Community College*
Cynthia Miller, *Georgia State University*
Pat Mower, *University of North Dakota*
James W. Newsom, *Tidewater Community College*
Linda Padilla, *Joliet Junior College*
Carolyn L. Pinchback, *The University of Central Arkansas*
Jane Pinnow, *Franklin University*
William Radulovich, *Florida Community College @ Jacksonville*
Janet Ritchie, *State University of New York, College @ Old Westbury*
Eddie Robinson, *Cedar Valley College*
Ken Seydel, *Skyline College*
Ann Skeath, *University of Pennsylvania*
Leslie Tanner, *The College of Idaho*
Steven Terry, *Ricks College*

Cheryl Roberts did an excellent job of working with the galley proofs and page proofs, insuring that they be as error-free as possible. Assisting her were Richard Semmler, Ara Sullenberger, and Fran Hopf. Joe May and Susan Friedman provided the answers and solutions, and contributed to the overall accuracy as well.

Finally, I would like to thank Laurie Golson, Steve Deitmer, and Christine Peckaitis for their invaluable contributions; production editors Virginia Huebner, Ed Thomas, and Barbara Grasso Mack; and executive editor Priscilla McGeehon, who is always there when I need her. Paul Banks's efforts as marketing manager, even before the book was published, are much appreciated.

Intermediate Algebra

CHAPTER **1**

Review of Real Numbers

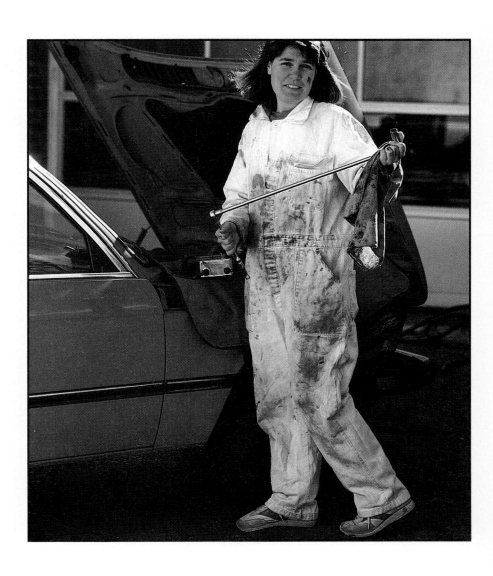

To buy or not to buy is the consumer's constant question. You may ask, for example, whether it is better to suffer the pings and rattles of an older car or take a loan against a new car. (See Critical Thinking, page 24.)

INTRODUCTION

In arithmetic, we add, subtract, multiply, divide, raise to powers, and take roots of **numbers.** In algebra, we add, subtract, multiply, divide, raise to powers, and take roots of **variables.** Letters, such as x, used to represent numbers are called variables. Understanding algebraic operations on variables depends on your proficiency and understanding of arithmetic operations. This chapter reviews these operations, beginning with a review of the sets of real numbers to which arithmetic operations apply.

1.1
Sets of Numbers

OBJECTIVES

Tape 1

1   Use roster and set builder notation to write sets.

2   Define equality of sets and subset of a set.

3   Identify common sets of numbers.

---

1   A **set** is a collection of objects. Individual objects in the set are called **elements** of the set, and **braces** { } are used to enclose the list of elements in the set. Capital letters are often used to name sets. Here are some examples of sets:

$$A = \{1, 2, 3, 4\}, \qquad B = \{\text{shirt, socks, shoes}\}, \qquad C = \{a, b, d, e, g, h, i\}$$

These sets above are called **finite sets** since they have a finite number of elements. Sets can also have an infinite number of elements. For example, the set of **natural numbers** N is an **infinite set:**

$$N = \{1, 2, 3, 4, 5, \ldots\}$$

The three dots, called an ellipsis, mean to continue in the same pattern. The symbol $\in$ is used to denote that an element is in a particular set. The symbol $\in$ is read as "is an element of." For example, the true statement

$$3 \text{ is an element of } \{1, 2, 3, 4, 5\}$$

can be written in symbols as:

$$3 \in \{1, 2, 3, 4, 5\}$$

The symbol $\notin$ is read as "is not an element of." In symbols, we write the true statement "p is not an element of $\{a, 5, g, j, q\}$" as

$$p \notin \{a, 5, g, j, q\}$$

HELPFUL HINT

If A is a set containing the number 5, it is true that $5 \in A$, read as "5 is an element of set A." It is **not** true that $A \in 5$. $A \in 5$ reads as "set A is an element of 5," which is false.

1

EXAMPLE 1   Determine whether each statement is true or false.

   a. $3 \in \{1, 2, 3\}$ **b.** $7 \notin \{1, 2, 3\}$ **c.** $\{2\} \in \{1, 2, 3\}$

Solution: **a.** True, since 3 is an element of the set $\{1, 2, 3\}$.

   b. True, since 7 is not an element of the set $\{1, 2, 3\}$.

   c. False. The **set** $\{2\}$ is not an **element** of the set.   ∎

All sets mentioned so far have had their elements listed. When the elements of a set are listed, the set is written in **roster** form. A set can also be written in **set builder notation,** which describes the elements but does not list them. For example, the set $P = \{x \mid x \text{ is a continent}\}$ is written in set builder notation. This set is read as

$$P = \{ \underline{x} \mid \underline{x \text{ is a continent}} \}$$

"the set of all x such that x is a continent"

The same set P in roster form is

$P = \{\text{Africa, Antarctica, Asia, Australia, North America, South America, Europe}\}$

The set $\{x \mid x \text{ is a natural number less than 3}\}$ written in roster form is $\{1, 2\}$. When a set contains no elements at all, we call it the **empty set** or the **null set.** The symbol $\varnothing$ or $\{\ \}$ can be used to denote the empty set. For example,

$$\{x \mid x \text{ is a month with 32 days}\} \text{ is } \varnothing$$

because no month has 32 days. The set has no elements.

HELPFUL HINT

Use $\{\ \}$ or $\varnothing$ to write the empty set. $\{0\}$ is **not** the empty set because it has one element: 0.

EXAMPLE 2   List the elements in each set.

   a. $\{x \mid x \text{ is a natural number between 1 and 6}\}$

   b. $\{x \mid x \text{ is a natural number greater than 100}\}$

Solution: **a.** $\{2, 3, 4, 5\}$

   b. $\{101, 102, 103, \ldots\}$ ∎

2   Two sets are **equal** if they have exactly the same elements. The order in which the elements of a set are written does not matter. If $D = \{3, 5, 7\}$ and $F = \{7, 3, 5\}$, then

$$D = F$$

Set X is a **subset** of set Y, written as $X \subseteq Y$, if all the elements of X are also elements of Y. For example, $\{2, 3\}$ is a subset of $\{2, 3, 5, 6\}$, and we can write this in symbols as

$$\{2, 3\} \subseteq \{2, 3, 5, 6\}$$

is a subset of

The symbol $\nsubseteq$ means "is not a subset of." The set $\{2, 3, 4, 6\}$ is not a subset of the set $\{3, 4, 5, 6, 7\}$ since 2 is an element of $\{2, 3, 4, 6\}$, but is not an element of $\{3, 4, 5, 6, 7\}$.

We can write this in symbols as

$$\{2, 3, 4, 6\} \nsubseteq \{3, 4, 5, 6, 7\}$$

is not a subset of

Note that the empty set is a subset of every set. This is because there is no element in the empty set, so "every element" of the empty set is an element of any other set. It is also true that every set is a subset of itself.

EXAMPLE 3   Determine whether the following statements are true or false.

    a. $\{1, 2\} \subseteq \{1, 2, 3\}$ **b.** $\{1, 2, 3\} \subseteq \{1, 2, 3\}$ **c.** $\varnothing \subseteq \{1, 2, 3\}$

    d. $\{2, 3\} = \{3, 2\}$ **e.** $3 \subseteq \{1, 2, 3\}$ **f.** $\{1, 2, 3, 4\} \subseteq \{2, 3\}$

Solution: **a.** True, since all elements of $\{1, 2\}$ are also elements of $\{1, 2, 3\}$.

    b. True, since every set is a subset of itself.

    c. True, since the empty set is a subset of any other set.

    d. True.

    e. False, since 3 is not a set. A true statement is "3 is an element of $\{1, 2, 3\}$" or, in symbols, $3 \in \{1, 2, 3\}$.

    f. False, since 1 and 4 are elements of $\{1, 2, 3, 4\}$ but are not elements of $\{2, 3\}$. ■

3   Sets of numbers can be visualized on a number line. To construct a number line, draw a line and label a point 0 with which we associate the number 0. This point is called the **origin.** Choose a point to the right of 0 and label it 1. The distance from 0 to 1 is called the **unit distance** and can be used to locate more points. The positive numbers lie to the right of the origin, and the negative numbers lie to the left of the origin. The number 0 is neither positive nor negative.

Zero

Negative numbers   Positive numbers

1 unit   1 unit   1 unit   1 unit   1 unit   1 unit

-3 -2 -1 0 1 2 3

A number is graphed on a number line by shading the point on the number line that corresponds to the number. Some common sets of numbers and their graphs include:

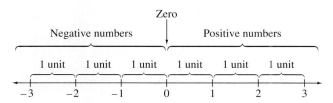

Real numbers: $\{x \mid x \text{ is a point on the number line}\}$

Natural numbers: $\{1, 2, 3, \ldots\}$

Whole numbers: $\{0, 1, 2, 3, \ldots\}$

Integers: $\{\ldots, -3, -2, -1, 0, 1, 2, 3, \ldots\}$

Rational numbers: $\left\{\dfrac{a}{b} \mid a \text{ and } b \text{ are integers and } b \neq 0\right\}$

Irrational numbers: $\{x \mid x \text{ is a real number and } x \text{ is not a rational number}\}$

Notice that integers are rational numbers since each can be written as a fraction with a denominator of 1.

$$3 = \frac{3}{1}, \qquad 0 = \frac{0}{1}, \qquad -8 = \frac{-8}{1}$$

Some square roots are rational numbers and some are irrational numbers. For example, $\sqrt{2}$, $\sqrt{3}$, and $\sqrt{7}$ are irrational numbers while $\sqrt{25}$ is a rational number because $\sqrt{25} = 5$. The number π is an irrational number. To help you make the distinction between rational and irrational numbers, here are a few examples of each.

Rational	*Irrational*
$-\dfrac{2}{3}$	$\sqrt{5}$
$\sqrt{36} \left(\sqrt{36} = \dfrac{6}{1}\right)$	$\dfrac{\sqrt{6}}{7}$
$5 \left(5 = \dfrac{5}{1}\right)$	$-\sqrt{13}$
$0 \left(0 = \dfrac{0}{1}\right)$	$\dfrac{-\sqrt{17}}{5}$
$1.2 \left(1.2 = \dfrac{12}{10}\right)$	π
$3\dfrac{7}{8}$	

Every rational number can be written as a decimal that either repeats or terminates. For example,

$$\frac{1}{2} = 0.5 \qquad\qquad \frac{5}{4} = 1.25$$

$$\frac{2}{3} = 0.6666666\ldots \qquad\qquad \frac{1}{11} = 0.090909\ldots$$

The natural numbers, whole numbers, integers, rational numbers, and irrational numbers are each a subset of the set of real numbers. The relationships between these sets of numbers are shown in the following diagram.

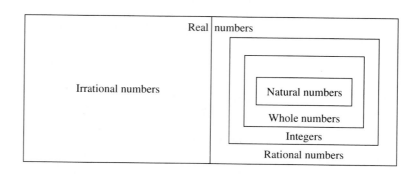

EXAMPLE 4   Determine whether the following statements are true or false.

    a. 3 is a real number.   **b.** $\frac{1}{5}$ is an irrational number.

    c. Every rational number is an integer.

Solution: **a.** True.

    b. False. The number $\frac{1}{5}$ is a rational number, since it is in the form $\frac{a}{b}$ with a and b integers and $b \neq 0$.

    c. False. The number $\frac{2}{3}$, for example, is a rational number, but it is not an integer. ∎

EXERCISE SET 1.1

Use the following sets for Exercises 1 through 21: $A = \{3, 4, 5\}$, $B = \{2, 4, 8\}$, $C = \{1, 2, 3, \ldots\}$, $D = \{2, 4, 6, \ldots\}$, $E = \{\ \}$, $F = \{x \mid x$ is an integer$\}$, $G = \{x \mid x$ is a positive even integer$\}$. Place $\in$ or $\notin$ in the space provided to make each statement true. See Example 1.

1. 5 A **2.** 8 B **3.** 10 C
4. 0 E **5.** $\varnothing$ E **6.** $\{7\}$ F

Place $\subseteq$ or $\not\subseteq$ in the space provided to make each statement true. See Example 3.

7. A C **8.** B C **9.** C A
10. D G **11.** G D **12.** E F

Determine whether each statement is true or false for the sets A, B, C, D, E, F, and G defined above. See Examples 1 and 3.

13. $14 \in D$ **14.** $-25 \in F$ **15.** $\{3, 0.5, -7\} \subseteq F$ **16.** $G \subseteq C$
17. $G \subseteq F$ **18.** $\{0\} = E$ **19.** $\{\varnothing\} = E$ **20.** $D = G$
21. $C = F$

List the elements in each set. See Example 2.

22. $\{x \mid x$ is a natural number less than 6$\}$
23. $\{x \mid x$ is a natural number greater than 6$\}$
24. $\{x \mid x$ is a natural number between 10 and 17$\}$
25. $\{x \mid x$ is an odd natural number$\}$
26. $\{x \mid x$ is a whole number that is not a natural number$\}$
27. $\{x \mid x$ is a natural number less than 1$\}$
28. $\{x \mid x$ is an even whole number less than 9$\}$
29. $\{x \mid x$ is an odd whole number less than 9$\}$

Determine whether the following statements are true or false. See Example 4.

30. Every integer is a rational number.
31. The number 0 is a natural number.
32. Every irrational number is also a real number.
33. Every natural number is also a whole number.
34. The number $\frac{2}{3}$ is an irrational number.
35. Some rational numbers are also irrational numbers.

Graph each set on a number line.

36. $\{0, 2, 4, 6\}$ **37.** $\{1, 3, 5, 7\}$
38. $\{-1, -2, -3\}$ **39.** $\{-2, -6, -10\}$
40. $\left\{\frac{1}{2}, \frac{2}{3}\right\}$ **41.** $\left\{\frac{1}{4}, \frac{1}{3}\right\}$

List the elements of the set $\{3, 0, \sqrt{7}, \sqrt{36}, \frac{2}{5}, -134\}$ *that are also elements of the given set.*

42. Whole numbers

43. Integers

44. Natural numbers

45. Rational numbers

46. Irrational numbers

47. Real numbers

Place ∈ *or* ∉ *in the space provided to make each statement true.*

48. -11 ___ $\{x \mid x \text{ is an integer}\}$

49. -6 ___ $\{2, 4, 6, \ldots\}$

50. 0 ___ $\{x \mid x \text{ is a positive integer}\}$

51. 12 ___ $\{1, 2, 3, \ldots\}$

52. 12 ___ $\{1, 3, 5, \ldots\}$

53. $\frac{1}{2}$ ___ $\{x \mid x \text{ is an irrational number}\}$

54. 0 ___ $\{1, 2, 3, \ldots\}$

55. 0 ___ $\{x \mid x \text{ is a natural number}\}$

Place ⊆ *or* ⊄ *in the space provided to make each statement true.*

56. $\{2, 4, 6\}$ ___ $\{6, 2, 4\}$

57. $\{ \ \}$ ___ $\{-3, -2, -1\}$

58. $\{2, 4, 8\}$ ___ $\{3, 4, 5\}$

59. $\{3, 4, 5\}$ ___ $\{2, 4, 8\}$

60. $\{13, 14, 15\}$ ___ $\{1, 2, 3, \ldots\}$

61. $\{12, 14, 18\}$ ___ $\{2, 4, 6, \ldots\}$

62. $\{0, 16, 82\}$ ___ $\{x \mid x \text{ is a positive even integer}\}$

63. $\{0\}$ ___ $\{x \mid x \text{ is an integer}\}$

Writing in Mathematics

64. In your own words, explain why every natural number is also a rational number, but not every rational number is a natural number.

65. Explain why the empty set is a subset of every set.

66. Explain why every set is a subset of itself.

1.2
Properties of Real Numbers

Tape 1

OBJECTIVES

1   Use the basic operation symbols.

2   Use properties of real numbers.

1   The four basic operations on real numbers are addition, subtraction, multiplication, and division. Symbols and their meanings for these operations are summarized next.

Operation	Expression	Symbols
Addition:	The sum of a and b	$a + b$
Subtraction:	The difference of a and b	$a - b$
Multiplication:	The product of a and b	$a \times b, \ a \cdot b, \ ab, \ a(b), \ (a)b$
Division:	The quotient of a and b	$\frac{a}{b}, \ a \div b$

EXAMPLE 1   Write each sentence using mathematical symbols.

    a. The sum of x and 5 is 20.

    b. Two times the sum of 3 and y is 4.

    c. Subtract 8 from x and the difference is the product of 2 and x.

    d. The quotient of z and 9 is 3 times the difference of z and 5.

Solution: **a.** The sum of x and 5 can be written as "$x + 5$" and "is" means "equals" in this sentence, so we write $x + 5 = 20$.

    b. $2(3 + y) = 4$

    c. $x - 8 = 2x$

    d. $\dfrac{z}{9} = 3(z - 5)$ ∎

2   To review properties of real numbers, we begin with properties of equality.

Properties of Equality

For all real numbers a, b, and c:

    1. $a = a$ Reflexive property

    2. If $a = b$, then $b = a$ Symmetric property

    3. If $a = b$ and $b = c$, then $a = c$ Transitive property

---

Example	*Property*
$5p = 5p$	Reflexive property
If $2 = x$, then $x = 2$	Symmetric property
If $x = y$ and $y = 7$, then $x = 7$	Transitive property

If we want to write in symbols that two numbers are not equal, we can use the symbol $\neq$, which means "is not equal to." For example,

$$3 \neq 2$$

If two real numbers are not equal, then one is larger than the other. The symbol $>$ means "is greater than." Since 3 is greater than 2, we write

$$3 > 2$$

Also, the symbol $<$ means "is less than." Since 2 is less than 3, we can write

$$2 < 3$$

In general, we can use the number line to compare real numbers. For two real numbers a and b, we say **a is less than b** or **$a < b$** if on the number line the point representing a lies to the left of the point representing b. Also, if b is to the right of a on the number line, then **b is greater than a, or $b > a$.**

Notice that if $a < b$, then also $b > a$.

The **trichotomy property** for real numbers assures us that, when given two real numbers a and b, they will either be equal or one will be greater than the other.

Trichotomy Property

If a and b are real numbers, then exactly one of these statements is true:
$$a = b, \quad a < b, \quad \text{or} \quad a > b$$

EXAMPLE 2   Insert $<$, $>$, or $=$ between each pair of numbers to form a true statement.

a. $-1 \quad -2$ **b.** $\dfrac{12}{4} \quad 3$ **c.** $-5 \quad 0$

Solution: **a.** $-1 > -2$ since -1 lies to the right of -2 on the number line.

b. $\dfrac{12}{4} = 3$.

c. $-5 < 0$ since -5 lies to the left of 0 on the number line.   ■

HELPFUL HINT

When inserting the $>$ or $<$ symbol, think of the symbols as arrowheads and "point" toward the smaller number. The resulting inequality will be true.

In addition to $<$ and $>$, there are the inequality symbols $\le$ and $\ge$. The symbol
$$\le \text{ means "is less than or equal to"}$$
and the symbol
$$\ge \text{ means "is greater than or equal to"}$$
Also, $\ne$ means "is not equal to."

For example, the following are true statements:

$$10 \le 10 \quad \text{since} \quad 10 = 10$$
$$-8 \le 13 \quad \text{since} \quad -8 < 13$$
$$-5 \ge -5 \quad \text{since} \quad -5 = -5$$
$$-7 \ge -9 \quad \text{since} \quad -7 > -9$$

EXAMPLE 3   Write each sentence using mathematical symbols.
a. The sum of 5 and y is greater than or equal to 7.
b. 11 is not equal to z.
c. 20 is less than the difference of 5 and x.

Solution: **a.** $5 + y \ge 7$ **b.** $11 \ne z$ **c.** $20 < 5 - x$ ■

The properties of real numbers, summarized next, are a collection of statements that are true for all real numbers.

Properties of Real Numbers

Let a, b, and c be real numbers.

Closure property: $a + b$ and $a \cdot b$ are also real numbers

	Addition	*Multiplication*
Commutative properties:	$a + b = b + a$	$a \cdot b = b \cdot a$
Associative properties:	$a + (b + c) = (a + b) + c$	$a \cdot (b \cdot c) = (a \cdot b) \cdot c$
Identities:	The additive identity is 0. $a + 0 = 0 + a = a$	The multiplicative identity is 1. $a \cdot 1 = 1 \cdot a = a$
Inverse properties:	For every number a, there is a number $-a$ called the **additive inverse** or **opposite** of a such that $a + (-a) = -a + a = 0$	For every nonzero number a, there is a number $1/a$ called the **multiplicative inverse** or the **reciprocal** of a such that $a \cdot \dfrac{1}{a} = \dfrac{1}{a} \cdot a = 1$

Distributive property: $a(b + c) = ab + ac$

The distributive property can also be extended as

$$a(b + c + d + \cdots + g) = ab + ac + ad + \cdots + ag$$

EXAMPLE 4   Name the property illustrated by each statement.

a. $(5 + 4) + y = 5 + (4 + y)$ **b.** $2 \cdot 3 = 3 \cdot 2$ **c.** $z + 0 = z$ **d.** $\dfrac{1}{3} \cdot 3 = 1$

e. $y \cdot 1 = y$ **f.** $2 + (-2) = 0$ **g.** $2(x + y) = 2x + 2y$

Solution: **a.** Associative property of addition   **b.** Commutative property of multiplication
c. Additive identity   **d.** Multiplicative inverse property
e. Multiplicative identity   **f.** Additive inverse property   **g.** Distributive property ∎

EXAMPLE 5   Write the reciprocal or multiplicative inverse of each.

a. 5   **b.** -2 **c.** $\dfrac{4}{7}$

Solution: **a.** The reciprocal of 5 is $\dfrac{1}{5}$.   **b.** The reciprocal of -2 is $\dfrac{1}{-2}$ or $-\dfrac{1}{2}$.

c. The reciprocal of $\dfrac{4}{7}$ is $\dfrac{1}{\frac{4}{7}} = 1 \cdot \dfrac{7}{4} = \dfrac{7}{4}$. ∎

Another property of real numbers has to do with the number 0. The multiplication property of 0 states that the product of 0 and any real number a is 0.

> **Multiplication Property of Zero**
>
> $$a \cdot 0 = 0 \cdot a = 0$$

For example, $5 \cdot 0 = 0$, $0 \cdot x = 0$, and $-3 \cdot 0 = 0$.

Two numbers that are the same distance from 0 on the number line but on opposite sides of 0 are called **opposites** or **additive inverses.**

The opposite of 6 is -6.

The opposite of $\dfrac{2}{3}$ is $-\dfrac{2}{3}$.

The opposite of -4 is 4.

We stated earlier that the opposite or additive inverse of a number a is $-a$. This means that the opposite of -4 is $-(-4)$. But we stated above that the opposite of -4 is 4. This means that $-(-4) = 4$ and leads us to the following **double negative property.**

> **Double Negative Property**
>
> If a is a real number, then $-(-a) = a$.

EXAMPLE 6 Write the additive inverse or the opposite of each.

a. 8 **b.** $\dfrac{1}{5}$ **c.** -9

Solution: **a.** The opposite of 8 is -8. **b.** The opposite of $\dfrac{1}{5}$ is $-\dfrac{1}{5}$.

c. The opposite of -9 is $-(-9) = 9$. ∎

EXERCISE SET 1.2

Write each sentence using mathematical symbols. See Example 1.

1. The product of 4 and c is 7.

2. The sum of 10 and x is -12.

3. 3 times the sum of x and 1 is 7.

4. 9 times the difference of 4 and m is 1.

5. The quotient of n and 5 is 4.

6. The quotient of 8 and y is 3.

Insert $<$, $>$, or $=$ in the space provided to form a true statement. See Example 2.

7. 0 _____ -2

8. -5 _____ 0

9. $\dfrac{12}{3}$ _____ $\dfrac{8}{2}$

10. $\dfrac{20}{5}$ _____ $\dfrac{20}{4}$

11. -7 _____ -8

12. -14 _____ -11

Write each statement using mathematical symbols. See Example 3.

13. The product of 7 and x is less than or equal to -21.

14. 10 subtracted from x is greater than 0.

15. The sum of -2 and x is not equal to 10.

16. The quotient of y and 3 is less than or equal to y.

17. Twice the difference of x and 6 is greater than the reciprocal of 11.

18. Four times the sum of 5 and x is not equal to the opposite of 15.

Name the property illustrated. See Example 4.

19. $7 \cdot 1 = 7$

20. $7 + (-7) = 0$

21. $7 + m = m + 7$

22. $zw = wz$

23. $(3 + x) + 5 = 5 + (3 + x)$

24. $(ab)c = c(ab)$

25. $2(x + y + 5) = 2x + 2y + 10$

26. $b + 0 = b$

27. $4 \cdot \dfrac{1}{4} = 1$

28. $-\dfrac{2}{3} + \dfrac{2}{3} = 0$

Find the reciprocal (or multiplicative inverse) of each number if one exists. See Example 5.

29. 5

30. -8

31. $-\dfrac{1}{4}$

32. $\dfrac{1}{9}$

33. 0

34. $\dfrac{0}{6}$

35. $-\dfrac{7}{8}$

36. $\dfrac{23}{5}$

Find the opposite (or additive inverse) of each number. See Example 6.

37. -6

38. -7.8

39. $\dfrac{4}{7}$

40. $\dfrac{9}{5}$

41. $-\dfrac{2}{3}$

42. $-\dfrac{14}{3}$

43. 0

44. 10.3

Write each sentence using mathematical symbols.

45. 7 subtracted from y is 6.

46. The sum of z and w is 12.

47. The product of 3 and y is less than -17.

48. 5 less x is not equal to 1.

49. Twice the difference of x and 6 is -27.

50. 5 times the sum of 6 and y is -35.

51. Subtract 4 from x and the difference is greater than or equal to 3 times x.

52. 4 times x subtracted from 10 is less than or equal to the opposite of x.

53. 6 subtracted from twice y is the reciprocal of 8.

54. 7 subtracted from the product of 5 and n is the opposite of n.

55. The sum of n and 5, divided by 2, is greater than twice n.

56. The product of 8 and x, divided by 5, is less than 3 more than x.

Use the distributive property to find the product.

57. $3(x + 5)$

58. $7(y + 2)$

59. $8(a + b)$

60. $9(c + d)$

61. $x(a - b)$

62. $b(d - 7)$

63. $3(x + y - 8)$

64. $6(b - c + 2)$

Complete the equation to illustrate the given property.

65. $3x + 6 = $ _____ Commutative property

66. $8 + 0 = $ _____ Additive identity property

67. $\dfrac{2}{3} + \left(-\dfrac{2}{3}\right) = $ _____ Additive inverse property

68. $4(x + 3) = $ _____ Distributive property

69. $7 \cdot 1 = $ _____ Multiplicative identity property

70. $0 \cdot (-5.4) = $ _____ Multiplication property of zero

Writing in Mathematics

71. Name the only real number that has no reciprocal, and explain why this is so.

1.3
Exponents, Roots, and Order of Operations

OBJECTIVES

Tape 1

> **1** Define and find the absolute value of a number.
>
> **2** Simplify expressions containing exponents.
>
> **3** Find roots of numbers.
>
> **4** Use the order of operations.
>
> **5** Find the value of an algebraic expression given replacement values.

1 The **absolute value** of a real number a, written as $|a|$, is the distance between a and zero on the number line. Since distance is always positive or zero, $|a|$ is always positive or zero. Using the number line, we see that

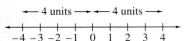

$$|4| = 4$$

because the distance between 4 and 0 is 4 units. Also,

$$|-4| = 4$$

because the distance between -4 and 0 is 4 units also.

An equivalent definition of absolute value is given next.

Absolute Value

The absolute value of a, written as $|a|$, is

$$|a| = \begin{cases} a \text{ if } a \text{ is } 0 \text{ or a positive number} \\ -a \text{ if } a \text{ is a negative number} \end{cases}$$

EXAMPLE 1 Find each absolute value.

    a. $|3|$ **b.** $|-5|$ **c.** $-|2|$ **d.** $-|-8|$ **e.** $|0|$

Solution: **a.** $|3| = 3$ since 3 is located 3 units from zero on the number line.

    b. $|-5| = 5$ since -5 is 5 units from zero on the number line.

    c. $-|2| = -2$. The negative sign outside the absolute value bars means to take the opposite of the absolute value of 2.

    d. $-|-8| = -8$. Since $|-8|$ is 8, $-|-8| = -8$.

    e. $|0| = 0$ since 0 is located 0 units from 0 on the number line. ∎

2 Recall that when two numbers are multiplied they are called **factors.** For example, in $3 \cdot 5 = 15$, the 3 and 5 are called factors.

An **exponent** is a shorthand notation for repeated multiplication of the same factor. This repeated factor is called the **base,** and the number of times it is used as a factor is indicated by the **exponent.** For example,

$$4^3 = \underbrace{4 \cdot 4 \cdot 4}_{} = 64$$
4 is a factor 3 times.

Also,

$$2^5 = \underbrace{2 \cdot 2 \cdot 2 \cdot 2 \cdot 2}_{} = 32$$
2 is a factor 5 times.

Exponents

If a is a real number and n is a natural number, then

$$a^n \text{ exponent, base} = \underbrace{a \cdot a \cdot a \cdot a \cdots a}_{a \text{ is a factor } n \text{ times}}$$

It is not necessary to write an exponent of 1. For example, 3 is assumed to be 3^1.

HELPFUL HINT

Remember that $2^3 = 2 \cdot 2 \cdot 2 = 8$, **not** $2 \cdot 3 = 6$.

EXAMPLE 2   Evaluate the following.

a. 4^2 **b.** $\left(\frac{1}{5}\right)^2$ **c.** $\left(\frac{2}{3}\right)^3$

Solution: **a.** $4^2 = 4 \cdot 4 = 16$ **b.** $\left(\frac{1}{5}\right)^2 = \frac{1}{5} \cdot \frac{1}{5} = \frac{1}{25}$ **c.** $\left(\frac{2}{3}\right)^3 = \frac{2}{3} \cdot \frac{2}{3} \cdot \frac{2}{3} = \frac{8}{27}$ ∎

3   The opposite of squaring a number is taking the **square root** of a number. For example, since the square of 4, or 4^2, is 16, we say that a square root of 16 is 4. The notation $\sqrt{a}$ is used to denote the **positive** or **principal square root** of a nonnegative number a. We then have in symbols that

$$\sqrt{16} = 4$$

EXAMPLE 3   Find the following square roots.

a. $\sqrt{9}$ **b.** $\sqrt{25}$ **c.** $\sqrt{\frac{1}{4}}$

Solution: **a.** $\sqrt{9} = 3$ since 3 is positive and $3^2 = 9$.
b. $\sqrt{25} = 5$ since $5^2 = 25$.

c. $\sqrt{\frac{1}{4}} = \frac{1}{2}$ since $\left(\frac{1}{2}\right)^2 = \frac{1}{4}$.   ∎

We can find roots other than square roots. Since 2 cubed, written as 2^3, is 8, we say that the cube root of 8 is 2. This is written as

$$\sqrt[3]{8} = 2$$

Also,

$$\sqrt[4]{81} = 3, \qquad \text{since } 3^4 = 81$$

EXAMPLE 4 Find the following roots.

a. $\sqrt[3]{27}$ **b.** $\sqrt[5]{1}$ **c.** $\sqrt[4]{16}$

Solution: **a.** $\sqrt[3]{27} = 3$ since $3^3 = 27$.

b. $\sqrt[5]{1} = 1$ since $1^5 = 1$.

c. $\sqrt[4]{16} = 2$ since $2^4 = 16$.   ∎

4   In expressions containing more than one operation, we follow an order of operations to get a unique answer to the problem. For example, when simplifying $2 + 3 \cdot 2$, should we add first or multiply first? To answer this question, we follow a fixed **order of operations.**

Order of Operations

Simplify expressions using the following order. If grouping symbols such as parentheses are present, simplify expressions within those first, starting with the innermost set. If fraction bars are present, simplify the numerator and denominator separately.

1. Simplify exponents or take roots in order from left to right.
2. Perform multiplications or divisions in order from left to right.
3. Perform additions or subtractions in order from left to right.

EXAMPLE 5 Simplify.

a. $2 + 3 \cdot 2$ **b.** $2 \cdot 3^3 + 5^2 - (3 + 4)$ **c.** $\dfrac{6^2 - 4}{2 + 2 \cdot 3}$ **d.** $\dfrac{|-2|^3}{7^1 - \sqrt{4}}$

Solution: **a.** First multiply; then add.

$$2 + 3 \cdot 2 = 2 + 6 = 8$$

b. First simplify inside parentheses.

$$2 \cdot 3^3 + 5^2 - (3 + 4) = 2 \cdot \boxed{3^3} + \boxed{5^2} - 7$$

$$= 2 \cdot \boxed{27} + \boxed{25} - 7 \qquad \text{Write } 3^3 \text{ as 27 and } 5^2 \text{ as 25.}$$

$$= 54 + 25 - 7 \qquad \text{Multiply.}$$

$$= 79 - 7 \qquad \text{Add or subtract from left to right.}$$

$$= 72 \qquad \text{Subtract.}$$

c. Simplify the numerator and the denominator separately; then divide.

$$\frac{6^2 - 4}{2 + 2 \cdot 3} = \frac{36 - 4}{2 + 6}$$ Square 6 in the numerator and multiply $2 \cdot 3$ in the denominator.

$$= \frac{32}{8}$$ Simplify the numerator and the denominator separately.

$$= 4$$ Divide.

d. Simplify the numerator and the denominator separately; then divide.

$$\frac{|-2|^3}{7^1 - \sqrt{4}} = \frac{2^3}{7 - 2}$$ Write $|-2|$ as 2 and $\sqrt{4}$ as 2.

$$= \frac{8}{5}$$ Simplify the numerator and the denominator.   ∎

5   An **algebraic expression** is formed by numbers and variables connected by the operations of addition, subtraction, multiplication, division, raising to powers, or taking roots. These expressions have different values depending on replacement numbers for the variables. If numbers are substituted for the variables in an algebraic expression and this expression is simplified, the result is called the **value** of the expression for the given replacement numbers.

EXAMPLE 6   Find the value of each algebraic expression when $x = 0$, $y = 2$, and $z = 5$.

a. $\dfrac{2x + y}{z}$ **b.** $|y^3| - (z + x)$ **c.** $z^3 + 3y^4$

Solution: Replace x with 0, y with 2, and z with 5 in each algebraic expression.

a. $\dfrac{2x + y}{z} = \dfrac{2(0) + 2}{5}$

$$= \frac{0 + 2}{5}$$ Multiply.

$$= \frac{2}{5}$$ Add.

b. $|y^3| - (z + x) = |2^3| - (5 + 0)$ Let $x = 0$, $y = 2$, and $z = 5$.

$$= |8| - 5$$ Find 2^3 and simplify inside parentheses.

$$= 8 - 5$$ Write $|8|$ as 8.

$$= 3$$ Subtract.

c. $z^3 + 3y^4 = 5^3 + 3 \cdot 2^4$ Replace y with 2 and z with 5.

$$= 125 + 3 \cdot 16$$ Write 5^3 as 125 and 2^4 as 16.

$$= 125 + 48$$ Multiply.

$$= 173$$ Add.   ∎

CALCULATOR BOX

Exponents:

To evaluate an exponential expression on a calculator, find the key marked $\boxed{y^x}$. To evaluate 2^7, press the keys

$$\boxed{2}\ \boxed{y^x}\ \boxed{7}\ \boxed{=}$$

The display should read $\boxed{128}$.

Taking Square Roots:

To take square roots on your calculator, find the key marked $\boxed{\sqrt{\ }}$. To find $\sqrt{289}$, press the keys

$$\boxed{289}\ \boxed{\sqrt{\ }}$$

The display should read $\boxed{17}$. (Square roots of numbers other than perfect squares will be covered in Chapter 7.)

Order of Operations:

To see whether your calculator has order of operations built in, evaluate $3 + 4 \cdot 5$ by pressing keys

$$\boxed{3}\ \boxed{+}\ \boxed{4}\ \boxed{\times}\ \boxed{5}\ \boxed{=}$$

The display should read $\boxed{23}$. Although order of operations is probably a built-in feature of your calculator, parentheses must sometimes be inserted. For example, to evaluate $\dfrac{16-4}{9-7}$, press the keys

$$\boxed{(}\ \boxed{16}\ \boxed{-}\ \boxed{4}\ \boxed{)}\ \boxed{\div}\ \boxed{(}\ \boxed{9}\ \boxed{-}\ \boxed{7}\ \boxed{)}\ \boxed{=}$$

The display should read $\boxed{6}$. Use a calculator to evaluate each expression.

Use a calculator to evaluate each expression.

1. 9^7
2. $\sqrt{961}$
3. $\sqrt{529} + 5^6$
4. $(\sqrt{625} + \sqrt{121})^2$
5. $\dfrac{7^4 - 4^4}{15}$
6. $\dfrac{(16-5)^3}{\sqrt{484}}$

EXERCISE SET 1.3

Find each absolute value. See Example 1.

1. $-|2|$ **2.** $|8|$ **3.** $|-4|$ **4.** $|-6|$
5. $|0|$ **6.** $|-1|$ **7.** $-|-3|$ **8.** $-|-11|$

Evaluate the following. See Example 2.

9. 3^2 **10.** 5^2 **11.** 1^4 **12.** 4^3

13. $|-2|^3$ **14.** $|-3|^3$ **15.** $\left(\dfrac{2}{3}\right)^2$ **16.** $\left(\dfrac{3}{4}\right)^3$

Find the following roots. See Examples 3 and 4.

17. $\sqrt{49}$ **18.** $\sqrt{81}$ **19.** $\sqrt{\dfrac{1}{9}}$ **20.** $\sqrt{\dfrac{1}{25}}$

21. $\sqrt[3]{64}$ **22.** $\sqrt[5]{32}$ **23.** $\sqrt[4]{81}$ **24.** $\sqrt[3]{1}$

Simplify the following. See Example 5.

25. $|-18| - |-7|$ **26.** $|-4| + |10|$ **27.** $2 \cdot 5^2$ **28.** $3 \cdot 2^3$

29. $45 - 6^2 \div 3^2 + \sqrt{1}$ **30.** $(12 \div 3)^2 - 10 \div 5$ **31.** $\dfrac{|-6|^2}{|-8| - 2^2}$ **32.** $\dfrac{|-4|^3}{3^2 + \sqrt{49}}$

Find the value of each expression when x = 3, y = 0, and z = 4. See Example 6.

33. $x^2 - y^2$ **34.** $z^2 - x^2$ **35.** $5(2z - y - x)$ **36.** $3(2x - z + y)$

37. $\dfrac{3y}{2x}$ **38.** $\dfrac{5y + z}{2}$ **39.** $\sqrt{z} + x^3$ **40.** $(y + x + z)^2$

Simplify the following.

41. $3^2 - 2^3$ **42.** $4^2 - 2^4$ **43.** $\sqrt{\dfrac{1}{4}} + \sqrt{\dfrac{9}{4}}$ **44.** $\sqrt{\dfrac{25}{9}} - \sqrt{\dfrac{1}{9}}$

45. $|7| - |-7|$ **46.** $|-8| - |-1|$ **47.** $\sqrt{100} - \sqrt{36}$ **48.** $\sqrt[3]{27} + \sqrt{64}$

49. $2 \cdot 3 + 4 \cdot 5$ **50.** $6 \cdot 2 - 2 \cdot 5$ **51.** $(15 - 11)(20 - 4)$ **52.** $(7 + 3)(35 - 2)$

53. $\dfrac{10 - 4}{5 - 1}$ **54.** $\dfrac{28 - 8}{7 - 2}$ **55.** $3 \cdot \sqrt{49} + 14 \div 2$ **56.** $2 \cdot \sqrt{121} - 100 \div 20$

57. $\dfrac{8(7 - 5)}{2^4}$ **58.** $\dfrac{3(10 - 7)}{15 \div 3 - 2}$ **59.** $\dfrac{7 - 3 \cdot 3}{1 + 2 \cdot 2}$ **60.** $\dfrac{4 - 2 \cdot 7}{4 + 3 \cdot 2}$

Find the value of each expression when a = 2, b = 1, and c = 5.

61. $\sqrt{b} + |-c|$ **62.** $a^3 + c^2$ **63.** $c^2 - b^5$ **64.** $6c - 4b$

65. $|a| + |b| + |c|$ **66.** $7b - a$ **67.** $4(b + c - a)$ **68.** $3(2a + b - c)$

69. $\dfrac{10c - 8a}{a}$ **70.** $\dfrac{|c| + a^3}{3c - a}$ **71.** $\dfrac{2a - 4}{b^4}$ **72.** $\dfrac{a + b + c}{a + b + c}$

Writing in Mathematics

73. Explain why $-(-2)$ and $-|-2|$ represent different numbers.

74. The boxed definition of an absolute value states that $|a| = -a$ if a is a negative number. Explain why $|a|$ is always positive, even though $|a| = -a$ for some values of a.

1.4
Operations on Real Numbers

OBJECTIVES

1   Add and subtract real numbers.

2   Multiply and divide real numbers.

3   Find the value of algebraic expressions given replacement values.

Tape 1

1   The addition of two real numbers may be summarized by the following rules.

Adding Real Numbers

1. To add two numbers with the same sign, add their absolute values and attach their common sign.

2. To add two numbers with different signs, subtract the smaller absolute value from the larger absolute value and attach the sign of the number with the larger absolute value.

For example, to add $-5 + (-7)$, first add their absolute values.

$$|-5| = 5, \quad |-7| = 7 \quad \text{and} \quad 5 + 7 = 12$$

Next, attach their common negative sign.

$$-5 + (-7) = -12$$

To find $(-4) + 3$, first subtract their absolute values.

$$|-4| = 4, \quad |3| = 3, \quad \text{and} \quad 4 - 3 = 1$$

Next, attach the sign of the number with the larger absolute value.

$$(-4) + 3 = -1$$
$$|-4| > |3|$$

EXAMPLE 1   Find each sum.

a. $-3 + (-11)$ **b.** $3 + (-7)$ **c.** $-10 + 15$ **d.** $-8 + (-1)$ **e.** $-\dfrac{1}{4} + \dfrac{1}{2}$

f. $-\dfrac{2}{3} + \dfrac{3}{7}$

Solution: **a.** $-3 + (-11) = -(3 + 11) = -14$ **b.** $3 + (-7) = -4$

c. $-10 + 15 = 5$ **d.** $-8 + (-1) = -9$ **e.** $-\dfrac{1}{4} + \dfrac{1}{2} = -\dfrac{1}{4} + \dfrac{2}{4} = \dfrac{1}{4}$

f. $-\dfrac{2}{3} + \dfrac{3}{7} = -\dfrac{14}{21} + \dfrac{9}{21} = -\dfrac{5}{21}$ ■

Subtraction of two real numbers is defined by addition.

Subtracting Real Numbers

If a and b are real numbers, then the difference of a and b, written $a - b$, is defined by

$$a - b = a + (-b)$$

In other words, to subtract two numbers, find the sum of the first number and the opposite of the second number.

EXAMPLE 2   Find each difference.

a. $2 - 8$ **b.** $-8 - (-1)$ **c.** $-11 - 5$ **d.** $10 - (-9)$ **e.** $\dfrac{2}{3} - \dfrac{1}{2}$

f. $1 - 0.06$ **g.** Subtract 7 from 4.

Solution:

Add the opposite.   Add the opposite.

a. $2 - 8 = 2 + (-8) = -6$ **b.** $-8 - (-1) = -8 + (1) = -7$

c. $-11 - 5 = -11 + (-5) = -16$ **d.** $10 - (-9) = 10 + (9) = 19$

e. $\dfrac{2}{3} - \dfrac{1}{2} = \dfrac{4}{6} + \left(-\dfrac{3}{6}\right) = \dfrac{1}{6}$ **f.** $1 - 0.06 = 1 + (-0.06) = 0.94$

g. $4 - 7 = 4 + (-7) = -3$ ■

To add or subtract three or more real numbers, follow the order of operations and add or subtract from left to right.

EXAMPLE 3   Simplify the following expressions.

a. $11 + 2 - 7$ **b.** $-5 - 4 + 2$ **c.** $-5 - (4 + 2)$ **d.** $|2 - 6| + |5 - 4|$

e. $(2 - 6) + (5 - 4)$

Solution: **a.** $11 + 2 - 7 = 13 - 7 = 6$ **b.** $-5 - 4 + 2 = -9 + 2 = -7$

c. Simplify inside parentheses first.

$$-5 - (4 + 2) = -5 - (6) = -5 + (-6) = -11$$

d. Since absolute value symbols are present, simplify within them first in order from left to right.

$|2 - 6| + |5 - 4| = |-4| + |1|$ Simplify within absolute value bars.

$\qquad\qquad\qquad\;\; = 4 + 1$ Find the absolute values.

$\qquad\qquad\qquad\;\; = 5$ Add.

e. Simplify inside parentheses first in order from left to right.

$(2 - 6) + (5 - 4) = (-4) + (1)$ Simplify inside parentheses.

$\qquad\qquad\qquad\;\; = -3$ Add.   ■

2   Sign rules for multiplying two real numbers may be summarized as follows.

Multiplying Two Real Numbers

The product of two numbers having the same sign is positive.
The product of two numbers having different signs is negative.

EXAMPLE 4   Find each product.

a. $(-8)(-1)$ **b.** $(-2)\dfrac{1}{6}$ **c.** $3(-3)$ **d.** $(0)(11)$ **e.** $\left(\dfrac{1}{5}\right)\left(-\dfrac{10}{11}\right)$

f. $(7)(1)(-2)(-3)$ **g.** $8(-2)(0)$

Solution: **a.** Since the signs of the two numbers are the same, the product is positive. Thus $(-8)(-1) = +8$ or 8.

b. Since the signs of the two numbers are different or unlike, the product is negative. Thus $(-2)\dfrac{1}{6} = -\dfrac{2}{6} = -\dfrac{1}{3}$.

c. $3(-3) = -9$ **d.** $0(11) = 0$ **e.** $\left(\frac{1}{5}\right)\left(-\frac{10}{11}\right) = -\frac{10}{55} = -\frac{2}{11}$

f. $(7)(1)(-2)(-3) = 7(-2)(-3)$
$$= -14(-3)$$
$$= 42$$

g. Since zero is a factor, the product is zero.

$$(8)(-2)(0) = 0 \quad \blacksquare$$

HELPFUL HINT

The following sign rules may be helpful when multiplying.

1. An odd number of negative factors gives a negative product.
2. An even number of negative factors gives a positive product.

EXAMPLE 5   Simplify each expression.

a. 3^2 **b.** $\left(\frac{1}{2}\right)^4$ **c.** -5^2 **d.** $(-5)^2$ **e.** -5^3 **f.** $(-5)^3$

Solution: **a.** $3^2 = 3 \cdot 3 = 9$ **b.** $\left(\frac{1}{2}\right)^4 = \left(\frac{1}{2}\right)\left(\frac{1}{2}\right)\left(\frac{1}{2}\right)\left(\frac{1}{2}\right) = \frac{1}{16}$

c. $-5^2 = -(5 \cdot 5) = -25$ **d.** $(-5)^2 = (-5)(-5) = 25$
e. $-5^3 = -(5 \cdot 5 \cdot 5) = -125$ **f.** $(-5)^3 = (-5)(-5)(-5) = -125$ $\blacksquare$

HELPFUL HINT

Be very careful when simplifying expressions such as -5^2 and $(-5)^2$.

$$-5^2 = -(5 \cdot 5) = -25 \quad \text{and} \quad (-5)^2 = (-5)(-5) = 25$$

Without parentheses, only 5 is squared, not -5.

The sign rules for division are the same as for multiplication.

Dividing Two Real Numbers

The quotient of two numbers with the same signs is positive.
The quotient of two numbers with different signs is negative.

EXAMPLE 6   Find each quotient.

a. $\dfrac{20}{-4}$ **b.** $\dfrac{-9}{-3}$ **c.** $-\dfrac{3}{8} \div 3$ **d.** $\dfrac{-40}{10}$ **e.** $\dfrac{\frac{-1}{10}}{\frac{-2}{5}}$

Solution: **a.** Since the signs are different or unlike, the quotient is negative and $\dfrac{20}{-4} = -5$.

b. Since the signs are the same, the quotient is positive and $\dfrac{-9}{-3} = 3$.

c. $-\dfrac{3}{8} \div 3 = -\dfrac{3}{8} \cdot \dfrac{1}{3} = -\dfrac{1}{8}$ **d.** $\dfrac{-40}{10} = -4$

e. $\dfrac{\dfrac{-1}{10}}{\dfrac{-2}{5}} = -\dfrac{1}{10} \cdot -\dfrac{5}{2} = \dfrac{1}{4}$ ■

With sign rules for division, we can understand why the positioning of the negative sign in a fraction does not change the value of the fraction. For example,

$$\frac{-12}{3} = -4, \quad \frac{12}{-3} = -4, \quad \text{and} \quad -\frac{12}{3} = -4$$

Since all fractions equal -4, we can say that

$$\frac{-12}{3} = \frac{12}{-3} = -\frac{12}{3}$$

In general, the following holds true:

If a and b are real numbers, $b \ne 0$, then $\dfrac{a}{-b} = \dfrac{-a}{b} = -\dfrac{a}{b}$.

Next, we practice simplifying expressions that contain more than one operation. Remember to use the order of operations.

EXAMPLE 7   Simplify the following expressions.

a. $2(1-4)^2$ **b.** $\dfrac{5-7}{-2}$ **c.** $\dfrac{-7}{0}$ **d.** $\dfrac{(6+2)-(-4)}{2-(-3)}$

Solution: **a.** $2(1-4)^2 = 2(-3)^2$ Simplify inside grouping symbols first.

$\qquad\qquad\qquad\ = 2(9)$ Write $(-3)^2$ as 9.

$\qquad\qquad\qquad\ = 18$ Multiply.

b. $\dfrac{5-7}{-2} = \dfrac{-2}{-2} = 1$

c. $\dfrac{-7}{0}$ is not a real number since no real number times zero gives a product of -7.

d. $\dfrac{(6+2)-(-4)}{2-(-3)} = \dfrac{8-(-4)}{2-(-3)}$ Simplify inside grouping symbols first.

$\qquad\qquad\qquad\ = \dfrac{8+4}{2+3}$ Write subtraction as equivalent addition.

$\qquad\qquad\qquad\ = \dfrac{12}{5}$ Add in both the numerator and denominator.   ■

3

EXAMPLE 8   Find the value of each algebraic expression when $x = 2$, $y = -1$, and $z = -3$.

a. $z - y$ **b.** z^2 **c.** $\dfrac{2x + y}{z}$

Solution: **a.** $z - y = -3 - (-1) = -3 + 1 = -2$ **b.** $z^2 = (-3)^2 = 9$

c. $\dfrac{2x + y}{z} = \dfrac{2(2) + (-1)}{-3} = \dfrac{4 + (-1)}{-3} = \dfrac{3}{-3} = -1$ ∎

Sometimes variables such as x_1 and x_2 will be used in this book. The small 1 and 2 are called **subscripts.** The variable x_1 can be read as "x sub 1" and the variable x_2 can be read as "x sub 2." The important thing to remember is that they are two different variables. For example, if $x_1 = -5$ and $x_2 = 7$, then

$$x_1 - x_2 = -5 - 7 = -12$$

CALCULATOR BOX

Entering Negative Numbers

To enter a negative number on a calculator, find a key marked $\boxed{+/-}$.
(On some calculators, this key is marked $\boxed{\text{CHS}}$ for "change sign.")
To enter -15, press $\boxed{15}$ $\boxed{+/-}$. The display will read $\boxed{\qquad -15}$.

Perform indicated operations.

1. $-57 + (-295)$ **2.** $-466 + 89$
3. $(-45)^2$ **4.** -19^2
5. $\dfrac{(-37)(20) + (-10)}{-5}$ **6.** $\dfrac{-462}{6} + \dfrac{738}{-9}$

EXERCISE SET 1.4

Find each sum or difference. See Examples 1 through 3.

1. $-3 + 8$ **2.** $-5 + (-9)$ **3.** $-14 + (-10)$
4. $12 + (-7)$ **5.** $-4 - 6$ **6.** $-8 - (-6)$
7. $13 - 17$ **8.** $15 - (-1)$ **9.** $(-2 - 3) + (-8 - 10)$
10. $-4 - (5 - 9)$ **11.** $19 - 10 - 11$ **12.** $-13 - 4 + 9$

Find each product or quotient. See Examples 4 and 6.

13. $(-5)(12)$ **14.** $6(-3)$ **15.** $(-8)(-10)$ **16.** $7(0)$
17. $\dfrac{-12}{-4}$ **18.** $\dfrac{60}{-6}$ **19.** $\dfrac{0}{-2}$ **20.** $\dfrac{-2}{0}$
21. $(-4)(-2)(-1)$ **22.** $5(-3)(-2)$ **23.** $\dfrac{-6}{7} \div 2$ **24.** $\dfrac{-9}{13} \div (-3)$

Evaluate. See Example 5.

25. -7^2

26. $(-7)^2$

27. $(-6)^2$

28. -6^2

29. $(-2)^3$

30. -2^3

Simplify each expression. See Example 7.

31. $3(5-7)^4$

32. $7(3-8)^2$

33. $-3^2 + 2^3$

34. $-5^2 - 2^4$

35. $\dfrac{3-(-12)}{-5}$

36. $\dfrac{-4-(-8)}{-4}$

37. $\dfrac{-6+|3-5|}{2}$

38. $\dfrac{-9-|7-10|}{-6}$

Find the value of each expression when $y_1 = -3$, $y_2 = 4$, and $y_3 = 0$. See Example 8.

39. $(y_1)^2$

40. $y_2 - y_1$

41. $y_1 + y_2 + y_3$

42. $\dfrac{y_3}{3y_1}$

43. $\dfrac{6y_2}{y_1}$

44. $y_1 - y_2$

Simplify each expression.

45. $-4 + 7$

46. $-9 + 15$

47. $-9 + (-3)$

48. $-17 + (-2)$

49. $-4 - (-19)$

50. $-5 - (-17)$

51. $6 - 18$

52. $15 - 21$

53. $(-4)(-7)(0)$

54. $(-9)(0)(-14)$

55. $\left(-\dfrac{2}{3}\right)\left(\dfrac{6}{4}\right)$

56. $\left(\dfrac{5}{6}\right)\left(\dfrac{-12}{15}\right)$

57. $-14 - 7$

58. $-6 - 31$

59. $-\dfrac{4}{5} - \left(\dfrac{-3}{10}\right)$

60. $-\dfrac{5}{2} - \left(-\dfrac{2}{3}\right)$

61. Subtract 14 from 8

62. Subtract 9 from -3

63. $-\dfrac{34}{2}$

64. $\dfrac{48}{-3}$

65. $16 - 8 - 9$

66. $-14 - 3 + 6$

67. $|4 - 7| + |4 + 7|$

68. $|7 - 3| - |2 - 19|$

69. $3|5 - 7|^4$

70. $7|3 - 8|^2$

71. $\dfrac{(3-9)-(-5)}{-3}$

72. $\dfrac{-14-(2-7)}{-15}$

73. $\dfrac{|3-9|-|-5|}{-3}$

74. $\dfrac{|-14|-|2-7|}{-15}$

75. $(-3)^2 + 2^3$

76. $(-15)^2 - 2^4$

77. $(7-2)^2$

78. $7^2 - 2^2$

79. $\dfrac{\left(\dfrac{-3}{10}\right)}{\left(\dfrac{42}{50}\right)}$

80. $\dfrac{\left(\dfrac{-5}{21}\right)}{\left(\dfrac{-6}{42}\right)}$

Find the value of each expression when $x = -2$, $y = -5$, and $z = 3$.

81. $y^2 + x^2 + z^2$

82. $y - 5x$

83. $\dfrac{3z + y}{2x}$

84. $\dfrac{5x - z}{-2y + z}$

85. $2x^2 - 3y^2$

86. $z^3 - y^2$

Writing in Mathematics

87. Explain why -2^4 and $(-2)^4$ represent different numbers.　　　**88.** Explain why -2^3 and $(-2)^3$ represent the same number.

CRITICAL THINKING　After putting nearly 60,000 miles on your car, you are faced with a decision. The car needs some major repairs, and you aren't sure the car is worth repairing. The mechanic estimates that the combined cost of parts and labor will be $750. He assures you, however, that the car's mileage rate of 24 miles per gallon will increase by nearly 15%. This increase in mileage rate will lower the amount of money you spend on gas, offsetting the cost of repairs.

On the other hand, this could be just the excuse you need to buy the $12,000 new car you've had your eye on. After your trade-in and down payment, you figure you could afford the monthly loan payment. You estimate you would be paying nearly $500 annually in interest for the next 4 years, but you reason that the new car will get 32 miles per gallon, and lower gas costs will offset interest payments.

On the basis of the facts described so far, should you buy a new car or repair the old one? What additional facts, financial and otherwise, might you consider to reach a good decision?

CHAPTER 1 GLOSSARY

The **absolute value** of a number, written as $|a|$, is the distance between a and 0 on the number line.

An **algebraic expression** is formed by numbers and variables connected by the operations of addition, subtraction, multiplication, division, raising to powers, or taking roots.

Individual objects in a set are called the **elements** of the set.

The **opposite** or **additive inverse** of a number a is $-a$.

The **positive** or **principal square root** of a nonnegative number a, written as $\sqrt{a}$, is the positive number whose square is a.

The **reciprocal** or **multiplicative inverse** of a nonzero number a is $1/a$.

A **set** is a collection of objects.

Set X is a **subset** of set Y, written as $X \subseteq Y$, if all elements of X are also elements of Y.

Letters used to represent numbers are called **variables**.

CHAPTER 1 SUMMARY

Real numbers:　$\{x \,|\, x \text{ is a point on the number line}\}$

Natural numbers:　$\{1, 2, 3, \ldots\}$

Whole numbers:　$\{0, 1, 2, 3, \ldots\}$

Integers:　$\{\ldots -3, -2, -1, 0, 1, 2, 3, \ldots\}$

Rational numbers:　$\left\{ \dfrac{a}{b} \,\middle|\, a \text{ and } b \text{ are integers and } b \neq 0 \right\}$

Irrational numbers:　$\{x \,|\, x \text{ is a real number and } x \text{ is not a rational number}\}$

ORDER OF OPERATIONS (1.3)

Simplify expressions using the following order. If grouping symbols such as parentheses are present, simplify expressions within those first, starting with the innermost set.

1. Simplify exponents or take roots in order from left to right.
2. Perform multiplications or divisions in order from left to right.
3. Perform additions or subtractions in order from left to right.

TO ADD TWO REAL NUMBERS WITH THE SAME SIGN (1.4)

Add the absolute values of the two numbers and use their common sign as the sign of the sum.

TO ADD TWO REAL NUMBERS WITH DIFFERENT SIGNS

Find the difference of their absolute values. Use the sign of the number with the larger absolute value as the sign of the difference.

TO SUBTRACT TWO REAL NUMBERS

Find the opposite of the number being subtracted, and then add the opposite.

$$a - b = a + (-b)$$

MULTIPLYING TWO REAL NUMBERS

The product of two numbers having the same sign is positive.
The product of two numbers having different signs is negative.

DIVIDING TWO REAL NUMBERS

The quotient of two numbers with the same sign is positive.
The quotient of two numbers with different signs is negative.

CHAPTER 1 REVIEW

(1.1) *Write each set in roster form.*

1. $\{x \mid x$ is an odd integer between -2 and $4\}$
2. $\{x \mid x$ is an even integer between -3 and $7\}$
3. $\{x \mid x$ is a negative whole number$\}$
4. $\{x \mid x$ is a natural number that is not a rational number $\}$
5. $\{x \mid x$ is a whole number greater than $5\}$
6. $\{x \mid x$ is an integer less than $3\}$

Determine whether each statement is true or false if $A = \{6, 10, 12\}$, $B = \{5, 9, 11\}$, $C = \{\ldots -3, -2, -1, 0, 1, 2, 3, \ldots\}$, $D = \{2, 4, 6, \ldots 16\}$, $E = \{x \mid x$ is a rational number$\}$, $F = \{\ \}$, $G = \{x \mid x$ is an irrational number$\}$, and $H = \{x \mid x$ is a real number$\}$.

7. $10 \in D$
8. $B \in 9$
9. $\sqrt{169} \notin G$
10. $0 \notin F$
11. $\pi \in E$
12. $\pi \in H$
13. $\sqrt{4} \in G$
14. $-9 \in E$
15. $A \subseteq D$
16. $C \not\subseteq B$
17. $C \not\subseteq E$
18. $F \subseteq H$
19. $B \subseteq B$
20. $D \subseteq C$
21. $C \subseteq H$
22. $G \subseteq H$
23. $\{5\} \in B$
24. $\{5\} \subseteq B$

List the elements of the set $\left\{5, -\dfrac{2}{3}, \dfrac{8}{2}, \sqrt{9}, 0.3, \sqrt{7}, 1\dfrac{5}{8}, -1, \pi\right\}$ that are also elements of each given set.

25. Whole numbers
26. Natural numbers
27. Rational numbers
28. Irrational numbers
29. Real numbers
30. Integers

(1.2) *Write each sentence as a mathematical statement.*

31. Twelve is the product of x and negative 4.

32. The sum of n and twice n is negative fifteen.

33. Four times the sum of y and three is -1.

34. The difference of t and five, multiplied by six is four.

35. Seven subtracted from z is six.

36. Ten less than the product of x and nine is five.

37. The difference of x and 5 is at least 12.

38. The opposite of four is less than the product of y and seven.

39. Two-thirds is not equal to twice the sum of n and one-fourth.

40. The sum of t and six is not more than negative twelve.

Name the property illustrated.

41. $(M + 5) + P = M + (5 + P)$

43. $(-4) + 4 = 0$

45. $(XY)Z = (YZ)X$

47. $T \cdot 0 = 0$

49. $A + 0 = A$

42. $5(3x - 4) = 15x - 20$

44. $(3 + x) + 7 = 7 + (3 + x)$

46. $\left(-\dfrac{3}{5}\right) \cdot \left(-\dfrac{5}{3}\right) = 1$

48. $(ab)c = a(bc)$

50. $8 \cdot 1 = 8$

Find the additive inverse or opposite.

51. $-\dfrac{3}{4}$

52. 0.6

53. 0

54. 1

Find the multiplicative inverse or reciprocal.

55. $-\dfrac{3}{4}$

56. 0.6

57. 0

58. 1

Complete the equation using the given property.

59. $5x - 15z = $ _____ Distributive property

60. $(7 + y) + (3 + x) = $ _____ Commutative property

61. $0 = $ _____ Additive inverse property

62. $1 = $ _____ Multiplicative inverse property

63. $[(3.4)(0.7)]5 = $ _____ Associative property

64. $7 = $ _____ Additive identity property

Insert $<$, $>$, or $=$ to make each statement true.

65. $-9 \quad\quad -12$

68. $7 \quad\quad |-7|$

66. $0 \quad\quad -6$

69. $-5 \quad\quad -(-5)$

67. $-3 \quad\quad -1$

70. $-(-2) \quad\quad -2$

(1.3) *Simplify.*

71. $3 + 2 \cdot 10 \div 4$

74. $8^2 \div 4^2 - 4$

77. $\dfrac{\sqrt{25}}{4 + 3 \cdot 7}$

80. $\dfrac{2 \cdot 5^2 - |15|}{35 - 50 \div 5}$

72. $7 + 24 \div 6 \cdot 4$

75. $(5 + 3)^2 - 3^3 - |7 - 4|$

78. $\dfrac{\sqrt{64}}{24 - 8 \cdot 2}$

73. $4 \cdot 6^2 \div 9 - 16$

76. $(6 - 3) + 7^2 - |8 - 7|$

79. $\dfrac{2(6^2 - 5^2)}{40 - 21 \div 3}$

(1.4) *Simplify.*

81. $5(-0.4)$

84. $9 - (-4.3)$

82. $(-3.1)(-0.1)$

85. $(-6)(-4)(0)(-3)$

83. $-7 - (-15)$

86. $(-12)(0)(-1)(-5)$

87. $(-24) \div 0$

90. $(60) \div (-12)$

93. $\left(-\dfrac{4}{5}\right) - \left(-\dfrac{2}{3}\right)$

96. $6(7 - 10)^2$

99. $\dfrac{-\dfrac{6}{15}}{\dfrac{8}{25}}$

102. $5(-2) - (-3) - \dfrac{1}{6} + \dfrac{2}{3}$

105. $(2^3 - 3^2) - (5 - 7)$

108. $\dfrac{(2 + 4)^2 + (-1)^5}{12 \div 2 \cdot 3 - 3}$

88. $0 \div (-45)$

91. $-5 + 7 - 3 - (-10)$

94. $\left(\dfrac{5}{4}\right) - \left(-2\dfrac{3}{4}\right)$

97. $\left(-\dfrac{8}{15}\right) \cdot \left(-\dfrac{2}{3}\right)^2$

100. $\dfrac{\dfrac{4}{9}}{-\dfrac{8}{45}}$

103. $|2^3 - 3^2| - |5 - 7|$

106. $(5^2 - 2^4) + [9 \div (-3)]$

109. $\dfrac{(4 - 9) + 4 - 9}{10 - 12 \div 4 \cdot 8}$

89. $(-36) \div (-9)$

92. $8 - (-3) + (-4) + 6$

95. $3(4 - 5)^4$

98. $\left(-\dfrac{3}{4}\right)^2 \cdot \left(-\dfrac{10}{21}\right)$

101. $-\dfrac{3}{8} + 3(2) \div 6$

104. $|5^2 - 2^2| + |9 \div (-3)|$

107. $\dfrac{(8 - 10)^3 - (-4)^2}{2 + 8(2) \div 4}$

110. $\dfrac{3 - 7 - (7 - 3)}{15 + 30 \div 6 \cdot 2}$

CHAPTER 1 TEST

Determine whether each statement is true or false.

1. $-2.3 > -2.33$

4. $(-2)(-3)(0) = \dfrac{(-4)}{0}$

2. $-6^2 = (-6)^2$

5. All natural numbers are integers.

3. $-5 - 8 = -(5 - 8)$

6. All rational numbers are integers.

Simplify.

7. $5 - 12 \div 3(2)$

10. $(4 - 9)^3 - |-4 - 6|^2$

13. $\dfrac{6(7 - 9)^3 + (-2)}{(-2)(-5)(-5)}$

8. $4^2 + 24 \div 2(3)$

11. $[2(6 - 3)^2 + 2] \div 5$

14. $\dfrac{-3(2 - 7)^2 + 5(2^3)}{-|10 - (-10)| - 5}$

9. $|4 - 6|^3 - (1 - 6^2)$

12. $[3|4 - 5|^5 - (-9)] \div (-6)$

Write each sentence as a mathematical statement.

15. Twice the absolute value of the sum of x and five is 30.

16. The square of the difference of six and y, divided by seven, is less than -2.

17. The product of nine and z, divided by the absolute value of -12, is not equal to 10.

18. Three times the quotient of n and five is the opposite of n.

19. Twenty is equal to 6 subtracted from twice x.

20. Negative two is equal to x divided by the sum of x and five.

Name each property illustrated.

21. $6(x - 4) = 6x - 24$

23. $(-7) + 7 = 0$

22. $(4 + x) + z = 4 + (x + z)$

24. $(-18)(0) = 0$

CHAPTER **2**

Solving Equations and Inequalities

Among the many environmental problems threatening our planet, pollutants dumped into waterways are one of the most sinister. Local authorities must evaluate the extent of pollution from the Chemlife Chemical Company to save a lake downstream. (See Critical Thinking, page 79.)

INTRODUCTION

Mathematics is a tool for solving problems in such diverse fields as transportation, engineering, economics, medicine, business, and biology. We solve problems using mathematics by modeling real-world phenomena with mathematical equations or inequalities. Our ability to solve problems using mathematics, then, depends in part on our ability to solve equations and inequalities. In this chapter, we solve linear equations and inequalities in one variable and graph their solutions on number lines.

2.1
Linear Equations

OBJECTIVES

Tape 2

1. Simplify algebraic expressions.
2. Define linear equations.
3. Solve linear equations.
4. Recognize identities and equations with no solution.

---

1 We begin this section with a review of algebraic expressions and adding or subtracting like terms. Recall that an **algebraic expression** is formed by numbers and variables connected by the operations of addition, subtraction, multiplication, division, raising to powers, or taking roots. For example,

$$2x + 3, \quad \frac{x + 5}{6} - \frac{z^5}{y^2}, \quad \text{and} \quad \sqrt{y}$$

are algebraic expressions or, more simply, expressions.

A **term** is a number or the product of a number and variables raised to powers.

Expression	Terms
$-2x + y$	$-2x, \quad y$
$3x^2 - \dfrac{y}{5} + 7$	$3x^2, \quad -\dfrac{y}{5}, \quad 7$

Terms with the same variable(s) raised to the same power are called **like terms.** We can add or subtract like terms by using the distributive property.

29

EXAMPLE 1 Use the distributive property to simplify each expression.

a. $3x - 5x$ **b.** $7y + y$ **c.** $4z + 6$

Solution: **a.** $3x - 5x = (3 - 5)x$. Apply the distributive property.
$$= -2x$$

b. $7y + y = (7 + 1)y = 8y$

c. $4z + 6$ cannot be simplified further since $4z$ and 6 are not like terms.   ■

The distributive property can also be used to multiply. For example,

$$-2(x + 3) = -2(x) + (-2)(3) = -2x - 6$$

The associative and commutative properties may sometimes be needed when simplifying expressions in order to rearrange and group like terms.

$$-7x + 5 + 3x - 2 = -7x + 3x + 5 - 2$$
$$= (-7 + 3)x + (5 - 2)$$
$$= -4x + 3$$

EXAMPLE 2 Simplify each expression.

a. $3x - 2x + 5 - 7 + x$ **b.** $7x + 3 - 5(x - 4)$ **c.** $(2x - 5) - (-x - 5)$

Solution: **a.** $3x - 2x + 5 - 7 + x = 3x - 2x + x + 5 - 7$ Apply the commutative property.

$$= (3 - 2 + 1)x + (5 - 7) \text{ Apply the distributive property.}$$

$$= 2x - 2 \qquad \text{Simplify.}$$

b. $7x + 3 - 5(x - 4) = 7x + 3 - 5x + 20$ Apply the distributive property.

$$= 2x + 23 \qquad \text{Simplify.}$$

c. Think of $-(-x - 5)$ as $-1(-x - 5)$ and use the distributive property.

$$(2x - 5) - 1(-x - 5) = 2x - 5 + x + 5$$

$$= 3x \qquad \text{Simplify.}   ■$$

2   An **equation** is a statement that two expressions are equal. In this section we concentrate on solving **linear equations** in one variable.

Linear Equations in One Variable

$$3x = -15 \qquad 7 - y = 3y \qquad 4n - 9n + 6 = 0 \qquad z = -2$$

Linear equations are also called **first-degree equations** since the exponent on the variable is 1.

Linear Equation in One Variable

A linear equation in one variable is an equation that can be written in the form

$$ax + b = c$$

where a, b, and c are real numbers and $a \neq 0$.

When a variable in an equation is replaced by a number and the resulting equation is true, then that number is called a **solution** of the equation. For example, 1 is a solution of the equation $3x + 4 = 7$, since $3(1) + 4 = 7$ is a true statement. But 2 is not a solution of this equation, since $3(2) + 4 = 7$ is not a true statement. The **solution set** of an equation is the set of solutions of the equation. For example, the solution set of $3x + 4 = 7$ is $\{1\}$.

3   To **solve an equation** is to find the solution set of an equation. Equations with the same solution set are called **equivalent equations.** For example

$$3x + 4 = 7 \qquad 3x = 3 \qquad x = 1$$

are equivalent equations because they all have the same solution set, namely $\{1\}$. To solve an equation in x, we start with the given equation and write a series of simpler equivalent equations until we obtain an equation of the form

$$x = \textbf{number}$$

The following properties are used to write equivalent equations.

Properties for Solving Equations

If a, b, and c, are real numbers, then the **addition property of equality** says that

$$a = b \quad \text{and} \quad a + c = b + c$$

are equivalent equations.

Also, if $c \neq 0$, then the **multiplication property of equality** says that

$$a = b \quad \text{and} \quad ac = bc$$

are equivalent equations.

In other words, the addition property of equality states that the same number may be added to (or subtracted from) both sides of an equation, and the result is an equivalent equation. The multiplication property of equality states that both sides of an equation may be multiplied by (or divided by) the same nonzero number, and the result is an equivalent equation.

For example, to solve $2x + 5 = 9$, use the addition and multiplication properties of equality to isolate x, that is, to write an equivalent equation of the form

$$x = \text{number}$$

EXAMPLE 3   Solve for x: $2x + 5 = 9$.

Solution: First, use the addition property of equality and subtract 5 from both sides.

$$2x + 5 = 9$$
$$2x + 5 \;-\; 5 = 9 \;-\; 5$$
$$2x = 4 \qquad\qquad \text{Simplify.}$$

To finish solving for x, use the multiplication property of equality and divide both sides by 2.

$$\frac{2x}{2} = \frac{4}{2}$$

$$x = 2 \qquad \text{Simplify.}$$

To check, replace x in the original equation with 2.

$$2x + 5 = 9 \qquad \text{Original equation.}$$

$$2(2) + 5 = 9 \qquad \text{Let } x = 2.$$

$$4 + 5 = 9$$

$$9 = 9 \qquad \text{True.}$$

The solution set is $\{2\}$. ■

EXAMPLE 4   Solve for x: $-6x - 1 + 5x = 3$.

Solution: First, combine like terms $-6x$ and $5x$. Then use the addition property of equality and add 1 to both sides of the equation.

$$-6x - 1 + 5x = 3$$

$$-x - 1 = 3 \qquad \text{Combine like terms.}$$

$$-x - 1\ +1\ = 3\ +1 \qquad \text{Add 1 to both sides of the equation.}$$

$$-x = 4 \qquad \text{Simplify.}$$

Notice that this equation is not solved for x since we have $-x$ or $-1x$, not x. To solve for x, divide both sides by -1.

$$\frac{-x}{-1} = \frac{4}{-1}$$

$$x = -4 \qquad \text{Simplify.}$$

Check to see that the solution set is $\{-4\}$. ■

If an equation contains parentheses, use the distributive property to remove them.

EXAMPLE 5   Solve for x: $2(x - 3) = 5x - 9$.

Solution: First, use the distributive property.

$$2(x - 3) = 5x - 9$$

$$2x - 6 = 5x - 9$$

Next, get variable terms on the same side of the equation by subtracting $5x$ from both sides.

$$2x - 6\ -5x\ = 5x - 9\ -5x$$

$$-3x - 6 = -9 \qquad \text{Simplify.}$$

$$-3x - 6 \boxed{+ 6} = -9 \boxed{+ 6}$$ Add 6 to both sides.

$$-3x = -3$$ Simplify.

$$\frac{-3x}{\boxed{-3}} = \frac{-3}{\boxed{-3}}$$ Divide both sides by -3.

$$x = 1$$

Let $x = 1$ in the original equation to see that $\{1\}$ is the solution set.   ■

If an equation contains fractions, we first clear the equation of fractions by multiplying both sides of the equation by the least common denominator (LCD) of all fractions in the equation.

EXAMPLE 6   Solve for x: $\dfrac{x}{3} - \dfrac{x}{4} = \dfrac{1}{6}$.

Solution: First, clear the equation of fractions by multiplying both sides of the equation by 12, the LCD of denominators 3, 4, and 6.

$$\frac{x}{3} - \frac{x}{4} = \frac{1}{6}$$

$$\boxed{12}\left(\frac{x}{3} - \frac{x}{4}\right) = \boxed{12}\left(\frac{1}{6}\right)$$ Multiply both sides by the LCD 12.

$$12\left(\frac{x}{3}\right) - 12\left(\frac{x}{4}\right) = 2$$ Apply the distributive property.

$$4x - 3x = 2$$ Simplify.

$$x = 2$$ Simplify.

To check, let $x = 2$ in the original equation.

$$\frac{x}{3} - \frac{x}{4} = \frac{1}{6}$$ Original equation.

$$\frac{2}{3} - \frac{2}{4} = \frac{1}{6}$$ Let $x = 2$.

$$\frac{8}{12} - \frac{6}{12} = \frac{1}{6}$$ Write fractions with LCD.

$$\frac{2}{12} = \frac{1}{6}$$ Subtract.

$$\frac{1}{6} = \frac{1}{6}$$ Simplify.

This is a true statement, so the solution set is $\{2\}$.   ■

As a general guideline, the following steps may be used to solve a linear equation in one variable.

> **To Solve a Linear Equation in One Variable**
>
> *Step 1* Clear the equation of fractions by multiplying both sides of the equation by the lowest common denominator (LCD) of all denominators in the equation.
>
> *Step 2* Use the distributive property to remove grouping symbols such as parentheses.
>
> *Step 3* Combine like terms on each side of the equation.
>
> *Step 4* Use the addition property of equality to rewrite the equation as an equivalent equation, with variable terms on one side and numbers on the other side.
>
> *Step 5* Use the multiplication property of equality to isolate the variable.
>
> *Step 6* Check the proposed solution in the original equation.

EXAMPLE 7 Solve $\dfrac{x + 5}{2} + \dfrac{1}{2} = 2x - \dfrac{x - 3}{8}$.

Solution: Multiply both sides of the equation by 8, the LCD of 2 and 8.

$$8 \left(\frac{x + 5}{2} + \frac{1}{2} \right) = 8 \left(2x - \frac{x - 3}{8} \right) \qquad \text{Multiply both sides by 8.}$$

$$4(x + 5) + 4 = 16x - (x - 3) \qquad \text{Apply the distributive property.}$$

$$4x + 20 + 4 = 16x - x + 3 \qquad \begin{array}{l}\text{Use the distributive property} \\ \text{to remove parentheses.}\end{array}$$

$$4x + 24 = 15x + 3 \qquad \text{Combine like terms.}$$

$$-11x + 24 = 3 \qquad \text{Subtract } 15x \text{ from both sides.}$$

$$-11x = -21 \qquad \text{Subtract 24 from both sides.}$$

$$\frac{-11x}{-11} = \frac{-21}{-11} \qquad \text{Divide both sides by } -11.$$

$$x = \frac{21}{11} \qquad \text{Simplify.}$$

To check, verify that replacing x with $\dfrac{21}{11}$ makes the original equation true. The solution set is $\left\{ \dfrac{21}{11} \right\}$. ∎

4 So far, each linear equation that we have solved has had a single solution. A linear equation in one variable that has exactly one solution is called a **conditional equation.** We will now look at two other types of equations: contradictions and identities.

An equation in one variable that has no solutions is called a **contradiction,** and an equation in one variable that has every number (for which the equation is defined) as a solution is called an **identity.** The next examples show how to recognize contradictions and identities.

EXAMPLE 8   Solve for x: $3x + 5 = 3(x + 2)$.

Solution: First, use the distributive property and remove parentheses.

$$3x + 5 = 3(x + 2)$$

$$3x + 5 = 3x + 6$$

$$3x + 5\ -\ 3x = 3x + 6\ -\ 3x \qquad \text{Subtract } 3x \text{ from both sides.}$$

$$5 = 6$$

The equation $5 = 6$ is a false statement no matter what value the variable x might have. Thus the original equation has no solution. Its solution set is written either as $\{\ \}$ or $\varnothing$. This equation is a contradiction.   ∎

EXAMPLE 9   Solve for x: $6x - 4 = 2 + 6(x - 1)$.

Solution: First, use the distributive property and remove parentheses.

$$6x - 4 = 2 + 6(x - 1)$$

$$6x - 4 = 2 + 6x - 6$$

$$6x - 4 = 6x - 4 \qquad \text{Combine like terms.}$$

At this point we might notice that both sides of the equation are the same, so replacing x by any real number gives a true statement. Thus the solution set of this equation is the set of real numbers and the equation is an identity. Continuing to "solve" $6x - 4 = 6x - 4$, we eventually arrive at the same conclusion.

$$6x - 4\ +\ 4 = 6x - 4\ +\ 4 \qquad \text{Add 4 to both sides.}$$

$$6x = 6x$$

$$6x\ -\ 6x = 6x\ -\ 6x \qquad \text{Subtract } 6x \text{ from both sides.}$$

$$0 = 0 \qquad \text{Simplify.}$$

Since $0 = 0$ is a true statement for every value of x, the solution set is the set of all real numbers or $\{x \mid x \text{ is a real number}\}$, and the equation is called an identity.   ∎

Before we solve applications, we will first practice writing mathematical expressions.

EXAMPLE 10   Write the following as mathematical expressions.

a. A piece of board of length x feet is cut from a 10-foot board. Write the length of the remaining piece of board as an expression in x.

b. Two angles are complementary if the sum of their measures is 90°. If the measure of one angle is $3x - 10$ degrees, express the measure of its complement in terms of x.

c. If x is the first of two consecutive integers, write their sum as an expression in x.

Solution: **a.** The length of the remaining board is $(10 - x)$ feet.
b. The measure of its complement is $90 - (3x - 10)$ degrees.
c. If x is the first of two consecutive integers, the next consecutive integer is one greater, or $x + 1$. Their sum is $x + (x + 1)$. ∎

MENTAL MATH

Simplify each expression. See Example 1.

1. $3x + 5x + 6 + 15$

2. $8y + 3y + 7 + 11$

3. $5n + n + 3 - 10$

4. $m + 2m + 4 - 8$

5. $8x - 12x + 5 - 6$

6. $4x - 10x + 13 - 16$

EXERCISE SET 2.1

Simplify each expression. See Examples 1 and 2.

1. $-9x + 4x + 18 - 10x$

2. $5y - 14 + 7y - 20y$

3. $5k - (3k - 10)$

4. $-11c - (4 - 2c)$

5. $(3x + 4) - (6x - 1)$

6. $(8 - 5y) - (4 + 3y)$

7. $3(x - 2) + x + 15$

8. $-4(y + 3) - 7y + 1$

Solve for the variable. See Examples 3 and 4.

9. $5x - 4 = 26$

10. $2y - 3 = 11$

11. $-4 - 7z = 3$

12. $10 - 6x = -2$

13. $5y + 12 = 2y - 3$

14. $4x + 14 = 6x + 8$

15. $8x - 5x + 3 = x - 7 + 10$

16. $6 + 3x + x = -x + 2 - 26$

Solve for the variable. See Examples 5 and 7.

17. $5x + 12 = 2(2x + 7)$

18. $2(x + 3) = x + 5$

19. $3(x - 6) = 5x$

20. $6x = 4(5 + x)$

21. $-2(5y - 1) - y = -4(y - 3)$

22. $-3(2w - 7) - 10 = 9 - 2(5w + 4)$

Solve for the variable. See Examples 6 and 7.

23. $\dfrac{x}{2} + \dfrac{2}{3} = \dfrac{3}{4}$

24. $\dfrac{x}{2} + \dfrac{x}{3} = \dfrac{5}{2}$

25. $\dfrac{3t}{4} - \dfrac{t}{2} = 1$

26. $\dfrac{4r}{5} - 7 = \dfrac{r}{10}$

27. $\dfrac{n - 3}{4} + \dfrac{n + 5}{7} = \dfrac{5}{14}$

28. $\dfrac{2 + h}{9} + \dfrac{h - 1}{3} = \dfrac{1}{3}$

29. $\dfrac{3x - 1}{9} + x = \dfrac{3x + 1}{3} + 4$

30. $\dfrac{2z + 7}{8} - 2 = z + \dfrac{z - 1}{2}$

Solve the following. See Examples 8 and 9.

31. $4(n + 3) = 2(6 + 2n)$ $\{n \mid n \text{ is a real number}\}$

32. $6(4n + 4) = 8(3 + 3n)$

33. $3(x - 1) + 5 = 3x + 2$

34. $5x - (x + 4) = 4 + 4(x - 2)$

Write the following expressions. See Example 10.

35. A 25-centimeter length of string is cut into three pieces. If one piece is x centimeters long and a second piece is $7x$ centimeters long, express the third piece as an expression in x.

36. If a pocket contains x nickels and $3x$ quarters, express their total value as an expression in x.

37. Express the sum of three odd consecutive integers as an expression in x. Let x be the first odd integer.

38. If the width of a pad of paper is x inches and its length is 3 more than twice the width, express the perimeter as an expression in x.

Simplify each expression.

39. $-(n + 5) + (5n - 3)$

40. $-(8 - t) + (2t - 6)$

41. $4(6n - 3) - 3(8n + 4)$

42. $5(2z - 6) + 10(3 - z)$

43. $3x - 2(x - 5) + x$

44. $7n + 3(2n - 6) - 2$

Solve the following.

45. $-5x + 1 = -19$

46. $-3x - 4 = 11$

47. $x - 10 = -6x + 4$

48. $4x - 7 = 2x - 7$

49. $3x - 4 - 5x = x + 4 + x$

50. $13x - 15x + 8 = 4x + 2 - 24$

51. $5(y + 4) = 4(y + 5)$

52. $6(y - 4) = 3(y - 8)$

53. $0.6x - 10 = 1.4x - 14$

54. $0.3x + 2.4 = 0.1x + 4$

55. $6x - 2(x - 3) = 4(x + 1) + 4$

56. $10x - 2(x + 4) = 8(x - 2) + 6$

57. $\dfrac{3}{8} + \dfrac{b}{3} = \dfrac{5}{12}$

58. $\dfrac{a}{2} + \dfrac{7}{4} = 5$

59. $z + 3(2 + 4z) = 6(z + 1) + 5z$

60. $4(m - 6) - m = 8(m - 3) - 5m$

61. $\dfrac{3t + 1}{8} = \dfrac{5 + 2t}{7} + 2$

62. $4 - \dfrac{2z + 7}{9} = \dfrac{7 - z}{12}$

63. $\dfrac{m - 4}{3} - \dfrac{3m - 1}{5} = 1$

64. $\dfrac{n + 1}{8} - \dfrac{2 - n}{3} = \dfrac{5}{6}$

65. $\dfrac{x}{5} - \dfrac{x}{4} = \dfrac{1}{2}(x - 2)$

66. $\dfrac{y}{3} + \dfrac{y}{5} = \dfrac{1}{10}(y + 3)$

67. $5(x - 2) + 2x = 7(x + 4)$

68. $3x + 2(x + 4) = 5(x + 1) + 3$

69. $y + 0.2 = 0.6(y + 3)$

70. $-(w + 0.2) = 0.3(4 - w)$

71. $2y + 5(y - 4) = 4y - 2(y - 10)$

72. $9c - 3(6 - 5c) = c - 2(3c + 9)$

73. $2(x - 8) + x = 3(x - 6) + 2$

74. $4(x + 5) = 3(x - 4) + x$

75. $\dfrac{5x - 1}{6} - 3x = \dfrac{1}{3} + \dfrac{4x + 3}{9}$

76. $\dfrac{2r - 5}{3} - \dfrac{r}{5} = 4 - \dfrac{r + 8}{10}$

77. $-2(b - 4) - (3b - 1) = 5b + 3$

78. $4(t - 3) - 3(t - 2) = 2t + 8$

79. $1.5(4 - x) = 1.3(2 - x)$

80. $2.4(2x + 3) = -0.1(2x + 3)$

81. $\dfrac{1}{4}(a + 2) = \dfrac{1}{6}(5 - a)$

82. $\dfrac{1}{3}(8 + 2c) = \dfrac{1}{5}(3c - 5)$

83. If the money box in a drink machine contains x nickels, $5x$ dimes, and $30x - 1$ quarters, express their total **value** as an expression in x.

84. A plot of land is in the shape of a triangle. If one side is x meters, a second side is $2x - 3$ meters, and a third side is $3x - 5$ meters, express the perimeter of the lot as an expression in x.

85. Terri Santa Coloma must complete a project by the end of September. If x days in September have already passed, express the number of days left to complete the project as an expression in x.

86. If x is the first of four consecutive even integers, write their sum as an expression in x.

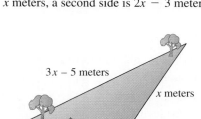

$3x - 5$ meters

x meters

$2x - 3$ meters

A Look Ahead

Solve the following. See the following example.

> EXAMPLE Solve for x: $5x(x - 1) + 14 = x(4x - 3) + x^2$.
>
> Solution: $5x^2 - 5x + 14 = 4x^2 - 3x + x^2$
>
> $$5x^2 - 5x + 14 = 5x^2 - 3x$$
> $$-5x + 14 = -3x$$
> $$14 = 2x$$
> $$7 = x$$ ∎

87. $x(x - 6) + 7 = x(x + 1)$

88. $7x^2 + 2x - 3 = 6x(x + 4) + x^2$

89. $3x(x + 5) - 12 = 3x^2 + 10x + 3$

90. $x(x + 1) + 16 = x(x + 5)$

Writing in Mathematics

91. Explain the difference between solving an equation for a variable and simplifying an expression.

92. Explain why the multiplication property of equality does not include multiplying both sides of an equation by 0.

Skill Review

Find the value of the following expressions for the given values. See Section 1.4.

93. $2a + b - c$; $a = 5$, $b = -1$, and $c = 3$

94. $-3a + 2c - b$; $a = -2$, $b = 6$, and $c = -7$

95. $4ab - 3bc$; $a = -5$, $b = -8$, and $c = 2$

96. $ab + 6bc$; $a = 0$, $b = -1$, and $c = 9$

97. $n^2 - m^2$; $n = -3$ and $m = -8$

98. $2n^2 + 3m^2$; $n = -2$ and $m = 7$

2.2
Formulas

OBJECTIVES

Tape 2

   1   Use formulas to model and solve problems.

   2   Solve a formula for a specified variable.

Equations that describe known relationships among real-life phenomena, such as time, area, and gravity, are called **formulas.** Examples of formulas are the following:

> *Formulas*
>
> $y = mx + b$ Slope–intercept form of a linear equation in two variables
>
> $A = P + Prt$ Banking formula used for loans
>
> $A = l \cdot w$ Formula for finding area of a rectangle
>
> $d = r \cdot t$ Physics formula used for distance.

1   In the next examples, we see how formulas are used as models to solve problems. The following steps may be helpful when using formulas to solve problems.

> **To Use Formulas to Solve Problems**
>
> *Step 1*   Determine the appropriate formula. See the inside front cover if needed for a list of common formulas.
> *Step 2*   Replace variables with their known measurement.
> *Step 3*   Solve the formula for the remaining variable.
> *Step 4*   State the solution to the problem.
> *Step 5*   Check the solution in the originally stated problem.

EXAMPLE 1   Find the amount of principal that must be invested to earn $240 in a passbook savings account in 1 year at 8% simple interest.

Solution: *Step 1*   Find the appropriate formula.
The formula for simple interest is $I = PRT$, where I is interest, P is principal, R is rate, and T is time, in years.

Step 2   Replace variables with their known measurement.
Substitute 240 for I, 0.08 for R, and 1 for T.

$$I = P \cdot R \cdot T$$
$$240 = P \cdot (0.08)(1) \qquad \text{Let } I = 240, R = 0.08, \text{ and } T = 1.$$
$$240 = 0.08P \qquad \text{Simplify.}$$

Step 3   Solve.
To solve for P, divide both sides by 0.08.

$$\frac{240}{0.08} = \frac{0.08P}{0.08}$$
$$3000 = P$$

Step 4   State the solution.
Thus $3000 must be invested to obtain $240 in interest.
Step 5   Check.
If $3000 is invested at 8% simple interest for 1 year, the interest is

$$(\$3000)(0.08)(1) = \$240, \text{ the given interest.} \quad\blacksquare$$

EXAMPLE 2   The distance from Atlanta, Georgia, to Savannah, Georgia, is about 240 miles. How long will it take to make this trip if the average speed is 50 miles per hour?

Solution: *Step 1*   Find the appropriate formula.
The appropriate formula is the distance formula: $d = rt$, where d is distance, r is rate, and t is time.

Step 2   Replace variables with their known measurement.
Replace *d* with 240 and *r* with 50.

$$d = rt$$

$$240 = (50)t \qquad \text{Let } d = 240, r = 50.$$

Step 3   Solve.

$$240 = 50t$$

$$\frac{240}{50} = \frac{50t}{50} \qquad \text{Divide both sides by 50.}$$

$$4.8 = t \qquad \text{Simplify.}$$

Step 4   State the solution.
It takes the driver 4.8 hours or 4 hours and 48 minutes to make the trip from Atlanta to Savannah.

Step 5   Check.
The diagram shows that this answer is reasonable. To check, notice that the distance traveled after 4.8 hours at a speed of 50 mph is 4.8(50) or 240 miles.

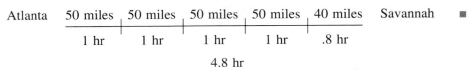

Atlanta	50 miles	50 miles	50 miles	50 miles	40 miles	Savannah
	1 hr	1 hr	1 hr	1 hr	.8 hr	

4.8 hr

EXAMPLE 3   A gallon of sealer covers 480 square feet. How many gallon containers of sealer should be bought to cover a rectangular driveway 24 feet wide by 90 feet long?

Solution: We will first find the area of the driveway. The area formula for a rectangle is $A = lw$. Let $l = 90$ feet and $w = 24$ feet.

$$A = lw$$

$$= (90 \text{ feet})(24 \text{ feet}) \qquad \text{Let } l = 90 \text{ feet and } w = 24 \text{ feet.}$$

$$= 2160 \text{ square feet}$$

To find how many gallon containers of sealer are needed, divide the area of the driveway, 2160 square feet, by the number of square feet that each gallon of sealer will cover, or 480 square feet.

$$\text{number of gallons needed} = \frac{2160 \text{ square feet}}{480 \text{ square feet}}$$

$$= 4.5$$

Since the question asks how many gallon containers should be bought, the answer is 5 one-gallon containers of sealer.   ■

2   Many times we want to solve an equation for a specified variable. In other words, we want to express the specified variable in terms of the other variables. To isolate the specified variable, we use steps similar to those for solving linear equations.

> **To Solve Equations for a Specified Variable**
>
> *Step 1.* Clear the equation of fractions by multiplying each side of the equation by the lowest common denominator.
>
> *Step 2.* Use the distributive property to remove grouping symbols such as parentheses.
>
> *Step 3.* Combine like terms on each side of the equation.
>
> *Step 4.* Use the addition property of equality to rewrite the equation as an equivalent equation, with terms containing the specified variable on one side and all other terms on the other side.
>
> *Step 5.* Use the distributive property and the multiplication property of equality to isolate the specified variable.

EXAMPLE 4   Solve $3y - 2x = 7$ for y.

Solution: This is a linear equation in two variables. Often an equation such as this is solved for y in order to reveal some properties of this equation, which we will learn more about in Chapter 3. Since there are no fractions or grouping symbols, we begin with step 4 and isolate the term containing the specified variable y by adding $2x$ to both sides of the equation.

$$3y - 2x = 7$$
$$3y - 2x \;+ 2x = 7 \;+ 2x$$
$$3y = 7 + 2x$$

To solve for y, divide both sides by 3.

$$\frac{3y}{3} = \frac{7 + 2x}{3}$$

$$y = \frac{2x + 7}{3} \quad \text{or} \quad y = \frac{2x}{3} + \frac{7}{3} \quad \blacksquare$$

EXAMPLE 5   Solve $A = \dfrac{1}{2}(B + b)h$ for b.

Solution: Since this formula for finding the area of a trapezoid contains fractions, we begin by multiplying both sides of the equation by the LCD 2.

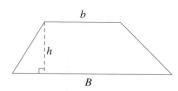

$$A = \frac{1}{2}(B + b)h$$

$$2 \cdot A = 2 \cdot \frac{1}{2}(B + b)h$$

$$2A = (B + b)h$$

Next, use the distributive property and remove parentheses.

$$2A = (B + b)h$$
$$2A = Bh + bh$$

$$2A - Bh = bh \qquad\qquad \text{Isolate the term containing } b$$
by subtracting Bh from both sides.

$$\frac{2A - Bh}{h} = \frac{bh}{h} \qquad\qquad \text{Divide both sides by } h.$$

$$\frac{2A - Bh}{h} = b \quad \text{or} \quad b = \frac{2A - Bh}{h} \qquad \blacksquare$$

HELPFUL HINT

Remember that we may isolate the specified variable on either side of the equation.

 CALCULATOR BOX

Evaluate Formulas:

A calculator is a useful aid in evaluating formulas, especially if the replacement numbers make the calculations tedious.

     The formula $I = PRT$ is used to find simple interest. Use a calculator to find the simple interest earned on depositing a principal P of \$1200 in an account paying a rate R of 6.25% simple interest for a time T of 5.5 years.

     Using the formula $I = PRT$, we have

$$I = (1200)(6.25\%)(5.5)$$

To evaluate this product, press the keys

$$\boxed{1200} \;\; \boxed{\times} \;\; \boxed{.0625} \;\; \boxed{\times} \;\; \boxed{5.5} \;\; \boxed{=}$$

The display will read $\boxed{412.5}$, which means that there will be \$412.50 earned in interest after 5.5 years.

Use a calculator to evaluate each formula with the given replacement values. If necessary, round your answer to three decimal places.

1. Evaluate $V = s^3$ when $s = 4.7$

2. Evaluate $m = \dfrac{y_2 - y_1}{x_2 - x_1}$ if $y_2 = 1.1$, $y_1 = 2.1$, $x_2 = -6.7$, and

    $x_1 = 5.9$

3. Evaluate $S = \dfrac{rl - a}{r - l}$ when $r = 2$, $l = 40{,}000$, and $a = 43$.

4. Evaluate $C = \dfrac{5}{9}(F - 32)$ when $F = 87$.

5. Evaluate $F = \dfrac{9C + 160}{5}$ when $C = 87$.

6. Evaluate $P = 2l + 2w$ when $l = 6.8$ and $w = 3.3$

MENTAL MATH

Solve each equation for the specified variable. See Examples 4 and 5.

1. $2x + y = 5$; for y

2. $7x - y = 3$; for y

3. $a - 5b = 8$; for a

4. $7r + s = 10$; for s

5. $5j + k - h = 6$; for k

6. $w - 4y + z = 0$; for z

EXERCISE SET 2.2

Solve the following. See Example 1.

1. Find how much interest $4000 earns in $2\frac{1}{2}$ years in a passbook savings account paying 8% simple interest annually.

2. Allan wants to win the Publisher's Clearinghouse Sweepstakes and live off the interest earned by placing it into a savings account paying 8% simple interest annually. Find how much he needs to win to have $40,000 annually to live on.

3. Frank needs $1050 for fall tuition next year. If he places $985 in a passbook savings account that pays 7% simple interest, will he have enough money in this account after 1 year to pay for fall tuition?

4. Find the amount of principal that must be invested to earn $400 interest in 2 years in a savings account that pays 5% simple interest.

Solve the following. See Example 2.

5. Omaha, Nebraska, is about 90 miles from Lincoln, Nebraska. Irania must go to the law library in Lincoln to get a document for the law firm she works for. Find how long it takes her to drive **round-trip** if she averages 50 mph.

6. It took Tanya $5\frac{1}{2}$ hours round-trip to drive from her house to her beach house 154 miles away. Find her average speed.

*Delta Airlines awards "Frequent Flyer" miles equal to the number of miles traveled rounded **up** to the nearest thousand. Use this for Exercises 7 and 8.*

7. A 5.5-hour nonstop flight from Orlando, Florida, to San Francisco, California, averages 470 mph. Find the "Frequent Flyer" miles earned on this flight.

8. A 45-minute $\left(\frac{3}{4}\text{-hour} \right)$ nonstop flight from New Orleans, Louisiana, to Houston, Texas, averages 500 mph. Find the "Frequent Flyer" miles earned on this flight.

Solve the following. See Example 3.

9. An aquarium is 30 inches long, 16 inches wide, and 20 inches tall. Find the volume of water in the aquarium if the water depth is 18 in.

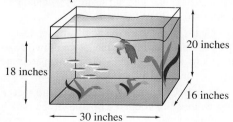

10. Piranha fish require 1.5 cubic feet of water per fish to maintain a healthy environment. Find the maximum number of piranhas you could put in a tank measuring 8 feet by 3 feet by 6 feet.

11. Find how much rope is needed to wrap around the earth at the equator, if the radius of the earth is 4000 miles. (Use $\pi = 3.14$)

$C = 2\pi r$

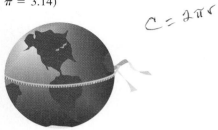

12. If the length of a rectangular flower bed is 16 meters and its width is 3.5 meters, find the amount of fencing needed to enclose the flower bed.

Solve each equation for the specified variable. See Example 4.

13. $P = 2L + 2W$; for W

14. $A = 3M - 2N$; for N

15. $J = AC - 3$; for A

16. $y = mx + b$; for x

Solve each equation for the specified variable. See Example 5.

17. $E = I(r + R)$; for r

18. $A = P(1 + rt)$; for t

19. $s = \dfrac{n}{2}(a + L)$; for L

20. $\dfrac{3}{4}(b - 2c) = a$; for b

Solve the following.

21. A package of floor tiles contains 24 one-foot-square tiles. Find how many packages should be bought to cover a square ballroom floor whose side measures 64 feet.

64 feet

22. One-foot-square ceiling tiles are sold in packages of 50. Find how many packages must be bought for a rectangular ceiling 18 feet by 12 feet.

23. The day's high temperature in Phoenix, Arizona, was recorded as 104°F. Write 104°F as degrees Celsius.

24. The annual low temperature in Nome, Alaska, was recorded as −15°C. Write −15°C as Fahrenheit degrees.

25. Find how much interest $10,000 earns in 2 years in a certificate of deposit paying 8.5% simple interest.

26. Bryan, Eric, Mandy, and Melissa would like to go to Disneyland in 3 years. Their total cost should be $3500. If each invests $800 in a savings account paying 5.5% interest, will they have enough in 3 years?

27. Find how long it takes Mark to drive 135 miles on I-10 if he merges onto I-10 at 10A.M. and drives nonstop with his cruise control set on 60 mph.

28. Beaumont, Texas, is about 150 miles from Toledo Bend. If Leo Miller leaves Beaumont at 4 A.M. and averages 45 mph, when should he arrive at Toledo Bend?

29. Find how many goldfish you can put in a cylindrical tank whose diameter is 8 meters and whose height is 3 meters, if each goldfish needs 2 cubic meters of water.

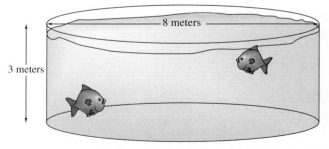

8 meters

3 meters

30. If the area of a triangular kite is 18 square feet and its base is 4 feet, find the height of the kite.

31. A gallon of latex paint can cover 500 square feet. Find how many gallons should be bought to paint two coats on each wall of a rectangular room whose dimensions are 14 feet by 16 feet (assume 8-foot ceilings).

32. A gallon of enamel paint can cover 300 square feet. Find how many gallons are needed to paint three coats on a wall measuring 21 feet by 8 feet.

33. Normal room temperature is about 72°F. Write this temperature as degrees Celsius.

34. Find the temperature at which the Celsius measurement and Fahrenheit measurement are the same number.

35. Maria's Pizza sells one 16-inch cheese pizza or two 10-inch cheese pizzas for $9.99. Determine which size gives more pizza.

— 16 inches —   — 10 inches —   — 10 inches —

36. A lawn is in the shape of a trapezoid whose height is 50 meters and whose bases are 75 meters and 105 meters. Find how many bags of fertilizer are needed to cover the lawn if each bag covers 600 square meters.

37. A bag of grass seed covers 500 square feet. Find how many bags are needed to cover a lawn in the shape of a parallelogram whose base is 90 feet and height is 65 feet.

Solve each equation for the specified variable.

38. $D = rt$; for t

39. $W = gh$; for g

40. $I = PRT$; for R

41. $V = LWH$; for L

42. $9x - 4y = 16$; for y

43. $2x + 3y = 17$; for y

44. $W = gh - 3gt^2$; for g

45. $A = Prt + P$; for P

46. $T = C(2 + AB)$; for B

47. $A = 5H(b + B)$; for b

48. $C = 2\pi r$; for r

49. $S = 2\pi r^2 + 2\pi rh$; for h

50. $\dfrac{1}{u} - \dfrac{1}{v} = \dfrac{1}{w}$; for w

51. $\dfrac{1}{r_1} + \dfrac{1}{r_2} = \dfrac{1}{R}$; for R

52. $N = 3st^4 - 5sv$; for v

53. $L = a + (n - 1)d$; for d

54. $S = \dfrac{a}{1 - r}$; for r

55. $m = \dfrac{y_2 - y_1}{x_2 - x_1}$; for y_1

56. $S = 2LW + 2LH + 2WH$; for H

57. $T = 3vs - 4ws + 5vw$; for v

Skill Review

Translate the phrase into an algebraic expression. See Section 1.2.

58. Four divided by the product of a number and 5.

59. Three times the sum of a number and four.

60. Double the sum of six and four times the number.

61. Twice a number multiplied by three times the opposite of the number.

Find the value of each expression when $x = 5$ and $y = 2$.

62. $\dfrac{x^2 - y^2}{3}$

63. $\dfrac{3x + 2y - 1}{6}$

64. $\dfrac{12(x - y)}{4 + x}$

65. $\dfrac{10x + 5y}{xy}$

2.3
Applications of Linear Equations

OBJECTIVE   **1**   Solve word problems.

Tape 3

1   In the previous section we solved problems using known formulas as models. In this section, we solve problems using equation models we create. The problem-solving steps given next may be helpful.

> **To Solve a Word Problem**
>
> *Step 1*   Read and then reread the problem. Choose a variable to represent one unknown quantity.
> *Step 2*   Use this variable to represent any other unknown quantities.
> *Step 3*   Draw a diagram if possible to visualize the known facts.
> *Step 4*   Translate the word problem into an equation.
> *Step 5*   Solve the equation.
> *Step 6*   Answer the question asked and check to see if the answer is **reasonable.**
> *Step 7*   Check the solution in the originally stated problem.

Although it may seem that every word problem is different, with practice you will learn to recognize some basic types and the strategies used to solve them. We will begin with an example involving consecutive integers. The following table may be helpful in reviewing consecutive integers and their variable representation.

	Variable Representation	*Examples*
Consecutive integers	$x, x + 1, x + 2, \ldots$	5, 6, 7 21, 22, 23 $-10, -9, -8$
Consecutive even integers	$x, x + 2, x + 4, \ldots$	8, 10, 12 $-52, -50, -48$ 74, 76, 78
Consecutive odd integers	$x, x + 2, x + 4, \ldots$	$-7, -5, -3$ 11, 13, 15 69, 71, 73

EXAMPLE 1   Find three consecutive odd integers such that the sum of the first and the third is eleven more than the second integer.

Solution: *Step 1*   Let a variable represent one unknown.
Let x represent the first of three consecutive odd integers.

Step 2   Use this variable to represent other unknown quantities.
Since

$$x = \text{first odd integer, then}$$
$$x + 2 = \text{second odd integer and}$$
$$x + 4 = \text{third odd integer}$$

Step 4   Write an equation from the stated problem.
Then

$$\boxed{\text{first}} + \boxed{\text{third}} = \boxed{\text{second}} + \boxed{11}$$

or

$$x + (x + 4) = (x + 2) + 11$$

Step 5   Solve the equation.

$$x + (x + 4) = (x + 2) + 11$$
$$2x + 4 = x + 13$$
$$x + 4 = 13$$
$$x = 9$$

Step 6   State the solution.
If $x = 9$, then $x + 2 = 11$ and $x + 4 = 13$. The three consecutive odd integers are 9, 11, and 13.

Step 7   Check the solution in the originally stated problem.
To check, notice that the sum of the first and the third consecutive odd integers is $9 + 13$ or 22, and this sum is 11 more than the second odd integer, 11. The integers are 9, 11, and 13.   ■

Geometric problems involve geometric figures and one or more geometric formulas. Neatly drawn diagrams help organize these word problems.

EXAMPLE 2   A square and an equilateral triangle have the same perimeter. Each side of the triangle is 6 centimeters longer than each side of the square. Find the length of each side of the triangle.

Solution: *Step 1*   Let a variable represent one unknown.
Let x represent the length of each side of the square.

Step 2   Use this variable to represent other unknown quantities.
Since each side of the triangle is 6 centimeters longer than each side of the square, let $x + 6$ represent the length of each side of the triangle.

Step 3   Draw a diagram.

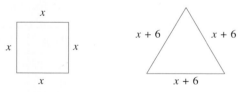

Step 4   Write an equation from the stated problem.
Recall that the perimeter of a polygon such as a square or a triangle is the sum of the lengths of its sides. We are given that the

$$\boxed{\text{square's perimeter}} \;=\; \boxed{\text{triangle's perimeter}}$$

or
$$x + x + x + x = (x + 6) + (x + 6) + (x + 6)$$
$$4x = 3x + 18$$

Step 5   Solve the equation.

$$4x = 3x + 18$$
$$x = 18$$

Step 6 State the solution in words.

Since the length of each side of the square is 18 centimeters, the length of each side of the triangle is $x + 6 = 18 + 6 = 24$ centimeters.

Step 7 Check the solution in the originally stated problem.

To check, see that the perimeter of the square equals the perimeter of the triangle. The perimeters are the same and the side of the triangle is 24 centimeters.

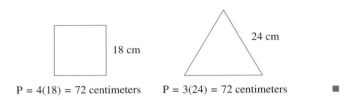

P = 4(18) = 72 centimeters   P = 3(24) = 72 centimeters ■

Our next example is a motion problem using the formula $d = r \cdot t$.

EXAMPLE 3 Marie rode her bicycle at an average speed of 18 mph on level roads and then slowed down to 10 mph on the hilly roads of the trip. The entire trip covered 98 miles. How long did the entire trip take if traveling the level roads took the same time as traveling the hilly roads?

Solution: *Step 1* Let a variable represent one unknown.

We are looking for the length of the entire trip. Begin by letting x represent the time spent on level roads.

Step 2 Use this variable to represent other unknown quantities.

Since both parts of the trip took the same time, we will let x also represent the time spent on hilly roads.

Step 3 Draw a diagram.

We summarize the information from the problem in the following table. Fill in the rates given and use the formula $d = r \cdot t$ to fill in the distance column.

Trip	Rate	Time	Distance ($= r \cdot t$)
Level	18	x	$18x$
Hilly	10	x	$10x$

Step 4 Write an equation from the stated problem.

Since the entire trip covered 98 miles, we have that

total distance	=	level distance	+	hilly distance
98	=	$18x$	+	$10x$

$$98 = 28x$$

Step 5 Solve the equation.

$$\frac{98}{28} = \frac{28x}{28}$$

$$3.5 = x$$

Step 6   State the solution in words.

Recall that $x = 3.5$ hours is the time of the level portion of the trip and $x = 3.5$ hours is also the time of the hilly portion. The time of the entire trip is then 3.5 hours + 3.5 hours or 7 hours.

Step 7   Check the solution in the originally stated problem.

If Marie rides for 3.5 hours at 18 mph, her distance is $18(3.5) = 63$ miles. If Marie rides for 3.5 hours at 10 mph, her distance is $10(3.5) = 35$ miles. The total distance 63 miles + 35 miles is the required distance, 98 miles.   ■

Collection problems like the following one involve a combination of objects. For example, finding the number of different priced tickets sold to a concert is a collection problem. This information can be valuable in deciding how to optimize profit.

EXAMPLE 4   The bottom of Jim's backpack contains 20 coins in nickels and dimes. If the coins have a total value of $1.85, find the number of each type of coin.

Solution: *Step 1*   Let a variable represent one unknown.

Let $x = $ the number of nickels.

Step 2   Use this variable to represent other unknown quantities.

Since there are 20 coins in all, $20 - x = $ number of dimes.

Step 3   Draw a diagram.

If we have x nickels, we have $x(0.05)$ money in nickels, since each nickel is worth $0.05. If we have $(20 - x)$ dimes, we have $(20 - x)(0.10)$ money in dimes. This is organized in a table as follows:

Coin	Number of Coins	Value of Coins	Total Value
Nickels	x	0.05	$0.05x$
Dimes	$20 - x$	0.10	$0.10(20 - x)$
Total	20		1.85

Step 4   Write an equation from the stated problem.

The total amount of money, $1.85, must equal the money in nickels plus the money in dimes.

money in nickels   money in dimes   total money

$$0.05x \quad + \quad 0.10(20 - x) \quad = \quad 1.85$$

Step 5   Solve the equation.

Use the multiplication property of equality and multiply both sides of this equation by 100. This has the effect of moving decimal points two places to the right, or

$$0.05x + 0.10(20 - x) = 1.85$$
$$100[0.05x + 0.10(20 - x)] = 100(1.85)$$
$$5x + 10(20 - x) = 185$$
$$5x + 200 - 10x = 185$$
$$-5x = -15$$
$$x = 3$$

Step 6 State the solution in words.

The number of nickels is 3 and the number of dimes is $20 - 3$ or 17.

Step 7 Check the solution in the originally stated problem.

The value of 3 nickels is ($0.05)3 or $0.15. The value of 17 dimes is ($0.10)17 or $1.70. The total value is their sum, $0.15 + $1.70 = $1.85, the given total. ∎

Mixture problems involve two or more different quantities being combined to form a new mixture. These applications range from Dow Chemical's need to form a chemical mixture of a required strength to Planter's Peanut Company's need to find the correct mixture of peanuts and cashews, given taste and price constraints.

EXAMPLE 5 A chemist working on his doctorate degree needs 12 liters of a 50% acid solution for a lab experiment. The stockroom has only 40% and 70% solutions. How much of each solution should be mixed together to form 12 liters of a 50% solution?

Solution: Let x = number of liters of 40% solution, then $12 - x$ = number of liters of 70% solution since a total of 12 liters is needed. The following table summarizes the information given.

	No. of liters	Strength	Amount of Acid
First solution	x	40%	0.40x
Second solution	$12 - x$	70%	0.70(12 − x)
Mixture needed	12	50%	0.50(12)

The amount of acid in each solution is found by multiplying the strength of each solution by the number of liters. The amount of acid in the final mixture is the sum of the amounts of acid in the two solutions.

acid in first solution	+	acid in second solution	=	acid in mixture
0.40x	+	0.70(12 − x)	=	0.50(12)

Now solve the equation for x.

$$0.40x + 0.70(12 - x) = 0.50(12)$$

$$0.4x + 8.4 - 0.7x = 6$$

$$-0.3x + 8.4 = 6$$

$$-0.3x = -2.4$$

$$x = 8$$

If $x = 8$, then $12 - x = 12 - 8 = 4$. Thus, if 8 liters of the 40% solution is mixed with 4 liters of the 70% solution, the result is 12 liters of a 50% solution. ∎

The next example is an investment problem.

EXAMPLE 6   Juan invested part of his $25,000 bonus in a stock that made a 12% profit and the rest in stock that suffered a 15% **loss.** In 1 year, his overall loss for both investments was $1590. Find the amount of each investment.

Solution: Let $x =$ amount of money invested that received a 12% profit. Then $25,000 - x =$ amount of money invested that received a 15% loss. Juan made a 12% profit in x amount of stock. This is 12% of x or $0.12x$. Juan also had a 15% loss in $(25,000 - x)$ amount of stock. This loss can be represented by 15% of $(25,000 - x)$ or $0.15(25,000 - x)$.

	Amount Invested	Percent Gain or Loss	Amount Profited or Lost
Profit stock	x	12%	$0.12x$
Loss stock	$25,000 - x$	15%	$0.15(25,000 - x)$

Represent Juan's overall loss of $1,590 by -1590. Then Juan's profit minus his loss is his overall loss, or

profit	$-$	loss	$=$	overall loss

or

$$0.12x - 0.15(25,000 - x) = -1590$$
$$0.12x - 3750 + 0.15x = -1590$$
$$0.27x = 2160$$
$$x = 8000$$

Thus $8000 was invested at a 12% profit and $25,000 - x = \$25,000 - \$8000 = \$17,000$ was invested with a 15% loss. To check, a 12% profit on $8000 = 0.12(\$8000) = \$960$. A 15% loss on $\$17,000 = 0.15(\$17,000) = \$2550$, and $960 - 2550 = -1590$, a loss of $1590. ∎

EXERCISE SET 2.3

Solve. See Example 1.

1. Find three consecutive even integers such that the sum of the first integer and three times the third integer is 64 more than the second integer.

2. Find three consecutive odd integers such that four times the first minus the third is the first increased by 10.

3. The sum of three consecutive integers is 13 more than twice the smallest integer. Find the integers.

4. Determine whether there are two consecutive odd integers such that 7 times the first exceeds 5 times the second by 54.

Solve. See Example 2.

5. The perimeter of an equilateral triangle is 7 inches more than the perimeter of a square, and the side of the triangle is 5 inches longer than the side of the square. Find the side of the triangle.

6. A square pig pen and a pen shaped like an equilateral triangle have equal perimeters. Find the side of each pen if the side of the triangular pen is 10 feet less than twice the side of the square pen.

7. The length of a rectangular sign is 2 feet less than three times its width. Find the dimensions if the perimeter is 28 feet.

x feet

8. In a blueprint of a rectangular room, the length is to be 2 centimeters greater than twice its width. Find the dimensions if the perimeter is to be 40 centimeters.

Solve. See Example 3.

9. A motorcyclist drove 200 miles in 4 hours. He kept a steady speed for the first 3 hours of the trip. Due to rain, he reduced his speed by 20 mph for the last hour of the trip. Find the two speeds.

10. A plane traveled to Los Angeles at 200 mph and returned at 300 mph. If the total traveling time was 2 hours, find how far away Los Angeles is.

11. Jose traveled to San Diego at an average rate of 50 mph. Carlos made the same trip in 1 hour less time at an average rate of 60 mph. Find the distance of the trip.

12. Two trains leave Chicago at the same time and travel in opposite directions. At the end of 4 hours they are 660 miles apart. One train is 15 mph slower than the other. Find their speeds.

Solve. See Example 4.

13. Amy's purse contains 75 coins in nickels and dimes. Find the number of each type of coin if their value is $5.95.

14. A stamp collection contains 15-cent stamps and 25-cent stamps with a total face value of $142.50. If the number of 25-cent stamps is 50 more than twice the number of 15-cent stamps, find the number of 25-cent stamps.

15. Eight hundred tickets to a play were sold for a total of $2000. If the adult tickets cost $4 each and student tickets cost $2 each, find how many of each kind of ticket was sold.

16. The Little League had a fund raiser at which they washed cars for $2 and washed bikes for 75 cents. If $450 was collected from 350 people, find how many bikes were washed.

Solve. See Example 5.

17. A pharmacist needs 6 ounces of a 30% codeine solution, but he only has bottles of 20% and 50% codeine solutions. Find how many ounces from each bottle should be mixed for this prescription.

18. A lab researcher has 4 liters of a 10% acid solution. Find how much pure acid should be added to increase the strength to 25%.

19. Find how many gallons of a 15% saline solution should be mixed with 10 gallons of a 20% saline solution to produce a 16% saline solution.

20. A doctor needs to dilute 10 centiliters of a 50% glucose solution by adding water to get a 20% glucose solution. Find how much water he should add.

Solve. See Example 6.

21. Lilian invested $24,000 in two accounts: a mutual fund paying 8% annual interest and a CD paying 9% interest. If her annual interest was $2020, find how much she invested in each account.

22. Julio invested money in a passbook savings account paying 6% annual interest and $3200 more in a bond paying 8%. If his total yearly income from both investments was $1656, find how much he invested in the bond.

23. Lin made two investments totaling $50,000. After 1 year, on one investment she made an 18% profit, but took an 11% loss on the other investment. Find the amount of each investment if her net gain was $4360.

24. Mr. Goldberg invested his $40,000 bonus in two accounts. He took a 4% loss on one investment and made a 12% profit on the other investment, but ended up breaking even. Find how much he invested at each rate.

Solve.

25. Find three consecutive integers such that the sum of the first integer and twice the second integer is 54 more than the third integer.

26. Find four consecutive integers such that their sum is −54.

27. Two cars leave Tuscon at the same time traveling in opposite directions, one driving 55 mph and the other at 65 mph. Find how long Donald and Howard are able to talk on their car phones if the phones have a 300-mile range.

28. Two joggers are 12 miles apart traveling toward each other. Find how long it takes them to meet if one jogger runs at 6 mph and the other runs 2 mph slower.

29. How many pounds of mint tea, which costs $2.25 per pound, must be mixed with 40 pounds of camomile tea, which costs $6.00 per pound, to have a blend that costs $3.50 per pound?

30. When Mr. Whipple mixes $6.00-per-pound coffee with $5.80-per-pound coffee, he sells the blend for $5.85 per pound. Find how much of each type of coffee he uses to make 100 pounds of the blend.

31. Find measures of the angles of a triangle if one angle is twice another angle, and the third angle is 105°.

32. One angle is twice its complement increased by 30°. Find the two complementary angles.

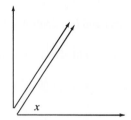

33. Irene Boesky invested a certain amount of money at 9% annually, twice that amount at 10% annually, and three times that amount at 11% annually. Find the amount invested at each rate if her total yearly income from the investments was $8370.

34. Hank Williams put $7000 in an account paying 10%. Find how much he should deposit at 7% annual interest so that the average return on the two investments is 9% annually.

35. On a 225-mile trip on Germany's autobahn system, Emil traveled at an average speed of 70 mph, stopped for gas, and then averaged 60 mph for the remainder of the trip. If the entire trip took 3.75 hours and the gas stop took 15 minutes, find how far Emil drove before the gas stop.

36. Two hikers are 21 miles apart on the Appalachian Trail walking toward each other. They meet in 3 hours. Find the rate of each hiker if one hiker walks 2 mph faster than the other.

37. A 6-gallon radiator contains a 40% antifreeze solution. Find how much needs to be drained and replaced by pure antifreeze to get a 60% solution.

38. A 4-gallon radiator contains a 50% antifreeze solution. Find how much needs to be drained and replaced by water to obtain a 40% antifreeze solution.

39. James Dean cashed a $170 check and got 10 bills, some of them $20 bills and some of them $10 bills. Find how many bills of each type he got.

40. Ben's Grocery Store begins the day with $845 in one-, five-, ten-, and twenty-dollar bills in the cash drawer. There are twice as many ones as twenties, five times as many tens as twenties, and the number of fives is half the number of tens. Find how many bills of each type are in the drawer.

41. The sum of the angles in a triangle is 180°. Find the angles of a triangle whose two base angles are equal, and whose third angle is 10° less than three times a base angle.

42. Find an angle such that its supplement is equal to twice its complement increased by 50°.

43. Carla rented a canoe for 4 hours. She can paddle upstream at 2 mph and downstream at 6 mph. Find how far upstream she should paddle if she wants to return in exactly 4 hours.

44. Jonathan can row a boat at a rate of 3 mph. He plans to row up a river with a current of 1.5 mph, turn around, and row downstream to his starting point in a total of 4 hours. Find how many hours he rows upstream.

45. A welder has a rectangular piece of sheet metal 16 feet by 12 feet. He plans to make an open box by cutting squares from each of the corners and folding up the sides. Find how large a square he should cut so that the length of the finished box is three times its width.

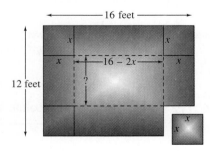

46. Kristin has a 10-inch by 15-inch piece of cardboard with which to make an open box by cutting squares from the corners and folding up the sides. Find how large a square she should cut so that the width of the finished box is three times the height of the box.

47. A butcher mixes ground sirloin worth $2.25 per pound with hamburger worth $1.00 per pound to make 500 pounds of a blend to sell for $1.50 per pound. Find how much ground sirloin and how much hamburger he uses.

48. Determine whether there are three consecutive integers such that their sum is three times the second integer.

49. To break even in a manufacturing business, revenue (income) must equal the cost of production. The cost C to produce x number of skateboards is $C = 100 + 20x$. The skateboards are sold wholesale for $24 each, so revenue R is given by $R = 24x$. Find how many skateboards the manufacturer needs to produce and sell to break even.

50. The revenue R from selling x number of computer boards is given by $R = 60x$, and the cost C of producing them is given by $C = 50x + 5000$. Find how many boards must be sold to break even. Find how much money is needed to produce the break-even number of boards.

51. The cost C of producing x number of paperback books is given by $C = 4.50x + 2400$. Income R from these books is given by $R = 7.50x$. Find how many books should be produced and sold to break even.

52. Find the break-even quantity for a company that makes x number of computer monitors at a cost C given by $C = 870 + 70x$, and receives revenue R given by $R = 105x$.

Writing in Mathematics

53. Problems 49 through 52 involve finding the break-even point for manufacturing. Discuss what happens if a company makes and sells fewer products than the break-even point. Discuss what happens if more products than the break-even point are made and sold. If fewer products are sold a loss results; if more are sold a profit results.

Skill Review

Translate each sentence into an equation. See Section 1.2.

54. Twice the sum of a number and three is forty-two.

55. Half of the difference of a number and one is thirty-seven.

56. Five times the opposite of a number is the number plus sixty.

57. If three times the sum of a number and 2 is divided by 5, the quotient is 0.

58. If the sum of a number and 9 is subtracted from 50, the result is 0.

59. Forty is the sum of twice a number and one, multiplied by eight.

60. If twice the sum of 4 and a number is added to 10, the sum is 12.

61. If the product of 8 and a number is subtracted from 20, the difference is −4.

2.4
Absolute Value Equations

OBJECTIVE | **1** Solve absolute value equations.

Tape 3

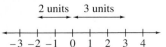

In Chapter 1, we defined the absolute value of a number as its distance on a number line from 0.

$$|-2| = 2 \text{ and } |3| = 3$$

1   Let us now consider an equation containing the absolute value of a variable: $|x| = 3$. The solution set of this equation will contain all numbers whose distance from 0 is 3 units. There are two numbers 3 units away from 0 on the number line: 3 and -3:

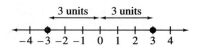

Thus the solution set of the equation $|x| = 3$ is $\{3, -3\}$. This suggests the following:

Solving Equations of the Form $|x| = a$

If a is a positive number, then $|x| = a$ is equivalent to $x = a$ or $x = -a$.

EXAMPLE 1   Solve $|p| = 2$.

Solution: Since 2 is positive, we may use the above rule and $|p| = 2$ is equivalent to $p = 2$ or $p = -2$.

To check, let $p = 2$ and then $p = -2$ in the original equation.

$\|p\| = 2$	Original equation.	$\|p\| = 2$	Original equation.
$\|2\| = 2$	Let $p = 2$.	$\|-2\| = 2$	Let $p = -2$.
$2 = 2$	True.	$2 = 2$	True.

The solution set is $\{2, -2\}$.   ∎

If the expression inside the absolute value bars is more complicated than the single variable x, replace x in the preceding rule with the expression inside the absolute value bars and solve the resulting equations.

EXAMPLE 2   Solve $|5w + 3| = 7$.

Solution: Here the expression inside the absolute value bars is $5w + 3$. If we think of the expression $5w + 3$ as x, we have that $|x| = 7$ is equivalent to

$$x = 7 \quad \text{or} \quad x = -7$$

Then substitute $5w + 3$ for x, and we have

$$5w + 3 = 7 \quad \text{or} \quad 5w + 3 = -7$$

Solve these two equations for w.

$$5w + 3 = 7 \quad \text{or} \quad 5w + 3 = -7$$
$$5w = 4 \quad \text{or} \quad 5w = -10$$
$$w = \frac{4}{5} \quad \text{or} \quad w = -2$$

To check, let $w = -2$ and then $w = \frac{4}{5}$ in the original equation.

$$\text{Let } w = -2 \qquad\qquad \text{Let } w = \frac{4}{5}$$

$$|5(-2) + 3| = 7 \qquad\qquad \left|5\left(\frac{4}{5}\right) + 3\right| = 7$$

$$|-10 + 3| = 7 \qquad\qquad |4 + 3| = 7$$

$$|-7| = 7 \qquad\qquad |7| = 7$$

$$7 = 7 \quad \text{True.} \qquad\qquad 7 = 7 \quad \text{True.}$$

Both solutions check and the solution set is $\left\{-2, \dfrac{4}{5}\right\}$. ∎

EXAMPLE 3   Solve $\left|\dfrac{x}{2} - 1\right| = 11$.

Solution: $\left|\dfrac{x}{2} - 1\right| = 11$ is equivalent to

$$\frac{x}{2} - 1 = 11 \qquad \text{or} \qquad \frac{x}{2} - 1 = -11$$

$$2\left(\frac{x}{2} - 1\right) = 2(11) \qquad \text{or} \qquad 2\left(\frac{x}{2} - 1\right) = 2(-11) \qquad \text{Clear fractions.}$$

$$x - 2 = 22 \qquad \text{or} \qquad x - 2 = -22 \qquad \text{Apply the distributive}$$
$$x = 24 \qquad \text{or} \qquad x = -20 \qquad \text{property.}$$

The solution set is $\{24, -20\}$. ∎

To apply the absolute value rule, first make sure that the absolute value expression is isolated.

HELPFUL HINT

If the equation has a single absolute value expression containing variables, isolate the absolute value expression first.

EXAMPLE 4   Solve $|2x| + 5 = 7$.

Solution: We want the absolute value expression alone on one side of the equation, so begin by subtracting 5 from both sides. Then proceed as usual.

$$|2x| + 5 = 7$$
$$|2x| = 2 \qquad \text{Subtract 5 from both sides.}$$
$$2x = 2 \quad \text{or} \quad 2x = -2$$
$$x = 1 \quad \text{or} \quad x = -1$$

The solution set is $\{-1, 1\}$. ∎

EXAMPLE 5   Solve $|y| = 0$.

Solution: We are looking for all numbers whose distance from 0 is zero units. The only number is 0. The solution set is $\{0\}$.   ■

EXAMPLE 6   Solve $|g| = -1$.

Solution: The absolute value of a number is never negative, so this equation has no solution. The solution set is $\{\ \}$ or $\varnothing$.   ■

EXAMPLE 7   Solve $\left|\dfrac{3x + 1}{2}\right| = -2$.

Solution: Again, the absolute value of any expression is never negative, so no solution exists. The solution set is $\{\ \}$ or $\varnothing$.   ■

Given two absolute value expressions, we might ask, when are the absolute values of two expressions equal? To see the answer, notice that

$$|2| = |2|, \quad |-2| = |-2|, \quad |-2| = |2|, \quad \text{and} \quad |2| = |-2|$$
$$\underset{\text{same}}{\nwarrow\nearrow} \qquad \underset{\text{same}}{\nwarrow\nearrow} \qquad \underset{\text{opposites}}{\uparrow\quad\uparrow} \qquad \underset{\text{opposites}}{\uparrow\quad\uparrow}$$

Two absolute value expressions are equal when the expressions inside the absolute value bars are equal or are opposites of each other.

EXAMPLE 8   Solve $|3x + 2| = |5x - 8|$.

Solution: This equation is true if the expressions inside the absolute value bars are equal or are opposites of each other.

$$3x + 2 = 5x - 8 \quad \text{or} \quad 3x + 2 = -(5x - 8)$$

Next, solve each equation.

$$3x + 2 = 5x - 8 \quad \text{or} \quad 3x + 2 = -5x + 8$$
$$-2x + 2 = -8 \quad \text{or} \quad 8x + 2 = 8$$
$$-2x = -10 \quad \text{or} \quad 8x = 6$$
$$x = 5 \quad \text{or} \quad x = \frac{3}{4}$$

The solution set is $\left\{\dfrac{3}{4}, 5\right\}$.   ■

EXAMPLE 9   Solve $|x - 3| = |5 - x|$.

Solution:
$$x - 3 = 5 - x \quad \text{or} \quad x - 3 = -(5 - x)$$
$$2x - 3 = 5 \quad \text{or} \quad x - 3 = -5 + x$$
$$2x = 8 \quad \text{or} \quad x - 3 - x = -5 + x - x$$
$$x = 4 \quad \text{or} \quad -3 = -5 \qquad \text{False.}$$

Recall from Section 2.1 that when an equation simplifies to a false statement the equation has no solution. Thus the only solution for x is 4, and the solution set is $\{4\}$. ∎

The following box summarizes the methods shown for solving absolute value equations.

> **Absolute Value Equations**
>
> $|x| = a$ $\begin{cases} \textbf{1. } \text{If } a \text{ is positive, then solve } x = a \text{ or } x = -a. \\ \textbf{2. } \text{If } a \text{ is 0, solve } x = 0. \\ \textbf{3. } \text{If } a \text{ is negative, the equation } |x| = a \text{ has no solution.} \end{cases}$
>
> $|x| = |y|$ Solve $x = y$ or $x = -y$.

MENTAL MATH

Simplify each expression.

1. $|-7|$
2. $|-8|$
3. $-|5|$
4. $-|10|$
5. $-|-6|$
6. $-|-3|$
7. $|-3| + |-2| + |-7|$
8. $|-1| + |-6| + |-8|$

EXERCISE SET 2.4

Solve each absolute value equation. See Examples 1 through 3.

1. $|x| = 7$
2. $|y| = 15$
3. $|3x| = 12$
4. $|6n| = 12$

5. $|2x - 5| = 9$
6. $|6 + 2n| = 4$
7. $\left|\dfrac{x}{2} - 3\right| = 1$
8. $\left|\dfrac{n}{3} + 2\right| = 4$

Solve. See Example 4.

9. $|z| + 4 = 9$
10. $|x| + 1 = 3$
11. $|3x| + 5 = 14$

12. $|2x| - 6 = 4$
13. $|2x| = 0$
14. $|7z| = 0$

Solve. See Examples 5 through 7.

15. $|4n + 1| + 10 = 4$
16. $|3z - 2| + 8 = 1$
17. $|5x - 1| = 0$
18. $|3y + 2| = 0$

Solve. See Examples 8 and 9.

19. $|5x - 7| = |3x + 11|$
20. $|9y + 1| = |6y + 4|$

21. $|z + 8| = |z - 3|$
22. $|2x - 5| = |2x + 5|$

Solve each absolute value equation.

23. $|x| = 4$
24. $|x| = 1$
25. $|y| = 0$

26. $|y| = 8$
27. $|z| = -2$
28. $|y| = -9$

29. $|7 - 3x| = 7$

30. $|4m + 5| = 5$

31. $|6x| - 1 = 11$

32. $|7z| + 1 = 22$

33. $|4p| = -8$

34. $|5m| = -10$

35. $|x - 3| + 3 = 7$

36. $|x + 4| - 4 = 1$

37. $\left|\dfrac{z}{4} + 5\right| = -7$

38. $\left|\dfrac{c}{5} - 1\right| = -2$

39. $|9v - 3| = -8$

40. $|1 - 3b| = -7$

41. $|8n + 1| = 0$

42. $|5x - 2| = 0$

43. $|1 + 6c| - 7 = -3$

44. $|2 + 3m| - 9 = -7$

45. $|5x + 1| = 11$

46. $|8 - 6c| = 1$

47. $|4x - 2| = |-10|$

48. $|3x + 5| = |-4|$

49. $|5x + 1| = |4x - 7|$

50. $|3 + 6n| = |4n + 11|$

51. $|6 + 2x| = -|-7|$

52. $|4 - 5y| = -|-3|$

53. $|2x - 6| = |10 - 2x|$

54. $|4n + 5| = |4n + 3|$

55. $\left|\dfrac{2x - 5}{3}\right| = 7$

56. $\left|\dfrac{1 + 3n}{4}\right| = 4$

57. $2 + |5n| = 17$

58. $8 + |4m| = 24$

59. $\left|\dfrac{2x - 1}{3}\right| = |-5|$

60. $\left|\dfrac{5x + 2}{2}\right| = |-6|$

61. $|2y - 3| = |9 - 4y|$

62. $|5z - 1| = |7 - z|$

63. $\left|\dfrac{3n + 2}{8}\right| = |-1|$

64. $\left|\dfrac{2r - 6}{5}\right| = |-2|$

65. $|x + 4| = |7 - x|$

66. $|8 - y| = |y + 2|$

67. $\left|\dfrac{8c - 7}{3}\right| = -|-5|$

68. $\left|\dfrac{5d + 1}{6}\right| = -|-9|$

Skill Review

Find the opposite or additive inverse of each. See Section 1.2.

69. 5

70. -12

71. -8

72. $-(-2)$

Find the reciprocal or multiplicative inverse of each. See Section 1.2.

73. -3

74. $\dfrac{2}{3}$

75. $\dfrac{5}{9}$

76. $-\dfrac{1}{7}$

2.5
Linear Inequalities

OBJECTIVES		
	1	Use interval notation.
	2	Solve linear inequalities using the addition property of inequality.
	3	Solve linear inequalities using the multiplication property of inequality.

Tape 4

1 Recall that the set $\{x \mid x > 2\}$ can be graphed on the number line by sketching an open circle at 2 and shading all points to the right of 2.

We are now going to change our graphing notation to help us understand **interval notation.** Instead of an open circle, we use a parenthesis; instead of a closed circle, we use a bracket. With this new notation, the graph of $\{x \mid x > 2\}$ now looks like

and can be represented in interval notation as $(2, \infty)$. The symbol ∞ is read "infinity" and indicates that the interval includes **all** numbers greater than 2. The left parenthesis indicates that 2 **is not** included in the interval. Using a left bracket would indicate that 2 **is** included in the interval. The following table shows three forms of describing intervals: in set notation, as a graph, and in interval notation.

Set Notation	Graph	Interval Notation
$\{x \mid x < a\}$		$(-\infty, a)$
$\{x \mid x > a\}$		(a, ∞)
$\{x \mid x \leq a\}$		$(-\infty, a]$
$\{x \mid x \geq a\}$		$[a, \infty)$
$\{x \mid a < x < b\}$		(a, b)
$\{x \mid a \leq x \leq b\}$		$[a, b]$
$\{x \mid a < x \leq b\}$		$(a, b]$
$\{x \mid a \leq x < b\}$		$[a, b)$

HELPFUL HINT

Notice that a parenthesis is always used to enclose ∞ and $-\infty$.

EXAMPLE 1 Graph each set on a number line and then write in interval notation.

a. $\{x \mid x \geq 2\}$ **b.** $\{x \mid x < -1\}$ **c.** $\{x \mid 0.5 < x \leq 3\}$

Solution: **a.** $[2, \infty)$

b. $(-\infty, -1)$

c. $(0.5, 3]$

2   Interval notation can be used to write solutions of linear inequalities. A linear inequality is similar to a linear equation except that the equality symbol is replaced with an inequality symbol such as $<$, $>$, $\leq$, or $\geq$.

Linear Inequalities in One Variable

$$3x + 5 \geq 4 \qquad 2y < 0 \qquad 4n \geq n - 3 \qquad 3(x - 4) < 5x \qquad \frac{x}{3} \leq 5$$

Linear Inequality in One Variable

A linear inequality in one variable is an inequality that can be written in the form

$$ax + b < c$$

where a, b, and c are real numbers and $a \neq 0$.

In this section, when we make definitions, state properties, or list steps about an inequality containing the symbol $<$, we mean that the definition, property, or steps apply to an inequality containing the symbols $>$, $\leq$, and $\geq$, also.

A solution of an inequality is a value of the variable that makes the inequality a true statement. To solve a linear inequality, we use a process similar to the one used to solve a linear equation. We use properties of inequalities to write equivalent inequalities until the variable is isolated.

Addition Property of Inequality

If a, b, and c are real numbers, then

$$a < b \quad \text{and} \quad a + c < b + c$$

are equivalent inequalities.

In other words, we may add the same real number to both sides of an inequality and the resulting inequality will have the same solution set. This property also allows us to subtract the same real number from both sides.

EXAMPLE 2   Solve for x: $3x + 4 \geq 2x - 6$. Graph the solution and write it in interval notation.

Solution:

$$3x + 4 \geq 2x - 6$$

$$3x + 4 \;-2x\; \geq 2x - 6 \;-2x\; \qquad \text{Subtract } 2x \text{ from both sides.}$$

$$x + 4 \geq -6 \qquad \text{Combine like terms.}$$

$$x + 4 \;-4\; \geq -6 \;-4\; \qquad \text{Subtract 4 from both sides.}$$

$$x \geq -10 \qquad \text{Simplify.}$$

The solution set is $\{x \mid x \geq -10\}$, which in interval notation is $[-10, \infty)$. The graph of the solution set is

$$[-10, \infty]$$

Thus, **all** real numbers greater than or equal to -10 are solutions of the linear inequality $3x + 4 \geq 2x - 6$. For example -10, 11, 38, and 1,000,000 are a few solutions. To see this, replace x in $3x + 4 \geq 2x - 6$ by each of the numbers and see that the result is a true inequality.

3   Next, we introduce and use the multiplication property of inequality to solve linear inequalities. To understand this property, let's start with the true statement $-3 < 7$ and multiply both sides by 2.

$$-3 < 7$$

$$-3(2) < 7(2) \qquad \text{Multiply by 2.}$$

$$-6 < 14 \qquad \text{True.}$$

The statement remains true.

Notice what happens if both sides of $-3 < 7$ are multiplied by -2.

$$-3 < 7$$

$$-3(-2) < 7(-2)$$

$$6 < -14 \qquad \textbf{False.}$$

The inequality $6 < -14$ is a false statement. However, **if the direction of the inequality sign is reversed,** the result is $6 > -14$, a true statement. These examples suggest the following property.

Multiplication Property of Inequality

If a, b, and c are real numbers, and c is **positive,** then $a < b$ and $ac < bc$ are equivalent inequalities.

If a, b, and c are real numbers and c is **negative,** then $a < b$ and $ac > bc$ are equivalent inequalities.

In other words, we may multiply both sides of an inequality by the same positive real number and the result is an equivalent inequality.

We may also multiply both sides of an inequality by the same **negative number** and **reverse the direction of the inequality symbol,** and the result is an equivalent inequality. The multiplication property holds for division also, since division is defined in terms of multiplication.

HELPFUL HINT

Whenever both sides of an inequality are multiplied or divided by a negative number, the direction of the inequality symbol **must be** reversed to form an equivalent inequality.

EXAMPLE 3   Solve for x.

    a. $2x < -6$

    b. $-2x < 6$

Solution: **a.** $2x < -6$

$$\frac{2x}{2} < \frac{-6}{2} \qquad \text{Divide both sides by 2.}$$

$$x < -3 \qquad \text{Simplify.}$$

The solution set is $\{x \mid x < -3\}$, which in interval notation is $(-\infty, -3)$. The graph of the solution set is

    b. $-2x < 6$

$$\frac{-2x}{-2} > \frac{6}{-2} \qquad \begin{array}{l}\text{Divide both sides by } -2 \text{ and reverse}\\ \text{the inequality symbol.}\end{array}$$

$$x > -3 \qquad \text{Simplify.}$$

The solution set is $\{x \mid x > -3\}$, which is $(-3, \infty)$ in interval notation. The graph of the solution set is

    To solve linear inequalities in general, we follow steps similar to those for solving linear equations.

To Solve a Linear Inequality in One Variable

Step 1   Clear the equation of fractions by multiplying both sides of the inequality by the lowest common denominator (LCD) of all fractions in the inequality.

Step 2   Use the distributive property to remove grouping symbols such as parentheses.

Step 3   Combine like terms on each side of the inequality.

Step 4   Use the addition property of inequality to write the inequality as an equivalent inequality with variable terms on one side and numbers on the other side.

Step 5   Use the multiplication property of inequality to isolate the variable.

EXAMPLE 4   Solve for x: $-(x - 3) + 2 \leq 3(2x - 5) + x$.

Solution:

$$-(x - 3) + 2 \leq 3(2x - 5) + x$$

$-x + 3 + 2 \leq 6x - 15 + x$	Apply the distributive property.
$5 - x \leq 7x - 15$	Combine like terms.
$5 - x + x \leq 7x - 15 + x$	Add x to both sides.
$5 \leq 8x - 15$	Combine like terms.
$5 + 15 \leq 8x - 15 + 15$	Add 15 to both sides.
$20 \leq 8x$	Combine like terms.
$\dfrac{20}{8} \leq \dfrac{8x}{8}$	Divide both sides by 8.

$$\frac{5}{2} \leq x \quad \text{or} \quad x \geq \frac{5}{2}$$

The solution set written in interval notation is $\left[\dfrac{5}{2}, \infty\right)$ and its graph is

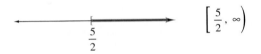

$$\left[\frac{5}{2}, \infty\right)$$

EXAMPLE 5   Solve for x: $\dfrac{2}{5}(x - 6) \geq x - 1$.

Solution:

$$\frac{2}{5}(x - 6) \geq x - 1$$

$5\left[\dfrac{2}{5}(x - 6)\right] \geq 5\,(x - 1)$	Multiply both sides by 5 to eliminate fractions.
$2x - 12 \geq 5x - 5$	Use the distributive property.
$-3x - 12 \geq -5$	Subtract $5x$ from both sides.
$-3x \geq 7$	Add 12 to both sides.
$\dfrac{-3x}{-3} \leq \dfrac{7}{-3}$	Divide both sides by -3 and reverse the inequality symbol.
$x \leq -\dfrac{7}{3}$	Simplify.

The solution is graphed on a number line and written in interval notation, as $\left(-\infty, -\dfrac{7}{3}\right]$

$$\left(-\infty, -\frac{7}{3}\right]$$

EXAMPLE 6   Solve for x: $2(x + 3) > 2x + 1$.

Solution:
$$2(x + 3) > 2x + 1$$
$$2x + 6 > 2x + 1 \qquad \text{Distribute on left side.}$$
$$2x + 6 - 2x > 2x + 1 - 2x \qquad \text{Subtract } 2x \text{ from both sides.}$$
$$6 > 1 \qquad \text{Simplify.}$$

$6 > 1$ is a true statement for all values of x, so this inequality and the original inequality are true for all numbers. The solution set is $\{x \mid x \text{ is a real number}\}$, or $(-\infty, \infty)$ in interval notation, and its graph is

$(-\infty, \infty)$ ∎

Application problems containing words such as "at least," "at most," "between," "no more than," and "no less than" usually indicate that an inequality be solved instead of an equation. In solving applications involving linear inequalities, we use the same procedure as when we solved applications involving linear equations.

EXAMPLE 7   A salesperson earns $600 per month plus 20% of sales. Find the minimum amount of sales needed to receive a salary of at least $1500 per month.

Solution: The phrase "at least" implies that an inequality be used. Let $x =$ amount of sales. To write an inequality, we know that

$$\boxed{600} \; + \; \boxed{\begin{array}{c} 20\% \text{ of} \\ \text{sales} \end{array}} \; \geq \; \boxed{1500}$$

or

$$600 + 0.20x \geq 1500$$
$$600 + 0.20x - 600 \geq 1500 - 600$$
$$0.20x \geq 900$$
$$x \geq 4500$$

The minimum amount of sales needed for the salesperson to earn at least $1500 per month is $4500.   ∎

EXERCISE SET 2.5

Graph the solution set of each inequality on a number line, and write the solution set in interval notation. See Example 1.

1. $\{x \mid x < -3\}$

2. $\{x \mid x \geq -7\}$

3. $\{x \mid x \geq 0.3\}$

4. $\{x \mid x < -0.2\}$

5. $\{x \mid 5 < x\}$

6. $\{x \mid -7 \geq x\}$

7. $\{x \mid -2 < x < 5\}$

8. $\{x \mid -5 \leq x \leq -1\}$

9. $\{x \mid 5 > x > -1\}$

10. $\{x \mid -3 \geq x \geq -7\}$

Solve each inequality for the variable. Graph the solution set. See Example 2.

11. $5x + 3 > 2 + 4x$

12. $7x - 1 \geq 6x - 1$

13. $8x - 7 \leq 7x - 5$

14. $12x + 14 < 11x - 2$

Solve each inequality for the variable. Graph the solution set. See Example 3.

15. $5x < -20$

16. $-4x > -12$

17. $-3x \geq 24$

18. $7x \leq -21$

Solve each inequality for the variable. Graph the solution set. See Examples 4 and 5.

19. $15 + 2x \geq 4x - 7$

20. $20 + x < 6x$

21. $\dfrac{3x}{4} \geq 2$

22. $\dfrac{5}{6}x \geq -8$

23. $3(x - 5) < 2(2x - 1)$

24. $5(x + 4) \leq 4(2x + 3)$

25. $\dfrac{1}{2} + \dfrac{2}{3} \geq \dfrac{x}{6}$

26. $\dfrac{3}{4} - \dfrac{2}{3} > \dfrac{x}{6}$

Solve each inequality for the variable. Graph the solution set. See Example 6.

27. $4(x - 1) \geq 4x - 8$

28. $3x + 1 < 3(x - 2)$

29. $7x < 7(x - 2)$

30. $8(x + 3) \leq 7(x + 5) + x$

Solve. See Example 7.

31. A small plane's maximum take-off weight is 2000 pounds. Six passengers weigh an average of 160 pounds each. Use an inequality to find the maximum weight of luggage and cargo the plane can carry.

32. A clerk must use the elevator to move boxes of paper. The elevator's weight limit is 1500 pounds. If each box of paper weighs 66 pounds and the clerk weighs 147 pounds, use an inequality to find the maximum number of boxes she can move on the elevator at one time.

33. To mail an envelope first class, the U.S. Post Office charges 29 cents for the first ounce and 23 cents per ounce for each additional ounce. Use an inequality to find the maximum weight that can be mailed for $4.20.

34. A parking garage charges $1 for the first half-hour and 60 cents for each additional half-hour or a portion of a half-hour. Use an inequality to find how long you can park if you have only $4.00 in cash.

Solve each inequality for the variable. Graph the solution set.

35. $5x > 10$

36. $9x < 45$

37. $-4x \leq 32$

38. $-6x \geq 42$

39. $-2x + 7 \geq 9$

40. $8 - 5x \leq 23$

41. $4(2x + 1) > 4$

42. $6(2 - x) \geq 12$

43. $\dfrac{x + 7}{5} > 1$

44. $\dfrac{2x - 4}{3} \leq 2$

45. $\dfrac{-5x + 11}{2} \le 7$

46. $\dfrac{4x - 8}{7} < 0$

47. $8x - 16 \le 10x + 2$

48. $18x - 24 < 10x + 64$

49. $2(x - 3) > 70$

50. $3(5x + 6) \ge -12$

51. $-5x + 4 \le -4(x - 1)$

52. $-6x + 2 < -3(x + 4)$

53. $\dfrac{1}{4}(x - 7) \ge x + 2$

54. $\dfrac{3}{5}(x + 1) \le x + 1$

55. $\dfrac{2}{3}(x + 2) < \dfrac{1}{5}(2x + 7)$

56. $\dfrac{1}{6}(3x + 10) > \dfrac{5}{12}(x - 1)$

57. $4(x - 6) + 2x - 4 \ge 3(x - 7) + 10x$

58. $7(2x + 3) + 4x \le 7 + 5(3x - 4)$

59. $\dfrac{5x + 1}{7} - \dfrac{2x - 6}{4} \ge -4$

60. $\dfrac{1 - 2x}{3} + \dfrac{3x + 7}{7} > 1$

61. $\dfrac{-x + 2}{2} - \dfrac{1 - 5x}{8} < -1$

62. $\dfrac{3 - 4x}{6} - \dfrac{1 - 2x}{12} \le -2$

63. $0.8x + 0.6x \ge 4.2$

64. $0.7x - x > 0.45$

65. $\dfrac{x + 5}{5} - \dfrac{3 + x}{8} \ge \dfrac{-3}{10}$

66. $\dfrac{x - 4}{2} - \dfrac{x - 2}{3} > \dfrac{5}{6}$

67. $\dfrac{x + 3}{12} + \dfrac{x - 5}{15} < \dfrac{2}{3}$

68. $\dfrac{3x + 2}{18} - \dfrac{1 + 2x}{6} \le -\dfrac{1}{2}$

69. Shureka has scores of 72, 67, 82, and 79 on her algebra tests. Use an inequality to find the minimum score she can make on the final exam to pass the course with an average of 60 or higher, given that the final exam counts as two tests.

70. In a Winter Olympics speed-skating event, Hans scored times of 3.52, 4.04, and 3.87 minutes on his first three trials. Use an inequality to find the maximum time he can score on his last trial so that his average time is under 4.0 minutes.

71. Northeast Telephone Company offers two billing plans for local calls. Plan 1 charges $25 per month for unlimited calls, and plan 2 charges $13 per month plus 6 cents per call. Use an inequality to find the number of monthly calls for which plan 1 is more economical than plan 2.

72. A car rental company offers two subcompact rental plans. Plan A charges $32 per day for unlimited mileage, and plan B charges $24 per day plus 15 cents per mile. Use an inequality to find the number of daily miles for which plan A is more economical than plan B.

PLAN B > PLAN A
$24 + .15x > 32$

Writing in Mathematics

73. Explain how solving a linear inequality is similar to solving a linear equation.

74. Explain how solving a linear inequality is different from solving a linear equation.

Skill Review

Find the following roots. See Section 1.3.

75. $\sqrt{25}$

76. $\sqrt{49}$

77. $\sqrt[3]{8}$

78. $\sqrt[3]{27}$

79. $\sqrt[4]{16}$

80. $\sqrt[4]{81}$

81. $-\sqrt{4}$

82. $-\sqrt{36}$

2.6
Compound Inequalities

1   Solve compound inequalities.

Tape 4

1   Two inequalities joined by the words **and** or **or** are called **compound inequalities.**

Compound Inequalities

$$x + 3 < 8 \quad \text{and} \quad x > 2$$

$$\frac{2x}{3} \geq 5 \quad \text{or} \quad -x + 10 < 7$$

The solution set of a compound inequality formed by the word **and** is the intersection of the solution sets of the two inequalities. The intersection of two sets, denoted by $\cap$, is the set of elements common to both sets. In other words, a solution of a compound inequality formed by the word **and** is a number that makes both inequalities true.

The graph of the compound inequality $x \leq 5$ and $x \geq 3$ contains all numbers that make the inequality $x \leq 5$ a true statement **and** the inequality $x \geq 3$ a true statement. The first graph shown next is the graph of $x \leq 5$, the second graph is the graph of $x \geq 3$, and the third graph shows the intersection of the two graphs. It is the graph of $x \leq 5$ **and** $x \geq 3$.

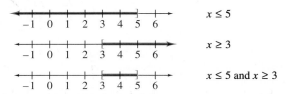

In interval notation, the set $\{x \mid x \leq 5 \text{ and } x \geq 3\}$ is written as [3, 5].

EXAMPLE 1   Solve for x: $x - 7 < 2$ and $2x + 1 < 9$.

Solution: First solve each inequality separately.

$$x - 7 < 2 \quad \text{and} \quad 2x + 1 < 9$$
$$x < 9 \quad \text{and} \quad 2x < 8$$
$$x < 9 \quad \text{and} \quad x < 4$$

Graph the two intervals on two number lines and find their intersection.

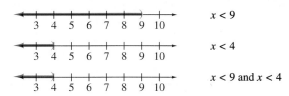

The solution set written as an interval is $(-\infty, 4)$.   ∎

Compound inequalities containing the word **and** can be written in a more compact form. The compound inequality $2 \leq x$ and $x \leq 6$ can be written as

$$2 \leq x \leq 6$$

The graph of $2 \leq x \leq 6$ is all numbers between 2 and 6, including 2 and 6.

The set $\{x \mid 2 \leq x \leq 6\}$ written in interval notation is $[2, 6]$.

To solve a compound inequality like $2 < 4 - x < 7$, we isolate x "on the middle side." Since a compound inequality is really two inequalities in one statement, we must perform the same operation to all three "sides" of the inequality.

EXAMPLE 2   Solve $2 < 4 - x < 7$.

Solution: To isolate x, first subtract 4 from all three sides.

$$2 < 4 - x < 7$$

$$2 - 4 < 4 - x - 4 < 7 - 4 \qquad \text{Subtract 4 from all three sides.}$$

$$-2 < -x < 3 \qquad \text{Simplify.}$$

$$\frac{-2}{-1} > \frac{-x}{-1} > \frac{3}{-1} \qquad \begin{array}{l}\text{Divide all three sides by } -1 \text{ and reverse} \\ \text{the inequality symbols.}\end{array}$$

$$2 > x > -3$$

This is equivalent to $-3 < x < 2$ and its graph is shown at the left.

The solution set in interval notation is $(-3, 2)$.   ■

EXAMPLE 3   Solve for x: $-1 \leq \dfrac{2x}{3} + 5 \leq 2$.

Solution: First, clear the inequality of fractions by multiplying all three sides by the LCD of 3.

$$-1 \leq \frac{2x}{3} + 5 \leq 2$$

$$3(-1) \leq 3\left(\frac{2x}{3} + 5\right) \leq 3(2) \qquad \text{Multiply by the LCD of 3.}$$

$$-3 \leq 2x + 15 \leq 6 \qquad \text{Apply the distributive property and multiply.}$$

$$-3 - 15 \leq 2x + 15 - 15 \leq 6 - 15 \quad \text{Subtract 15 from all three sides.}$$

$$-18 \leq 2x \leq -9 \qquad \text{Simplify.}$$

$$\frac{-18}{2} \leq \frac{2x}{2} \leq \frac{-9}{2} \qquad \text{Divide all three sides by 2.}$$

$$-9 \leq x \leq -\frac{9}{2} \qquad \text{Simplify.}$$

The graph of the solution is shown at the left.

The solution set in interval notation is $\left[-9, -\dfrac{9}{2}\right]$.   ■

The solution set of a compound inequality formed by the word **or** is the **union** of the solution sets of the two inequalities. The union of two sets, denoted by ∪, is the set of elements that belong to either of the sets. In other words, a solution of a compound inequality formed by the word **or** is a number that makes either inequality true.

The graph of the compound inequality $x \leq 1$ or $x \geq 3$ contains all numbers that make the inequality $x \leq 1$ a true statement **or** the inequality $x \geq 3$ a true statement.

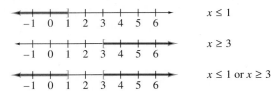

In interval notation, the set $\{x \mid x \leq 1 \text{ or } x \geq 3\}$ is written as $(-\infty, 1] \cup [3, \infty)$.

EXAMPLE 4   Solve $5x - 3 \leq 10$ or $x + 1 \geq 5$.

Solution: Solve each inequality separately.

$$5x - 3 \leq 10 \qquad x + 1 \geq 5$$
$$5x \leq 13$$
$$x \leq \frac{13}{5} \quad \text{or} \qquad x \geq 4$$

Graph each interval on a number line. Then find their union.

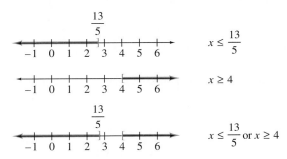

The solution set in interval notation is $\left(-\infty, \dfrac{13}{5}\right] \cup [4, \infty)$. ■

EXERCISE SET 2.6

Solve each compound inequality. Graph the solution set. See Example 1.

1. $x < 5$ and $x > -2$

2. $x \leq 7$ and $x \leq 1$

3. $x + 1 \geq 7$ and $3x - 1 \geq 5$

4. $-2x < -8$ and $x - 5 < 5$

5. $4x + 2 \leq -10$

6. $x + 4 > 0$ and $4x > 0$

Solve each compound inequality. Graph the solution set. See Examples 2 and 3.

7. $5 < x - 6 < 11$

8. $-2 \le x + 3 \le 0$

9. $-2 \le 3x - 5 \le 7$

10. $1 < 4 + 2x < 7$

11. $1 \le \frac{2}{3}x + 3 \le 4$

12. $-2 < \frac{1}{2}x - 5 < 1$

13. $-5 \le \frac{x + 1}{4} \le -2$

14. $-4 \le \frac{2x + 5}{3} \le 1$

Solve each compound inequality. Graph the solution set. See Example 4.

15. $x < -1$ or $x > 0$

16. $x \le 1$ or $x \le -3$

17. $-2x \le -4$ or $5x - 20 \ge 5$

18. $x + 4 < 0$ or $6x > -12$

19. $3(x - 1) < 12$ or $x + 7 > 10$

20. $5(x - 1) \ge -5$ or $5 - x \le 11$

Solve each compound inequality. Graph the solution set.

21. $x < 2$ and $x > -1$

22. $x < 5$ and $x < 1$

23. $x < 2$ or $x > -1$

24. $x < 5$ or $x < 1$

25. $x \ge -5$ and $x \ge -1$

26. $x \le 0$ or $x \ge -3$

27. $x \ge -5$ or $x \ge -1$

28. $x \le 0$ and $x \ge -3$

29. $0 \le 2x - 3 \le 9$

30. $3 < 5x + 1 < 11$

31. $\frac{1}{2} < x - \frac{3}{4} < 2$

32. $\frac{2}{3} < x + \frac{1}{2} < 4$

33. $x + 3 \ge 3$ and $x + 3 \le 2$

34. $2x - 1 \ge 3$ and $-x > 2$

35. $3x \ge 5$ or $-x - 6 < 1$

36. $\frac{3}{8}x + 1 \le 0$ or $-2x < -4$

37. $0 < \frac{5 - 2x}{3} < 5$

38. $-2 < \frac{-2x - 1}{3} < 2$

39. $-6 < 3(x - 2) \le 8$

40. $-5 < 2(x + 4) < 8$

41. $-x + 5 > 6$ and $1 + 2x \le -5$

42. $5x \le 0$ and $-x + 5 < 8$

43. $3x + 2 \le 5$ or $7x > 29$

44. $-x < 7$ or $3x + 1 < -20$

45. $5 - x > 7$ and $2x + 3 \geq 13$

46. $-2x < -6$ or $1 - x > -2$

47. $-\dfrac{1}{2} \leq \dfrac{4x - 1}{6} < \dfrac{5}{6}$

48. $-\dfrac{1}{2} \leq \dfrac{3x - 1}{10} < \dfrac{1}{2}$

49. $\dfrac{1}{15} < \dfrac{8 - 3x}{15} < \dfrac{4}{5}$

50. $-\dfrac{1}{4} < \dfrac{6 - x}{12} < -\dfrac{1}{6}$

51. $0.3 < 0.2x - 0.9 < 1.5$

52. $-0.7 \leq 0.4x + 0.8 < 0.5$

53. In Oslo, the temperature ranges from $-10°$ to $18°$ Celsius. Use a compound inequality to convert these temperatures to the Fahrenheit scale.

54. Wendy has scores of 80, 90, 82, and 75 on her chemistry tests. Use a compound inequality to find the range of

scores she can make on her final exam in order to receive a B in the course. The final exam counts as two tests, and a B is received if the final course average is from 80 to 89.

x = final score (EXAM)
2x = twice the final score.

A Look Ahead

Solve each compound inequality for x. See the following example.

EXAMPLE   Solve: $x - 6 < 3x < 2x + 5$

Solution: $x - 6 < 3x$ and $3x < 2x + 5$

$\qquad\qquad -6 < 2x$ and $x < 5$

$\qquad\qquad -3 < x$

$\qquad\qquad x > -3$ and $x < 5$

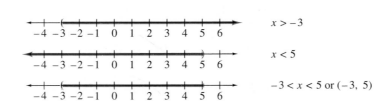

$x > -3$

$x < 5$

$-3 < x < 5$ or $(-3, 5)$

55. $2x - 3 < 3x + 1 < 4x - 5$

56. $x + 3 < 2x + 1 < 4x + 6$

57. $-3(x - 2) \leq 3 - 2x \leq 10 - 3x$

58. $7x - 1 \leq 7 + 5x \leq 3(1 + 2x)$

59. $5x - 8 < 2(2 + x) < -2(1 + 2x)$

60. $1 + 2x < 3(2 + x) < 1 + 4x$

Skill Review

Find each sum or difference. See Section 1.4.

61. $-7 + (-8)$

62. $9 - 14$

63. $-10 - (-5)$

64. $-20 + 26$

65. $14 + (-17)$

66. $-3 - 12$

67. $4 - (-6)$

68. $8 + (-15)$

2.7
Absolute Value Inequalities

Tape 5

OBJECTIVES		
	1	Solve absolute value inequalities of the form $|x| < a$.
	2	Solve absolute value inequalities of the form $|x| > a$.

1 The solution set of an absolute value inequality such as $|x| < 2$ will contain all numbers whose distance from 0 is less than 2 units, as shown next.

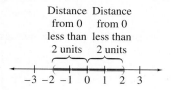

The solution set is $\{x \mid -2 < x < 2\}$ or $(-2, 2)$ in interval notation.

EXAMPLE 1 Solve $|x| \leq 3$.

Solution: The solution set of this inequality contains all numbers whose distance from 0 is less than or equal to 3. Thus 3, -3, and all numbers between 3 and -3 are in the solution set.

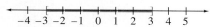

The solution set is $[-3, 3]$. ∎

In general, we have the following:

> **Solving Absolute Value Inequalities of the Form $|x| < a$**
>
> If a is a positive number, then $|x| < a$ is equivalent to $-a < x < a$.

This statement also holds true for the inequality symbol $\leq$.

EXAMPLE 2 Solve for m: $|m - 6| < 2$.

Solution: Replace x with $m - 6$ and a with 2 in the preceding rule and we have that

$$|m - 6| < 2 \quad \text{is equivalent to} \quad -2 < m - 6 < 2$$

Solve this compound inequality for m by adding 6 to all three sides.

$$-2 < m - 6 < 2$$
$$-2 + 6 < m - 6 + 6 < 2 + 6 \qquad \text{Add 6 to all three sides.}$$
$$4 < m < 8 \qquad\qquad\qquad \text{Simplify.}$$

The solution set is $(4, 8)$, and its graph is shown at the left. ∎

> **HELPFUL HINT**
>
> Isolate the absolute value expression on one side of the inequality before applying the absolute value inequality rule.

EXAMPLE 3 Solve for x: $|5x + 1| + 1 \leq 10$.

Solution: First, isolate the absolute value expression by subtracting 1 from both sides.

$$|5x + 1| + 1 \leq 10$$

$$|5x + 1| \leq 10 - 1 \qquad \text{Subtract 1 from both sides.}$$

$$|5x + 1| \leq 9 \qquad \text{Simplify.}$$

Since 9 is positive, we apply the absolute value inequality rule for $|x| \leq a$.

$$-9 \leq 5x + 1 \leq 9$$

$$-9 - 1 \leq 5x + 1 - 1 < 9 - 1 \qquad \text{Subtract 1 from all three sides.}$$

$$-10 \leq 5x \leq 8 \qquad \text{Simplify.}$$

$$-2 \leq x \leq \frac{8}{5} \qquad \text{Divide all three sides by 5.}$$

The solution set is $\left[-2, \dfrac{8}{5} \right]$ and the graph is shown at the left. ■

EXAMPLE 4 Solve for x: $\left| 2x - \dfrac{1}{10} \right| < -13$.

Solution: The absolute value of a number is always nonnegative and will never be less than -13. Thus this absolute value inequality has no solution. The solution set is $\{ \ \}$ or $\varnothing$. ■

2 Let us now solve an absolute value inequality of the form $|x| > a$, such as $|x| \geq 3$. The solution set contains all numbers whose distance from 0 is 3 or more units. Thus the graph of the solution set contains 3 and all points to the right of 3 on the number line or -3 and all points to the left of -3 on the number line.

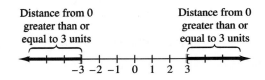

This solution set is written as $\{x | x \leq -3 \text{ or } x \geq 3\}$. In interval notation, the solution set is $(-\infty, -3] \cup [3, \infty)$, since "or" means "union." In general, we have the following.

> **Solving Absolute Value Inequalities of the Form $|x| > a$**
>
> If a is a positive number, than $|x| > a$ is equivalent to $x < -a$ or $x > a$.

This rule also holds true for the inequality symbol $\geq$.

EXAMPLE 5   Solve for y: $|y - 3| > 7$.

Solution: Since 7 is positive, we apply the preceding rule.

$$|y - 3| > 7 \quad \text{is equivalent to} \quad y - 3 < -7 \text{ or } y - 3 > 7$$

Next, solve the compound inequality.

$$y - 3 < -7 \qquad \text{or} \quad y - 3 > 7$$
$$y - 3 + 3 < -7 + 3 \quad \text{or} \quad y - 3 + 3 > 7 + 3 \qquad \text{Add 3 to both sides.}$$
$$y < -4 \qquad \text{or} \quad y > 10 \qquad\qquad\qquad \text{Simplify.}$$

The solution set is $(-\infty, -4) \cup (10, \infty)$ and its graph is shown at the left.   ∎

EXAMPLE 6   Solve $|2x + 9| + 5 > 3$.

Solution: First isolate the absolute value expression by subtracting 5 from both sides.

$$|2x + 9| + 5 > 3$$
$$|2x + 9| + 5 - 5 > 3 - 5 \qquad \text{Subtract 5 from both sides.}$$
$$|2x + 9| > -2 \qquad\qquad \text{Simplify.}$$

The absolute value of any number is always nonnegative and thus is always greater than -2. This inequality and the original inequality are true for all values of x. The solution set is $\{x \mid x \text{ is a real number}\}$ or $(-\infty, \infty)$, and its graph is at the left.   ∎

EXAMPLE 7   Solve $\left|\dfrac{x}{3} - 1\right| - 7 \geq -5$.

Solution: First, isolate the absolute value expression by adding 7 to both sides.

$$\left|\frac{x}{3} - 1\right| - 7 \geq -5$$
$$\left|\frac{x}{3} - 1\right| - 7 + 7 \geq -5 + 7 \qquad \text{Add 7 to both sides.}$$
$$\left|\frac{x}{3} - 1\right| \geq 2 \qquad\qquad \text{Simplify.}$$

Next, write the absolute value inequality as an equivalent compound inequality and solve.

$$\frac{x}{3} - 1 \leq -2 \qquad \text{or} \qquad \frac{x}{3} - 1 \geq 2$$
$$3\left(\frac{x}{3} - 1\right) \leq 3(-2) \quad \text{or} \quad 3\left(\frac{x}{3} - 1\right) \geq 3(2) \qquad \text{Clear the inequalities of fractions.}$$
$$x - 3 \leq -6 \qquad \text{or} \qquad x - 3 \geq 6 \qquad \text{Apply the distributive property.}$$
$$x \leq -3 \qquad \text{or} \qquad x \geq 9 \qquad\qquad \text{Add 3 to both sides.}$$

The solution set is $(-\infty, -3] \cup [9, \infty)$ and its graph is shown at the left.   ∎

EXAMPLE 8   Solve for x: $\left| \dfrac{2(x + 1)}{3} \right| \leq 0$.

Solution: Recall that " $\leq$ " means "less than or equal to." The absolute value of any expression will never be less than 0, but it may be equal to 0. Thus, to solve $\left| \dfrac{2(x + 1)}{3} \right| \leq 0$, we solve $\dfrac{2(x + 1)}{3} = 0$.

$$\frac{2(x + 1)}{3} = 0$$

$$3\left[\frac{2(x + 1)}{3} \right] = 3(0) \quad \text{Clear the equation of fractions.}$$

$$2x + 2 = 0 \quad \text{Apply the distributive property.}$$

$$2x = -2 \quad \text{Subtract 2 from both sides.}$$

$$x = -1 \quad \text{Divide both sides by 2.}$$

The solution set is $\{-1\}$.   ∎

The following box summarizes the types of absolute value equations and inequalities.

Solving Absolute Value Equations and Inequalities with $a > 0$

Algebraic Solution	Solution Graph
$\lvert x \rvert = a$ is equivalent to $x = a$ or $x = -a.$	
$\lvert x \rvert < a$ is equivalent to $-a < x < a.$	
$\lvert x \rvert > a$ is equivalent to $x < -a$ or $x > a.$	

EXERCISE SET 2.7

Solve each inequality. Then graph the solution set. See Examples 1 and 2.

1. $\lvert x \rvert \leq 4$ **2.** $\lvert x \rvert < 6$

3. $\lvert x - 3 \rvert < 2$ **4.** $\lvert y \rvert \leq 5$

5. $\lvert x + 3 \rvert < 2$ **6.** $\lvert x + 4 \rvert < 6$

7. $\lvert 2x + 7 \rvert \leq 13$ **8.** $\lvert 5x - 3 \rvert \leq 18$

Solve each inequality. Graph the solution set. See Examples 3 and 4.

9. $\lvert x \rvert + 7 \leq 12$ **10.** $\lvert x \rvert + 6 \leq 7$ **11.** $\lvert 3x - 1 \rvert < -5$

12. $|8x - 3| < -2$ **13.** $|x - 6| - 7 \le -1$ **14.** $|z + 2| - 7 < -3$

Solve each inequality. Graph the solution set. See Example 5.

15. $|x| > 3$ **16.** $|y| \ge 4$

17. $|x + 10| \ge 14$ **18.** $|x - 9| \ge 2$

Solve each inequality. Graph the solution set. See Examples 6 and 7.

19. $|x| + 2 > 6$ **20.** $|x| - 1 > 3$ **21.** $|5x| > -4$

22. $|4x - 11| > -1$ **23.** $|6x - 8| + 3 > 7$ **24.** $|10 + 3x| + 1 > 2$

Solve each inequality. Graph the solution set. See Example 8.

25. $|x| \le 0$ **26.** $|x| \ge 0$

27. $|8x + 3| > 0$ **28.** $|5x - 6| < 0$

Solve each inequality. Graph the solution set.

29. $|x| \le 2$ **30.** $|z| < 6$

31. $|y| > 1$ **32.** $|x| \ge 10$

33. $|x - 3| < 8$ **34.** $|-3 + x| \le 10$

35. $|6x - 8| > 4$ **36.** $|1 + 0.3x| \ge 0.1$

37. $5 + |x| \le 2$ **38.** $8 + |x| < 1$

39. $|x| > -4$ **40.** $|x| \le -7$

41. $|2x - 7| \le 11$ **42.** $|5x + 2| < 8$

43. $|x + 5| + 2 \ge 8$ **44.** $|-1 + x| - 6 > 2$

45. $|x| > 0$ **46.** $|x| < 0$

47. $9 + |x| > 7$ **48.** $5 + |x| \ge 4$

49. $6 + |4x - 1| \le 9$ **50.** $-3 + |5x - 2| \le 4$

51. $\left| \dfrac{2}{3}x + 1 \right| > 1$ **52.** $|5x - 1| \ge 2$

53. $|5x + 3| < -6$ **54.** $|4 + 9x| \ge -6$

55. $|8x + 3| \ge 0$ **56.** $|5x - 6| \le 0$

57. $|1 + 3x| + 4 < 5$

58. $|7x - 3| - 1 \le 10$

59. $|x| - 3 \ge -3$

60. $|x| + 6 < 6$

61. $|8x| - 10 > -2$

62. $|6x| - 13 \ge -7$

63. $\left|\dfrac{x + 6}{3}\right| > 2$

64. $\left|\dfrac{7 + x}{2}\right| \ge 4$

65. $|2(3 + x)| > 6$

66. $|5(x - 3)| \ge 10$

67. $\left|\dfrac{5(x + 2)}{3}\right| < 7$

68. $\left|\dfrac{6(3 + x)}{5}\right| \le 4$

69. $-15 + |2x - 7| \le -6$

70. $-9 + |3 + 4x| < -4$

71. $\left|2x + \dfrac{3}{4}\right| - 7 \le -2$

72. $\left|\dfrac{3}{5} + 4x\right| - 6 < -1$

Solve each equation or inequality for x.

73. $|2x - 3| < 7$

74. $|2x - 3| > 7$

75. $|2x - 3| = 7$

76. $|5 - 6x| = 29$

77. $|x - 5| \ge 12$

78. $|x + 4| \ge 20$

79. $|9 + 4x| = 0$

80. $|9 + 4x| \ge 0$

81. $|2x + 1| + 4 < 7$

82. $8 + |5x - 3| \ge 11$

83. $|3x - 5| + 4 = 5$

84. $|8x| = -5$

85. $|x + 11| = -1$

86. $|4x - 4| = -3$

87. $\left|\dfrac{2x - 1}{3}\right| = 6$

88. $\left|\dfrac{6 - x}{4}\right| = 5$

89. $\left|\dfrac{3x - 5}{6}\right| > 5$

90. $\left|\dfrac{4x - 7}{5}\right| < 2$

Skill Review

List the elements in each set. See Section 1.1.

91. $\{x \mid x \text{ is an odd natural number less than } 10\}$

92. $\{x \mid x \text{ is a negative integer greater than } 1\}$

93. $\{x \mid x \text{ is a natural number that is not a whole number}\}$

94. $\{x \mid x \text{ is an even integer greater than } 5\}$

Simplify the following. See Section 1.4.

95. $(-3)^2 + (-5)^2$

96. $|-3|^2 + |-5|^2$

97. $\dfrac{2 - (-7)}{3^2}$

98. $\dfrac{-8 + (-4)}{2^2}$

CRITICAL
THINKING

The Chemlife Chemical Company recently discovered that for some time the company had unknowingly been pumping polluted water into a stream, owing apparently to incorrect diverting of wastewater at the plant. By the time the flow of pollutants was stopped, the lake downstream had been seriously affected. In fact, 2% of the lakewater now is pollutants, and until this pollution level is reduced to 0.1%, the lake must be declared off limits to all recreational activity.

The streams flowing into and out of the lake exchange with the lake about 100 gallons of water per minute, so local authorities should be able to estimate when the lake can be reopened to the public, assuming the streams are now pollutant free and assuming the authorities know how much water is actually in the lake. The lake is nearly 1200 feet long and is roughly kidney shaped, its width ranging from 400 to 700 feet. At no point is the lake deeper than 27 feet, and the vast majority of the lake is much shallower than 27 feet.

What strategy can you suggest for estimating the volume of water in the lake? Making certain reasonable assumptions, can you predict when the lake can be reopened?

CHAPTER 2 GLOSSARY

Two inequalities joined by the words **and** or **or** are called **compound inequalities.**

A linear equation in one variable that has exactly one solution is called a **conditional equation.**

An equation in one variable that has no solutions is called a **contradiction.**

An **equation** is a statement that two expressions are equal.

Equivalent equations are equations that have the same solution set.

Formulas are equations that describe relationships among variables that represent real-life phenomena such as time, area, and gravity.

A **linear equation** in one variable is an equation that can be written in the form $ax + b = c$, where a, b, and c are real numbers and $a \neq 0$.

An equation in one variable that has every number (for which the equation is defined) as a solution is called an **identity.**

A **solution** of an equation is a replacement value for the variable that makes the resulting equation true.

The **solution set** of an equation is the set of solutions of the equation.

To **solve an equation** is to find the solution set of an equation.

A **term** is a number or the product of a number and variables raised to powers.

CHAPTER 2 SUMMARY

PROPERTIES FOR SOLVING EQUATIONS (2.1)

If a, b, and c, are real numbers, then the **addition property of equality** says that

$$a = b \quad \text{and} \quad a + c = b + c$$

are equivalent equations. Also, if $c \neq 0$, then the **multiplication property of equality** says that

$$a = b \quad \text{and} \quad ac = bc$$

are equivalent equations.

TO SOLVE A LINEAR EQUATION IN ONE VARIABLE (2.1)

Step 1 Clear the equation of fractions by multiplying both sides of the equation by the lowest common denominator (LCD) of all denominators in the equation.

Step 2 Use the order of operations and the distributive property to remove grouping symbols such as parentheses.

Step 3 Combine like terms on each side of the equation.

Step 4 Use the addition property of equality to rewrite the equation as an equivalent equation, with variable terms on one side and numbers on the other side.

Step 5 Use the multiplication property of equality to isolate the variable.

Step 6 Check the proposed solution in the original equation.

TO SOLVE A WORD PROBLEM (2.3)

Step 1 Read and then reread the problem. Choose a variable to represent one unknown quantity.

Step 2 Use this variable to represent any other unknown quantities.

Step 3 Draw a diagram if possible to visualize the known facts.

Step 4 Translate the word problem into an equation.

Step 5 Solve the equation.

Step 6 Answer the question asked and check to see if the answer is **reasonable.**

Step 7 Check the solution in the originally stated problem.

ADDITION PROPERTY OF INEQUALITY (2.5)

If a, b, and c are real numbers, then

$$a < b \quad \text{and} \quad a + c < b + c$$

are equivalent inequalities.

MULTIPLICATION PROPERTY OF INEQUALITY (2.5)

If a, b, and c are real numbers, and c is **positive,** then $a < b$ and $ac < bc$ are equivalent inequalities.

If a, b, and c are real numbers and c is **negative,** then $a < b$ and $ac > bc$ are equivalent inequalities.

SOLVING ABSOLUTE VALUE EQUATIONS AND INEQUALITIES WITH $a > 0$ (2.4 and 2.7)

Algebraic Solution	Solution Graph
$|x| = a$ is equivalent to $x = a$ or $x = -a$.	
$|x| < a$ is equivalent to $-a < x < a$.	
$|x| > a$ is equivalent to $x < -a$ or $x > a$.	

CHAPTER 2 REVIEW

(2.1) *Solve each linear equation.*

1. $4(x - 5) = 2x - 14$

2. $x + 7 = -2(x + 8)$

3. $3(2y - 1) = -8(6 + y)$

4. $-(z + 12) = 5(2z - 1)$

5. $n - (8 + 4n) = 2(3n - 4)$

6. $4(9v + 2) = 6(1 + 6v) - 10$

7. $0.3(x - 2) = 1.2$

8. $1.5 = 0.2(c - 0.3)$

9. $-4(2 - 3h) = 2(3h - 4) + 6h$

10. $6(m - 1) + 3(2 - m) = 0$

11. $6 - 3(2g + 4) - 4g = 5(1 - 2g)$

12. $20 - 5(p + 1) + 3p = -(2p - 15)$

13. $\dfrac{x}{3} - 4 = x - 2$

14. $\dfrac{9}{4}y = \dfrac{2}{3}y$

15. $\dfrac{3n}{8} - 1 = 3 + \dfrac{n}{6}$

16. $\dfrac{z}{6} + 1 = \dfrac{z}{2} + 2$

17. $\dfrac{y}{4} - \dfrac{y}{2} = -8$

18. $\dfrac{2x}{3} - \dfrac{8}{3} = x$

19. $\dfrac{b - 2}{3} = \dfrac{b + 2}{5}$

20. $\dfrac{2t - 1}{3} = \dfrac{3t + 2}{15}$

21. $\dfrac{2(t + 1)}{3} = \dfrac{2(t - 1)}{3}$

22. $\dfrac{3a - 3}{6} = \dfrac{4a + 1}{15} + 2$

23. $\dfrac{x - 2}{5} + \dfrac{x + 2}{2} = \dfrac{x + 4}{3}$

24. $\dfrac{2z - 3}{4} - \dfrac{4 - z}{2} = \dfrac{z + 1}{3}$

(2.2) *Solve each equation for the specified variable.*

25. $V = LWH$; W

26. $C = 2\pi r$; r

27. $5x - 4y = -12$; y

28. $5x - 4y = -12$; x

29. $y - y_1 = m(x - x_1)$; m

30. $y - y_1 = m(x - x_1)$; x

31. $E = I(R + r)$; r

32. $S = vt + gt^2$; g

33. $T = gr + gvt$; g

34. $I = Prt + P$; P

35. $A = \dfrac{h}{2}(B + b)$; B

36. $V = \dfrac{1}{3}\pi r^2 h$; h

37. $R = \dfrac{r_1 + r_2}{2}$; r_1

38. $\dfrac{V_1}{T_1} = \dfrac{V_2}{T_2}$; T_2

39. $\dfrac{1}{a} + \dfrac{1}{b} = \dfrac{1}{c}$; b

40. $\dfrac{2}{x} - \dfrac{3}{y} = \dfrac{1}{z}$; y

41. $R = \dfrac{R_1 R_2}{R_1 + R_2}$; R_2

42. $C = \dfrac{2AB}{A - B}$; A

43. $\dfrac{x - y}{5} + \dfrac{y}{4} = \dfrac{2x}{3}$; y

44. $\dfrac{b + c}{d} - \dfrac{b}{c} = \dfrac{5}{c}$; d

Solve (see Appendix for a list of common geometric formulas).

45. Find the amount of principal that must be invested at 8% interest per year for 2 years to earn $720 in interest.

46. One-square-foot floor tiles come 24 to a package. Find how many packages are needed to cover a rectangular floor 18 feet by 21 feet.

47. Determine which container holds more ice cream, an 8 inch x 5 inch x 3 inch box or a cylinder with radius of 3 inches and height of 6 inches.

48. Angie left Los Angeles at 11 A.M. and drove nonstop to San Diego, 130 miles away. If she arrived at 1:15 P.M., find her average speed, rounded to the nearest mile per hour.

(2.3) *Solve.*

49. A car rental company charges $19.95 per day for a compact car plus 12 cents per mile for every mile over 100 miles driven per day. If Mr. Woo's bill for 2 days use is $46.86, find how many miles he drove.

50. A mother spent $12.70 for her three children to bowl. If her sons bowled three games each, her daughter bowled two games, and shoes rent for $0.50 per pair, find the price of each game.

51. Find four consecutive integers such that twice the first subtracted from the sum of the other three integers is sixteen.

52. Determine whether there are two consecutive odd integers such that 5 times the first exceeds 3 times the second by 54.

53. LaTonya invested part of her $25,000 book royalty at 8% annual interest and the rest at 9% annual interest. If her total yearly interest from both accounts is $2135, find the amount invested at each rate.

54. Manuel invested part of his $10,000 commission check in a mutual fund, which paid an 11% profit annually, and invested the rest in a stock that suffered a 4% annual loss. Find the amount of each investment if his overall net profit was $650.

55. Two hikers leave simultaneously from the same point traveling in opposite directions, one walking at 4 mph and the other jogging at 5 mph. Find how long a time they can talk on walkie-talkies that have a 15-mile range?

56. On a 320-mile trip, a car traveled at an average speed of 70 mph, stopped for a speeding ticket, and then traveled at 60 mph for the remainder of the trip. If the entire trip took 5.5 hours and the speeding ticket stop took 30 minutes, find how long the car traveled at 70 mph.

57. A van traveled on a level road for 3 hours at an average speed of 20 mph faster than it traveled on a hilly road. The time spent on the hilly road was 4 hours. Find the average speed on the level road if the entire trip was 305 miles.

58. Alex can row upstream at 5 mph and downstream at 11 mph. If he starts rowing upstream until he gets tired and then rows downstream to his starting point, find how far he rowed upstream if the entire trip took 4 hours.

59. A 100-seat theater sold out for a recent concert for which floor seats cost $18 each and balcony seats cost $12 each. Gate receipts totaled $1440. Find how many seats are in the balcony.

60. While going to college, Jamal works part-time at school and earns $4 per hour. He also works in his family's catering business and earns $7.50 per hour. Last week he worked 16 hours and made $85. Determine whether he made more money at school or at his catering job.

61. Ian wants to combine a 15% glycerin solution with a 45% glycerin solution to form 150 centiliters of a 35% glycerin solution. Find how much of each solution he should use.

62. Connie needs to add pure bleach to 100 cubic centimeters of a 25% bleach solution to increase the strength to 40%. Find how much pure bleach she should add.

63. Find how many gallons of water must evaporate from 200 gallons of a 4% saline solution to strengthen it to a 6% saline solution.

64. A coffee merchant blends Colombian roast beans selling for $4.00 per pound with French roast beans selling for $6.50 per pound to make 300 pounds of a blend to sell for $4.75 per pound. Find how many pounds of each bean the merchant uses.

65. The length of a rectangular playing field is 5 meters less than twice its width. If 230 meters of fencing goes around the field, find the dimensions of the field.

66. Angie has a photograph in which the length is 2 inches longer than the width. If she increases each dimension by 4 inches, the area is increased by 88 square inches. Find the original dimensions.

67. The cost C of producing x number of scientific calculators is given by $C = 4.50x + 3000$, and the revenue R from selling them is given by $R = 16.50x$. Find the number of calculators that must be sold to break even. (In this case, revenue = cost.)

68. An entrepreneur can sell her musically vibrating plants for $40 each, while her cost C to produce x number of plants is given by $C = 20x + 100$. Find her break-even point. Find her revenue if she sells exactly that number of plants.

(2.4) *Solve each absolute value equation.*

69. $|x - 7| = 9$

70. $|8 - x| = 3$

71. $|2x + 9| = 9$

72. $|-3x + 4| = 7$

73. $|3x - 2| + 6 = 10$

74. $5 + |6x + 1| = 5$

75. $-5 = |4x - 3|$

77. $|7x| - 26 = -5$

79. $\left|\dfrac{3x - 7}{4}\right| = 2$

81. $|6x + 1| = |15 + 4x|$

76. $|5 - 6x| + 8 = 3$

78. $-8 = |x - 3| - 10$

80. $\left|\dfrac{9 - 2x}{5}\right| = -3$

82. $|x - 3| = |7 + 2x|$

(2.5) *Solve each linear inequality.*

83. $3(x - 5) > -(x + 3)$

85. $4x - (5 + 2x) < 3x - 1$

87. $24 \geq 6x - 2(3x - 5) + 2x$

89. $\dfrac{x}{3} + \dfrac{1}{2} > \dfrac{2}{3}$

91. $\dfrac{x - 5}{2} \leq \dfrac{3}{8}(2x + 6)$

84. $-2(x + 7) \geq 3(x + 2)$

86. $3(x - 8) < 7x + 2(5 - x)$

88. $48 + x \geq 5(2x + 4) - 2x$

90. $x + \dfrac{3}{4} < \dfrac{-x}{2} + \dfrac{9}{4}$

92. $\dfrac{3(x - 2)}{5} > \dfrac{-5(x - 2)}{3}$

Solve.

93. George Boros can pay his housekeeper \$15 per week to do his laundry, or he can have the laundromat do it at a cost of 50 cents per pound for the first 10 pounds and 40 cents for each additional pound. Use an inequality to find the weight at which it is more economical to use the housekeeper than the laundromat.

94. Ceramic firing temperatures usually range from 500° to 1000° Fahrenheit. Use a compound inequality to convert this range to the Celsius scale. Round to the nearest degree.

95. In the Olympic ice dancing competition, Nana must score 9.65 to win the silver medal. Seven of the eight

judges have reported scores of 9.5, 9.7, 9.9, 9.7, 9.7, 9.6, and 9.5. Use an inequality to find the minimum score that the last judge can give so that Nana wins the silver medal.

96. Carol would like to pay cash for a car when she graduates from college and estimates that she can afford a car that costs between \$4000 and \$8000. She has saved \$500 so far and plans to earn the rest of the money by working the next two summers. If Carol plans to save the same amount each summer, use a compound inequality to find the range of money she must save each summer to buy the car.

(2.6) *Solve each inequality.*

97. $1 \leq 4x - 7 \leq 3$

99. $-3 < 4(2x - 1) < 12$

101. $\dfrac{1}{6} < \dfrac{4x - 3}{3} \leq \dfrac{4}{5}$

103. $x \leq 2$ and $x > -5$

105. $3x - 5 > 6$ or $-x < -5$

98. $-2 \leq 8 + 5x < -1$

100. $-6 < x - (3 - 4x) < -3$

102. $0 \leq \dfrac{2(3x + 4)}{5} \leq 3$

104. $x \leq 2$ or $x > -5$

106. $-2x \leq 6$ and $-2x + 3 < -7$

(2.7) *Solve each absolute value inequality. Graph the solution set and write in interval notation.*

107. $|5x - 1| < 9$

109. $|3x| - 8 > 1$

111. $|6x - 5| \leq -1$

113. $\left|3x + \dfrac{2}{5}\right| \geq 4$

115. $\left|\dfrac{x}{3} + 6\right| - 8 > -5$

108. $|6 + 4x| \geq 10$

110. $) + |5x| < 24$

112. $|6x - 5| \geq -1$

114. $\left|\dfrac{4x - 3}{5}\right| < 1$

116. $\left|\dfrac{4(x - 1)}{7}\right| + 10 < 2$

CHAPTER 2 TEST

Solve each equation.

1. $8x + 14 = 5x + 44$

2. $3(x + 2) = 11 - 2(2 - x)$

3. $3(y - 4) + y = 2(6 + 2y)$

4. $7n - 6 + n = 2(4n - 3)$

5. $\dfrac{z}{2} + \dfrac{z}{3} = 10$

6. $\dfrac{7w}{4} + 5 = \dfrac{3w}{10} + 1$

7. $|6x - 5| = 1$

8. $|8 - 2t| = -6$

Solve each equation for the specified variable.

9. $3x - 4y = 8$; y

10. $4(2n - 3m) - 3(5n - 7m) = 0$; n

11. $S = gt^2 + gvt$; g

12. $F = \dfrac{9}{5}C + 32$; C

Solve each inequality.

13. $3(2x - 7) - 4x > -(x + 6)$

14. $8 - \dfrac{x}{2} \le 7$

15. $-3 < 2(x - 3) \le 4$

16. $|3x + 1| > 5$

17. $x \ge 5$ and $x \ge 4$

18. $x \ge 5$ or $x \ge 4$

19. $-x > 1$ and $3x + 3 \ge x - 3$

20. $6x + 1 > 5x + 4$ or $1 - x > -4$

Solve.

21. A circular dog pen has a circumference of 78.5 feet. Approximate π by 3.14 and estimate how many hunting dogs could be safely kept in the pen if each dog needs at least 60 square feet of room.

22. The company that makes Photoray sunglasses figures that the cost C to make x number of sunglasses weekly is given by $C = 3910 + 2.8x$, and the weekly revenue R is given by $R = 7.4x$. Use an inequality to find the number of sunglasses that must be made and sold to make a profit. (Revenue must exceed cost in order to make a profit.)

23. Sedric invested an amount of money in Amoxil stock that earned an annual 10% return and then invested twice the original amount in IBM stock that earned an annual 12% return. If his total return from both investments was $2890, find how much he invested in each stock?

24. Two cars leave from the same point at the same time traveling in opposite directions, one driving at 55 mph and the other at 65 mph. Find how long a time they can talk on their car phones if the phones have a 250-mile range?

25. A bank teller cashed a $300 check and gave the customer 19 bills using twenty-dollar bills and ten-dollar bills. Find how many bills of each type the teller gave.

26. Find how many liters of a 12% acid solution must be added to 10 liters of a 25% acid solution to get a 20% acid solution.

CHAPTER 2 CUMULATIVE REVIEW

1. Determine whether the following statements are true or false.
  a. $\{1, 2\} \subseteq \{1, 2, 3\}$
  b. $\{1, 2, 3,\} \subseteq \{1, 2, 3\}$
  c. $\varnothing \subseteq \{1, 2, 3\}$
  d. $\{2, 3\} = \{3, 2\}$
  e. $3 \subseteq \{1, 2, 3\}$
  f. $\{1, 2, 3, 4\} \subseteq \{2, 3\}$

2. Insert $<$, $>$, or $=$ to make each statement true.
  a. -1 ____ -2 **b.** $\dfrac{12}{4}$ ____ 3
  c. -5 ____ 0

3. Find each absolute value.
  a. $|3|$ **b.** $|-5|$ **c.** $-|2|$
  d. $-|-8|$ **e.** $|0|$

4. Find the following square roots.

   a. $\sqrt{9}$ **b.** $\sqrt{25}$ **c.** $\sqrt{\dfrac{1}{4}}$

5. Find the value of each algebraic expression when $x = 0$, $y = 2$, and $z = 5$.

   a. $\dfrac{2x + y}{z}$ **b.** $|y^3| - (z + x)$ **c.** $z^3 + 3y^4$

6. Find each difference.

   a. $2 - 8$ **b.** $-8 - (-1)$ **c.** $-11 - 5$

   d. $10 - (-9)$ **e.** $\dfrac{2}{3} - \dfrac{1}{2}$ **f.** $1 - 0.06$

7. Simplify the following expressions.

   a. $2(1 - 4)^2$ **b.** $\dfrac{5 - 7}{-2}$ **c.** $\dfrac{-7}{0}$

   d. $\dfrac{(6 + 2) - (-4)}{2 - (-3)}$

8. Simplify the following expressions by combining like terms.

   a. $3x - 2x + 5 - 7 + x$
   b. $7x + 3 - 5(x - 4)$
   c. $(2x - 5) - (-x - 5)$

9. Solve for x: $2(x - 3) = 5x - 9$.

10. Solve for x: $\dfrac{x}{3} - \dfrac{x}{4} = \dfrac{1}{6}$.

11. A gallon of sealer covers 480 square feet. How many gallon containers of sealer should be bought to cover a rectangular driveway 24 feet wide by 90 feet long?

12. The bottom of Jim's backpack contains 20 coins in nickels and dimes. If the coins have a total value of $1.85, find the number of each type of coin.

13. Solve $|5w + 3| = 7$.

14. Solve $\left|\dfrac{3x + 1}{2}\right| = -2$.

15. Solve $|x - 3| = |5 - x|$.

16. Solve for x.
   a. $2x < -6$
   b. $-2x < 6$

17. Solve for x: $2(x + 3) > 2x + 1$.

18. Solve for x: $x - 7 < 2$ and $2x + 1 < 9$.

19. Solve for m: $|m - 6| < 2$.

20. Solve $\left|\dfrac{x}{3} - 1\right| - 7 \geq -5$.

CHAPTER **3**

Graphing Linear Equations and Inequalities

Financial wizards once scoffed at cable television as a concept with limited potential. Today, however, even these doubters see the future possibilities, in part, by analyzing information such as the increase in the amount of money Americans spend on cable television.

INTRODUCTION

The linear equations we explored in Chapter 2 are statements about a single variable. This chapter examines statements about two variables: linear equations and inequalities in two variables. We focus particularly on graphs of these equations and inequalities, which lead in this chapter to the notion of relation and to the notion of function, perhaps the single most important and useful notion in all of mathematics.

3.1
The Cartesian Coordinate System

OBJECTIVES

Tape 6

1 Plot ordered pairs.

2 Use the distance formula.

3 Use the midpoint formula.

1   A versatile tool for graphing equations in two variables is the **Cartesian** or **rectangular** coordinate system, named after its inventor, René Déscartes. The Cartesian coordinate system consists of two number lines that intersect at right angles at their 0 coordinates. We position these axes on paper so that one number line is horizontal and the other number line is then vertical. The horizontal number line is called the **x-axis** (or the axis of the **abscissa**), and the vertical number line is called the **y-axis** (or the axis of the **ordinate**). The point of intersection of these axes is named the **origin.**

Notice that the axes divide the plane into four regions. These regions are called **quadrants,** with the top-right region being quadrant I. Quadrants II, III, and IV are numbered counterclockwise from the first quadrant as shown. The x-axis and the y-axis are not in any quadrant.

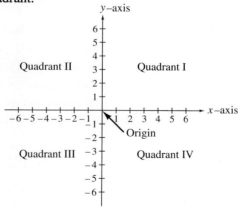

Each point in the plane can be located or **plotted** by describing its position in terms of distances along each axis from the origin. An **ordered pair,** represented by the notation (x, y), records these distances.

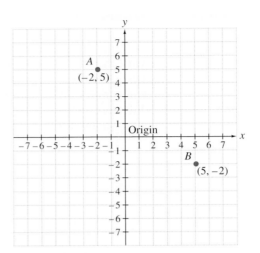

For example, the position of point A is described as 2 units to the left of the origin and 5 units upward. Thus we identify point A with the ordered pair $(-2, 5)$. Notice that the order of these numbers is critical. The x-value -2 is called the **x-coordinate** and is associated with the x-axis. The y-value 5 is called the **y-coordinate** and is associated with the y-axis. Compare the position of point A with the position of point B, which corresponds to the ordered pair $(5, -2)$. The x-coordinate 5 indicates that we move 5 units to the right of the origin. The y-coordinate -2 indicates that we move 2 units down. Point A is in a different position than point B. Two ordered pairs are considered equal and correspond to the same point if and only if their x-coordinates are equal and their y-coordinates are equal.

Keep in mind that **each ordered pair corresponds to exactly one point in the real plane and that each point in the plane corresponds to exactly one ordered pair.**

EXAMPLE 1   Plot each ordered pair on a Cartesian coordinate system and name the quadrant in which the point is located.

   a. $A(2, -1)$ **b.** $B(0, 5)$ **c.** $C(-3, 5)$ **d.** $D(-2, 0)$ **e.** $\left(-\dfrac{1}{2}, -4\right)$

Solution: The five points are graphed as shown:
   a. Point A lies in quadrant IV.
   b. Point B is not in any quadrant.
   c. Point C lies in quadrant II.
   d. Point D is not in any quadrant.
   e. Point E is in quadrant III.

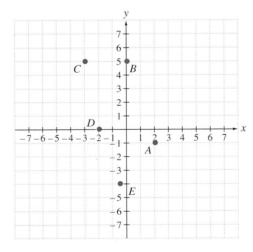

Notice that the y-coordinate of any point of the x-axis is 0. For example, the coordinates of point D are $(-2, 0)$. Also, the x-coordinate of any point on the y-axis is 0. For example, the coordinates of point B are $(0, 5)$.

2   Plotting points on the Cartesian coordinate system helps us to visualize a distance between points. To find the distance between two points, we use the **distance formula,** which is derived from the Pythagorean theorem. Recall that the Pythagorean theorem states that for any **right triangle** whose legs have length a and b and whose hypotenuse has length d, it must be true that

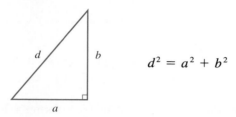

$$d^2 = a^2 + b^2$$

For example, in the following right triangle, we have that

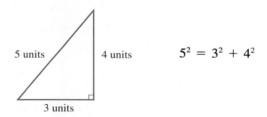

$$5^2 = 3^2 + 4^2$$

or

$$25 = 9 + 16$$

To find the distance d between two points (x_1, y_1) and (x_2, y_2), we use the Pythagorean theorem.

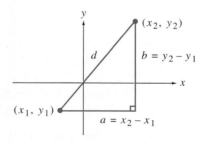

The length of leg a is $x_2 - x_1$ and the length of leg b is $y_2 - y_1$. Thus, the Pythagorean theorem tells us that

$$d^2 = a^2 + b^2$$

or

$$d^2 = (x_2 - x_1)^2 + (y_2 - y_1)^2$$

or

$$d = \sqrt{(x_2 - x_1)^2 + (y_2 - y_1)^2}$$

This formula gives us the distance between any two points on the real plane.

Distance Formula

The distance between two points (x_1, y_1) and (x_2, y_2) is given by

$$d = \sqrt{(x_2 - x_1)^2 + (y_2 - y_1)^2}$$

EXAMPLE 2   Find the distance between the following pairs of points.
a. $(-3, 3)$ and $(1, 0)$ **b.** $(2, -5)$ and $(1, -4)$

Solution: **a.** Use the distance formula. It makes no difference which point we call (x_1, y_1) and which point we call (x_2, y_2). Let the point $(-3, 3) = (x_1, y_1)$ and the point $(1, 0) = (x_2, y_2)$.

$$
\begin{aligned}
d &= \sqrt{(x_2 - x_1)^2 + (y_2 - y_1)^2} \\
&= \sqrt{[1 - (-3)]^2 + (0 - 3)^2} \qquad \text{Let } x_1 = -3,\ x_2 = 1,\ y_1 = 3,\ \text{and } y_2 = 0. \\
&= \sqrt{(4)^2 + (-3)^2} \\
&= \sqrt{16 + 9} \\
&= \sqrt{25} \\
&= 5
\end{aligned}
$$

The distance between the two points is 5 units.

b. Let $(2, -5) = (x_1, y_1)$ and $(1, -4) = (x_2, y_2)$.

$$
\begin{aligned}
d &= \sqrt{(x_2 - x_1)^2 + (y_2 - y_1)^2} \\
&= \sqrt{(1 - 2)^2 + [-4 - (-5)]^2} \\
&= \sqrt{(-1)^2 + (1)^2} \\
&= \sqrt{1 + 1} \\
&= \sqrt{2}
\end{aligned}
$$

The distance between the two points is $\sqrt{2}$ units.   ■

3   The **midpoint** of a line segment is the **point** located exactly halfway between the two end points of the line segment. On the following graph, the point M is the midpoint of line segment PQ. The distance between M and P equals the distance between M and Q.

The x-coordinate of M is half the distance between the x-coordinates of P and Q, and the y-coordinate of M is half the distance between the y-coordinates of P and Q.

Midpoint Formula

The midpoint of the line segment whose end points are (x_1, y_1) and (x_2, y_2) is the point

$$\left(\frac{x_1 + x_2}{2}, \frac{y_1 + y_2}{2}\right)$$

EXAMPLE 3   Find the midpoint of the line segment joining points $P(-3, 3)$ and $Q(1, 0)$.

Solution: Use the midpoint formula. It makes no difference which point we call (x_1, y_1) or which point we call (x_2, y_2). Let $(-3, 3) = (x_1, y_1)$ and $(1, 0) = (x_2, y_2)$.

$$\begin{aligned}
\text{midpoint} &= \left(\frac{x_1 + x_2}{2}, \frac{y_1 + y_2}{2}\right) \\[2mm]
&= \left(\frac{-3 + 1}{2}, \frac{3 + 0}{2}\right) \\[2mm]
&= \left(\frac{-2}{2}, \frac{3}{2}\right) \\[2mm]
&= \left(-1, \frac{3}{2}\right)
\end{aligned}$$

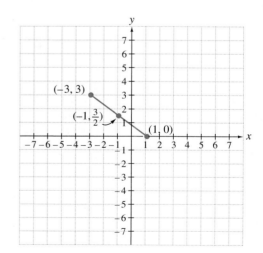

The midpoint of the segment is $\left(-1, \dfrac{3}{2}\right)$.   ∎

MENTAL MATH

Determine the coordinates of each point on the graph.

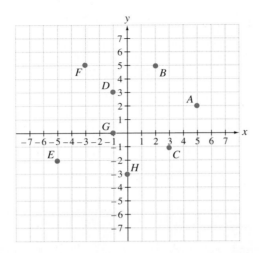

1. Point A

2. Point B

3. Point C

4. Point D

5. Point E

6. Point F

7. Point G

8. Point H

EXERCISE SET 3.1

Plot each point and name the quadrant or axis in which the point lies. See Example 1.

1. $(3, 2)$

2. $(2, -1)$

3. $(-5, 3)$

4. $(-3, -1)$

5. $(5, -4)$

6. $(2, 3)$

7. $(0, 3)$

8. $(-2, 4)$

9. $(-2, -4)$

10. $(5, 0)$

Find the distance between each pair of points. See Example 2.

11. (5, 1) and (8, 5)

12. (2, 3) and (14, 8)

13. (−8, 1) and (16, 8)

14. (2, 1) and (−4, 9)

15. (−3, 2) and (1, −3)

16. (3, −2) and (−4, 1)

17. (5, −3) and (2, −8)

18. (−1, 8) and (−5, 7)

Find the midpoint of the line segment whose end points are given. See Example 3.

19. (6, −8), (2, 4)

20. (3, 9), (7, 11)

21. (−3, 0), (7, 8)

22. (2, −7), (0, 5)

23. (−2, −1), (−8, 6)

24. (−3, −4), (6, −8)

25. (2, −5), (10, −8)

26. (−3, −6), (−9, 3)

Find the distance between each pair of points.

27. (6, 3) and (2, 4)

28. (8, 10) and (5, 8)

29. (−2, 1) and (−2, −9)

30. (0, −3) and (1, 5)

31. (4, −9) and (1, −8)

32. (3, −2) and (−6, −6)

33. (−4, −1) and (−7, −3)

34. (−2, −1) and (−1, 5)

35. (2, −7) and (−3, −3)

36. (10, −14) and (5, −11)

37. (x, 3) and (x, 7)

38. (2, y) and (−5, y)

Find the midpoint of the line segment whose end points are given.

39. (4, 9), (6, 5)

40. (10, 8), (4, 12)

41. (−3, 4), (−1, −2)

42. (7, 3), (−1, −3)

43. (−2, 5), (−1, 6)

44. (3, 0), (−3, 0)

45. $\left(-\dfrac{1}{2}, 3\right)$, (2, 6)

46. (2, −3), $\left(9, \dfrac{1}{2}\right)$

47. Determine the quadrant(s) for which the *x*-coordinates are greater than 0 and the *y*-coordinates are less than zero.

48. Determine the quadrant(s) for which both the *x*- and *y*-coordinates are negative.

49. Determine the quadrant(s) for which the *x*-coordinates are positive.

50. Determine the quadrant(s) for which the *y*-coordinates are positive.

51. Two surveyors need to find the distance across a lake. They placed a reference pole at point *A* in the diagram. Point *B* is 3 meters east and 1 meter north of the reference

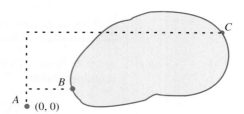

point *A*. Point *C* is 19 meters east and 13 meters north of point *A*. Find the distance across the lake.

52. Determine whether the triangle with vertices (2, 6), (0, −2), and (5, 1) is an isosceles triangle.

Skill Review

Solve the following equations. See Section 2.1.

53. 3(x − 2) + 5x = 6x − 16

54. 5 + 7(x + 1) = 12 + 10x

55. $3x + \dfrac{2}{5} = \dfrac{1}{10}$

56. $\dfrac{1}{6} + 2x = \dfrac{2}{3}$

Solve the following inequalities. See Section 2.5.

57. 3x ≤ −15

58. −3x > 18

59. 2x − 5 > 4x + 3

60. 9x + 8 ≤ 6x − 4

3.2
Graphing Linear Equations

OBJECTIVES

Tape 6

1   Find ordered pairs that satisfy a linear equation.

2   Graph lines.

3   Identify x- and y-intercepts.

In Chapter 2, we solved linear equations in one variable, such as $3x - 5 = 12$. In this section, we graph the solutions of **linear equations in two variables,** such as $3x - y = 12$.

Linear Equation in Two Variables

A linear equation in two variables is an equation that can be written in the form

$$ax + by = c$$

where a, b, and c are real numbers, and a and b are not both 0.

A linear equation written in the form $ax + by = c$ is said to be written in **standard form.**

Examples of Linear Equations in Standard Form

$$3x + 5y = 6$$
$$-2x + 3y = 0$$

1   We will write solutions of linear equations in two variables as ordered pairs of numbers. For example, a solution of $3x - y = 12$ is $(2, -6)$ because if x is replaced with 2 and y with -6, the resulting statement is true.

$$3x - y = 12$$

$$3(2) - (-6) = 12 \qquad \text{Let } x = 2 \text{ and } y = -6.$$

$$12 = 12 \qquad \text{True.}$$

To find an ordered pair that satisfies a linear equation, we can choose any value for one of the variables, replace the variable by the chosen value in the equation, and solve for the other variable. Example 1 illustrates this technique.

EXAMPLE 1   Complete the ordered pair solutions for the equation $3x - y = 12$.
   a. (, 0)   **b.** (1,)   **c.** (0,)

Solution: **a.** Let $y = 0$ in the linear equation and find the corresponding value for x.

$$3x - y = 12$$

$$3x - 0 = 12 \qquad \text{Let } y = 0.$$

$$3x = 12$$

$$x = 4 \qquad \text{Solve.}$$

Since $x = 4$ when $y = 0$, the ordered pair $(4, 0)$ is a solution of $3x - y = 12$.

b. Let $x = 1$ in the linear equation and solve for y.

$$3x - y = 12$$

$$3(1) - y = 12 \qquad \text{Let } x = 1.$$

$$3 - y = 12$$

$$-y = 9 \qquad \text{Solve.}$$

$$y = -9$$

The ordered pair is $(1, -9)$.

c. Let $x = 0$ in the linear equation and solve for y.

$$3x - y = 12$$

$$3(0) - y = 12 \qquad \text{Let } x = 0.$$

$$0 - y = 12$$

$$-y = 12 \qquad \text{Solve.}$$

$$y = -12$$

The ordered pair is $(0, -12)$.   ■

2   A few more ordered pairs that satisfy $3x - y = 12$ are $(2, -6)$, $(5, 3)$ and $(3, -3)$. These ordered pair solutions along with the ordered pair solutions from Example 1 are plotted on the following graph. The graph of $3x - y = 12$ is the single

x	y
4	0
1	-9
0	-12
2	-6
5	3
3	-3

line containing these points. Every ordered pair solution of the equation corresponds to a point on this line, and every point on this line corresponds to an ordered pair solution. In general, **the graph of every linear equation in two variables is a line.**

3   The line representing the solutions of $3x - y = 12$ intersects the x-axis at $(4, 0)$, so the number 4 is called the **x-intercept.** The line also intersects the y-axis at $(0, -12)$, and the number -12 is called the **y-intercept.**

Finding x- and y-Intercepts

To find the x-intercept, let $y = 0$ and solve for x.
To find the y-intercept, let $x = 0$ and solve for y.

EXAMPLE 2   Find the x- and y-intercepts of $2x + 3y = 5$ and graph the equation.

Solution: To find the x-intercept, let $y = 0$.

$$2x + 3y = 5$$
$$2x + 3(0) = 5 \qquad \text{Let } y = 0.$$
$$2x + 0 = 5 \qquad \text{Multiply.}$$
$$2x = 5 \qquad \text{Simplify.}$$
$$x = \frac{5}{2} \qquad \text{Divide by 2.}$$

The x-intercept is $\frac{5}{2}$ and the ordered pair solution is $\left(\frac{5}{2}, 0\right)$.

To find the y-intercept, let $x = 0$.

$$2x + 3y = 5$$
$$2(0) + 3y = 5 \qquad \text{Let } x = 0.$$
$$0 + 3y = 5 \qquad \text{Multiply.}$$
$$3y = 5$$
$$y = \frac{5}{3} \qquad \text{Divide by 3.}$$

The y-intercept is $\frac{5}{3}$ and the ordered pair solution is $\left(0, \frac{5}{3}\right)$.

Plot the points $\left(\frac{5}{2}, 0\right)$ and $\left(0, \frac{5}{3}\right)$. The graph of $2x + 3y = 5$ is the line containing these points.

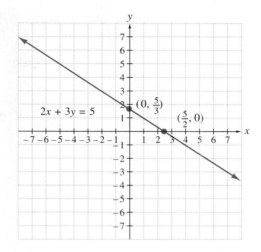

Although two points determine a line, when plotting points to graph a linear equation, we generally plot three points. The third point serves as a check to make sure our computations are correct, when we see that all three points lie on the same line.   ∎

EXAMPLE 3   Graph $y = 4x$.

Solution: Begin by finding the x- and y-intercepts. To find the x-intercept, let $y = 0$.

$$y = 4x$$
$$0 = 4x \qquad \text{Let } y = 0.$$
$$0 = x \qquad \text{Solve for } x.$$

The x-intercept is 0 and the ordered pair is $(0, 0)$. To find the y-intercept, let $x = 0$.

$$y = 4x$$
$$y = 4(0) \qquad \text{Let } x = 0.$$
$$y = 0$$

The y-intercept is also 0 and the ordered pair is $(0, 0)$. The x- and y-intercepts lead to the same ordered pair, so we will find two more points on the graph. Let $x = 2$ and $x = 1$ to obtain two other points.

$$y = 4x \qquad\qquad\qquad y = 4x$$
$$y = 4(2) \quad \text{Let } x = 2. \qquad y = 4(1) \quad \text{Let } x = 1.$$
$$y = 8 \quad \text{Multiply.} \qquad\quad y = 4 \quad \text{Multiply.}$$

The ordered pair is $(2, 8)$.   The ordered pair is $(1, 4)$.

Finally, plot the points $(0, 0)$, $(2, 8)$, and $(1, 4)$. The line through these points is the graph of $y = 4x$.

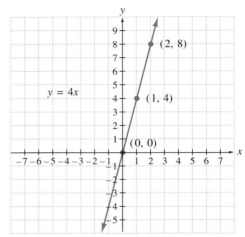

x	y
0	0
2	8
1	4

HELPFUL HINT

When graphing a linear equation, it is a good idea to find and plot three points. The third point serves as a check, when we see that a single line contains all three points.

EXAMPLE 4   Graph $x = 1$.

Solution: The line representing the equation $x = 1$ contains all points whose x-coordinate is 1. A few points are $(1, 0)$, $(1, 1)$, and $(1, 2)$. The graph of $x = 1$ is the line through these points.

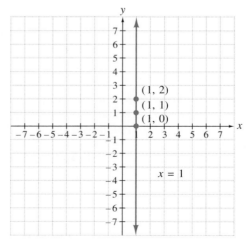

In general, **the graph of an equation of the form $x = c$ is a vertical line with x-intercept c.** ■

EXAMPLE 5   Graph $y = -2$.

Solution: The line representing the equation $y = -2$ contains all points whose y-coordinate is -2. A few points are $(1, -2)$, $(2, -2)$, $(3, -2)$. The graph of $y = -2$ is the line through these points.

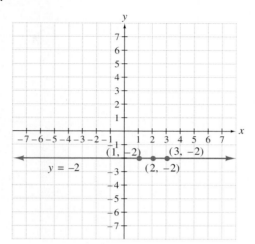

In general, **the graph of an equation of the form $y = b$ is a horizontal line with y-intercept b.**   ■

EXERCISE SET 3.2

Complete the ordered pair solutions of $4x + 3y = 15$. *See Example 1.*

1. $(0, \quad)$ **2.** $(\quad , 0)$ **3.** $(\quad , 1)$ **4.** $(-3, \quad)$

Complete the ordered pair solutions of $5x - 2y = 12$. *See Example 1.*

5. $(\quad , 0)$ **6.** $(0, \quad)$ **7.** $(-2, \quad)$ **8.** $(\quad , -2)$

Complete the ordered pair solutions of $6x - 10y = 15$. *See Example 1.*

9. $(0, \quad)$ **10.** $(\quad , 0)$ **11.** $(\quad , 3)$ **12.** $(4, \quad)$

Find the x- and y-intercepts for each equation, and use the intercepts to graph the equation. See Examples 2 and 3.

13. $x + y = 5$ **14.** $x + y = 2$ **15.** $2x + y = 4$ **16.** $x = 3y - 3$

17. $x = 2y - 4$ **18.** $-2x + y = 4$ **19.** $-3x + y = 6$ **20.** $4x - 3y = -12$

Graph each equation. See Examples 4 and 5.

21. $x = 5$ **22.** $y = 7$ **23.** $y = -2$ **24.** $x = -4$

25. $x = 0$ **26.** $y = 0$

Find the intercepts and graph each equation.

27. $x + 2y = 8$ **28.** $3x + 2y = -6$ **29.** $2x + 5y = -10$ **30.** $2x - 4y = -8$

31. $y = -3$ **32.** $x = -5$ **33.** $4y = 9 - 6x$ **34.** $2y = 5 - 10x$

35. $y = 3x$

36. $y = -2x$

37. $x = 10$

38. $y = 1$

39. $3x - 2y = -4$

40. $2y - 4x = -5$

41. $2x + 7y = 14$

42. $9x + 2y = 18$

43. $x = y + 5$

44. $y = x - 3$

45. $y = x$

46. $x = -y$

47. The perimeter P of a rectangle whose width is a constant 3 inches and whose length is L inches is given by the equation

$$P - 2L = 6$$

  a. Draw a graph of this equation. Use the horizontal axis for length L.
  b. Read from the graph the perimeter P of a rectangle whose length L is 4 inches.

48. The distance d traveled in a train moving at a constant speed of 50 miles per hour is given by the equation

$$d = 50t$$

where t is the time in hours traveled.

  a. Draw a graph of this equation. Use the horizontal axis for time t.
  b. Read from the graph the distance d traveled after 6 hours.

49. Broyhill Furniture found that it takes 2 hours to manufacture each table for one of its special dining room sets. Each chair takes 3 hours to manufacture. A total of 1500 hours is available to produce tables and chairs of this style. The linear equation that models this situation is $2x + 3y = 1500$, where x represents the number of tables produced and y the number of chairs produced.

  a. Complete the ordered pair solution $(0, \quad)$ of this equation. Describe the manufacturing situation this solution corresponds to.
  b. Complete the ordered pair $(\quad, 0)$ for this equation. Describe the manufacturing situation this solution corresponds to.
  c. If 50 tables are produced, find the greatest number of chairs they can make.

50. While manufacturing two different camera models, Kodak found that the basic model costs $55 to produce, while the deluxe model costs $75. The weekly budget for these two models is limited to $33,000 in production costs. The linear equation that models this situation is $55x + 75y = 33,000$, where x represents the number of basic models and y the number of deluxe models.

a. Complete the ordered pair solution $(0, \quad)$ of this equation. Describe the manufacturing situation this solution corresponds to.

b. Complete the ordered pair solution $(\quad, 0)$ of this equation. Describe the manufacturing situation this solution corresponds to.

c. If 350 deluxe models are produced, find the greatest number of basic models that can be made in one week.

A Look Ahead

The graph of an equation in two variables is not always a line. To see this, graph each equation. See the following example.

EXAMPLE   Graph $y = x^2$ by plotting ordered pair solutions and drawing a smooth curve through the points.

Solution: For $y = x^2$:
If $x = 0$, then $y = 0^2$ or 0.
If $x = 1$, then $y = 1^2$ or 1.
If $x = -1$, then $y = (-1)^2$ or 1.
If $x = 2$, then $y = 2^2$ or 4.
If $x = -2$, then $y = (-2)^2$ or 4.

x	y
0	0
1	1
-1	1
2	4
-2	4

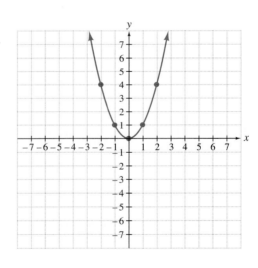

51. $y = |x|$

52. $y = x^3$

53. $y = \sqrt{x}$ (Let x = 0, 1, 4, and 9.)

54. $y = \sqrt{x + 1}$ (Let x = −1, 0, 3, and 8.)

55. $y = 2x^2$

56. $y = -x^2$ (Recall that $-x^2$ means $-1 \cdot x^2$)

Writing in Mathematics

57. Explain why we replace y by 0 to find the x-intercept.

58. Explain why we generally find three points to graph a line, when only two points are needed.

Skill Review

Solve the following. See Sections 2.4 and 2.7.

59. $|x - 3| = 6$

60. $|x + 2| < 4$

61. $|2x + 5| > 3$

62. $|5x| = 10$

63. $|3x - 4| \le 2$

64. $|7x - 2| \ge 5$

3.3
The Slope of a Line

Tape 7

OBJECTIVES

1   Find the slope of a line given two points on the line.

2   Find the slope of a vertical or a horizontal line.

3   Determine whether two lines are parallel or perpendicular.

---

1   Anyone who has experienced the plunging descent of a roller coaster or the rapid climb of a jet-liner has some awareness of the steepness of such rides. In mathematics, the steepness or tilt of a line is also known as its **slope.** We measure the slope of a line as a ratio of **vertical change** to **horizontal change.** Slope is usually designated by the letter m.

Suppose we want to measure the slope of a line through the points (2, 1) and (4, 5).

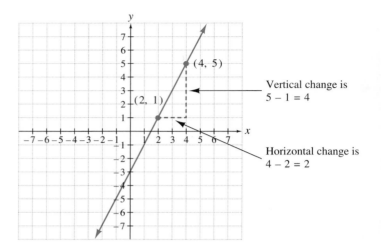

The vertical change is the distance between the y-coordinate 1 and the y-coordinate 5, a distance of $5 - 1 = 4$ units. The horizontal change is the distance between the x-coordinate 2 and the x-coordinate 4, a distance of $4 - 2 = 2$ units. Then

$$m = \frac{\text{change in } y \text{ (vertical change)}}{\text{change in } x \text{ (horizontal change)}} = \frac{4}{2} = 2$$

In general, the slope of the line through points (x_1, y_1) and (x_2, y_2) is the ratio of the change in y-coordinates, $y_2 - y_1$, to the change in x-coordinates, $x_2 - x_1$.

$$m = \frac{y_2 - y_1}{x_2 - x_1}$$

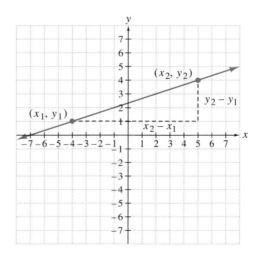

Slope of a Line

Given a line passing through points (x_1, y_1) and (x_2, y_2), the slope m of the line is

$$m = \frac{y_2 - y_1}{x_2 - x_1}$$

as long as $x_2 \neq x_1$.

EXAMPLE 1   Find the slope of the line containing the points $(0, 3)$ and $(2, 5)$.

Solution: Use the slope formula. It does not matter which point we call (x_1, y_1) and which point we call (x_2, y_2). Let $(0, 3) = (x_1, y_1)$ and $(2, 5) = (x_2, y_2)$.

$$m = \frac{y_2 - y_1}{x_2 - x_1}$$

$$= \frac{5 - 3}{2 - 0} = \frac{2}{2} = 1$$

Notice in this example that the slope is positive and that the graph of the line containing $(0, 3)$ and $(2, 5)$ moves upward as we go from left to right.

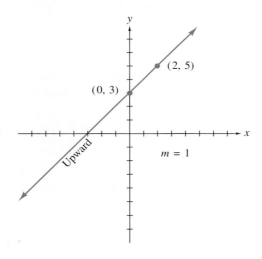

Positive slope   ∎

EXAMPLE 2   Find the slope of the line containing the points $(5, -7)$ and $(-3, 6)$.

Solution: Use the slope formula. Let $(5, -7) = (x_1, y_1)$ and $(-3, 6) = (x_2, y_2)$.

$$m = \frac{y_2 - y_1}{x_2 - x_1}$$

$$= \frac{6 - (-7)}{-3 - 5} = \frac{13}{-8} = -\frac{13}{8}$$

Notice in this example that the slope is negative and that the graph of the line through $(5, -7)$ and $(-3, 6)$ moves downward as we go from left to right.

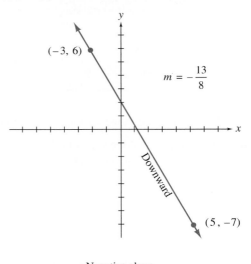

Negative slope ■

2   Two special cases of linear equations and their slopes are horizontal and vertical lines.

EXAMPLE 3   Find the slope of the line $x = -5$.

Solution: Find the coordinates of two points on the line, such as $(-5, 0)$ and $(-5, 3)$, and use these ordered pairs to find the slope.

$$m = \frac{3 - 0}{-5 - (-5)} = \frac{3}{0}, \qquad \text{which is undefined}$$

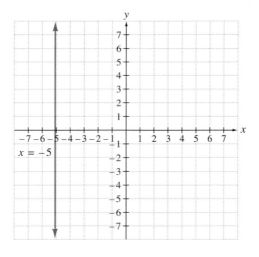

The slope of the line $x = -5$ is undefined. ■

It is true in general that **the slope of a vertical line is undefined.**

EXAMPLE 4   Find the slope of the line $y = 2$.

Solution: Find two points on the line, such as $(0, 2)$ and $(1, 2)$, and use these points to find the slope.

$$m = \frac{2 - 2}{1 - 0} = \frac{0}{1} = 0$$

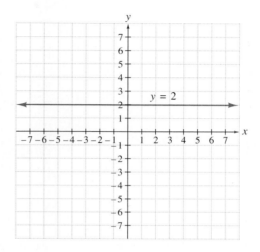

The slope of the line $y = 2$ is 0.   ■

It is true in general that **the slope of a horizontal line is 0.**
The following four graphs summarize the overall appearance of lines with positive, negative, zero, and undefined slopes.

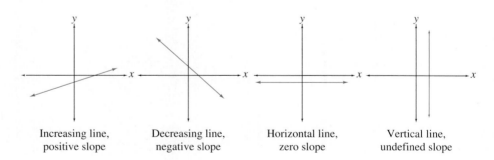

| Increasing line, positive slope | Decreasing line, negative slope | Horizontal line, zero slope | Vertical line, undefined slope |

EXAMPLE 5   Graph the line with slope $-\dfrac{2}{3}$ that contains the point $(-1, 3)$.

Solution: We need two points to graph a line. One point is $(-1, 3)$. A second point can be found by using the slope formula.

$$m = \frac{\text{change in } y}{\text{change in } x} = \frac{-2}{3}$$

From the point $(-1, 3)$, move down 2 units, then move right 3 units. A second point is $(2, 1)$. The line through $(2, 1)$ and $(-1, 3)$ will pass through $(-1, 3)$ and have slope of $\dfrac{-2}{3}$.

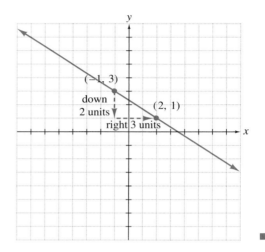

3 Two special cases of pairs of linear equations and their slopes are parallel lines and perpendicular lines.

Parallel lines have the same steepness, so it is reasonable to expect their slopes to be equal. Also, two lines whose slopes are equal are parallel.

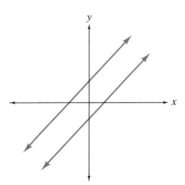

Parallel Lines

If the slopes of two lines are equal, then the lines are parallel. Also, nonvertical parallel lines have the same slope.

EXAMPLE 6 If L_1 is a line containing the points $(3, 2)$ and $(8, 5)$ and L_2 is a line containing the points $(5, 7)$ and $(0, 4)$, determine whether L_1 and L_2 are parallel.

Solution: To determine whether two lines are parallel, we see if their slopes are equal. The slope m_1 of line L_1 is

$$m_1 = \frac{5-2}{8-3} = \frac{3}{5}$$

The slope m_2 of line L_2 is

$$m_2 = \frac{4-7}{0-5} = \frac{-3}{-5} = \frac{3}{5}$$

The slopes are equal and so the lines are parallel.   ■

If two lines intersect to form right angles, we say that they are **perpendicular lines.** It can be shown that if the product of the slopes of two lines is -1 then the lines are perpendicular.

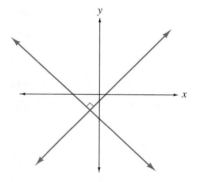

> ### Perpendicular Lines
>
> If the product of the slopes of two lines is -1, then the lines are perpendicular. Also, if two nonvertical lines are perpendicular, then the product of their slopes is -1.

For example, if line L_1 has a slope of 6 and line L_2 has a slope of $-\dfrac{1}{6}$, then these lines are perpendicular because the product of their slopes is

$$6\left(-\frac{1}{6}\right) = -1$$

HELPFUL HINT

Notice that the slopes of perpendicular lines are opposite reciprocals of each other. For example, the slope of a line perpendicular to a line with slope $\dfrac{3}{4}$ is

$$\frac{-1}{\frac{3}{4}} = -\frac{4}{3}.$$

EXAMPLE 7   Find the slope of any line perpendicular to the line through $(-2, -2)$ and $(3, -1)$.

Solution: First, determine the slope of the line through $(-2, -2)$ and $(3, -1)$.

$$m = \frac{-1 - (-2)}{3 - (-2)} = \frac{1}{5}$$

The slope of any line perpendicular to this line is -5, since $-5\left(\frac{1}{5}\right) = -1$.   ■

EXAMPLE 8   If L_1 is a line containing the points $(-5, 3)$ and $(1, -1)$ and L_2 is a line containing the points $(2, 5)$ and $(0, 2)$, determine whether L_1 and L_2 are perpendicular.

Solution: Find the slope of each line. The slope m_1 of line L_1 is

$$m_1 = \frac{-1 - 3}{1 - (-5)} = \frac{-4}{6} = \frac{-2}{3} = -\frac{2}{3}$$

The slope m_2 of line L_2 is

$$m_2 = \frac{2 - 5}{0 - 2} = \frac{-3}{-2} = \frac{3}{2}$$

The two lines have slopes of $-\frac{2}{3}$ and $\frac{3}{2}$. Since the product of these slopes is -1, the two lines are perpendicular.   ■

MENTAL MATH

Determine whether each line has positive, negative, 0, or undefined slope.

1.

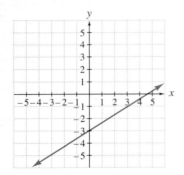

2.

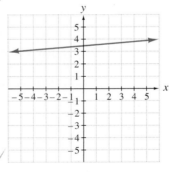

3.

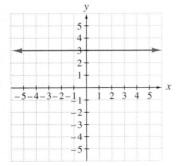

4.

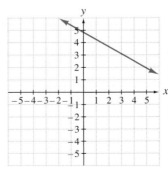

5.

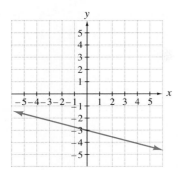

6.

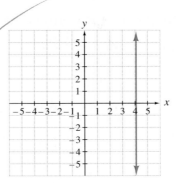

7.

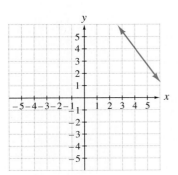

8.

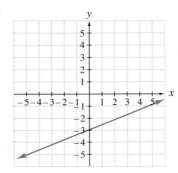

9.

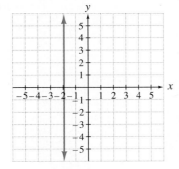

10.

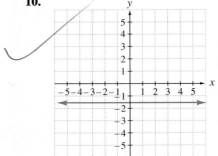

EXERCISE SET 3.3

Find the slope of the line containing each pair of points. See Examples 1 and 2.

1. $(3, 2)$, $(8, 11)$ **2.** $(1, 6)$, $(7, 11)$ **3.** $(3, 1)$, $(1, 8)$ **4.** $(2, 9)$, $(6, 4)$
5. $(-2, 8)$, $(4, 3)$ **6.** $(3, 7)$, $(-2, 11)$ **7.** $(-2, -6)$, $(4, -4)$ **8.** $(-3, -4)$, $(-1, 6)$
9. $(-3, -1)$, $(-12, 11)$ **10.** $(3, -1)$, $(-6, 5)$

Find the slope of the line containing each pair of points. See Examples 3 and 4.

11. $(-2, 5)$, $(3, 5)$ **12.** $(4, 2)$, $(4, 0)$ **13.** $(-1, 1)$, $(-1, -5)$ **14.** $(-2, -5)$, $(3, -5)$

Graph the line with given slope through the given point. See Example 5.

15. $m = \dfrac{1}{5}$, (2, 1) **16.** $m = \dfrac{3}{4}$, (0, 5) **17.** $m = 3$, (−1, −4) **18.** $m = 5$, (−2, 3)

19. $m = -\dfrac{2}{3}$, (3, 0) **20.** $m = -\dfrac{1}{6}$, (1, 5) **21.** $m = -2$, (2, −3) **22.** $m = -4$, (4, 2)

Determine whether lines L_1 and L_2 are parallel lines, perpendicular lines, or neither if they contain the given points. See Examples 6 through 8.

23. L_1: (2, 5), (6, 7)
  L_2: (−1, 5), (1, 6)

24. L_1: (6, 2), (8, 6)
  L_2: (1, −2), (2, 0)

25. L_1: (−1, 5), (3, 2)
  L_2: (2, 2), (5, 6)

26. L_1: (6, −2), (4, 6)
  L_2: (5, 2), (9, 1)

27. L_1: (6, −2), (2, 2)
  L_2: (−3, −2), (0, 1)

28. L_1: (3, −1), (5, −7)
  L_2: (−1, 6), (2, 7)

29. L_1: (2, −3), (6, −5)
  L_2: (−3, 1), (−1, 2)

30. L_1: (5, −2), (2, −3)
  L_2: (1, −2), (3, 4)

Find the slope of the line containing each pair of points.

31. (5, 1), (1, 0) **32.** (7, 9), (8, 7) **33.** (−4, 10), (−1, 7) **34.** (−6, −6), (2, −1)

35. (−3, 8), (−3, 1) **36.** (4, 0), (−5, 0) **37.** (−11, −4), (−9, 3) **38.** (−10, 6), (−5, 2)

39. (−5, −1), (−6, 2) **40.** (−1, −9), (1, −2)

Graph the line with given slope containing the given point.

41. $m = \dfrac{3}{2}$, (1, 5) **42.** $m = \dfrac{2}{5}$, (−2, 4) **43.** $m = -\dfrac{5}{4}$, (−4, −3) **44.** $m = -\dfrac{1}{3}$, (6, −7)

45. $m = 2, (-1, 0)$ **46.** $m = 1, (0, 6)$ **47.** $m = -3, (5, -2)$ **48.** $m = -6, (-3, -5)$

Determine whether lines L_1 and L_2 are parallel lines, perpendicular lines, or neither if they contain the given points.

49. L_1: $(-2, -3), (-7, -5)$
L_2: $(8, 10), (3, 8)$

50. L_1: $(0, -2), (1, 6)$
L_2: $(-6, 1), (10, -1)$

51. L_1: $(8, -10), (2, -7)$
L_2: $(-8, -6), (8, 3)$

52. L_1: $(-30, 0), (2, 1)$
L_2: $(2, -7), (3, -2)$

53. L_1: $(-1, -5), (8, 5)$
L_2: $(10, 8), (-10, -10)$

54. L_1: $(6, 1), (5, -2)$
L_2: $(2, 15), (-2, 3)$

55. L_1: $(-3, -9), (-10, -8)$
L_2: $(-2, -2), (-4, -16)$

56. L_1: $(2, 10), (-3, 11)$
L_2: $(-7, 5), (8, 2)$

57. Find the pitch or slope of the roof shown.

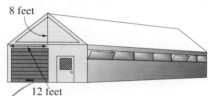

8 feet

12 feet

58. Upon take-off, a Delta Airlines jet climbs to 3 miles as it passes over 25 miles of land below it. Find the slope of its climb.

3 miles

25 miles

59. Driving down Bald Mountain in Wyoming, Dean finds that he descends 1600 feet in elevation by the time he is 2.5 miles (horizontally) away from the high point on the mountain road. Find the slope of his descent. (1 mile = 5280 feet)

60. Find the grade of slope of the road shown.

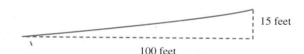

15 feet

100 feet

Writing in Mathematics

61. Explain why the slope of a vertical line is undefined.
62. Explain how merely looking at a line can tell us whether its slope is negative.

63. Explain why the graph of $y = b$ is a horizontal line.
64. Explain why it is reasonable that nonvertical parallel lines have the same slope.

Skill Review

Solve each equation for y. See Section 2.2.

65. $3x + 5y = 22$ **66.** $7x + 4y = 5$ **67.** $6x + 2y = 9$ **68.** $8x + 4y = 3$

Evaluate the following. See Section 1.2.

69. $|-6|$ **70.** $-|7|$ **71.** $-|-1|$ **72.** $|-3|$

3.4
Equations of Lines

OBJECTIVES

Tape 7

1	Find the equation of a line using the point–slope form.
2	Find the slope of a line using the slope–intercept form.
3	Find the equation of a line using the slope–intercept form.
4	Find the equation of a vertical line and a horizontal line.
5	Find the equation of parallel or perpendicular lines.

1   In the last section, we graphed a line given its slope and the coordinate of one point on the line. In this section, we write the equation of a line given its slope and the coordinate of one point on the line. Suppose the line passes through the point (x_1, y_1) and has slope m. For any other point (x, y) on that line, the slope of the line is

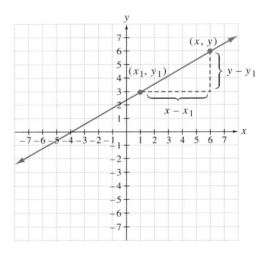

Multiply both sides by the common denominator $(x - x_1)$ to clear fractions. The result is

$$m(x - x_1) = y - y_1 \quad \text{or} \quad y - y_1 = m(x - x_1)$$

This is the **point–slope form** of the line.

Point–Slope Form

The point–slope form of a line is $y - y_1 = m(x - x_1)$, where m is the slope of the line and (x_1, y_1) is a point of the line.

EXAMPLE 1   Find the equation of the line with slope -3 containing the point $(1, -5)$.

Solution: Because we know the slope and a point of the line, we use the point–slope form with $m = -3$ and $(x_1, y_1) = (1, -5)$.

$$y - y_1 = m(x - x_1) \qquad \text{Point–slope form.}$$

$$y - (-5) = -3(x - 1) \qquad \text{Let } m = -3 \text{ and } (x_1, y_1) = (1, -5).$$

$$y + 5 = -3x + 3 \qquad \text{Remove parentheses.}$$

$$3x + y = -2 \qquad \text{Standard form.} \qquad \blacksquare$$

EXAMPLE 2   Find the equation of the line through points $(4, 0)$ and $(-4, -5)$.

Solution: First, find the slope of the line.

$$m = \frac{-5 - 0}{-4 - 4} = \frac{-5}{-8} = \frac{5}{8}$$

Next, make use of the point–slope form. Replace (x_1, y_1) by either $(4, 0)$ or $(-4, -5)$ in the point–slope equation. We will choose the point $(4, 0)$. The line through $(4, 0)$ with slope $\frac{5}{8}$ is

$$y - y_1 = m(x - x_1) \qquad \text{Point–slope form.}$$

$$y - 0 = \frac{5}{8}(x - 4) \qquad \text{Let } m = \frac{5}{8} \text{ and } (x_1, y_1) = (4, 0).$$

$$8y = 5(x - 4) \qquad \text{Multiply both sides by 8.}$$

$$8y = 5x - 20 \qquad \text{Remove parentheses.}$$

$$-5x + 8y = -20 \qquad \text{Subtract } 5x \text{ from both sides.}$$

The standard form of the line is $-5x + 8y = -20$ or, if we multiply through by -1, $5x - 8y = 20$. $\blacksquare$

2   One way to find the slope of the graph of an equation such as $x - 4y = -7$ is to find two ordered pair solutions and use the definition of slope. Another way is to use the **slope–intercept form.** This form can be derived from the point–slope form of the line by letting $(x_1, y_1) = (0, b)$, where b is the y-intercept of a line.

$$y - y_1 = m(x - x_1) \qquad \text{Point–slope form.}$$

$$y - b = m(x - 0) \qquad \text{Let } x_1 = 0 \text{ and } y_1 = b.$$

$$y - b = mx$$

$$y = mx + b \qquad \text{Solve for } y.$$

The expression $y = mx + b$ is known as the slope–intercept form.

> **Slope–Intercept Form**
>
> The slope–intercept form of a line is $y = mx + b$, where m is the slope of the line and b is the y-intercept of the line.

This means that if we write the equation $x - 4y = -7$ in the form $y = mx + b$, we can identify the slope, m and y-intercept, b, directly from the equation.

EXAMPLE 3 Find the slope and y-intercept of the graph of $x - 4y = -7$.

Solution: Write this equation in slope–intercept form. To write an equation in the form $y = mx + b$, solve the equation for y.

$$x - 4y = -7 \qquad \text{Original equation.}$$

$$-4y = -x - 7 \qquad \text{Subtract } x \text{ from both sides.}$$

$$y = \frac{-x - 7}{-4} \qquad \text{Divide by } -4.$$

$$y = \frac{-x}{-4} - \frac{7}{-4} \quad \text{or} \quad y = \frac{1}{4}x + \frac{7}{4}$$

This equation is now in the form $y = mx + b$. The slope m is $\frac{1}{4}$, the coefficient of x, and the y-intercept b is $\frac{7}{4}$. ∎

3 The slope-intercept form of a line can also be used to write the equation of a line.

EXAMPLE 4 Find the equation of the line with slope $\frac{1}{4}$ and y-intercept $-\frac{2}{3}$. Write the equation in standard form.

Solution: Because we are given the slope and the y-intercept, we use the slope–intercept form and replace m and b by their known values.

$$y = mx + b \qquad \text{Slope–intercept form.}$$

$$y = \frac{1}{4}x - \frac{2}{3} \qquad \text{Let } m = \frac{1}{4} \text{ and } b = -\frac{2}{3}.$$

The slope–intercept form of the line is $y = \frac{1}{4}x - \frac{2}{3}$. To write the equation in standard form, begin by multiplying both sides of the equation by 12, the LCD.

$$y = \frac{1}{4}x - \frac{2}{3}$$

$$12(y) = 12\left(\frac{1}{4}x - \frac{2}{3}\right) \qquad \text{Multiply both sides by 12 to clear fractions.}$$

$$12y = 3x - 8 \qquad \text{Remove parentheses.}$$

$$-3x + 12y = -8 \qquad \text{Subtract } 3x \text{ from both sides.}$$

The standard form of the line is $-3x + 12y = -8$ or, if we multiply through by -1, $3x - 12y = 8$. ∎

4   A few special types of linear equations are linear equations whose graphs are vertical and horizontal lines.

EXAMPLE 5   Find the equation of the horizontal line containing the point (2, 3).

Solution: Recall that a horizontal line has an equation of the form $y = b$. Since the line contains the point (2, 3), the equation is $y = 3$.   ☒

EXAMPLE 6   Find the equation of the line containing the point (2, 3) with undefined slope.

Solution: Since the line has undefined slope, the line must be vertical. A vertical line has an equation of the form $x = c$, and since the line contains the point (2, 3), the equation is $x = 2$.   ■

EXAMPLE 7   Are the graphs of $x + 3y = -1$ and $2x = 6y - 1$ parallel lines, perpendicular lines, or neither?

Solution: Write both equations in slope–intercept form and compare their slopes.

$$\text{Line 1:} \quad x + 3y = -1$$

$$3y = -x - 1$$

$$y = \frac{-x - 1}{3} \qquad \text{Divide both sides by 3.}$$

$$y = -\frac{1}{3}x - \frac{1}{3} \qquad \text{Slope–intercept form.}$$

The slope of the line is $-\dfrac{1}{3}$.

$$\text{Line 2:} \quad 2x = 6y - 1$$

$$2x + 1 = 6y \qquad \text{Add 1 to both sides.}$$

$$\frac{2x + 1}{6} = y \qquad \text{Divide by 6.}$$

$$\frac{2}{6}x + \frac{1}{6} = y$$

$$\text{or} \quad y = \frac{1}{3}x + \frac{1}{6} \qquad \text{Slope–intercept form.}$$

The slope of the line is $\dfrac{1}{3}$. The two slopes are not the same, so the graphs of the lines are not parallel. The product of the two slopes is not -1, so the graphs of the lines are not perpendicular. They are neither.   ■

5   We can use the fact that parallel lines have the same slope and perpendicular lines have slopes whose product is -1 to find equations of lines.

EXAMPLE 8   Find the equation of the line containing the point (4, 4) and parallel to the line $2x + 3y = -6$.

Solution: Because the line we want to find is **parallel** to the line $2x + 3y = -6$, the two lines must have equal slopes. Find the slope of $2x + 3y = -6$ by writing it in the form $y = mx + b$.

$$2x + 3y = -6$$
$$3y = -2x - 6 \qquad \text{Subtract } 2x \text{ from both sides.}$$
$$y = \frac{-2x - 6}{3} \qquad \text{Divide by 3.}$$
$$y = -\frac{2}{3}x - 2 \qquad \text{Slope–intercept form.}$$

The slope of this line is $-\frac{2}{3}$. Thus a line parallel to this line will also have a slope of $-\frac{2}{3}$. The equation we are asked to find describes a line containing the point (4, 4) with a slope of $-\frac{2}{3}$. We use the point–slope form.

$$y - y_1 = m(x - x_1)$$
$$y - 4 = -\frac{2}{3}(x - 4) \qquad \text{Let } m = -\frac{2}{3},\ x_1 = 4,\ \text{and } y_1 = 4.$$
$$3(y - 4) = -2(x - 4) \qquad \text{Multiply both sides by 3.}$$
$$3y - 12 = -2x + 8 \qquad \text{Apply the distributive property.}$$
$$2x + 3y = 20 \qquad \text{Standard form.} \quad \blacksquare$$

EXAMPLE 9   Find the equation of the line containing the point (0, −5) and perpendicular to the line $x = 2y$.

Solution: Find the slope of the line $x = 2y$ by solving it for y in order to write it in the form $y = mx + b$.

$$x = 2y$$
$$\frac{x}{2} = y \quad \text{or} \quad y = \frac{1}{2}x$$

The slope of this line is $\frac{1}{2}$. A line perpendicular to the line $y = \frac{1}{2}x$ will have slope -2. (Recall that perpendicular lines have slopes that are negative reciprocals of one another.) Since the line contains the point (0, −5), the y-intercept b of the desired line is −5. To find an equation of the line with slope −2 and with y-intercept −5 we use slope-intercept form.

$$y = mx + b \qquad \text{Slope–intercept form.}$$
$$y = -2x - 5 \qquad \text{Let } m = -2 \text{ and } b = -5.$$
$$2x + y = -5 \qquad \text{Standard form.} \quad \blacksquare$$

In Example 9, we could have used the point–slope form at the end of the problem instead of the slope–intercept form to find the equation of the line.

EXAMPLE 10   Find the equation of the line containing the point $(1, 0)$ and perpendicular to the line $y = 9$.

Solution: The given line, $y = 9$, is a horizontal line. Therefore, a line perpendicular to this line is vertical and has the form $x = c$. The vertical line containing $(1, 0)$ is $x = 1$.   ∎

Notice in Example 10 that we did not use the point–slope form to obtain the equation. Vertical lines have undefined slope, so this form cannot be used.

The following box gives a summary of the five forms for linear equations in two variables. The form you use depends on the information provided and the type of line described.

Five Forms of Equations of Lines

Standard form: $ax + by = c$, where a, b, and c are real numbers and a and b are not both 0.

Point–slope form: $y - y_1 = m(x - x_1)$, where m is the slope of the line and (x_1, y_1) is a point of the line.

Slope–intercept form: $y = mx + b$, where m is the slope and b is the y-intercept.

Vertical lines: $x = c$ and have undefined slope.

Horizontal lines: $y = b$ and have slope 0.

MENTAL MATH

Find the slope and the y-intercept of each of the following lines.

1. $y = 6x + 2$

2. $y = 5x + 3$

3. $y = -2x + 5$

4. $y = -3x + 6$

5. $y = \dfrac{4}{3}x - 3$

6. $y = \dfrac{2}{9}x - 7$

7. $y = -\dfrac{3}{5}x - \dfrac{7}{2}$

8. $y = -\dfrac{4}{7}x - \dfrac{5}{3}$

EXERCISE SET 3.4

Find the equation of the line with the given slope and containing the given point. Write the equation in standard form. See Example 1.

1. Slope 3; through $(1, 2)$

2. Slope 4; through $(5, 1)$

3. Slope -2; through $(1, -3)$

4. Slope -4; through $(2, -4)$

5. Slope $\dfrac{1}{2}$; through $(-6, 2)$

6. Slope $\dfrac{2}{3}$; through $(-9, 4)$

7. Slope $-\dfrac{9}{10}$; through $(-3, 0)$

8. Slope $-\dfrac{1}{5}$; through $(4, -6)$

Find the equation of the line passing through the given points. Write the equation in standard form. See Example 2.

9. $(2, 0)$, $(4, 6)$ **10.** $(3, 0)$, $(7, 8)$ **11.** $(-2, 5)$, $(-6, 13)$ **12.** $(7, -4)$, $(2, 6)$

13. $(-2, -4)$, $(-4, -3)$ **14.** $(-9, -2)$, $(-3, 10)$ **15.** $(-3, -8)$, $(-6, -9)$ **16.** $(8, -3)$, $(4, -8)$

Find the slope and the y-intercept of each line. See Example 3.

17. $3x + y = 8$ **18.** $4x + y = 2$ **19.** $2x - y = 3$

20. $6x - y = -3$ **21.** $2x = 3y + 9$ **22.** $5x = 2y + 8$

Find the equation of the line with the given slope and y-intercept. Write the equation in standard form. See Example 4.

23. Slope 2; y-intercept 5 **24.** Slope 7; y-intercept 1 **25.** Slope $\frac{4}{5}$; y-intercept -3

26. Slope -6; y-intercept $\frac{3}{10}$ **27.** Slope $-\frac{1}{8}$; y-intercept $-\frac{3}{2}$ **28.** Slope $-\frac{5}{9}$; y-intercept 0

Write the equation of each line. See Examples 5 and 6.

29. Vertical; through $(2, 6)$ **30.** Slope 0; through $(-2, -4)$ **31.** Horizontal; through $(-3, 1)$

32. Vertical; through $(4, 7)$ **33.** Undefined slope; through $(0, 5)$ **34.** Horizontal; through $(0, 5)$

Determine whether the graphs of each pair of equations are parallel lines, perpendicular lines, or neither. See Example 7.

35. $x - 5y = -2$ **36.** $3x = 2y$ **37.** $x - y = -2$
    $-2x + 10y = 0$ $4x - y = 3$ $3x + 3y = 4$

38. $-4x + y = 10$ **39.** $5x + 2y = 3$ **40.** $y = -\frac{2}{3}x + 4$
    $2x + 8y = 10$ $4y = 3x$ $6x + 9y = 2$

Find the equation of each line. Write the equation in standard form. See Examples 8 and 9.

41. Through $(3, 8)$; parallel to $y = 4x - 2$ **42.** Through $(1, 5)$; parallel to $y = 3x - 4$

43. Through $(2, -5)$; perpendicular to $3y = x - 6$ **44.** Through $(-4, 8)$; perpendicular to $2x - 3y = 1$

45. Through $(-2, -3)$; parallel to $3x + 2y = 5$ **46.** Through $(-2, -3)$; perpendicular to $3x + 2y = 5$

Find the equation of each line. Write the equation in standard form. See Example 10.

47. Through $(4, 6)$; parallel to the line $y = 2$ **48.** Through $(2, 9)$; parallel to the line $x = -3$

49. Through $(0, 5)$; perpendicular to the line $y = -2$ **50.** Through $(-1, -8)$; perpendicular to the line $x = 0$

51. Through $(3, -3)$; parallel to the line $x = 5$ **52.** Through $(-2, 0)$; perpendicular to the line $y = -10$

Find the equation of each line. Write the equation in standard form.

53. Slope 2; through $(-2, 3)$ **54.** Slope 3; through $(-4, 2)$

55. Through $(1, 6)$ and $(5, 2)$ **56.** Through $(2, 9)$ and $(8, 6)$

57. With slope $-\frac{1}{2}$; y-intercept 11 **58.** With slope -4; y-intercept $\frac{2}{9}$

59. Through $(-7, -4)$ and $(0, -6)$ **60.** Through $(2, -8)$ and $(-4, -3)$

61. Slope $-\frac{4}{3}$; through $(-5, 0)$ **62.** Slope $-\frac{3}{5}$; through $(4, -1)$

63. Vertical line; through $(-2, -10)$ **64.** Horizontal line; through $(1, 0)$

65. Through $(6, -2)$; parallel to the line $2x + 4y = 9$ **66.** Through $(8, -3)$; parallel to the line $6x + 2y = 5$

67. Slope 0; through $(-9, 12)$ **68.** Undefined slope; through $(10, -8)$

69. Through $(6, 1)$; parallel to the line $8x - y = 9$

71. Through $(5, -6)$; perpendicular to $y = 9$

73. Through $(2, -8)$ and $(-6, -5)$

75. The equation relating Celsius temperature and Fahrenheit temperature is linear. Find this linear equation if the freezing point of water is 0°C or 32°F, and the boiling point is 100°C or 212°F. [*Hint:* This is equivalent to saying that the line passes through points $(0, 32)$ and $(100, 212)$.]

76. The relationship between temperatures on the Kelvin scale and the Celsius scale is modeled by a linear equation. Since 273 K equals 0°C, and 373 K is the same as 100°C, find the linear equation that models the Kelvin–Celsius relationship.

77. The Whammo Company has learned that, by pricing a newly released Frisbee at $6, sales will reach 2000 per day. Raising the price to $8 will cause the sales to fall to 1500 per day. Assume that the ratio of change in price to change in daily sales is constant.

70. Through $(3, 5)$; perpendicular to the line $2x - y = 8$

72. Through $(-3, -5)$; parallel to $y = 9$

74. Through $(-4, -2)$ and $(-6, 5)$

a. Find the linear equation that models the price–sales relationship for this Frisbee. [*Hint:* The line must pass through $(6, 2000)$ and $(8, 1500)$.]

b. Predict the daily sales of Frisbees if the price is set at $7.50.

78. Del Monte Fruit Company recently released a new pineapple sauce. By the end of its first year, profits on this product amounted to $30,000. The anticipated profit for the end of the fourth year is $66,000. The ratio of change in time to change in profit is constant.

a. Find the linear equation that models the time–profit relationship. [*Hint:* The line must pass through the points $(1, 30{,}000)$ and $(4, 66{,}000)$.]

b. Predict the profit the company should anticipate for the ninth year.

A Look Ahead

Find an equation of the perpendicular bisector of the line segment whose end points are given. See the following example.

EXAMPLE   Find an equation of the perpendicular bisector of the line segment whose end points are $(2, 6)$ and $(0, -2)$.

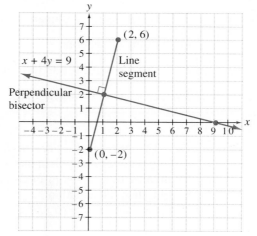

Solution: A perpendicular bisector is a line that contains the midpoint of the given segment and is perpendicular to the segment.

Step 1   The midpoint of the segment with endpoints $(2, 6)$ and $(0, -2)$ is $(1, 2)$.

Step 2   The slope of the segment containing points $(2, 6)$ and $(0, -2)$ is 4.

Step 3   A line perpendicular to this line segment will have slope of $-\dfrac{1}{4}$.

Step 4   The equation of the line through the midpoint $(1, 2)$ with a slope of $-\dfrac{1}{4}$ will be the equation of the perpendicular bisector. This equation in standard form is $x + 4y = 9$.

■

79. $(3, -1); (-5, 1)$

80. $(-6, -3); (-8, -1)$

81. $(-2, 6); (-22, -4)$

82. $(5, 8); (7, 2)$

83. $(2, 3); (-4, 7)$

84. $(-6, 8); (-4, -2)$

Writing in Mathematics

85. Discuss how you choose between the point–slope form and the slope–intercept form to find the equation of a line.

Skill Review

Find the value of each expression if $x = 1$, $y = -4$, and $z = 0$. See Section 1.4.

86. $\dfrac{x^2}{y^2 - zx}$

87. $\dfrac{2x}{zx - y^2}$

88. $\dfrac{3x - y}{2z - x^2}$

89. $\dfrac{4z - 3y}{2x - x^2}$

90. $\dfrac{|y| + z^2}{2x}$

91. $\dfrac{|x| + 3z}{2y}$

92. $\dfrac{3x^2 - y}{z}$

93. $\dfrac{3xy - 2x}{4x^2 - 3y}$

3.5
Introduction to Functions

OBJECTIVES

Tape 8

1   Define relation, domain, and range.

2   Identify functions.

3   Use the vertical line test for functions.

4   Find the domain and range of a function.

1   The amount of money inserted into a parking meter is related to the length of parking time received. Suppose that each quarter buys fifteen minutes. If x represents the number of quarters inserted in the meter and y represents the number of minutes bought, then the equation

$$y = 15x$$

models the relationship between quarters inserted and minutes bought. For the model $y = 15x$, the number of quarters, x, is called the **independent variable,** and the number of minutes, y, is called the **dependent variable.** This is because we can **independently** decide how many quarters to insert, but the time bought **depends** on the number of quarters inserted.

The set of possible values for the independent variable x is the **domain** of the **relation** $y = 15x$, and the set of values of the dependent variable y that correspond to some value of x is the **range** of the relation. The domain of this quarter–time relation is $\{0, 1, 2, 3, \ldots\}$, and the range is $\{0, 15, 30, 45, \ldots\}$.

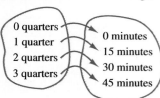

x
Quarters inserted

y
Time bought

0 quarters
1 quarter
2 quarters
3 quarters

0 minutes
15 minutes
30 minutes
45 minutes

> **Relation, Domain, and Range**
>
> A **relation** is a correspondence between two sets X and Y that assigns to each element of set X an element of set Y. The **domain** of the relation is the set X, and the **range** of the relation is the set of elements of Y that corresponds to some element of X.

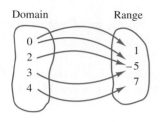

Domain   Range

An equivalent definition of relation is that a relation is a set of ordered pairs. For example, $\{(0, 1), (0, -5), (2, -5), (3, -5), (4, 7)\}$ is a relation whose domain is $\{0, 2, 3, 4\}$ and whose range is $\{1, -5, 7\}$. Notice that the domain is the set of first coordinates and the range is the set of second coordinates.

> **Alternate Definition of Relation, Domain, and Range**
>
> A **relation** is a set of ordered pairs. The **domain** of the relation is the set of first coordinates, and the **range** is the set of second coordinates.

EXAMPLE 1   Determine the domain and range of each relation. Write the domain and range in set notation.

a. $\{(2, 3), (2, 4), (0, -1), (3, -1)\}$

b.

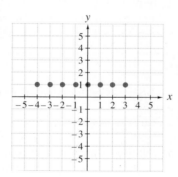

c.

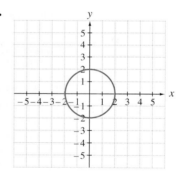

d.

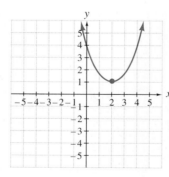

e. $3y = 5x - 2$

Solution: **a.** The domain is the set of all first coordinates of the ordered pairs, $\{2, 0, 3\}$. The range is the set of all second coordinates, $\{3, 4, -1\}$.

b. Ordered pairs are not listed here, but are given in graph form. The relation is $\{(-4, 1), (-3, 1), (-2, 1), (-1, 1), (0, 1), (1, 1), (2, 1), (3, 1)\}$. The domain is $\{-4, -3, -2, -1, 0, 1, 2, 3\}$.
The range is $\{1\}$.

c. From this graph, note that the x values lie between -2 and 2 inclusive, and the y values lie between -2 and 2 inclusive. The domain is $\{x \mid -2 \le x \le 2\}$, and the range is $\{y \mid -2 \le y \le 2\}$.

d. On this curve, x may take on any value. The domain then is the set of real numbers $\{x \mid x \text{ is a real number}\}$. Note that the lowest point on the graph has a y value of 1, so y is always greater than or equal to 1. The range then is $\{y \mid y \ge 1\}$.

e. The graph of the relation is a line. Its domain is the set of real numbers $\{x \mid x \text{ is a real number}\}$ and its range is the set of real numbers $\{y \mid y \text{ is a real number}\}$.

∎

2 Now we consider a special kind of relation called a function. A **function** is a relation for which each x-value corresponds to **exactly one** y-value.

Function, Domain, and Range

A **function** is a relation in which each first coordinate of an ordered pair is assigned to exactly one second coordinate.

A function can be thought of as a rule that assigns to each element of a set X a **unique** element of a set Y. In other words, a relation is also called a function if no ordered pairs have the same first coordinates but different second coordinates.

HELPFUL HINT

A function is a special type of relation, so all functions are relations, but not all relations are functions.

EXAMPLE 2   Determine whether the relation is also a function.

a. $\{(1, 1), (2, 7), (3, -1), (0, 0)\}$ **b.** $\{(2, 3), (2, 4), (7, 6)\}$

c. $\{(-2, 5), (-3, 5), (-1, 5)\}$ **d.** $y = x^2$ **e.** $y = \dfrac{2x + 3}{6}$

f.

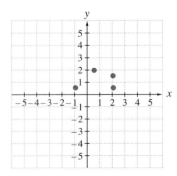

g. $\{(0, 0), (1, 15), (2, 30), (3, 45), (4, 60), \ldots\}$

Solution: **a.** This relation is a function because each first coordinate is paired with exactly one second coordinate.

b. This relation is not a function because 2 is paired with 3 and also 4. In a function, each first coordinate is paired with one and only one second coordinate.

c. This is a function. Each first coordinate is paired with exactly one second coordinate. It makes no difference if some second coordinates are the same.

d. This is a function because each value substituted for x produces exactly one value of y.

e. This is a function because each value of x yields exactly one value of y.

f. The ordered pairs graphed are $\left(-1, \frac{1}{2}\right)$, $\left(\frac{1}{2}, 2\right)$, $\left(2, \frac{1}{2}\right)$, and $\left(2, \frac{3}{2}\right)$. Since $\left(2, \frac{1}{2}\right)$ and $\left(2, \frac{3}{2}\right)$ have the same first coordinate but different second coordinates, the set of ordered pairs is not a function.

g. This relation is the quarters–parking time relation we modeled with the equation $y = 15x$. Since every value of x yields a unique value of y, this relation is a function. ■

3   In Example 2f, drawing a vertical line through $\left(2, \frac{1}{2}\right)$ also intersects $\left(2, \frac{3}{2}\right)$. In general, if a vertical line intersects a graph at more than one point, these points have the same first coordinates but different second coordinates. Thus the graph is **not** the graph of a function.

Vertical Line Test for Functions

If a vertical line intersects a graph at more than one point, the graph is **not** the graph of a function.

EXAMPLE 3   Use the vertical line test to determine whether the graph is the graph of a function.

a.

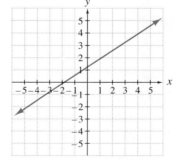

b.

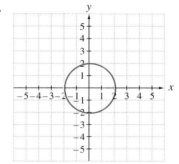

c.

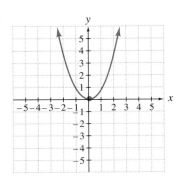

d.

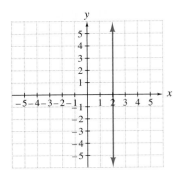

Solution: **a.** Every vertical line intersects this straight line only once, so this is the graph of a function.

b. Many vertical lines intersect this circle more than once, so this is not the graph of a function.

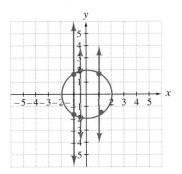

c. This is a function.

d. This graph is a vertical line, so a vertical line intersects this graph more than once and it is not the graph of a function. A vertical line is the only straight line that is not the graph of a function.   ■

4   Since a function is a special type of a relation, functions have domains and ranges also.

EXAMPLE 4   Find the domain and range of each function.

a.

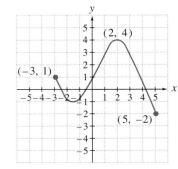

b. $y = \dfrac{1}{x}$ **c.** $y = 2x + 1$

Solution: **a.** From the graph, as x takes on values from -3 to 5 inclusive, y takes on values from -2 to 4 inclusive. Thus the domain is $\{x \mid -3 \leq x \leq 5\}$ and the range is $\{y \mid -2 \leq y \leq 4\}$.

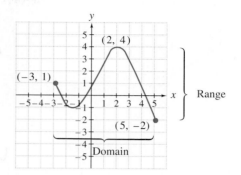

b. In this example, x cannot be 0 since the denominator of a fraction cannot be 0. And no matter what value x is, y will not equal 0 since the numerator of a fraction must be 0 for the fraction to be 0. Thus the domain is $\{x \mid x$ is a real number and $x \neq 0\}$, and the range is $\{y \mid y$ is a real number and $y \neq 0\}$.

c. The graph of this linear equation is a line. The domain is the set of all real numbers, and the range is the set of all real numbers. In set notation, the domain is $\{x \mid x$ is a real number$\}$, and the range is $\{y \mid y$ is a real number$\}$. ■

EXERCISE SET 3.5

Find the domain and range of each relation. Also determine whether the relation is a function. See Example 1.

1. $\{(-1, 7), (0, 6), (-2, 2), (5, 6)\}$

2. $\{(4, 9), (-4, 9), (2, 3), (10, -5)\}$

3. $\{(-2, 4), (6, 4), (-2, -3), (-7, -8)\}$

4. $\{(6, 6), (5, 6), (5, -2), (7, 6)\}$

5. $\{(1, 1), (1, 2), (1, 3), (1, 4)\}$

6. $\{(1, 1), (2, 1), (3, 1), (4, 1)\}$

7. $\left\{\left(\dfrac{3}{2}, \dfrac{1}{2}\right), \left(1\dfrac{1}{2}, -7\right), \left(0, \dfrac{4}{5}\right)\right\}$

8. $\{(\pi, 0), (0, \pi), (-2, 4), (4, -2)\}$

9. $\{(-3, -3), (0, 0), (3, 3)\}$

10. $\left\{\left(\dfrac{1}{2}, \dfrac{1}{4}\right), \left(0, \dfrac{7}{8}\right), (0.5, \pi)\right\}$

Find the domain and range of each relation. Also determine whether the relation is a function. See Examples 3 and 4.

11.

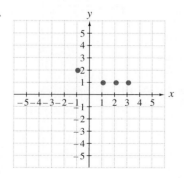

12.

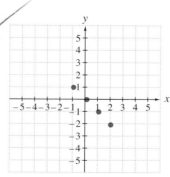

13.

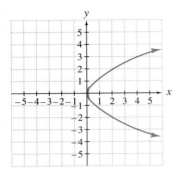

14.

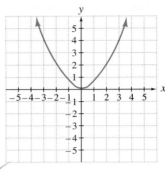

15.

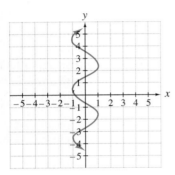

16.

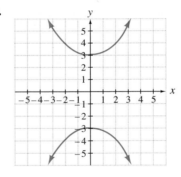

17.

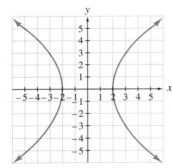

18.

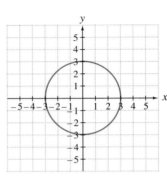

Find the domain of each relation. Also determine whether the relation is a function. See Examples 2 and 4.

19. $y = 5x - 12$ **20.** $y = \frac{1}{2}x + 4$ **21.** $x = y^2$ **22.** $x = |y|$

23. $y = 6$ **24.** $x = 4$ **25.** $y = \frac{5}{x + 4}$ **26.** $y = \frac{1}{x - 5}$

Find the domain and range of each relation. Also determine whether the relation is a function.

27. $\{(4, 7), (7, 4), (-4, 0)\}$ **28.** $\{(8, -6), (7, 2), (4, 0)\}$

29. $\{(5, 1), (-6, 1), (4, -8), (2, -8)\}$ **30.** $\{(3, 6), (-3, -6), (-6, -3), (-6, 0)\}$

31. $\{(0, 10), (2, 9), (5, 10)\}$ **32.** $\{(7, 7), (8, 6), (-8, 6), (4, 4)\}$

33.

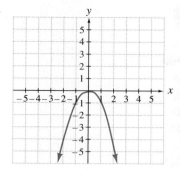

34.

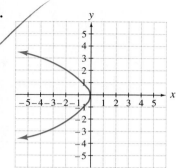

35.

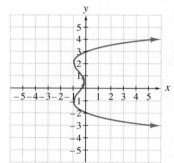

36.

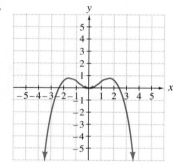

37.

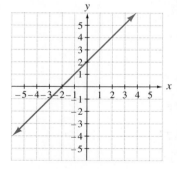

38.

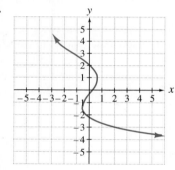

39.

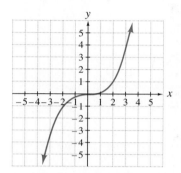

40.

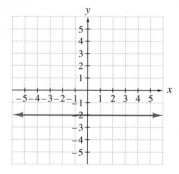

41.

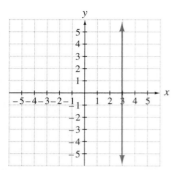

42.

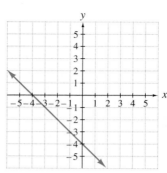

43. $2x - y = 7$

44. $5x + 2y = -10$

Find the domain of each function. Also determine whether the relation is a function.

45. $y = \sqrt{x}$

46. $x = \sqrt{y}$

47. $x^2 + y^2 = 16$

48. $x^2 + y^2 = 1$

49. $x = y^3$

50. $y = x^3$

51. $y - 7 = 0$

52. $x + 2 = 0$

Write an equation that describes each of the following relations. Then graph the equation and determine whether the relation is also a function.

53. The weight in ounces, y, of x boxes of oatmeal if each box weighs 12 ounces.

54. The number, y, if y is always 6 less than twice x.

55. The cost in cents, y, to buy candy that is 29 cents per pound.

56. The cost in dollars, y, of driving a rental car x miles if there is a flat fee of $12 and a charge of 19 cents per mile.

57. The property tax, y, on a property with assessed value x if the tax rate is 8% of assessed value.

58. The dosage in milligrams, y, of Ivermectin (a heartworm preventive) for a dog of weight x if 136 milligrams is required for each 25 pounds of weight.

Writing in Mathematics

59. Describe a function whose domain and range are sets of people.

60. Explain how the vertical line test accurately reveals whether a relation is a function.

Skill Review

Solve the following. See Section 2.6.

61. $x < 4$ and $x > -1$

62. $x > -3$ and $x > 0$

63. $-2x > -10$ and $x + 7 < 15$

64. $3x - 1 < 5$ and $2x + 7 > 3$

Solve the following. See Section 2.2.

65. If two angles are supplementary and one angle's measure is three times the measure of the other angle, find the measure of each angle.

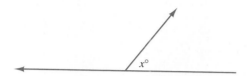

66. Is it possible to find the perimeter of the following geometric figure? If so, find the perimeter.

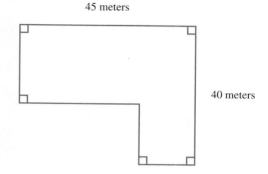

45 meters

40 meters

3.6
Graphing Linear Inequalities

OBJECTIVES

Tape 8

1 Graph linear inequalities.

2 Graph the intersection or union of two linear inequalities.

Recall that the graph of a linear equation in two variables is the graph of all ordered pairs that satisfy the equation, and we determined that the graph is a line. Here we graph **linear inequalities** in two variables; that is, we graph all the ordered pairs that satisfy the inequality.

If the equal sign in a linear equation in two variables is replaced with an inequality symbol, the result is a linear inequality in two variables.

Examples of Linear Inequalities

$$3x + 5y \geq 6 \qquad 2x - 4y < -3$$
$$4x > 2 \qquad y \leq 5$$

To graph the linear inequality $x + y \leq 5$, recall that this inequality means

$$x + y = 5 \quad \text{or} \quad x + y < 5$$

The graph of $x + y = 5$ is a line. This line is called a **boundary** because it separates the plane into two **half-planes.** All points "above" the boundary line $x + y = 5$ have coordinates that satisfy the inequality $x + y > 5$, and all points "below" the line have coordinates that satisfy the inequality $x + y < 5$. Thus the graph of $x + y \leq 5$ is the boundary line together with the half-plane below it.

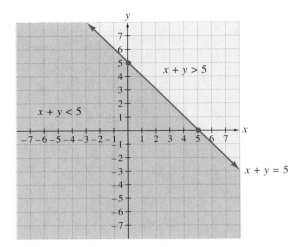

The following steps may be used to graph linear inequalities in two variables.

To Graph a Linear Inequality

Step 1   Graph the boundary line found by replacing the inequality sign with an equal sign. If the inequality sign is $<$ or $>$, graph a dashed line indicating that points on the line are not solutions of the inequality. If the inequality sign is $\leq$ or $\geq$, graph a solid line indicating that points on the line are solutions of the inequality.

Step 2   Choose a **test point not on the boundary line** and substitute the coordinates of this test point into the **original inequality.**

Step 3   If a true statement is obtained in *Step 2*, shade the half-plane that contains the test point. If a false statement is obtained, shade the half-plane that does not contain the test point.

EXAMPLE 1   Graph $2x - y < 6$.

Solution: First, the boundary line for this inequality is the graph of $2x - y = 6$. Graph a dashed boundary line because the inequality symbol is $<$. Next, choose a test point on either side of the boundary line. The point $(0, 0)$ is not on the boundary line so we use this point. Replacing x with 0 and y with 0 in the **original inequality** $2x - y < 6$ leads to the following:

$$2x - y < 6$$
$$2(0) - 0 < 6 \qquad \text{Let } x = 0 \text{ and } y = 0.$$
$$0 < 6 \qquad \text{True.}$$

Because $(0, 0)$ satisfies the inequality, so does every point on the same side of the boundary line as $(0, 0)$. Shade the half-plane that contains $(0, 0)$. The half-plane graph of the inequality is shown next.

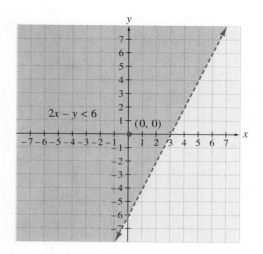

Every point in the shaded half-plane satisfies the original inequality.   ■

EXAMPLE 2   Graph $3x \geq y$.

Solution: First, graph the boundary line $3x = y$. Graph a solid boundary line because the inequality symbol is $\geq$. Test a point not on the boundary line to determine which half-plane contains points that satisfy the inequality. We choose $(0, 1)$ as our test point.

$$3x \geq y$$

$$3(0) \geq 1 \qquad \text{Let } x = 0 \text{ and } y = 1.$$

$$0 \geq 1 \qquad \text{False.}$$

This point does not satisfy the inequality, so the correct half-plane is on the opposite side of the boundary line from $(0, 1)$. The graph of $3x \geq y$ is the boundary line together with the shaded region shown next.

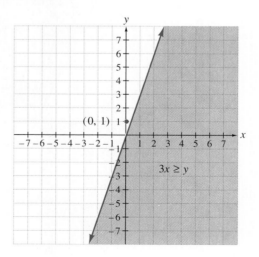

2   The intersection and the union of linear inequalities can be graphed also as shown in the next two examples.

EXAMPLE 3   Graph the intersection $x \geq 1$ and $y \geq 2x - 1$.

Solution: Graph each inequality. The intersection of the two graphs is all points common to both regions, as shown by the heaviest shading in the third graph.

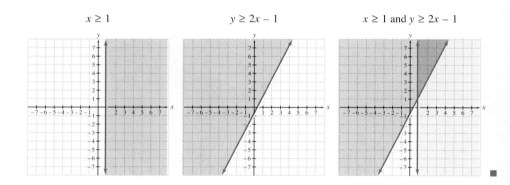

EXAMPLE 4   Graph the union $x + \frac{1}{2}y \geq -4$ or $y \leq -2$.

Solution: Graph each inequality. The union of the two inequalities is both shaded regions, including the solid boundary lines shown in the third graph.

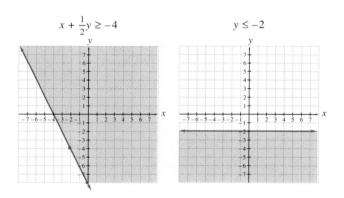

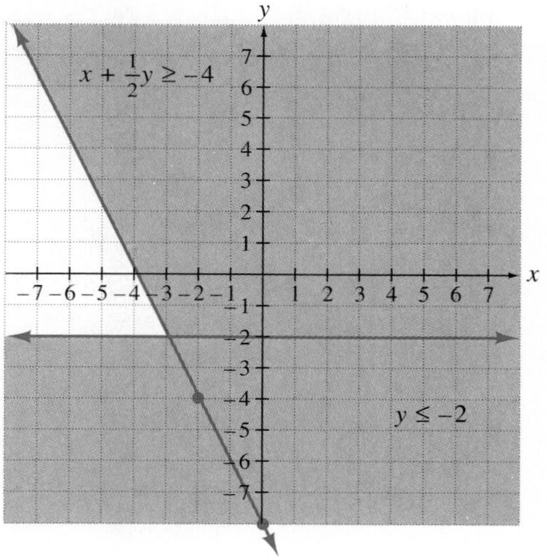

EXERCISE SET 3.6

Graph each inequality. See Examples 1 and 2.

1. $x < 2$

2. $x > -3$

3. $x - y \geq 7$

4. $3x + y \leq 1$

5. $3x + y > 6$

6. $2x + y > 2$

7. $y \leq -2x$

8. $y \leq 3x$

9. $2x + 4y \geq 8$

10. $2x + 6y \leq 12$

11. $5x + 3y > -15$

12. $2x + 5y < -20$

Graph each inequality. See Examples 3 and 4.

13. The intersection $x \geq 3$
and $y \leq -2$

14. The union $x \geq 3$
or $y \leq -2$

15. The union $x \leq -2$
or $y \geq 4$

16. The intersection $x \leq -2$
and $y \geq 4$

17. The intersection $x - y < 3$
and $x > 4$

18. The intersection $2x > y$
and $y > x + 2$

19. The union $x + y \leq 3$
or $x - y \geq 5$

20. The union $x - y \leq 3$
or $x + y > -1$

Graph each inequality.

21. $y \geq -2$

22. $y \leq 4$

23. $x - 6y < 12$

24. $x - 4y < 8$

25. $x > 5$

26. $y \geq -2$

27. $-2x + y \leq 4$

28. $-3x + y \leq 9$

29. $x - 3y < 0$

30. $x + 2y > 0$

31. $3x - 2y \leq 12$

32. $2x - 3y \leq 9$

33. The union $x - y \geq 2$ or $y < 5$

34. The union $x - y < 3$ or $x > 4$

35. The intersection $x + y \leq 1$ and $y \leq -1$

36. The intersection $y \geq x$ and $2x - 4y \geq 6$

37. The union $2x + y > 4$ or $x \geq 1$

38. The union $3x + y < 9$ or $y \leq 2$

39. The intersection of $x \geq -2$ and $x \leq 1$

40. The intersection of $x \geq -4$ and $x \leq 3$

41. The union $x + y \leq 0$ or $3x - 6y \geq 12$

42. The intersection $x + y \leq 0$ and $3x - 6y \geq 12$

43. The intersection $2x - y > 3$ and $x \geq 0$

44. The union $2x - y > 3$ or $x \geq 0$

Write the inequality whose graph is given.

45.
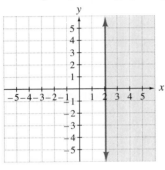

46.

47.

48.
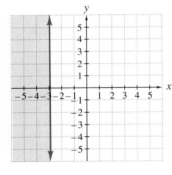

49.
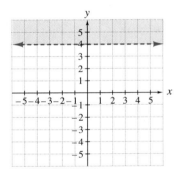

50.
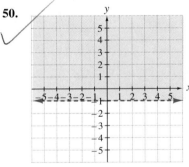

51.
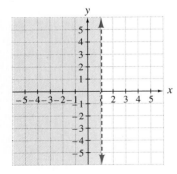

52.
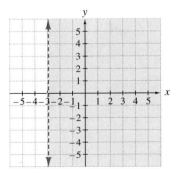

53. Rheem Abo-Zahrah decides that she will study at most 20 hours every week and that she must work at least 10 hours every week. Let *x* represent the hours studying and *y* represent the hours working. Write two inequalities that model this situation and graph their intersection.

54. The movie and TV critic for the *New York Times* spends between 2 and 6 hours daily reviewing movies, and less than 5 hours reviewing TV shows. Let *x* represent the hours watching movies and *y* represent the time spent watching TV. Write two inequalities that model this situation and graph their intersection.

55. Chris-Craft manufactures boats out of fiber glass and wood. Fiber-glass hulls require 2 hours work, while wood hulls require 4 hours work. Employees work less than 40 hours a week. The following inequalities model these restrictions, where *x* represents the number of fiber-glass hulls produced and *y* represents the number of wood hulls produced.

$$\begin{cases} x \geq 0 \\ y \geq 0 \\ 2x + 4y \leq 40 \end{cases}$$

Graph the intersection of these inequalities.

Writing in Mathematics

56. Explain a dashed boundary line in the graph of an inequality.

57. After the boundary line is sketched, explain why we test a point on either side of this boundary in the original inequality.

58. When the union of two inequalities is graphed, explain why all shaded regions are included, and not just the overlap of the shaded regions.

Skill Review

Evaluate each expression. See Sections 1.3 and 1.4.

59. 2^3

60. 3^2

61. -5^2

62. $(-5)^2$

63. $(-2)^4$

64. -2^4

65. $\left(\dfrac{3}{5}\right)^3$

66. $\left(\dfrac{2}{7}\right)^2$

CRITICAL THINKING   The graph shows the amount of money Americans spent on cable television each year from 1982 through 1992. Notice that this graph is roughly linear, that is, like a line.

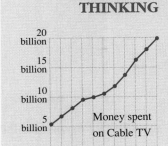

What does the linearity of this graph reveal to analysts about the growth of the cable industry beyond 1992? What is the rate of growth? What assumptions must analysts make to predict the amount of money Americans will spend on cable television in the year 2000? On the basis of these assumptions, what is your prediction? How might your prediction change under these conditions?

CHAPTER 3 GLOSSARY

The **Cartesian coordinate system** consists of two number lines that intersect at right angles at their 0 coordinates. The horizontal number line is called the **x-axis** (or the axis of the **abscissa**), and the vertical number line is called the **y-axis** (or the axis of the **ordinate**). The point of intersection of these axes is named the **origin.**

A **function** is a relation by which each first component is assigned to exactly one second component.

A **linear equation in two variables** is an equation that can be written in the form $ax + by = c$, where a, b, and c are real numbers, and a and b are not both 0.

A **linear inequality in two variables** is an inequality that can be written in the form $ax + by < c$ or $ax + by > c$, where a, b, and c are real numbers and a and b are not both 0.

The **midpoint** of a line segment is the point located exactly halfway between the two end points of that segment.

An **ordered pair** is a pair of values written as (x, y). The first value is the **x-coordinate** and the second value is the **y-coordinate.**

A **relation** is a correspondence that assigns to each element in set X an element in set Y. The **domain** of the relation is set X, and the **range** of the relation is the set of all values in Y that correspond to some element in X.

The **slope** of a line is a measure of its steepness.

The **x-intercept** of a line is the x-coordinate of the point where the line crosses the x-axis.

The **y-intercept** of a line is the y-coordinate of the point where the line crosses the y-axis.

CHAPTER 3 SUMMARY

(3.1)

The **distance** between two points (x_1, y_1) and (x_2, y_2) is

$$d = \sqrt{(x_2 - x_1)^2 + (y_2 - y_1)^2}$$

The **midpoint** of the line segment joining points (x_1, y_1) and (x_2, y_2) is the point

$$\left(\frac{x_1 + x_2}{2}, \frac{y_1 + y_2}{2} \right)$$

Given two points (x_1, y_1) and (x_2, y_2) on a line, the **slope of the line** (denoted by m) is $m = \dfrac{y_2 - y_1}{x_2 - x_1}$.

(3.2)

To find the x-intercept, let $y = 0$ and solve for x.
To find the y-intercept, let $x = 0$ and solve for y.

(3.3)

Parallel lines have the same slope.
If the product of the slopes of two lines is -1, then the two lines are **perpendicular.**

EQUATIONS OF LINES   **(3.4)**

Standard form: $ax + by = c$, where a, b, and c are real numbers and a and b are not both 0.

Point–slope form: $y - y_1 = m(x - x_1)$, where m is the slope of the line and (x_1, y_1) is a point on the line.

Slope–intercept form: $y = mx + b$, where m is the slope and b is the y-intercept.

Vertical lines: $x = c$ and have undefined slope.
Horizontal lines: $y = b$ and have slope 0.

VERTICAL LINE TEST FOR FUNCTIONS   (3.5)
If a vertical line intersects a graph at more than one point, the graph is not the graph of a function.

CHAPTER 3 REVIEW EXERCISES

(3.1) *Plot the points and name the quadrant in which each point lies.*

1. $A(2, -1), B(-2, 1), C(0, 3), D(-3, -5)$

2. $A(-3, 4), B(4, -3), C(-2, 0), D(-4, 1)$

Find the distance between each pair of points.

3. $(-6, 3)$ and $(8, 4)$

4. $(3, 5)$ and $(8, 9)$

5. $(-4, -6)$ and $(-1, 5)$

6. $(-1, 5)$ and $(2, -3)$

Find the midpoint of the line segment whose end points are given.

7. $(2, 6)$ and $(-12, 4)$

8. $(-3, 8)$ and $(11, 24)$

9. $(-6, -5)$ and $(-9, 7)$

10. $(4, -6)$ and $(-15, 2)$

(3.2) *Determine whether each ordered pair is a solution of the given equation.*

11. $7x + 6y = -2$
 a. $\left(0, -\dfrac{1}{3}\right)$ **b.** $(2, -2)$ **c.** $(-8, 9)$

12. $4x - 3y = 12$
 a. $(0, 4)$ **b.** $(6, -4)$ **c.** $(-3, -8)$

Graph each equation.

13. $4x + 5y = 20$

14. $3x - 2y = -9$

15. $4x - y = 3$

16. $2x + 6y = 9$

17. $y = 5$

18. $x = -2$

(3.3) *Find the slope of the line through each pair of points.*

19. $(2, 8)$ and $(6, -4)$

20. $(-3, 9)$ and $(5, 13)$

21. $(-7, -4)$ and $(-3, 6)$

22. $(7, -2)$ and $(-5, 7)$

Determine whether lines L_1 and L_2 are parallel lines, perpendicular lines, or neither if they pass through the given points.

23. Line L_1: $(2, -3), (5, 6)$; line L_2: $(4, -1), (7, 0)$

24. Line L_1: $(-6, 8), (4, -2)$; line L_2: $(2, 5), (-4, 11)$

(3.4) *Find an equation of the line satisfying the conditions given.*

25. Horizontal; through $(3, -1)$

26. Vertical; through $(-2, -4)$

27. Parallel to the line $x = 6$; through $(-4, -3)$

28. Slope 0; through $(2, 5)$

Find the slope and y-intercept of each line.

29. $6x - 15y = 20$

30. $4x + 14y = 21$

Find the standard form equation of each line satisfying the conditions given.

31. Through $(-3, 5)$; slope 3

32. Slope 2; through $(5, -2)$

33. Slope $-\dfrac{2}{3}$; y-intercept 4

34. Slope -1; y-intercept -2

35. Through $(2, -6)$; parallel to $6x + 3y = 5$

36. Through $(-4, -2)$; parallel to $3x + 2y = 8$

37. Through $(-6, -1)$; perpendicular to $4x + 3y = 5$

38. Through $(-4, 5)$; perpendicular to $2x - 3y = 6$

39. Through $(-6, -1)$ and $(-4, -2)$

40. Through $(-5, 3)$ and $(-4, -8)$

41. Through $(-2, 3)$; perpendicular to $x = 4$

42. Through $(-2, -5)$; parallel to $y = 8$

(3.5)   *Find the domain and range of each relation. Also determine whether the relation is a function.*

43. $\left\{\left(-\dfrac{1}{2}, \dfrac{3}{4}\right), (6, 0.75), (0, -12), (25, 25)\right\}$

44. $\left\{\left(\dfrac{3}{4}, -\dfrac{1}{2}\right), (0.75, 6), (-12, 0), (25, 25)\right\}$

45.

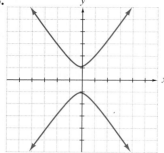

46.

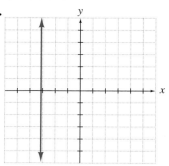

47.

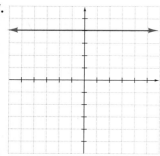

48.

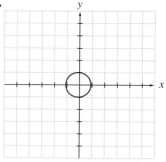

(3.6)   *Graph each linear inequality.*

49. $3x + y > 4$

50. $\dfrac{1}{2}x - y < 2$

51. $5x - 2y \le 9$

52. $3y \ge x$

53. $y < 1$

54. $x > -2$

55. Graph the union $y > 2x + 3$ or $x \leq -3$.

56. Graph the intersection $2x < 3y + 8$ and $y \geq -2$.

CHAPTER 3 TEST

1. Find the x- and y-intercepts of the line $4x - 6y = 12$.

2. Complete the ordered pair solution $(-6, \quad)$ of the equation $2y - 3x = 12$.

3. Plot the points, and name the quadrant in which each is located: $A(6, -2)$, $B(4, 0)$, $C(-1, 6)$.

Graph each line.

4. $2x - 3y = -6$

5. $4x + 6y = 7$

6. $y = \dfrac{2}{3}x$

7. $y = -3$

8. Find the distance between the points $(-6, 3)$ and $(-8, -7)$.

9. Find the midpoint of the line segment whose end points are $(-2, -5)$ and $(-6, 12)$.

10. Find the slope of the line that passes through $(5, -8)$ and $(-7, 10)$.

11. Find the slope and the y-intercept of the line $3x + 12y = 8$.

Find an equation of each line satisfying the conditions given.

12. Horizontal; through $(2, -8)$

13. Vertical; through $(-4, -3)$

14. Perpendicular to $x = 5$; through $(3, -2)$

15. Through $(4, -1)$; slope -3

16. Through $(0, -2)$; slope 5

17. Through $(4, -2)$ and $(6, -3)$

18. Through $(-1, 2)$; perpendicular to $3x - y = 4$

19. Parallel to $2y + x = 3$; through $(3, -2)$

20. Line L_1 has the equation $2x - 5y = 8$. Line L_2 passes through the points $(1, 4)$ and $(-1, -1)$. Determine whether these lines are parallel lines, perpendicular lines, or neither.

Graph each inequality.

21. $x \leq -4$

22. $y > -2$

23. $2x - y > 5$

24. The intersection $2x + 4y < 6$ and $y \leq -4$

Determine whether each equation or graph represents a function.

25. $y + 5 = 0$

26. $2x - 12y = 0$

27.

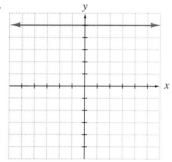

28.

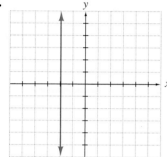

29.

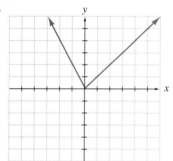

30.

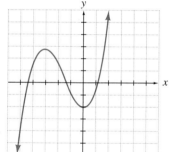

CHAPTER 3 CUMULATIVE REVIEW

1. Translate each sentence into a mathematical statement.
  a. The sum of x and 5 is 20.
  b. Two times the sum of 3 and y is 4.
  c. Subtract 8 from x and the difference is the product of 2 and x.
  d. The quotient of z and 9 is three times the difference of z and 5.

2. Write the reciprocal or multiplicative inverse of each.

  a. 5   **b.** -2 **c.** $\dfrac{4}{7}$

3. Evaluate the following.

  a. 4^2 **b.** $\left(\dfrac{1}{5}\right)^2$ **c.** $\left(\dfrac{2}{3}\right)^3$

4. Find each sum:
  a. $-3 + (-11)$ **b.** $3 + (-7)$
  c. $-10 + 15$ **d.** $-8 + (-1)$
  e. $-\dfrac{1}{4} + \dfrac{1}{2}$ **f.** $-\dfrac{2}{3} + \dfrac{3}{7}$

5. Simplify each expression:

  a. 3^2 **b.** $\left(\dfrac{1}{2}\right)^4$ **c.** -5^2 **d.** $(-5)^2$

  e. -5^3 **f.** $(-5)^3$

6. Find the value of each algebraic expression when $x = 2$, $y = -1$, and $z = -3$.

  a. $z - y$ **b.** z^2 **c.** $\dfrac{2x + y}{z}$

7. Solve for x: $-6x - 1 + 5x = 3$.

8. Solve for x: $6x - 4 = 2 + 6(x - 1)$.

9. The distance from Atlanta, Georgia, to Savannah, Georgia, is about 240 miles. How long will it take to make this trip if the average speed is 50 miles per hour?

10. Solve $|2x| + 5 = 7$.

11. Solve for x: $3x + 4 \geq 2x - 6$. Graph the solution and write it in interval notation.

12. Solve for x: $\dfrac{2}{5}(x - 6) \geq x - 1$.

13. Solve for x: $-1 \leq \dfrac{2x}{3} + 5 \leq 2$.

14. Solve for y: $|y - 3| > 7$.

15. Find the distance between the following pairs of points.
　a. $(-3, 3)$ and $(1, 0)$
　b. $(2, -5)$ and $(1, -4)$

16. Graph $y = 4x$.

17. Graph $y = -2$.

18. If L_1 is a line through $(-5, 3)$ and $(1, -1)$ and L_2 is a line through $(2, 5)$ and $(0, 2)$, determine whether L_1 and L_2 are perpendicular.

19. Find the equation of the line through $(2, 3)$ with undefined slope.

20. Use the vertical line test to determine whether the graph is the graph of a function.

a.

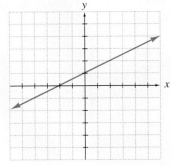

b.

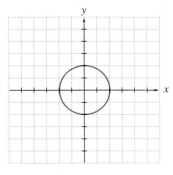

c.

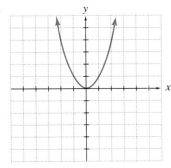

d.

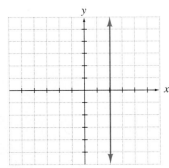

21. Graph $2x - y < 6$.

CHAPTER **4**

Exponents and Polynomials

Yee, an instructor at Florida Heights Community College, just bought a programmable calculator. She understands that programmable calculators are designed so that she can define and program special keys according to her needs. She is preparing to program a key so that the touch of this key will adjust her final exam scores.

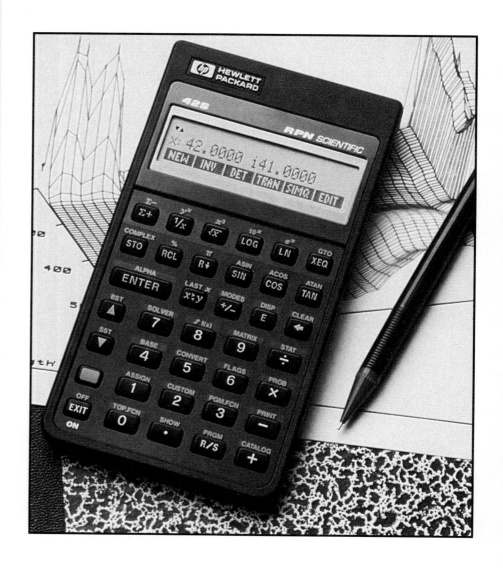

INTRODUCTION

Linear equations are important for solving problems. They are not sufficient, however, to solve all problems. If, for example, we want to express the area of a square field whose side has length s, a linear expression will not do. Many nonlinear expressions are written with exponents, a shorthand notation for repeated multiplication. The first two sections of this chapter refresh your skills working with exponents. The next sections are devoted to operations on polynomials. Polynomials model many real-world phenomena, and hence your ability to work with them can apply to solving many problems. We conclude the chapter with additional concepts relating to functions, particularly polynomial functions.

4.1
Exponents and Scientific Notation

Tape 9

OBJECTIVES

1. Define a^n.

2. Use the product rule for exponents.

3. Evaluate a raised to the 0 power.

4. Use the quotient rule for exponents.

5. Define a raised to the negative nth power.

6. Write numbers in scientific notation.

7. Convert numbers from scientific notation to standard notation.

---

1 In the product $3(7) = 21$, recall that 3 and 7 are called factors of 21. When a number is a factor more than once, as in $5 \cdot 5 \cdot 5$, for example, **exponents** may be used to write these repeated factors in shorthand notation.

$$\underbrace{5 \cdot 5 \cdot 5}_{\text{5 is a factor 3 times}} = 5^3$$

In the exponential expression 5^3, 5 is called the **base** and 3 is called the **exponent.** The base is the repeated factor and the exponent is the number of times the factor is repeated.

Exponents can also be used when the repeated factor is a variable. For example

$$\underbrace{y \cdot y \cdot y \cdot y \cdot y}_{\text{y is a factor 5 times}} = y^5$$

In the exponential expression y^5, y is the base and 5 is the exponent.

Definition of a^n

If a is a real number and n is a positive integer, then a raised to the nth power, written a^n, is the product of n factors of a.

$$a^n = \underbrace{a \cdot a \cdot a \cdot a \cdots a}_{n \text{ factors of } a}$$

EXAMPLE 1 Evaluate the following.

   a. 9^2 **b.** 2^3 **c.** 7^1 **d.** $\left(\dfrac{2}{3}\right)^3$ **e.** -2^4 **f.** $(-2)^4$ **g.** $2 \cdot 4^2$

Solution: **a.** $9^2 = 9 \cdot 9 = 81$; 9^2 is read "9 squared"
   b. $2^3 = 2 \cdot 2 \cdot 2 = 8$; 2^3 is read "2 cubed"
   c. $7^1 = 7$
   d. $\left(\dfrac{2}{3}\right)^3 = \dfrac{2}{3} \cdot \dfrac{2}{3} \cdot \dfrac{2}{3} = \dfrac{8}{27}$
   e. $-2^4 = -(2 \cdot 2 \cdot 2 \cdot 2) = -16$
   f. $(-2)^4 = (-2)(-2)(-2)(-2) = 16$
   g. $2 \cdot 4^2 = 2 \cdot 4 \cdot 4 = 32$ ∎

HELPFUL HINT

Notice in the preceding example that

   $-2^4 = -(2 \cdot 2 \cdot 2 \cdot 2) = -16$ and $(-2)^4 = (-2)(-2)(-2)(-2) = 16$

Without parentheses, the exponent 4 applies to the base of 2 only.

2 Exponential expressions can be multiplied, divided, added, subtracted, and themselves raised to powers. We review multiplication first.
   To multiply x^2 by x^3, use the definition of a^n.

$$x^2 \cdot x^3 = \underbrace{(x \cdot x)(x \cdot x \cdot x)}_{x \text{ is a factor 5 times}}$$

$$= x^5$$

Notice that the result is exactly the same if we add the exponents.

$$x^2 \cdot x^3 = x^{2+3} = x^5$$

This suggests the following.

Product Rule for Exponents

If m and n are positive integers and a is a real number, then

$$a^m \cdot a^n = a^{m+n}$$

In other words, the **product** of exponential expressions with a common base is the common base raised to a power equal to the **sum** of the exponents of the factors.

EXAMPLE 2   Find each product.

a. $2^2 \cdot 2^5$ **b.** $x^7 x^3$ **c.** $y \cdot y^2 \cdot y^4$

Solution: **a.** $2^2 \cdot 2^5 = 2^{2+5} = 2^7$

b. $x^7 x^3 = x^{7+3} = x^{10}$

c. $y \cdot y^2 \cdot y^4 = (y^1 \cdot y^2) \cdot y^4$

$$= y^3 \cdot y^4$$

$$= y^7 \quad \blacksquare$$

EXAMPLE 3   Find each product.

a. $(3x^6)(5x)$ **b.** $(-2x^3 p^2)(4xp^{10})$

Solution: **a.** $(3x^6)(5x) = 3(5)x^6 x^1 = 15x^7$

b. $(-2x^3 p^2)(4xp^{10}) = -2(4)x^3 x^1 p^2 p^{10} = -8x^4 p^{12}$ $\blacksquare$

3   The definition of a^n does not include the possibility that n might be 0. But if it did, then, by the product rule,

$$\underbrace{a^0(a^n)}_{} = a^{0+n} = a^n = \underbrace{(1)(a^n)}_{}.$$

From this, we reasonably define that $a^0 = 1$, as long as a does not equal 0.

Zero Exponent

If a does not equal 0, then $a^0 = 1$.

EXAMPLE 4   Evaluate the following.

a. 7^0 **b.** -7^0 **c.** $(2x + 5)^0$ **d.** $2x^0$

Solution: **a.** $7^0 = 1$

b. Without parentheses, only 7 is raised to the 0 power.

$$-7^0 = -(7^0) = -(1) = -1$$

c. $(2x + 5)^0 = 1$

d. $2x^0 = 2(1) = 2$ $\blacksquare$

4   To find quotients of exponential expressions, we again begin with the definition of a^n to simplify $\dfrac{x^9}{x^2}$. For example,

$$\frac{x^9}{x^2} = \frac{x \cdot x \cdot x \cdot x \cdot x \cdot x \cdot x \cdot x \cdot x}{x \cdot x} = x^7$$

(Assume that denominators containing variables are not 0.)

Notice that the result is exactly the same if we subtract the exponents.

$$\frac{x^9}{x^2} = x^{9-2} = x^7$$

This suggests the following.

Quotient Rule for Exponents

If a is a nonzero real number and n and m are integers, then

$$\frac{a^m}{a^n} = a^{m-n}$$

In other words, the **quotient** of exponential expressions with a common base is the common base raised to a power equal to the **difference** of the exponents.

EXAMPLE 5   Find each quotient.

a. $\dfrac{x^7}{x^4}$ b. $\dfrac{5^8}{5^2}$ c. $\dfrac{20x^6}{4x^5}$ d. $\dfrac{12y^{10}z^7}{14y^8z^7}$

Solution: a. $\dfrac{x^7}{x^4} = x^{7-4} = x^3$

b. $\dfrac{5^8}{5^2} = 5^{8-2} = 5^6$

c. $\dfrac{20x^6}{4x^5} = 5x^{6-5} = 5x^1$ or $5x$

d. $\dfrac{12y^{10}z^7}{14y^8z^7} = \dfrac{6}{7}y^{10-8} \cdot z^{7-7} = \dfrac{6}{7}y^2z^0 = \dfrac{6}{7}y^2$ or $\dfrac{6y^2}{7}$ ■

5   When the exponent of the denominator is larger than the exponent of the numerator, applying the quotient rule yields a negative exponent. For example,

$$\frac{x^3}{x^5} = x^{3-5} = x^{-2}$$

Using the definition of a^n, though, gives us

$$\frac{x^3}{x^5} = \frac{x \cdot x \cdot x}{x \cdot x \cdot x \cdot x \cdot x} = \frac{1}{x^2}$$

From this, we reasonably define $a^{-n} = \dfrac{1}{a^n}$.

Negative Exponents

If a is a real number other than 0 and n is a positive integer, then

$$a^{-n} = \frac{1}{a^n}$$

EXAMPLE 6 Write the following using only positive exponents. Simplify if possible.

a. 5^{-2} **b.** $2x^{-3}$ **c.** $(3x)^{-1}$ **d.** $\dfrac{m^5}{m^{15}}$ **e.** $\dfrac{3^3}{3^6}$ **f.** $2^{-1} + 3^{-2}$ **g.** $\dfrac{1}{t^{-5}}$

Solution: **a.** $5^{-2} = \dfrac{1}{5^2} = \dfrac{1}{25}$

b. Without parentheses, only x is raised to the -3 power.

$$2x^{-3} = 2 \cdot \frac{1}{x^3} = \frac{2}{x^3}$$

c. With parentheses, both 3 and x are raised to the -1 power.

$$(3x)^{-1} = \frac{1}{(3x)^1} = \frac{1}{3x}$$

d. $\dfrac{m^5}{m^{15}} = m^{5-15} = m^{-10} = \dfrac{1}{m^{10}}$

e. $\dfrac{3^3}{3^6} = 3^{3-6} = 3^{-3} = \dfrac{1}{3^3} = \dfrac{1}{27}$

f. $2^{-1} + 3^{-2} = \dfrac{1}{2^1} + \dfrac{1}{3^2} = \dfrac{1}{2} + \dfrac{1}{9} = \dfrac{9}{18} + \dfrac{2}{18} = \dfrac{11}{18}$

g. $\dfrac{1}{t^{-5}} = \dfrac{1}{\dfrac{1}{t^5}} = 1 \div \dfrac{1}{t^5} = 1 \cdot \dfrac{t^5}{1} = t^5$ ∎

HELPFUL HINT

Notice that when a factor containing an exponent is moved from the numerator to the denominator or from the denominator to the numerator, the sign of its exponent changes.

$$x^{-3} = \frac{1}{x^3}, \qquad\qquad 5^{-2} = \frac{1}{5^2} = \frac{1}{25}$$

$$\frac{1}{y^{-4}} = y^4, \qquad \frac{1}{2^{-3}} = 2^3 = 8$$

EXAMPLE 7 Simplify each expression. Write answers using positive exponents.

a. $\dfrac{x^{-9}}{x^2}$ **b.** $\dfrac{p^4}{p^{-3}}$ **c.** $\dfrac{2^{-3}}{2^{-1}}$ **d.** $\dfrac{2x^{-7}y^2}{10xy^{-5}}$ **e.** $\dfrac{(3x^{-3})(x^2)}{x^6}$

Solution: **a.** $\dfrac{x^{-9}}{x^2} = x^{-9-2} = x^{-11} = \dfrac{1}{x^{11}}$

b. $\dfrac{p^4}{p^{-3}} = p^{4-(-3)} = p^7$

c. $\dfrac{2^{-3}}{2^{-1}} = 2^{-3-(-1)} = 2^{-2} = \dfrac{1}{2^2} = \dfrac{1}{4}$

d. $\dfrac{2x^{-7}y^2}{10xy^{-5}} = \dfrac{x^{-7-1} \cdot y^{2-(-5)}}{5} = \dfrac{x^{-8}y^7}{5} = \dfrac{y^7}{5x^8}$

e. Simplify the numerator first.

$$\dfrac{(3x^{-3})(x^2)}{x^6} = \dfrac{3x^{-3+2}}{x^6} = \dfrac{3x^{-1}}{x^6} = 3x^{-1-6} = 3x^{-7} = \dfrac{3}{x^7} \quad \blacksquare$$

EXAMPLE 8   Simplify. Assume that a and t are nonzero integers and x is not 0.

a. $x^{2a} \cdot x^3$ **b.** $\dfrac{x^{2t-1}}{x^{t-5}}$

Solution: **a.** $x^{2a} \cdot x^3 = x^{2a+3}$ Use the product rule.

numerator exponent
minus
denominator exponent

b. $\dfrac{x^{2t-1}}{x^{t-5}} = x^{(2t-1)-(t-5)}$ Use the quotient rule.

$= x^{2t-1-t+5} = x^{t+4} \quad \blacksquare$

6   Very large and very small numbers occur frequently in nature. For example, the distance between the earth and the sun is approximately 150,000,000 kilometers. A helium atom has a diameter of 0.000 000 022 centimeters. It can be tedious to write down these very large and small numbers in standard notation like this. **Scientific notation** is a convenient shorthand notation for writing very large and very small numbers.

Scientific Notation
A positive number is written in **scientific notation** if it is written as the product of a number a, where $1 \le a < 10$, and an integer power r of 10: $a \times 10^r$.

The following are examples of numbers written in scientific notation:

$$2.03 \times 10^2, \qquad 7.362 \times 10^7, \qquad 8.1 \times 10^{-5}$$

To write the approximate distance between the earth and the sun in scientific notation, move the decimal point to the left until the number is between 1 and 10.

150,000,000.

The decimal point was moved 8 places left, so

$$150,000,000 = 1.5 \times 10^8$$

Next, to write the diameter of a helium atom in scientific notation, again move the decimal point until we have a number between 1 and 10.

0. 000 000 022

The decimal point was moved 8 places to the right, so

$$0.\,000\,000\,022 = 2.2 \times 10^{-8}$$

> **To Write a Number in Scientific Notation**
>
> *Step 1*   Move the decimal point in the original number until the new number has a value between 1 and 10.
>
> *Step 2*   Count the number of decimal places the decimal point was moved in *Step 1*. If the decimal point was moved to the left, the count is positive. If the decimal point was moved to the right, the count is negative.
>
> *Step 3*   Multiply the new number in *Step 1* by 10 raised to an exponent equal to the count found in *Step 2*.

EXAMPLE 9   Write each number in scientific notation.
a. 730,000   **b.** 0.00000104

Solution: **a.** *Step 1*   Move the decimal point until the number is between 1 and 10.

$$730{,}000.$$

Step 2   The decimal point is moved to the left 5 places, so the count is positive 5.
Step 3 $730{,}000 = 7.3 \times 10^5$.

b. *Step 1*   Move the decimal point until the number is between 1 and 10.

$$0.00000104$$

Step 2   The decimal point is moved to the right 6 places, so the count is -6.
Step 3 $0.00000104 = 1.04 \times 10^{-6}$.   ■

7   To write a scientific notation number in standard form, we reverse the preceding steps.

> **To Write a Scientific Notation Number in Standard Notation**
>
> Move the decimal point in the number the same number of places as the exponent on 10. If the exponent is positive, move the decimal point right. If the exponent is negative, move the decimal point left.

EXAMPLE 10   Write each number in standard notation.
a. 7.7×10^8 **b.** 1.025×10^{-3}

Solution: **a.** Since the exponent is positive, move the decimal point 8 places to the right. Add zeros as needed.

$$7.7 \times 10^8 = 770{,}000{,}000$$

b. Since the exponent is negative, move the decimal point 3 places to the left. Add zeros as needed.

$$1.025 \times 10^{-3} = 0.001025$$ ■

CALCULATOR BOX

Multiply 5,000,000 by 700,000 on your calculator. The display should read $\boxed{3.5 \qquad 12}$, which is the product written in scientific notation. This notation means 3.5×10^{12}.

To enter a number written in scientific notation on a calculator, find the key marked $\boxed{EE}$. (On some calculators, this key may be marked $\boxed{EXP}$.)

To enter 7.26×10^{13}, press the keys

$$\boxed{7.26}\ \boxed{EE}\ \boxed{13}$$

The display will read $\boxed{7.26 \qquad 13}$.

Use your calculator to perform each operation indicated.

1. Multiply 3×10^{11} and 2×10^{32}.
2. Divide 6×10^{14} by 3×10^{9}.
3. Multiply 5×10^{23} and 7×10^{4}. Did the answer surprise you?
4. Divide 4.38×10^{41} by 3×10^{17}.

MENTAL MATH

Identify the exponent and the base.

1. 3^5
2. $(-7)^4$
3. -4^5
4. 6^3
5. $3x^2$
6. $3 + 2^7$
7. $(2y + z)^0$
8. $-x^0$

EXERCISE SET 4.1

Evaluate the following. See Example 1.

1. 2^4
2. -4^3
3. -9^2
4. 3^4
5. $\left(\dfrac{3}{4}\right)^2$
6. $5 \cdot 2^2$
7. $4 \cdot 3^2$
8. $(-1)^5$

Find each product. See Examples 2 and 3.

9. $4^2 \cdot 4^3$
10. $3^3 \cdot 3^5$
11. $x^5 \cdot x^3$
12. $a^2 \cdot a^9$
13. $(4xy)(-5x)$
14. $(7xy)(7aby)$
15. $(-4x^3p^2)(4y^3x^3)$
16. $(-6a^2b^3)(-3ab^3)$

Evaluate the following. See Example 4.

17. -8^0
18. $(-9)^0$
19. $(4x + 5)^0$
20. $8x^0 + 1$

Find each quotient. See Example 5.

21. $\dfrac{a^5}{a^2}$
22. $\dfrac{x^9}{x^4}$
23. $\dfrac{x^9 y^6}{x^8 y^6}$
24. $\dfrac{a^{12}b^2}{a^9 b}$

25. $-\dfrac{26z^{11}}{2z^{7}}$

26. $\dfrac{16x^{5}}{8x}$

Write the following using positive exponents only. Simplify if possible. See Example 6.

27. 4^{-2}

28. 2^{-3}

29. $\dfrac{x^{7}}{x^{15}}$

30. $\dfrac{z}{z^{3}}$

31. $5a^{-4}$

32. $10b^{-1}$

Simplify each expression. Write answers using positive exponents. See Example 7.

33. $\dfrac{x^{-2}}{x^{5}}$

34. $\dfrac{y^{-6}}{y^{-9}}$

35. $\dfrac{8r^{4}}{2r^{-4}}$

36. $\dfrac{3s^{3}}{15s^{-3}}$

37. $\dfrac{x^{-9}x^{4}}{x^{-5}}$

38. $\dfrac{y^{-7}y}{y^{8}}$

Simplify. Assume that variables in the exponent represent nonzero integers and x and y are not 0. See Example 8.

39. $x^{5}\cdot x^{7a}$

40. $y^{2p}\cdot y^{9p}$

41. $\dfrac{x^{3t-1}}{x^{t}}$

42. $\dfrac{y^{4p-2}}{y^{3p}}$

Write each number in scientific notation. See Example 9.

43. 31,250,000

44. 678,000

45. 0.016

46. 0.007613

Write each number in standard notation, without exponents. See Example 10.

47. 3.6×10^{-9}

48. 2.7×10^{-5}

49. 9.3×10^{7}

50. 6.378×10^{8}

Simplify the following. Write answers with positive exponents.

51. -6^{2}

52. -3^{4}

53. 4^{-3}

54. 2^{-3}

55. $4^{-1} + 3^{-2}$

56. $1^{-3} - 4^{-2}$

57. $4x^{0} + 5$

58. $-5x^{0}$

59. $8 \cdot 2^{3}$

60. $-3 \cdot 5^{2}$

61. $x^{7} \cdot x^{8}$

62. $y^{6} \cdot y$

63. $\dfrac{z^{12}}{z^{15}}$

64. $\dfrac{x^{11}}{x^{20}}$

65. $\dfrac{y^{-3}}{y^{-7}}$

66. $\dfrac{z^{-12}}{z^{10}}$

67. $3x^{-1}$

68. $(4x)^{-1}$

69. $3^{0} - 3t^{0}$

70. $4^{0} + 4x^{0}$

71. $\dfrac{r^{4}}{r^{-4}}$

72. $\dfrac{x^{-5}}{x^{3}}$

73. $\dfrac{x^{-7}}{x^{2}}$

74. $\dfrac{10^{-9}}{10^{-3}}$

75. $\dfrac{2a^{-6}b^{2}}{18ab^{-5}}$

76. $\dfrac{18ab^{-6}}{3a^{-3}b^{6}}$

77. $\dfrac{(24x^{8})(x)}{20x^{-7}}$

78. $\dfrac{(30z^{2})(z^{5})}{55z^{-4}}$

Simplify the following. Assume that variables in the exponent represent nonzero integers and that all other variables are not 0.

79. $x^{4a} \cdot x^{7}$

80. $x^{9y} \cdot x^{-7y}$

81. $\dfrac{z^{6x}}{z^{7}}$

82. $\dfrac{y^{6}}{y^{4z}}$

83. $\dfrac{x^{3t} \cdot x^{4t-1}}{x^{t}}$

84. $\dfrac{z^{35} \cdot z^{-5+2}}{z^{25}}$

Write each number in scientific notation.

85. 67,413

86. 36,800,000

87. 0.0125

88. 0.00084

89. 0.000053

90. 98,700,000,000

Write each number in standard notation.

91. 1.278×10^{6}

92. 7.6×10^{4}

93. 7.35×10^{12}

94. 1.66×10^{-5}

95. 4.03×10^{-7}

96. 8.007×10^{8}

Express each number in scientific notation.

97. The approximate distance between Jupiter and Earth is 918,000,000 kilometers.

98. A computer can perform 48,000,000,000 arithmetic operations in one minute.

99. The number of millimeters in 50 kilometers is 50,000,000.

100. The approximate distance between the sun and the earth is 93,000,000 miles.

Writing in Mathematics

101. Explain how to convert a number from standard notation to scientific notation.

102. Explain how to convert a number from scientific notation to standard notation.

103. Explain whether 0.4×10^{-5} is written in scientific notation.

104. Explain why -3^2 does not equal $(-3)^2$.

Skill Review

Use the slope–intercept form of a line, $y = mx + b$, to find the slope of each line. See Section 3.4.

105. $y = -2x + 7$

106. $y = \dfrac{3}{2}x - 1$

107. $3x - 5y = 14$

108. $x + 7y = 2$

Use the vertical line test to determine which of the following are graphs of functions. See Section 3.5.

109.

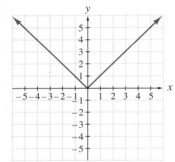

110.

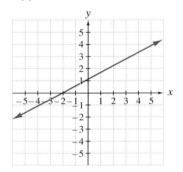

4.2
More Work with Exponents

Tape 9

OBJECTIVES

1 Use the power rule for exponents.

2 Use exponent rules and definitions to simplify exponential expressions.

3 Compute using scientific notation.

1 To simplify an expression like $(x^2)^3$, we use the definition of a^n. Then

$$(x^2)^3 = \underbrace{(x^2)(x^2)(x^2)}_{x^2 \text{ is a factor 3 times}} = x^{2+2+2} = x^6$$

Notice that the result is exactly the same if the exponents are multiplied.

$$(x^2)^3 = x^{2 \cdot 3} = x^6$$

This suggests that the power of an exponential expression raised to a power is the product of the exponents. Two additional power rules for exponents are given in the following box.

Power Rules for Exponents

If a and b are real numbers and m and n are integers, then

$$(a^m)^n = a^{m \cdot n}$$ Power of a power

$$(ab)^m = a^m b^m$$ Power of a product

$$\left(\frac{a}{b}\right)^n = \frac{a^n}{b^n} \quad (b \neq 0)$$ Power of a quotient

---

EXAMPLE 1   Simplify the following expressions. Write all answers using positive exponents.
a. $(x^5)^7$ **b.** $(2^2)^3$ **c.** $(5^{-1})^2$ **d.** $(y^{-3})^{-4}$

Solution: **a.** $(x^5)^7 = x^{5 \cdot 7} = x^{35}$

b. $(2^2)^3 = 2^{2 \cdot 3} = 2^6 = 64$

c. $(5^{-1})^2 = 5^{-1 \cdot 2} = 5^{-2} = \dfrac{1}{5^2} = \dfrac{1}{25}$

d. $(y^{-3})^{-4} = y^{-3(-4)} = y^{12}$ ■

EXAMPLE 2   Simplify the following. Write answers using positive exponents.

a. $(5x^2)^3$ **b.** $\left(\dfrac{2}{3}\right)^3$ **c.** $\left(\dfrac{3p^4}{q^5}\right)^2$ **d.** $\left(\dfrac{2^{-3}}{y}\right)^{-2}$ **e.** $(x^{-5}y^2z^{-1})^7$

Solution: **a.** $(5x^2)^3 = 5^3 \cdot (x^2)^3 = 5^3 \cdot x^{2 \cdot 3} = 125x^6$

b. $\left(\dfrac{2}{3}\right)^3 = \dfrac{2^3}{3^3} = \dfrac{8}{27}$

c. $\left(\dfrac{3p^4}{q^5}\right)^2 = \dfrac{(3p^4)^2}{(q^5)^2} = \dfrac{3^2 \cdot (p^4)^2}{(q^5)^2} = \dfrac{9p^8}{q^{10}}$

d. $\left(\dfrac{2^{-3}}{y}\right)^{-2} = \dfrac{(2^{-3})^{-2}}{y^{-2}}$

$$= \dfrac{2^6}{y^{-2}} = 64y^2 \qquad \text{Use the negative exponent rule.}$$

e. $(x^{-5}y^2z^{-1})^7 = (x^{-5})^7 \cdot (y^2)^7 \cdot (z^{-1})^7$

$$= x^{-35}y^{14}z^{-7} = \dfrac{y^{14}}{x^{35}z^7} \qquad ■$$

2   In the next few examples, we practice using several of the rules and definitions for exponents. The following is a summary of these rules and definitions.

Summary of Rules for Exponents

If a and b are real numbers and m and n are integers, then

Product rule for exponents	$a^m \cdot a^n = a^{m+n}$
Zero exponent	$a^0 = 1 \qquad (a \neq 0)$
Negative exponent	$a^{-n} = \dfrac{1}{a^n} \qquad (a \neq 0)$
Quotient rule	$\dfrac{a^m}{a^n} = a^{m-n} \qquad (a \neq 0)$
Power rules	$(a^m)^n = a^{m \cdot n}$
	$(ab)^m = a^m \cdot b^m$
	$\left(\dfrac{a}{b}\right)^m = \dfrac{a^m}{b^m} \qquad (b \neq 0)$

EXAMPLE 3   Simplify each expression. Write answers using positive exponents.

a. $(2x^0 y^{-3})^{-2}$ **b.** $\left(\dfrac{x^{-5}}{x^{-2}}\right)^{-3}$ **c.** $\left(\dfrac{2}{7}\right)^{-2}$ **d.** $\dfrac{5^{-2}x^{-3}y^{11}}{x^2 y^{-5}}$

Solution: **a.** $(2x^0 y^{-3})^{-2} = 2^{-2}(x^0)^{-2}(y^{-3})^{-2}$

$= 2^{-2} x^0 y^6$

$= \dfrac{1(y^6)}{2^2}$ Write x^0 as 1.

$= \dfrac{y^6}{4}$

b. $\left(\dfrac{x^{-5}}{x^{-2}}\right)^{-3} = \dfrac{(x^{-5})^{-3}}{(x^{-2})^{-3}} = \dfrac{x^{15}}{x^6} = x^{15-6} = x^9$

c. $\left(\dfrac{2}{7}\right)^{-2} = \dfrac{2^{-2}}{7^{-2}} = \dfrac{7^2}{2^2} = \dfrac{49}{4}$

d. $\dfrac{5^{-2}x^{-3}y^{11}}{x^2 y^{-5}} = \left(5^{-2}\right)\left(\dfrac{x^{-3}}{x^2}\right)\left(\dfrac{y^{11}}{y^{-5}}\right) = 5^{-2}x^{-3-2}y^{11-(-5)} = 5^{-2}x^{-5}y^{16}$

$= \dfrac{y^{16}}{5^2 x^5} = \dfrac{y^{16}}{25x^5}$ ■

EXAMPLE 4   Simplify each expression. Write answers using positive exponents.

a. $\left(\dfrac{3x^2 y}{y^{-9}z}\right)^{-2}$ **b.** $\left(\dfrac{3a^2}{2x^{-1}}\right)^3 \left(\dfrac{x^{-3}}{4a^{-2}}\right)^{-1}$

Solution: There is often more than one way to simplify exponential expressions. Here, we will simplify inside parentheses if possible before applying power rules for exponents.

a. $\left(\dfrac{3x^2y}{y^{-9}z}\right)^{-2} = \left(\dfrac{3x^2y^{10}}{z}\right)^{-2} = \dfrac{3^{-2}x^{-4}y^{-20}}{z^{-2}} = \dfrac{z^2}{3^2x^4y^{20}} = \dfrac{z^2}{9x^4y^{20}}$

b. $\left(\dfrac{3a^2}{2x^{-1}}\right)^3 \left(\dfrac{x^{-3}}{4a^{-2}}\right)^{-1} = \dfrac{27a^6}{8x^{-3}} \cdot \dfrac{x^3}{4^{-1}a^2}$

$$= \dfrac{27 \cdot 4 \cdot a^6 x^3 x^3}{8 \cdot a^2} = \dfrac{27a^4x^6}{2} \quad \blacksquare$$

EXAMPLE 5   Simplify the expression. Assume that a and b are integers and that x and y are not 0.

a. $x^{-b}(2x^b)^2$ **b.** $\dfrac{(y^{3a})^2}{y^{a-6}}$

Solution: **a.** $x^{-b}(2x^b)^2 = x^{-b}2^2x^{2b} = 4x^{-b+2b} = 4x^b$

b. $\dfrac{(y^{3a})^2}{y^{a-6}} = \dfrac{y^{2(3a)}}{y^{a-6}} = \dfrac{y^{6a}}{y^{a-6}} = y^{6a-(a-6)} = y^{6a-a+6} = y^{5a+6} \quad \blacksquare$

3

EXAMPLE 6   Use scientific notation to find each quotient in standard notation.

a. $\dfrac{1.2 \times 10^4}{3 \times 10^{-2}}$ **b.** $\dfrac{2000 \times 0.000021}{700}$

Solution: **a.** $\dfrac{1.2 \times 10^4}{3 \times 10^{-2}} = \left(\dfrac{1.2}{3}\right)\left(\dfrac{10^4}{10^{-2}}\right) = 0.4 \times 10^{4-(-2)}$

$$= 0.4 \times 10^6 = 400{,}000$$

b. $\dfrac{2000 \times 0.000021}{700} = \dfrac{(2 \times 10^3)(2.1 \times 10^{-5})}{7 \times 10^2} = \dfrac{2(2.1)}{7} \cdot \dfrac{10^3 \cdot 10^{-5}}{10^2}$

$$= 0.6 \times 10^{-4} = 0.00006 \quad \blacksquare$$

MENTAL MATH

Simplify. See Example 1.

1. $(x^4)^5$ **2.** $(5^6)^2$ **3.** $x^4 \cdot x^5$ **4.** $x^7 \cdot x^8$

5. $(y^6)^7$ **6.** $(x^3)^4$ **7.** $(z^4)^5$ **8.** $(z^3)^7$

9. $(z^{-6})^{-3}$ **10.** $(y^{-4})^{-2}$

EXERCISE SET 4.2

Simplify. Write each answer using positive exponents. See Example 1.

1. $(3^{-1})^2$ **2.** $(2^{-2})^2$ **3.** $(x^4)^{-9}$ **4.** $(y^7)^{-3}$

5. $(y)^{-5}$

6. $(z^{-1})^{10}$

Simplify. Write each answer using positive exponents. See Example 2.

7. $(3x^2y^3)^2$

8. $(4x^3yz)^2$

9. $\left(\dfrac{2x^5}{y^{-3}}\right)^4$

10. $\left(\dfrac{3a^{-4}}{b^7}\right)^3$

11. $(a^2bc^{-3})^{-6}$

12. $(6x^{-6}y^7z^0)^{-2}$

13. $\left(\dfrac{x^7y^{-3}}{z^{-4}}\right)^{-5}$

14. $\left(\dfrac{a^{-2}b^{-5}}{c^{-11}}\right)^{-6}$

Simplify. Write each answer using positive exponents. See Examples 3 and 4.

15. $\left(\dfrac{a^{-4}}{a^{-5}}\right)^{-2}$

16. $\left(\dfrac{x^{-9}}{x^{-4}}\right)^{-3}$

17. $\left(\dfrac{2ab^5}{4ab^5}\right)^{-2}$

18. $\left(\dfrac{5x^3y^4}{10x^3y^4}\right)^{-3}$

19. $\dfrac{4^{-1}x^2yz}{x^{-2}yz^3}$

20. $\dfrac{8^{-2}x^{-3}y^{11}}{x^2y^{-5}}$

Simplify the following. Assume that variables in the exponents represent nonzero integers and that all other variables are not 0. See Example 5.

21. $(x^{3a+6})^3$

22. $(x^{2b+7})^2$

23. $\dfrac{x^{4a}(x^{4a})^3}{x^{4a-2}}$

24. $\dfrac{x^{-5y+2}x^{2y}}{x}$

Use scientific notation to find each quotient. Express the quotient in standard notation without exponents. See Example 6.

25. $\dfrac{0.0069}{0.023}$

26. $\dfrac{0.00048}{0.0016}$

27. $\dfrac{18,200 \times 100}{91,000}$

28. $\dfrac{0.0003 \times 0.0024}{0.0006 \times 20}$

Simplify. Write each answer using positive exponents.

29. $(5^{-1})^3$

30. $(8^2)^{-1}$

31. $(x^7)^{-9}$

32. $(y^{-4})^5$

33. $\left(\dfrac{7}{8}\right)^3$

34. $\left(\dfrac{4}{3}\right)^2$

35. $(4x^2)^2$

36. $(-8x^3)^2$

37. $(-2^{-2}y)^3$

38. $(-4^{-6}y^{-6})^{-4}$

39. $\left(\dfrac{4^{-4}}{y^3x}\right)^{-2}$

40. $\left(\dfrac{7^{-3}}{ab^2}\right)^{-2}$

41. $\left(\dfrac{6p^6}{p^{12}}\right)^2$

42. $\left(\dfrac{4p^6}{p^9}\right)^3$

43. $(-8y^3xa^{-2})^{-3}$

44. $(-xy^0x^2a^3)^{-3}$

45. $\left(\dfrac{x^{-2}y^{-2}}{a^{-3}}\right)^{-7}$

46. $\left(\dfrac{x^{-1}y^{-2}}{5^{-3}}\right)^{-5}$

47. $\left(\dfrac{3x^5}{6x^4}\right)^4$

48. $\left(\dfrac{8^{-3}}{y^2}\right)^{-2}$

49. $\left(\dfrac{1}{4}\right)^{-3}$

50. $\left(\dfrac{1}{8}\right)^{-2}$

51. $\dfrac{(y^3)^{-4}}{y^3}$

52. $\dfrac{2(y^3)^{-3}}{y^{-3}}$

53. $\dfrac{8p^7}{4p^9}$

54. $\left(\dfrac{2x^4}{x^2}\right)^3$

55. $(4x^6y^5)^{-2}(6x^4y^3)$

56. $(5xy)^3(z^{-2})^{-3}$

57. $x^6(x^6bc)^{-6}$

58. $2(y^2b)^{-4}$

59. $\dfrac{2^{-3}x^2y^{-5}}{5^{-2}x^7y^{-1}}$

60. $\dfrac{7^{-1}a^{-3}b^5}{a^2b^{-2}}$

61. $\left(\dfrac{2x^2}{y^4}\right)^3 \cdot \left(\dfrac{2x^5}{y}\right)^{-2}$

62. $\left(\dfrac{3z^{-2}}{y}\right)^2 \cdot \left(\dfrac{9y^{-4}}{z^{-3}}\right)^{-1}$

Simplify. Assume that variables in the exponents represent nonzero integers and that all other variables are not 0.

63. $(b^{5x-2})^{2x}$

64. $(c^{2a+3})^3$

65. $\dfrac{(y^{2a})^8}{y^{a-3}}$

66. $\dfrac{(y^{4a})^7}{y^{2a-1}}$

67. $\left(\dfrac{2x^{3t}}{x^{2t-1}}\right)^{4}$

68. $\left(\dfrac{3y^{5a}}{y^{-a+1}}\right)^{2}$

Use scientific notation to find each quotient. Express the quotient in standard notation without exponents.

69. $\dfrac{6000 \times 0.006}{0.009 \times 400}$

70. $\dfrac{0.00016 \times 300}{0.064 \times 100}$

71. $\dfrac{0.00064 \times 2000}{16,000}$

72. $\dfrac{0.00072 \times 0.003}{0.00024}$

73. $\dfrac{66,000 \times 0.001}{0.002 \times 0.003}$

74. $\dfrac{0.0007 \times 11,000}{0.001 \times 0.0001}$

75. The fastest computer can add two numbers in about 10^{-8} second. Express in scientific notation how long it would take this computer to do this task 200,000 times.

76. The density D of an object is equivalent to the quotient of its mass M and volume V. Thus $D = \dfrac{M}{V}$. Express in scientific notation the density of an object whose mass is 500,000 pounds and whose volume is 250 cubic feet.

77. Each side of the cube shown is $\dfrac{2x^{-2}}{y}$ meters. Find its volume.

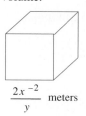

$\dfrac{2x^{-2}}{y}$ meters

78. The lot shown is in the shape of a parallelogram with base $\dfrac{3x^{-1}}{y^{-3}}$ feet and height $5x^{-7}$ feet. Find its area.

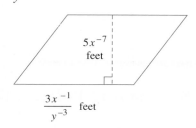

$5x^{-7}$ feet

$\dfrac{3x^{-1}}{y^{-3}}$ feet

A Look Ahead

Simplify each expression. Write answers using positive exponents. See the following example.

EXAMPLE   Simplify. Write using only positive exponents.

$$\frac{(3x^{2}yz)^{-2}(y^{-2}z^{-5})^{3}}{(x^{7}y^{-2})^{-3}(2^{-1}x^{-7})}$$

Solution: $\dfrac{(3x^{2}yz)^{-2}(y^{-2}z^{-5})^{3}}{(x^{7}y^{-2})^{-3}(2^{-1}x^{-7})} = \dfrac{3^{-2}x^{-4}y^{-2}z^{-2}\cdot y^{-6}z^{-15}}{x^{-21}y^{6}\cdot2^{-1}x^{-7}}$

$= \dfrac{3^{-2}x^{-4}y^{-8}z^{-17}}{2^{-1}x^{-28}y^{6}}$

$= \dfrac{2x^{24}y^{-14}z^{-17}}{3^{2}}$

$= \dfrac{2x^{24}}{9y^{14}z^{17}}$ ∎

79. $\left(\dfrac{27x^{-5}x^{5}}{18x^{-6}y^{2}}\right) \cdot \left(\dfrac{x^{4}y^{-1}}{x^{-2}y^{3}}\right)^{2}$

80. $\left(\dfrac{2x^{-3}y^{2}}{4xy^{-1}}\right)^{-2} \cdot \left(\dfrac{x^{-2}y^{3}}{x^{4}y^{-1}}\right)^{2}$

81. $\dfrac{(3x^{2}yz)^{2}(xy)^{-2}}{(x^{2}yz)^{-3}(3^{-1}x^{-7})}$

82. $\dfrac{(5x^2y^3)^2(8xy)^6}{4x^3(2x^3)^3}$

83. $\left(\dfrac{3^3x^0y^{-2}}{2^3x^3y^{-5}}\right)^{-1} \cdot \left(\dfrac{3^3x^{-1}y}{2^2x^2y^{-2}}\right)^2$

84. $\left(\dfrac{2^2x^2y^0}{8x^{-1}}\right)^{-2} \cdot \left(\dfrac{x^{-3}}{x^{-5}}\right)^3$

Skill Review

Simplify each expression. See Section 1.4.

85. $2 - (8^2 - 6)$

86. $7 - (4^2 - 3)$

See Section 3.4.

87. Find the equation of the line with slope 5 that contains the point $(3, -5)$.

88. Find the equation of the line that contains the points $(-2, 4)$ and $(0, 6)$.

4.3
Adding and Subtracting Polynomials

OBJECTIVES		
	1	Define term, constant, polynomial, monomial, binomial, and trinomial.
	2	Identify the degree of a term and of a polynomial.
	3	Add polynomials.
	4	Subtract polynomials.
	5	Use function notation for polynomials.

Tape 10

---

1   Recall from Section 2.1 that a **term** of an algebraic expression is a number or the product of a number and one or more variables raised to powers. The **numerical coefficient,** or simply the **coefficient,** is the numerical factor of a term.

Term	*Numerical Coefficient*
$-12x^5$	-12
x^3y	1
$-z$	-1
2	2

If a term contains only a number, it is called a **constant term,** or simply a **constant.**

A **polynomial** is a finite sum of terms in which all variables have exponents raised to nonnegative integer powers and no variables appear in the denominator.

Polynomials	*Not Polynomials*	
$4x^{5y} + 7xz$	$5x^{-3} + 2x$	(negative integer exponent)
$-5x^3 + 2x + \dfrac{2}{3}$	$\dfrac{6}{x^2} - 5x + 1$	(variable in denominator)

A polynomial that contains only one variable is called a **polynomial in one variable.** For example, $3x^2 - 2x + 7$ is a **polynomial in x.** This polynomial in x is written in **descending order** since the terms are listed in descending order of the variable's exponents. (The term 7 can be thought of as $7x^0$.) The following examples are polynomials in one variable written in **descending order.**

$$4x^3 - 7x^2 + 5, \qquad y^2 - 4, \qquad 8a^4 - 7a^2 + 4a$$

A **monomial** is a polynomial consisting of one term. A **binomial** is a polynomial consisting of two terms. A **trinomial** is a polynomial consisting of three terms.

Monomials	*Binomials*	*Trinomials*
ax^2	$x + y$	$x^2 + 4xy + y^2$
$-3x$	$6y^2 - 2$	$-x^4 + 3x^3 + 1$
4	$\frac{5}{7}z^3 - 2z$	$8y^2 - 2y - 10$

By definition, all monomials, binomials, and trinomials are also polynomials.

2   Each term of a polynomial has a **degree.**

Degree of a Term

The **degree of a term** is the sum of the exponents on the variables contained in the term.

EXAMPLE 1   Find the degree of each term.
a. $3x^2$ **b.** -2^3x^5 **c.** y **d.** $12x^2yz^3$ **e.** 5

Solution: **a.** The exponent on x is 2, so the degree of the term is 2.
b. The exponent on x is 5, so the degree of the term is 5.
c. The degree of y or y^1 is 1.
d. The degree is the sum of the exponents on the variables, or $2 + 1 + 3 = 6$.
e. The degree of 5, which can be written as $5x^0$, is 0.   ∎

From the preceding example, we can say that the degree of a constant is 0. Also, the term 0 has no degree.
    Each polynomial also has a degree.

Degree of a Polynomial

The **degree of a polynomial** is the largest degree of all its terms.

EXAMPLE 2   Find the degree of each polynomial and indicate whether the polynomial is a monomial, binomial, trinomial, or none of these.
a. $7x^3 - 3x + 2$ **b.** $-xyz$ **c.** $x^2 - 4$ **d.** $2xy + x^2y^2 - 5x^2 - 6$

Solution:

a. The degree of the trinomial $7x^3 - 3x + 2$ is 3, the largest degree of any of its terms.

b. The degree of the monomial $-xyz$ or $-x^1y^1z^1$ is 3.

c. The degree of the binomial $x^2 - 4$ is 2.

d. The degree of each term of the polynomial $2xy + x^2y^2 - 5x^2 - 6$ is:

Term	Degree
$2xy$ or $2x^1y^1$	$1 + 1 = 2$
x^2y^2	$2 + 2 = 4$
$-5x^2$	2
-6	0

The highest degree of any term is 4, so the degree of this polynomial is 4. ∎

3   To add and subtract polynomials, we **combine like terms.** Terms are considered to be **like terms** if they contain exactly the same variables raised to exactly the same powers.

Like Terms	*Unlike Terms*
$-5x^2, -x^2$	$4x^2, 3x$
$7xy^3z, -2xzy^3$	$12x^2y^3, -2xy^3$

We can **combine like terms** by using the distributive property. For example, by the distributive property,

$$5x + 7x = (5 + 7)x = 12x$$

EXAMPLE 3   Combine like terms.

a. $-12x^2 + 7x^2 - 6x$ **b.** $3xy - 2x + 5xy - x$

Solution: By the distributive property, we have

a. $-12x^2 + 7x^2 - 6x = (-12 + 7)x^2 - 6x = -5x^2 - 6x$

b. Use the associative and commutative properties to group together like terms; then combine.

$$3xy - 2x + 5xy - x = 3xy + 5xy - 2x - x$$
$$= (3 + 5)xy + (-2 - 1)x$$
$$= 8xy - 3x \quad ∎$$

Now we have the necessary skills to add polynomials.

To Add Polynomials

Combine all like terms.

EXAMPLE 4   Add.

a. $(7x^3y - xy^3 + 11) + (6x^3y - 4)$ **b.** $(3a^3 - b + 2a - 5) + (a + b + 5)$

Solution: **a.** To add, remove the parentheses and group like terms.

$$(7x^3y - xy^3 + 11) + (6x^3y - 4)$$

$$= 7x^3y - xy^3 + 11 + 6x^3y - 4$$

$$= 7x^3y + 6x^3y - xy^3 + 11 - 4 \qquad \text{Group like terms.}$$

$$= 13x^3y - xy^3 + 7 \qquad \text{Combine like terms.}$$

b. $(3a^3 - b + 2a - 5) + (a + b + 5)$

$$= 3a^3 - b + 2a - 5 + a + b + 5$$

$$= 3a^3 - b + b + 2a + a - 5 + 5 \qquad \text{Group like terms.}$$

$$= 3a^3 + 3a \qquad \text{Combine like terms.}$$

$\blacksquare$

EXAMPLE 5   Add $11x^3 - 12x^2 + x - 3$ and $x^3 - 10x + 5$.

Solution: $(11x^3 - 12x^2 + x - 3) + (x^3 - 10x + 5)$

$$= 11x^3 + x^3 - 12x^2 + x - 10x - 3 + 5 \qquad \text{Group like terms.}$$

$$= 12x^3 - 12x^2 - 9x + 2 \qquad \text{Combine like terms.} \quad \blacksquare$$

Sometimes it is more convenient to add polynomials vertically. To do this, line up like terms underneath one another and add like terms.

EXAMPLE 6   Add $11x^3 - 12x^2 + x - 3$ and $x^3 - 10x + 5$ vertically.

Solution:

$$
\begin{array}{l}
11x^3 - 12x^2 + x - 3 \\
\underline{x^3 - 10x + 5} \qquad \text{Line up like terms.} \\
12x^3 - 12x^2 - 9x + 2 \qquad \text{Combine like terms.}
\end{array}
$$

Notice that this example is the same as Example 5, only here we added vertically.

$\blacksquare$

4   The definition of subtraction of real numbers can be extended to apply to polynomials. To subtract a number, we add its opposite:

$$a - b = a + (-b)$$

To subtract a polynomial, we add its opposite. In other words, if P and Q are polynomials, then

$$P - Q = P + (-Q)$$

The polynomial $-Q$ is the **opposite** or **additive inverse** of the polynomial Q. We can find $-Q$ by changing the sign of each term of Q.

To Subtract Polynomials

To subtract polynomials, change the signs of the terms of the second polynomial; then add.

For example,

To subtract, change the signs; then

Add

$$(3x^2 + 4x - 7) - (3x^2 - 2x - 5) = (3x^2 + 4x - 7) + (-3x^2 + 2x + 5)$$
$$= 3x^2 + 4x - 7 - 3x^2 + 2x + 5$$
$$= 6x - 2 \qquad \text{Combine like terms.}$$

EXAMPLE 7 Subtract $(12z^5 - 12z^3 + z) - (-3z^4 + z^3 + 12z)$.

Solution: To subtract, change the sign of each term of the second polynomial and add the result to the first polynomial.

$$(12z^5 - 12z^3 + z) - (-3z^4 + z^3 + 12z)$$
$$= 12z^5 - 12z^3 + z + 3z^4 - z^3 - 12z \qquad \text{Change signs and add.}$$
$$= 12z^5 + 3z^4 - 12z^3 - z^3 + z - 12z \qquad \text{Group like terms.}$$
$$= 12z^5 + 3z^4 - 13z^3 - 11z \qquad \text{Combine like terms.}$$

EXAMPLE 8 Subtract $4x^3y^2 - 3x^2y^2 + 2y^2$ from $10x^3y^2 - 7x^2y^2$.

Solution: If we subtract 2 from 8, the difference is $8 - 2 = 6$. Notice the order of the numbers, and then write "Subtract $4x^3y^2 - 3x^2y^2 + 2y^2$ from $10x^3y^2 - 7x^2y^2$" as a mathematical expression.

$$(10x^3y^2 - 7x^2y^2) - (4x^3y^2 - 3x^2y^2 + 2y^2)$$
$$= 10x^3y^2 - 7x^2y^2 - 4x^3y^2 + 3x^2y^2 - 2y^2 \qquad \text{Remove parentheses.}$$
$$= 6x^3y^2 - 4x^2y^2 - 2y^2 \qquad \text{Combine like terms.}$$

EXAMPLE 9 Perform the subtraction $(10x^3y^2 - 7x^2y^2) - (4x^3y^2 - 3x^2y^2 + 2y^2)$ vertically.

Solution: Add the opposite of the second polynomial.

$$10x^3y^2 - 7x^2y^2$$
$$-(4x^3y^2 - 3x^2y^2 + 2y^2)$$

is equivalent to

$$10x^3y^2 - 7x^2y^2$$
$$-4x^3y^2 + 3x^2y^2 - 2y^2$$
$$6x^3y^2 - 4x^2y^2 - 2y^2$$

5 At times it is convenient to represent polynomials using **function notation** such as $P(x)$ (read as "P of x"). The symbol $P(x)$ represents a particular polynomial function in x. We may write, for example, $P(x)$ to represent the polynomial $3x^2 - 5x + 6$, or $R(x)$ to represent the polynomial $9x^5 + 7$. In symbols, this is

$$P(x) = 3x^2 - 5x + 6, \qquad R(x) = 9x^5 + 7$$

Sometimes we want to evaluate a polynomial at a given value. For example, we write $P(2)$ to represent the value of $P(x)$ when x is replaced by 2. Thus, if

$$P(x) = 3x^2 - 5x + 6$$

then

$$P(2) = 3(2)^2 - 5(2) + 6$$
$$= 3(4) - 10 + 6 = 8$$

Then $P(2) = 8$.

HELPFUL HINT

The symbol $P(x)$ **does not mean** P times x.

EXAMPLE 10   If $P(x) = 7x^2 - 3x + 1$ and $Q(x) = 3x - 2$, find the following.
       a. $P(1)$ **b.** $Q(1)$ **c.** $P(-2)$ **d.** $Q(0)$

Solution: **a.** Substitute 1 for x in $P(x) = 7x^2 - 3x + 1$ and simplify.

$$P(x) = 7x^2 - 3x + 1$$
$$P(1) = 7(1)^2 - 3(1) + 1 = 5$$

b. $Q(x) = 3x - 2$

$$Q(1) = 3(1) - 2 = 1$$

c. $P(x) = 7x^2 - 3x + 1$

$$P(-2) = 7(-2)^2 - 3(-2) + 1 = 35$$

d. $Q(x) = 3x - 2$

$$Q(0) = 3(0) - 2 = -2$$ ■

EXERCISE SET 4.3

Find the degree of each term. See Example 1.

1. 4

2. 7

3. $5x^2$

4. $-z^3$

5. $-3xy^2$

6. $12x^2z$

Find the degree of each polynomial and indicate whether the polynomial is a monomial, binomial, trinomial, or none of these. See Example 2.

7. $6x + 3$

8. $7x - 8$

9. $3x^2 - 2x + 5$

10. $5x^2 - 3x^2y - 2x^3$

11. $-xyz$

12. -9

13. $x^2y - 4xy^2 + 5x + y$

14. $-2x^2y - 3y^2 + 4x + y^5$

Simplify by combining like terms. See Example 3.

15. $5y + y$

16. $-x + 3x$

17. $4x + 7x - 3$

18. $-8y + 9y + 4y^2$

19. $-9xy + 7xy + 9x^2y$

20. $xyz + 2xy - xy$

21. $4xy + 2x - 3xy - 1$

22. $-8xy^2 + 4x - x + 2xy^2$

Add. See Examples 4, 5, and 6.

23. $(9y^2 - 8) + (9y^2 - 9)$

24. $(x^2 + 4x - 7) + (8x^2 + 9x - 7)$

25. $(x^2 + xy - y^2)$ and $(2x^2 - 4xy + 7y^2)$

26. $(4x^3 - 6x^2 + 5x + 7)$ and $(2x^2 + 6x - 3)$

27. $x^2 - 6x + 3$
$\underline{+ \quad (2x + 5)}$

28. $-2x^2 + 3x - 9$
$\underline{+ \qquad (2x - 3)}$

Subtract. See Examples 7, 8, and 9.

29. $(9y^2 - 7y + 5) - (8y^2 - 7y + 2)$

30. $(2x^2 + 3x + 12) - (5x - 7)$

31. $(6x^2 - 3x)$ from $(4x^2 + 2x)$

32. $(xy + x - y)$ from $(xy + x - 3)$

33. $\quad 3x^2 - 4x + 8$
$\underline{- \qquad (5x^2 - 7)}$

34. $-3x^2 - 4x + 8$
$\underline{- \qquad (5x + 12)}$

If $P(x) = x^2$ and $Q(x) = 5x - 1$, find the following. See Example 10.

35. $P(7)$

36. $Q(4)$

37. $Q(-10)$

38. $P(-4)$

39. $P(0)$

40. $Q(0)$

Perform the indicated operations.

41. $(5x - 11) + (-x - 2)$

42. $(3x^2 - 2x) + (5x^2 - 9x)$

43. $(7x^2 + x + 1) - (6x^2 + x - 1)$

44. $(4x - 4) - (-x - 4)$

45. $(7x^3 - 4x + 8) + (5x^3 + 4x + 8x)$

46. $(9xyz + 4x - y) + (-9xyz - 3x + y + 2)$

47. $(9x^3 - 2x^2 + 4x - 7) - (2x^3 - 6x^2 - 4x + 3)$

48. $(3x^2 + 6xy + 3y^2) - (8x^2 - 6xy - y^2)$

49. Add $(y^2 + 4y + 7)$ and $(-19y^2 + 7y + 7)$

50. Subtract $(x - 4)$ from $(3x^2 - 4x + 5)$

51. $(3x^3 - b + 2a - 6) + (-4x^3 + b + 6a - 6)$

52. $(5x^2 - 6) + (2x^2 - 4x + 8)$

53. $(4x^2 - 6x + 2) - (-x^2 + 3x + 5)$

54. $(5x^2 + x + 9) - (2x^2 - 9)$

55. $(-3x + 8) + (-3x^2 + 3x - 5)$

56. $(5y^2 - 2y + 4) + (3y + 7)$

57. $(-3 + 4x^2 + 7xy) + (2x^3 - x^2 + xy)$

58. $(-3xy + 4) - (-7xy - 8y)$

59. $\quad 6y^2 - 6y + 4$
$\underline{- \quad (-y^2 - 6y + 7)}$

60. $\quad -4x^3 + 4x^2 - 4x$
$\underline{- \quad (2x^3 - 2x^2 + 3x)}$

61. $\quad 3x^2 + 15x + \;\; 8$
$\underline{+ (2x^2 + \;\; 7x + \;\; 8)}$

62. $\quad 9x^2 + 9x - 4$
$\underline{+ (7x^2 - 3x - 4)}$

63. Find the sum of $(5q^4 - 2q^2 - 3q)$ and $(-6q^4 + 3q^2 + 5)$.

64. Find the sum of $(5y^4 - 7y^2 + x^2 - 3)$ and $(-3y^4 + 2y^2 + 4)$.

65. Subtract $(3x + 7)$ from the sum of $(7x^2 + 4x + 9)$ and $(8x^2 + 7x - 8)$.

66. Subtract $(9x + 8)$ from the sum of $(3x^2 - 2x - x^3 + 2)$ and $(5x^2 - 8x - x^3 + 4)$.

67. Find the sum of $(4x^4 - 7x^2 + 3)$ and $(2 - 3x^4)$.

68. Find the sum of $(8x^4 - 14x^2 + 6)$ and $(-12x^6 - 21x^4 - 9x^2)$.

If $P(x) = 3x + 3$, $Q(x) = 4x^2 - 6x + 3$, and $R(x) = 5x^2 - 7$, find the following.

69. $P(4)$

70. $Q(-1)$

71. $R(-3)$

72. $P(0)$

73. $Q(2)$

74. $R(1)$

75. $Q(0)$

76. $R(-2)$

77. A piece of cable $3x^2 + 2x - 5$ centimeters is cut from a cable of length $7x^2 - 4$ centimeters. Express the length of the remaining piece of cable as a polynomial in x.

$$\longleftarrow \qquad 7x^2 - 4 \text{ centimeters} \longrightarrow$$

$3x^2 + 2x - 5$?
$\leftarrow$ centimeters $\rightarrow$

78. Boards are to be placed around the border of a proposed driveway (see figure) before cement can be poured. To determine the number of feet of board strips to order, find the perimeter of the proposed driveway.

79. The polynomial function $P(x) = 45x - 100{,}000$ models the relationship between the number of lamps that Sherry's Lamp Shop sells and the profit the shop makes. If x represents the number of lamps sold, $45x - 100{,}000$ represents the profit from selling x lamps. Find $P(4000)$, the profit from selling 4000 lamps.

80. The polynomial $P(t) = -32t + 500$ models the relationship between the length of time a particle flies through space, beginning at a velocity of 500 feet per second, and its accrued velocity. If t represents seconds, then $-32t + 500$ represents accrued velocity. Find $P(3)$, the accrued speed after 3 seconds.

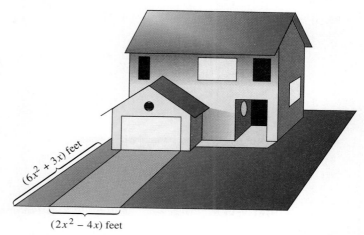

$(6x^2 + 3x)$ feet

$(2x^2 - 4x)$ feet

A Look Ahead

If $P(x)$ is the polynomial given, find $P(a)$, $P(-x)$, and $P(x + h)$. See the following example.

EXAMPLE   If $P(x) = 2x^2 - 3x + 5$, find the following:
a. $P(a)$ **b.** $P(-x)$

Solution: **a.** $P(x) = 2x^2 - 3x + 5$
$P(a) = 2a^2 - 3a + 5$

b. $P(x) = 2x^2 - 3x + 5$
$P(-x) = 2(-x)^2 - 3(-x) + 5$
$= 2x^2 + 3x + 5$ ∎

81. $P(x) = 2x - 3$ **82.** $P(x) = 8x + 3$ **83.** $P(x) = 3x^2 + 4x$ **84.** $P(x) = 9x^2 - 4$
85. $P(x) = 4x - 1$ **86.** $P(x) = 3x - 2$

Writing in Mathematics

87. Describe how to find the degree of a polynomial.

88. Explain why xyz is a monomial while $x + y + z$ is a trinomial.

Skill Review

Multiply. See Section 4.1.

89. $(x^2y)(3x)$

90. $(6x^4y^6)(9x^2y^3)$

91. $(x^2y^2)(-16xy)$

92. $(21x^2y)(-2xy)$

Graph each linear equation. See Section 3.2.

93. $x - y = 4$

94. $-x + 2y = 3$

95. $2x + y = 6$

96. $3x - 4y = 12$

4.4
Multiplying Polynomials

OBJECTIVES

Tape 10

1   Multiply two polynomials.

2   Multiply binomials.

3   Square a binomial.

4   Multiply the sum and difference of two terms.

1   Properties of real numbers and exponents are used continually in the process of multiplying polynomials. To multiply monomials, for example, we apply the commutative and associative properties of real numbers and the product rule for exponents.

EXAMPLE 1   Multiply:

     a. $(2x^3)(5x^6)$ **b.** $(7y^4z^4)(-xy^{11}z^5)$

Solution: Group like bases and apply the product rule for exponents.

     a. $(2x^3)(5x^6) = 2(5)(x^3)(x^6) = 10x^9$

     b. $(7y^4z^4)(-xy^{11}z^5) = 7(-1)x(y^4y^{11})(z^4z^5) = -7xy^{15}z^9$ ∎

To multiply a monomial by a polynomial other than a monomial, we use an expanded form of the distributive property:

$$a(b + c + d + \cdots + z) = ab + ac + ad + \cdots + az$$

Notice that the monomial a is multiplied by each term of the polynomial.

EXAMPLE 2   Find the following products.

     a. $2x(5x - 4)$ **b.** $-3x^2(4x^2 - 6x + 1)$ **c.** $-xy(7x^2y + 3xy - 11)$

Solution: Apply the distributive property.

     a. $2x \ (5x - 4) = \ 2x \ (5x) + \ 2x \ (-4)$

                 $= 10x^2 - 8x$

b. $-3x^2(4x^2 - 6x + 1) = -3x^2(4x^2) + (-3x^2)(-6x) + (-3x^2)(1)$

$$= -12x^4 + 18x^3 - 3x^2$$

c. $-xy(7x^2y + 3xy - 11) = -xy(7x^2y) + (-xy)(3xy) + (-xy)(-11)$

$$= -7x^3y^2 - 3x^2y^2 + 11xy \quad \blacksquare$$

To multiply any polynomial by a polynomial, we again use the distributive property, multiplying each term of one polynomial by each term of the other polynomial.

EXAMPLE 3   Multiply and simplify the product if possible.

a. $(x + 3)(2x + 5)$ **b.** $(2x^3 - 3)(5x^2 - 6x + 7)$

Solution: **a.** Multiply each term of $(x + 3)$ by $(2x + 5)$.

$$(x + 3)(2x + 5) = x(2x + 5) + 3(2x + 5) \qquad \text{Apply the distributive property.}$$

$$= 2x^2 + 5x + 6x + 15 \qquad \text{Apply the distributive property again.}$$

$$= 2x^2 + 11x + 15 \qquad \text{Combine like terms.}$$

b. Multiply each term of $(2x^3 - 3)$ by each term of $(5x^2 - 6x + 7)$.

$$(2x^3 - 3)(5x^2 - 6x + 7) = 2x^3(5x^2 - 6x + 7) + (-3)(5x^2 - 6x + 7)$$

$$= 10x^5 - 12x^4 + 14x^3 - 15x^2 + 18x - 21 \quad \blacksquare$$

Sometimes polynomials are easier to multiply vertically, in the same way we multiply real numbers. When multiplying vertically, line up like terms in the **partial products** vertically. This makes combining like terms easier.

EXAMPLE 4   Find the product of $(4x^2 + 7)$ and $(x^2 + 2x + 8)$.

Solution:

$$
\begin{array}{r}
x^2 + 2x + 8 \\
4x^2 + 7 \\
\hline
7x^2 + 14x + 56 \\
4x^4 + 8x^3 + 32x^2 \\
\hline
4x^4 + 8x^3 + 39x^2 + 14x + 56
\end{array}
$$

$7(x^2 + 2x + 8)$.
$4x^2(x^2 + 2x + 8)$.
Combine like terms. $\blacksquare$

EXAMPLE 5   Multiply $(x + 3)$ by $(2x + 5)$ vertically.

Solution:

$$
\begin{array}{r}
x + 3 \\
2x + 5 \\
\hline
5x + 15 \\
2x^2 + 6x \\
\hline
2x^2 + 11x + 15
\end{array}
$$

$5(x + 3)$.
$2x(x + 3)$.
Combine like terms. $\blacksquare$

2 When multiplying a binomial by a binomial, a special order of multiplying terms, called the **FOIL** order, may be used. The letters of FOIL stand for "**F**irst–**O**uter–**I**nner–**L**ast." To illustrate this method, multiply $(2x - 3)$ by $(3x + 1)$.

Multiply the **F**irst terms of each binomial.

$$(2x - 3)(3x + 1) \qquad \begin{matrix} \mathbf{F} \\ 2x(3x) = 6x^2 \end{matrix}$$

Multiply the **O**uter terms of each binomial.

$$(2x - 3)(3x + 1) \qquad \begin{matrix} \mathbf{O} \\ 2x(1) = 2x \end{matrix}$$

Multiply the **I**nner terms of each binomial.

$$(2x - 3)(3x + 1) \qquad \begin{matrix} \mathbf{I} \\ -3(3x) = -9x \end{matrix}$$

Multiply the **L**ast terms of each binomial.

$$(2x - 3)(3x + 1) \qquad \begin{matrix} \mathbf{L} \\ -3(1) = -3 \end{matrix}$$

Combine like terms.

$$6x^2 + 2x - 9x - 3 = 6x^2 - 7x - 3$$

EXAMPLE 6 Multiply $(x - 1)(x + 2)$. Use the FOIL order.

Solution:

$$\begin{matrix} & \text{First} & \text{Outer} & \text{Inner} & \text{Last} \\ & \downarrow & \downarrow & \downarrow & \downarrow \\ (x - 1)(x + 2) = & x \cdot x & + \ 2 \cdot x & + \ (-1)x & + \ (-1)(2) \end{matrix}$$

$$= x^2 + 2x - x - 2$$
$$= x^2 + x - 2 \qquad \text{Combine like terms.} \qquad \blacksquare$$

EXAMPLE 7 Multiply $(2x - 7)(3x - 4)$.

Solution:

$$\begin{matrix} & \text{First} & \text{Outer} & \text{Inner} & \text{Last} \\ & \downarrow & \downarrow & \downarrow & \downarrow \\ (2x - 7)(3x - 4) = & 2x(3x) & + \ 2x(-4) & + \ (-7)(3x) & + \ (-7)(-4) \end{matrix}$$

$$= 6x^2 - 8x - 21x + 28$$
$$= 6x^2 - 29x + 28 \qquad \blacksquare$$

3 The **square of a binomial** is a special case of the product of two binomials. Find $(a + b)^2$ by the FOIL order for multiplying two binomials.

$$(a + b)^2 = (a + b)(a + b)$$

$$\begin{matrix} \ \ \ \mathbf{F} & \ \mathbf{O} & \ \mathbf{I} & \ \mathbf{L} \\ = a^2 & + \ ab & + \ ba & + \ b^2 \end{matrix}$$

$$= a^2 + 2ab + b^2$$

We use this result as the basis of a quick method for squaring a binomial. The method has two forms: the sum of terms and the difference of terms. We call such quick methods **special products.**

> **Square of a Binomial**
> $$(a + b)^2 = a^2 + 2ab + b^2, \qquad (a - b)^2 = a^2 - 2ab + b^2$$

A binomial squared is the sum of the first term squared, twice the product of both terms, and the second term squared.

EXAMPLE 8   Find the following products.

   a. $(x + 5)^2$ **b.** $(x - 9)^2$ **c.** $(3x + 2z)^2$ **d.** $(4m^2 - 3n)^2$

Solution: **a.** $(x + 5)^2 = x^2 + 2 \cdot x \cdot 5 + 5^2 = x^2 + 10x + 25$

   b. $(x - 9)^2 = x^2 - 2 \cdot x \cdot 9 + (9)^2 = x^2 - 18x + 81$

   c. $(3x + 2z)^2 = (3x)^2 + 2(3x)(2z) + (2z)^2 = 9x^2 + 12xz + 4z^2$

   d. $(4m^2 - 3n)^2 = (4m^2)^2 - 2(4m^2)(3n) + (3n)^2 = 16m^4 - 24m^2n + 9n^2$ ■

> **HELPFUL HINT**
>
> Note that $(a + b)^2 = a^2 + 2ab + b^2$ and **not** $a^2 + b^2$. Also, $(a - b)^2 = a^2 - 2ab + b^2$ and **not** $a^2 - b^2$.

4   Another special product applies to the sum and difference of the same two terms. Multiply $(a + b)(a - b)$.

$$(a + b)(a - b) = a^2 - ab + ba - b^2$$
$$= a^2 - b^2$$

> **Product of the Sum and Difference of Two Terms**
> $$(a + b)(a - b) = a^2 - b^2$$

The product of the sum and difference of the same two terms is the difference of the first term squared and the second term squared.

EXAMPLE 9   Find the following products.

   a. $(x - 3)(x + 3)$ **b.** $(4y + 1)(4y - 1)$ **c.** $(x^2 + 2y)(x^2 - 2y)$

Solution: **a.** $(x - 3)(x + 3) = x^2 - 3^2 = x^2 - 9$

   b. $(4y + 1)(4y - 1) = (4y)^2 - 1^2 = 16y^2 - 1$

   c. $(x^2 + 2y)(x^2 - 2y) = (x^2)^2 - (2y)^2 = x^4 - 4y^2$ ■

EXAMPLE 10   Multiply $[3 + (2a + b)]^2$.

Solution: Think of 3 as the first term and $(2a + b)$ as the second term, and apply the method for squaring a binomial.

$$[3 + (2a + b)]^2 = \underset{\substack{\text{First term}\\\text{squared}}}{(3)^2} + \underset{\substack{\text{Twice the}\\\text{product of}\\\text{both terms}}}{2(3)(2a + b)} + \underset{\substack{\text{Last term}\\\text{squared}}}{(2a + b)^2}$$

$$= 9 + 6(2a + b) + (2a + b)^2$$

$$= 9 + 12a + 6b + (2a)^2 + 2(2a)(b) + b^2 \qquad \text{Square } (2a + b).$$

$$= 9 + 12a + 6b + 4a^2 + 4ab + b^2 \qquad ■$$

EXAMPLE 11   Multiply $[(5x - 2y) - 1][(5x - 2y) + 1]$.

Solution: Think of $(5x - 2y)$ as the first term and 1 as the second term, and apply the method for the product of the sum and difference of two terms.

$$[(5x - 2y) - 1][(5x - 2y) + 1] = \underset{\substack{\text{First term}\\\text{squared}}}{(5x - 2y)^2} - \underset{\substack{\text{Second term}\\\text{squared}}}{1^2}$$

$$= (5x - 2y)^2 - 1$$

$$= (5x)^2 - 2(5x)(2y) + (2y)^2 - 1 \qquad \substack{\text{Square}\\(5x - 2y).}$$

$$= 25x^2 - 20xy + 4y^2 - 1 \qquad ■$$

CALCULATOR BOX

Evaluate Polynomials:

The process of evaluating polynomials using given replacement values can be simplified by using a calculator. For example, to evaluate the polynomial $x^5 - 4x^3 + 2x$ when $x = 1.5$, press the keys

$$\boxed{1.5}\ \boxed{y^x}\ \boxed{5}\ \boxed{-}\ \boxed{4}\ \boxed{\times}\ \boxed{1.5}\ \boxed{y^x}\ \boxed{3}\ \boxed{+}\ \boxed{2}\ \boxed{\times}\ \boxed{1.5}\ \boxed{=}$$

The display should read $\boxed{-2.90625}$.

Evaluate each polynomial using the given replacement values.

1. $x^4 - 18$ when $x = \dfrac{1}{5}$

2. $3x^2 - 5x + 13$ when $x = 3.4$

3. $x^5 - x^3 - x$ when $x = 7$

4. $5y^3 - 5x^3$ when $x = 9.1$ and $y = 2.7$

5. $x^2y^2 - 4x + 2y + 6$ when $x = 5.4$ and $y = 3.1$ (round your answer to the nearest thousandths)

6. $2x^3 + x^2y + xy^2 + 2y^3$ when $x = -4$ and $y = 11$

EXERCISE SET 4.4

Multiply. See Example 1.

1. $(-4x^3)(3x^2)$

3. $(-xyz)(-9xy^2z^2)$

2. $(-6a)(4a)$

4. $(-4yt^2)(6yt^3z)$

Multiply. See Example 2.

5. $3x(4x + 7)$

7. $-6xy(4x + y)$

9. $-4ab(xa^2 + ya^2 - 3)$

6. $5x(6x - 4)$

8. $-8y(6xy + 4x)$

10. $-6b^2z(z^2a + baz - 3b)$

Multiply. See Example 3.

11. $(x - 3)(2x + 4)$

13. $(2x + 3)(x^3 - x + 2)$

15. $(x^2 + 2x - 1)^2$

12. $(y + 5)(3y - 2)$

14. $(a + 2)(3a^2 - a + 5)$

16. $(2x^2 - 3x + 2)^2$

Multiply vertically. See Examples 4 and 5.

17.
$$\begin{array}{r} 3x + 2 \\ x + 4 \\ \hline \end{array}$$

18.
$$\begin{array}{r} 2s - 3 \\ s + 2 \\ \hline \end{array}$$

19.
$$\begin{array}{r} 2x^2 - 4x + 2 \\ x - 5 \\ \hline \end{array}$$

20.
$$\begin{array}{r} -3b^2 + 2b - 4 \\ b - 3 \\ \hline \end{array}$$

21.
$$\begin{array}{r} a^2 - 4a - 3 \\ a^2 - 2a - 5 \\ \hline \end{array}$$

22.
$$\begin{array}{r} 3b^2 + 2b - 2 \\ -3b^2 + b - 4 \\ \hline \end{array}$$

23.
$$\begin{array}{r} 2a^2 + ab + b^2 \\ 3a^2 - ab + b^2 \\ \hline \end{array}$$

24.
$$\begin{array}{r} a^2 - 4ab + 3b^2 \\ 4a^2 + 4ab - 5 \\ \hline \end{array}$$

Multiply the binomials. See Examples 6 and 7.

25. $(x - 3)(x + 4)$

27. $(2x - 8)(2x - 4)$

29. $(3x - 1)(x + 3)$

31. $\left(3x + \dfrac{1}{2}\right)\left(3x - \dfrac{1}{2}\right)$

26. $(c - 3)(c + 1)$

28. $(3n - 9)(n + 7)$

30. $(5d - 3)(d + 6)$

32. $\left(2x - \dfrac{1}{3}\right)\left(2x + \dfrac{1}{3}\right)$

Multiply using special product methods. See Examples 8 and 9.

33. $(x + 4)^2$

34. $(x - 5)^2$

35. $(6y - 1)(6y + 1)$

36. $(x - 9)(x + 9)$

37. $(3x - y)^2$

38. $(4x - z)^2$

39. $(3b - 6y)(3b + 6y)$

40. $(2x - 4y)(2x + 4y)$

Multiply using special product methods. See Examples 10 and 11.

41. $[3 + (4b + 1)]^2$

42. $[5 - (3b - 3)]^2$

43. $[(2s - 3) - 1][(2s - 3) + 1]$

44. $[(2y + 5) + 6][(2y + 5) - 6]$

45. $[(xy + 4) - 6]^2$

46. $[(2a^2 + 4a) + 1]^2$

Multiply.

47. $(3ab)(-4b)$

50. $3x(6x^2 + 4y - 3)$

53. $(2x^3 + 5)(5x^2 + 4x + 1)$

56. $(4x + 1)(4x - 1)$

48. $(-6x^2y)(2x^3)$

51. $(3x + 1)(3x + 5)$

54. $(3y^3 - 1)(3y^3 - 6y + 1)$

57.
$$\begin{array}{r} 3x^2 + 4x - 4 \\ 3x + 6 \\ \hline \end{array}$$

49. $4y(x^2 + y + z)$

52. $(4x - 5)(5x + 6)$

55. $(7x - 3)(7x + 3)$

58.
$$\begin{array}{r} 6x^2 + 2x - 1 \\ 3x - 6 \\ \hline \end{array}$$

59. $\left(4x + \dfrac{1}{3}\right)\left(4x - \dfrac{1}{2}\right)$

60. $\left(4y - \dfrac{1}{3}\right)\left(3y - \dfrac{1}{8}\right)$

61. $(6x + 1)^2$

62. $(4x + 7)^2$

63. $(x^2 + 2y)(x^2 - 2y)$

64. $(3x + 2y)(3x - 2y)$

65. $-6a^2b^2(5a^2b^2 - 6a - 6b)$

66. $7x^2y^3(-3ax - 4xy + z)$

67. $(a - 4)(2a - 4)$

68. $(2x - 3)(x + 1)$

69. $(7ab + 3c)(7ab - 3c)$

70. $(3xy - 2b)(3xy + 2b)$

71. $[4 - (3x + 1)][4 + (3x + 1)]$

72. $[7 - (3x + 5)][7 + (3x + 5)]$

73. $(m - 4)^2$

74. $(x + 2)^2$

75. $[3 + (2y - c)]^2$

76. $[4 - (3x + 2y)]^2$

77. $[(5x - 2y) - 4][(5x - 2y) + 4]$

78. $[(4x - 2y) + 3][(4x - 2y) - 3]$

79. $(3x + 1)^2$

80. $(4x + 6)^2$

81. $(y - 4)(y - 3)$

82. $(c - 8)(c + 2)$

83. $[4 - (2x + y)][4 + (2x + y)]$

84. $[7 - (6x + b)][(7 + (6x + b)]$

85. $(x + y)(2x - 1)(x + 1)$

86. $(z + 2)(z - 3)(2z + 1)$

87. $(3x^2 + 2x - 1)^2$

88. $(4x^2 + 4x - 4)^2$

89. $(3x + 1)(4x^2 - 2x + 5)$

90. $(2x - 1)(5x^2 - x - 2)$

91. Find the volume of the refrigerator.

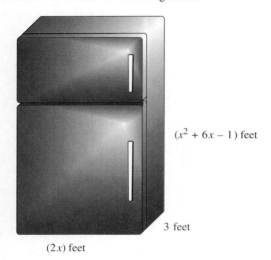

$(x^2 + 6x - 1)$ feet

3 feet

$(2x)$ feet

92. Find the area of the circle. Do not approximate π.

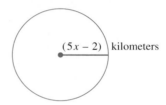

$(5x - 2)$ kilometers

A Look Ahead

Multiply. See the following example.

> **EXAMPLE** Multiply.
>
> $$(x^{-2} + y^{-1})(x^{-1} - y^3)$$
>
> **Solution:** $(x^{-2} + y^{-1})(x^{-1} - y^3) = x^{-2}x^{-1} - x^{-2}y^3 + x^{-1}y^{-1} - y^{-1}y^3$
>
> $$= x^{-3} - x^{-2}y^3 + x^{-1}y^{-1} - y^2 \quad \blacksquare$$

93. $5x^{-1}(x^{-2} + 3x^{-1} + 2)$

94. $x^{-2}(7x^2 + 2x - 6)$

95. $(x^{-2} + 2y)(x^{-2} - 2y)$

96. $(4x^{-3} + 6)(x + 1)$

97. $(3a + y^{-4})(a + y^{-2})$

98. $(6z^{-1} + 2y^{-3})(6z^{-1} - 2y^{-3})$

Multiply. Assume that all variables in the exponents represent integers and that all other variables are not 0. See the following example.

> **EXAMPLE**   Multiply.
>
> $$6a^n(3a^{2n} - 5)$$
>
> Solution: $6a^n(3a^{2n} - 5) = 6a^n(3a^{2n}) + 6a^n(-5)$
>
> $$= 18a^{3n} - 30a^n \quad \blacksquare$$

99. $6a^4(3a^{n+1} - 6)$

100. $4a^{n+1}(3a^{n-1} - 2)$

101. $(3x^y + 7)^2$

102. $(4y^{2c} - 5)(4y^{2c} + 5)$

Writing in Mathematics

103. Explain how to multiply a polynomial by a polynomial.   **104.** Explain why $(3x + 2)^2$ does not equal $9x^2 + 4$.

Skill Review

Simplify. See Sections 4.1 and 4.2.

105. $\dfrac{3x^3y^2}{12x}$

106. $\dfrac{-36xb^3}{9xb^2}$

107. $\dfrac{144x^5y^5}{-16x^2y}$

108. $\dfrac{48x^3y^2}{-4xy}$

Solve the following. See Section 2.4.

109. $|x - 5| = 9$

110. $|2y + 1| = 1$

4.5
Dividing Polynomials

OBJECTIVES	
1	Divide by a monomial.
2	Divide by a polynomial.

Tape 11

1   When dividing a **monomial** by a **monomial,** use the rules for exponents developed at the beginning of this chapter. In this section, assume that a variable in the denominator does not have a value that makes the denominator 0.

EXAMPLE 1   Simplify $\dfrac{2xz^4}{3x^5z^3}$.

Solution: $\dfrac{2xz^4}{3x^5z^3} = \dfrac{2}{3}x^{1-5}z^{4-3} = \dfrac{2}{3}x^{-4}z^1 = \dfrac{2z}{3x^4} \quad \blacksquare$

Dividing a **polynomial** by a **monomial** uses your knowledge of fractions, since a fraction, after all, is a quotient. Recall that two fractions that have a common denominator are added by adding the numerators:

$$\frac{a}{c} + \frac{b}{c} = \frac{a + b}{c}$$

If a, b, and c are monomials, we might read this equation from right to left and gain insight into dividing a polynomial by a monomial.

> **To Divide a Polynomial by a Monomial**
>
> Divide each term in the polynomial by the monomial.
>
> $$\frac{a + b}{c} = \frac{a}{c} + \frac{b}{c}, \qquad c \neq 0$$

EXAMPLE 2   Divide $10x^2 - 5x + 20$ by 5.

Solution: Divide each term of $10x^2 - 5x + 20$ by 5 and simplify.

$$\frac{10x^2 - 5x + 20}{5} = \frac{10x^2}{5} - \frac{5x}{5} + \frac{20}{5} = 2x^2 - x + 4 \qquad \blacksquare$$

To check, see that (quotient)(divisor) = dividend or

$$(2x^2 - x + 4)(5) = 10x^2 - 5x + 20$$

EXAMPLE 3   Find the quotient: $\dfrac{7a^2b - 2ab^2}{2ab^2}$.

Solution: Divide each term of the polynomial in the numerator by $2ab^2$.

$$\frac{7a^2b - 2ab^2}{2ab^2} = \frac{7a^2b}{2ab^2} - \frac{2ab^2}{2ab^2} = \frac{7a}{2b} - 1 \qquad \blacksquare$$

EXAMPLE 4   Find the quotient: $\dfrac{3x^5y^2 - 15x^3y - 6x}{6x^2y^3}$.

Solution: Divide each term in the numerator by $6x^2y^3$.

$$\frac{3x^5y^2 - 15x^3y - 6x}{6x^2y^3} = \frac{3x^5y^2}{6x^2y^3} - \frac{15x^3y}{6x^2y^3} - \frac{6x}{6x^2y^3} = \frac{x^3}{2y} - \frac{5x}{2y^2} - \frac{1}{xy^3} \qquad \blacksquare$$

2   To divide a polynomial by a polynomial other than a monomial, we use **long division.** Polynomial long division is similar to long division of real numbers. We review long division of real numbers by dividing 7 into 296.

Divisor:
$$\begin{array}{r} 42 \\ 7\overline{)296} \\ -28 \\ \hline 16 \\ -14 \\ \hline 2 \end{array}$$

$4(7) = 28$.

Subtract and bring down the next digit in the dividend.

$2(7) = 14$.

Subtract. The remainder is 2.

The quotient is $42\dfrac{2 \text{ (remainder)}}{7 \text{ (divisor)}}$. To check, notice that

$$42(7) + 2 = 296, \qquad \text{the dividend.}$$

This same division process can be applied to polynomials, as shown next.

EXAMPLE 5   Divide $2x^2 - x - 10$ by $x + 2$.

Solution: $2x^2 - x - 10$ is the dividend while $x + 2$ is the divisor.

Step 1   Divide $2x^2$ by x.

$$x + 2 \overline{)2x^2 - x - 10} \qquad \dfrac{2x^2}{x} = 2x \,,$$

with $2x$ above, so $2x$ is the first term of the quotient.

Step 2   Multiply $2x(x + 2)$.

$$
\begin{array}{r}
2x \\
x + 2 \overline{)2x^2 - x - 10} \\
2x^2 + 4x
\end{array}
\qquad 2x(x + 2)
$$

Like terms are lined up vertically.

Step 3   Subtract $(2x^2 + 4x)$ from $(2x^2 - x - 10)$ by changing the signs of $(2x^2 + 4x)$ and adding.

$$
\begin{array}{r}
2x \\
x + 2 \overline{)2x^2 - x - 10} \\
-2x^2 - 4x \\
\hline
-5x
\end{array}
$$

Step 4   Bring down the next term, -10, and start the process over.

$$
\begin{array}{r}
2x \\
x + 2 \overline{)2x^2 - x - 10} \\
-2x^2 - 4x \\
\hline
-5x - 10
\end{array}
$$

Step 5   Divide $-5x$ by x.

$$
\begin{array}{r}
2x - 5 \\
x + 2 \overline{)2x^2 - x - 10} \\
-2x^2 - 4x \\
\hline
-5x - 10
\end{array}
\qquad \dfrac{-5x}{x} = -5
$$

so -5 is the second term of the quotient.

Step 6   Multiply $-5(x + 2)$.

$$
\begin{array}{r}
2x - 5 \\
x + 2 \overline{)2x^2 - x - 10} \\
-2x^2 - 4x \\
\hline
-5x - 10 \\
-5x - 10
\end{array}
\qquad -5(x + 2)
$$

Like terms are lined up vertically.

Step 7 Subtract $(-5x - 10)$ from $(-5x - 10)$.

$$
\begin{array}{r}
2x - 5 \\
x + 2 \overline{)2x^2 - x - 10} \\
\underline{-2x^2 - 4x} \\
-5x - 10 \\
\underline{+5x + 10} \\
0
\end{array}
$$

Then $\dfrac{2x^2 - x - 10}{x + 2} = 2x - 5$. There is no remainder.

Check this result by multiplying $2x - 5$ by $x + 2$. Their product is

$$(2x - 5)(x + 2) = 2x^2 - x - 10, \text{ the dividend.} \quad \blacksquare$$

EXAMPLE 6 Find the quotient: $\dfrac{6x^2 - 19x + 12}{3x - 5}$.

Solution:

$$
\begin{array}{r}
2x \\
3x - 5 \overline{)6x^2 - 19x + 12} \\
\underline{6x^2 - 10x} \\
-9x + 12
\end{array}
$$

Divide: $\dfrac{6x^2}{3x} = \boxed{2x}$.

Multiply: $2x(3x - 5)$.
Subtract by changing the signs of $6x^2 - 10x$ and adding. Bring down $+12$.

$$
\begin{array}{r}
2x - 3 \\
3x - 5 \overline{)6x^2 - 19x + 12} \\
\underline{6x^2 - 10x} \\
-9x + 12 \\
\underline{-9x + 15} \\
-3
\end{array}
$$

Divide: $\dfrac{-9x}{3x} = \boxed{-3}$.

Multiply: $-3(3x - 5)$.
Subtract.

When checking, we call the **divisor** polynomial $3x - 5$. The **quotient** polynomial is $2x - 3$. The **remainder** polynomial is -3. See that

$$\textbf{dividend} = \textbf{divisor} \cdot \textbf{quotient} + \textbf{remainder}$$

or

$$6x^2 - 19x + 12 = (3x - 5)(2x - 3) + (-3)$$
$$= 6x^2 - 19x + 15 - 3$$
$$= 6x^2 - 19x + 12$$

The division checks, so

$$\frac{6x^2 - 19x + 12}{3x - 5} = 2x - 3 - \frac{3}{3x - 5} \quad \blacksquare$$

EXAMPLE 7 Divide $2x^3 + 3x^4 - 8x + 6$ by $x^2 - 1$.

Solution: Before dividing, we will write both the divisor and the dividend in descending order of exponents. Any "missing powers" can be represented by the product of 0 and the variable raised to the missing power. There is no x^2 term in the dividend, so include $0x^2$ to represent the missing term. Also, there is no x term in the divisor, so include $0x$ in the divisor.

$$
\begin{array}{r}
3x^2 + 2x + 3 \\
x^2 + 0x - 1\overline{)3x^4 + 2x^3 + 0x^2 - 8x + 6}
\end{array}
$$

$$\frac{3x^4}{x^2} = 3x^2.$$

$$3x^4 + 0x^3 - 3x^2$$
$$2x^3 + 3x^2 - 8x$$

$3x^2(x^2 + 0x - 1).$
Subtract. Bring down $-8x$.
$2x^3/x^2 = 2x$, a term of the quotient.

$$2x^3 + 0x^2 - 2x$$
$$3x^2 - 6x + 6$$

$2x(x^2 + 0x - 1).$
Subtract. Bring down 6.
$3x^2/x^2 = 3$, a term of the quotient.

$$3x^2 + 0x - 3$$
$$-6x + 9$$

$3(x^2 + 0x - 1).$
Subtract.

The division process is finished when the degree of the remainder polynomial is less than the degree of the divisor. Thus,

$$\frac{3x^4 + 2x^3 - 8x + 6}{x^2 - 1} = 3x^2 + 2x + 3 + \frac{-6x + 9}{x^2 - 1}$$ ■

To check, see that

$$3x^4 + 2x^3 - 8x + 6 = (x^2 - 1)(3x^2 + 2x + 3) + (-6x + 9).$$

EXAMPLE 8   Divide $27x^3 + 8$ by $2 + 3x$.

Solution: Write both the divisor and the dividend in descending order of exponents. Replace the missing terms in the dividend with $0x^2$ and $0x$.

$$
\begin{array}{r}
9x^2 - 6x + 4 \\
3x + 2\overline{)27x^3 + 0x^2 + 0x + 8}
\end{array}
$$

$$27x^3 + 18x^2$$
$$-18x^2 + 0x$$

$9x^2(3x + 2).$
Subtract. Bring down $0x$.

$$-18x^2 - 12x$$
$$12x + 8$$

$-6x(3x + 2).$
Subtract. Bring down 8.

$$12x + 8$$

$4(3x + 2).$

Thus, $\dfrac{27x^3 + 8}{3x + 2} = 9x^2 - 6x + 4.$ ■

EXERCISE SET 4.5

Find each quotient. See Example 1.

1. $\dfrac{50b^{10}}{25b^5}$

2. $\dfrac{x^3y^6}{x^6y^2}$

3. $\dfrac{26x^2y^3z^7}{13x^6bz^5}$

4. $\dfrac{-48ab^{10}}{32a^4b^3}$

5. Divide $-2x^4y^2$ by $6x^4y^4$.

6. Divide $x^{17}y^5$ by $-x^7y^{10}$.

Find each quotient. See Examples 2 through 4.

7. Divide $4a^2 + 8a$ by $2a$.

8. Divide $6x^4 - 3x^3$ by $3x^2$.

9. $\dfrac{12a^5b^2 + 16a^4b}{4a^4b}$

10. $\dfrac{4x^3y + 12x^2y^2 - 4xy^3}{4xy}$

11. $\dfrac{4x^2y^2 + 6xy^2 - 4y^2}{2y^2}$

12. $\dfrac{6x^5 + 74x^4 + 24x^3}{2x^3}$

13. $\dfrac{4x^2 + 8x + 4}{4}$

14. $\dfrac{15x^3 - 5x^2 + 10x}{5x^2}$

Find each quotient. See Examples 5 through 8.

15. $\dfrac{x^2 + 3x + 3}{x + 2}$

16. $\dfrac{y^2 + 7y + 10}{y + 5}$

17. $\dfrac{2x^2 - 6x - 8}{x + 1}$

18. $\dfrac{3x^2 + 19x + 20}{x + 5}$

19. Divide $2x^2 + 3x - 2$ by $2x + 4$

20. Divide $6x^2 - 17x - 3$ by $3x - 9$

21. $\dfrac{4x^3 + 7x^2 + 8x + 20}{2x + 4}$

22. $\dfrac{18x^3 + x^2 - 90x - 5}{9x^2 - 45}$

Find each quotient.

23. Divide $25a^2b^{12}$ by $10a^5b^7$

24. Divide $12a^2b^3$ by $8a^7b$

25. $\dfrac{x^6y^6 - x^3y^3}{x^3y^3}$

26. $\dfrac{25xy^2 + 75xyz + 125x^2yz}{-5x^2y}$

27. $\dfrac{a^2 + 4a + 3}{a + 1}$

28. $\dfrac{3x^2 - 14x + 16}{x - 2}$

29. $\dfrac{2x^2 + x - 10}{x - 2}$

30. $\dfrac{x^2 - 7x + 12}{x - 5}$

31. Divide $-16y^3 + 24y^4$ by $-4y^2$

32. Divide $-20a^2b + 12ab^2$ by $-4ab$

33. $\dfrac{2x^2 + 13x + 15}{x - 5}$

34. $\dfrac{2x^2 + 13x + 5}{2x + 3}$

35. $\dfrac{20x^2y^3 + 6xy^4 - 12x^3y^5}{2xy^3}$

36. $\dfrac{3x^2y + 6x^2y^2 + 3xy}{3xy}$

37. $\dfrac{6x^2 + 16x + 8}{3x + 2}$

38. $\dfrac{x^2 - 25}{x + 5}$

39. $\dfrac{2y^2 + 7y - 15}{2y - 3}$

40. $\dfrac{3x^2 - 4x + 6}{x - 2}$

41. Divide $4x^2 - 9$ by $2x - 3$.

42. Divide $8x^2 + 6x - 27$ by $4x + 9$.

43. Divide $2x^3 + 6x - 4$ by $x + 4$.

44. Divide $4x^3 - 5x$ by $2x - 1$.

45. Divide $3x^2 - 4$ by $x - 1$.

46. Divide $x^2 - 9$ by $x + 4$.

47. $\dfrac{-13x^3 + 2x^4 + 16x^2 - 9x + 20}{5 - x}$

48. $\dfrac{5x^2 - 5x + 2x^3 + 20}{4 + x}$

49. Divide $3x^5 - x^3 + 4x^2 - 12x - 8$ by $x^2 - 2$.

50. Divide $-8x^3 + 2x^4 + 19x^2 - 33x + 15$ by $x^2 - x + 5$.

51. $\dfrac{3x^3 - 5}{3x^2}$

52. $\dfrac{14x^3 - 2}{7x - 1}$

53. If the area of the rectangle is $15x^2 - 29x - 14$ square inches, find its width.

5x + 2 inches

54. A board of length $3x^4 + 6x^2 - 18$ meters is to be cut into

three pieces of the same length. Find the length of each piece.

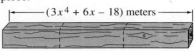

$(3x^4 + 6x - 18)$ meters

55. If the area of a parallelogram is $2x^2 - 17x + 35$ square centimeters and its base is $2x - 7$ centimeters, find its height.

A Look Ahead

Find each quotient. See the following example.

EXAMPLE   Divide $x^2 - \dfrac{7}{2}x + 4$ by $x + 2$.

Solution:

$$
\begin{array}{r}
x - \dfrac{11}{2} \\
x + 2 \overline{\smash{\big)}\, x^2 - \dfrac{7}{2}x + 4} \\
\underline{x^2 + 2x} \\
-\dfrac{11}{2}x + 4 \\
\underline{-\dfrac{11}{2}x - 11} \\
15
\end{array}
$$

The quotient is $x - \dfrac{11}{2} + \dfrac{15}{x + 2}$.   ∎

56. $\dfrac{x^4 + \dfrac{2}{3}x^3 + x}{x - 1}$

57. $\dfrac{2x^3 + \dfrac{9}{2}x^2 - 4x - 10}{x + 2}$

58. $\dfrac{3x^4 - x - x^3 + \dfrac{1}{2}}{2x - 1}$

59. $\dfrac{2x^4 + \dfrac{1}{2}x^3 + x^2 + x}{x - 2}$

60. $\dfrac{5x^4 - 2x^2 + 10x^3 - 4x}{5x + 10}$

61. $\dfrac{9x^5 + 6x^4 - 6x^2 - 4x}{3x + 2}$

Writing in Mathematics

62. Explain how to check polynomial long division.

63. Try performing the following division without changing the order of the terms. Describe why this makes the process more complicated. Then perform the division again after putting the terms in the dividend in descending order of exponents.

$$\dfrac{4x^2 - 12x - 12 + 3x^3}{x - 2}$$

Skill Review

Insert $<$, $>$, *or* $=$ *to make each statement true. See Section 4.1.*

64. 3^2 _____ $(-3)^2$ **65.** $(-5)^2$ _____ 5^2 **66.** -2^3 _____ $(-2)^3$ **67.** 3^4 _____ $(-3)^4$

Solve each inequality. See Section 2.7.

68. $|x + 5| < 4$ **69.** $|x - 1| \leq 8$

70. $|2x + 7| \geq 9$ **71.** $|4x + 2| > 10$

4.6
Synthetic Division

OBJECTIVES

Tape 11

1   Use synthetic division to divide a polynomial by a binomial.

2   Use the remainder theorem to evaluate polynomials.

1   When a polynomial is to be divided by a binomial of the form $x - c$, a shortcut process called **synthetic division** may be used. On the left is an example of long division, and on the right the same example showing the coefficients of the variables only.

$$
\begin{array}{r}
2x^2 + 5x + 2 \\
x - 3\overline{)2x^3 - x^2 - 13x + 1} \\
\underline{2x^3 - 6x^2} \\
5x^2 - 13x \\
\underline{5x^2 - 15x} \\
2x + 1 \\
\underline{2x - 6} \\
7
\end{array}
\qquad
\begin{array}{r}
2 \quad 5 \quad 2 \\
1 - 3\overline{)2 - 1 - 13 + 1} \\
\underline{2 - 6} \\
5 - 13 \\
\underline{5 - 15} \\
2 + 1 \\
\underline{2 - 6} \\
7
\end{array}
$$

Notice that as long as we keep coefficients of powers of x in the same column, we can perform division of polynomials by performing algebraic operations on the coefficients only. This shortcut process of dividing with coefficients only in a special format is called synthetic division. To find $(2x^3 - x^2 - 13x + 1) \div (x - 3)$ by synthetic division, follow the example shown.

EXAMPLE 1   Use synthetic division to divide $2x^3 - x^2 - 13x + 1$ by $x - 3$.

Solution: To use synthetic division, the divisor must be in the form $x - c$. Since we are dividing by $x - 3$, c is 3. Write down 3 and the coefficients of the dividend.

$$
\begin{array}{r|rrrr}
3 & 2 & -1 & -13 & 1 \\
& \downarrow & & & \\
\hline
& 2 & & &
\end{array}
$$

Next, draw a line and bring down the first coefficient of the dividend.

$$\begin{array}{r|rrrr} 3 & 2 & -1 & -13 & 1 \\ & & 6 & & \\ \hline & 2 & & & \end{array}$$

Multiply $3 \cdot 2$ and write down the product 6.

$$\begin{array}{r|rrrr} 3 & 2 & -1 & -13 & 1 \\ & & 6 & & \\ \hline & 2 & 5 & & \end{array}$$

Add $-1 + 6$. Write down the sum 5.

$$\begin{array}{r|rrrr} 3 & 2 & -1 & -13 & 1 \\ & & 6 & 15 & \\ \hline & 2 & 5 & 2 & \end{array}$$

$3 \cdot 5 = 15.$
$-13 + 15 = 2.$

$$\begin{array}{r|rrrr} 3 & 2 & -1 & -13 & 1 \\ & & 6 & 15 & 6 \\ \hline & 2 & 5 & 2 & 7 \end{array}$$

$3 \cdot 2 = 6.$
$1 + 6 = 7.$

The quotient is found in the bottom row. The numbers 2, 5, and 2 are the coefficients of the quotient polynomial, and the number 7 is the remainder. The degree of the quotient polynomial is one less than the degree of the dividend. In our example, the degree of the dividend is 3, so the degree of the quotient polynomial is 2. As we found when we performed the long division, the quotient is

$$2x^2 + 5x + 2, \qquad \text{remainder } 7$$

or

$$2x^2 + 5x + 2 + \frac{7}{x - 3} \qquad \blacksquare$$

EXAMPLE 2 Use synthetic division to divide $x^4 - 2x^3 - 11x^2 + 5x + 34$ by $x + 2$.

Solution: The divisor is $x + 2$, which we write in the form $x - c$ as $x - (-2)$. Thus c is -2. The dividend coefficients are 1, -2, -11, 5, and 34.

$$\begin{array}{r|rrrrr} -2 & 1 & -2 & -11 & 5 & 34 \\ & & -2 & 8 & 6 & -22 \\ \hline & 1 & -4 & -3 & 11 & 12 \end{array}$$

The dividend is a fourth-degree polynomial, so the quotient polynomial is a third-degree polynomial. The quotient is $x^3 - 4x^2 - 3x + 11$ with a remainder of 12. Thus

$$\frac{x^4 - 2x^3 - 11x^2 + 5x + 34}{x + 2} = x^3 - 4x^2 - 3x + 11 + \frac{12}{x + 2} \qquad \blacksquare$$

HELPFUL HINT

Before dividing by synthetic division, write the dividend in descending order of variable exponents. Any "missing powers" of the variable should be represented by 0 times the variable raised to the missing power.

EXAMPLE 3   If $P(x) = 2x^3 - 4x^2 + 5$, **(a)** find $P(2)$ by substitution and **(b)** use synthetic division to find the remainder when $P(x)$ is divided by $x - 2$.

Solution: **a.** $P(x) = 2x^3 - 4x^2 + 5$

$P(2) = 2(2)^3 - 4(2)^2 + 5$

$= 2(8) - 4(4) + 5 = 16 - 16 + 5 = 5$

Thus $P(2) = 5$.

b. The coefficients of $P(x)$ are 2, -4, 0, and 5. The number 0 is a coefficient of the missing power of x^1. The divisor is $x - 2$, so c is 2.

$$
\begin{array}{r|rrrr}
2 & 2 & -4 & 0 & 5 \\
  &   & 4 & 0 & 0 \\
\hline
  & 2 & 0 & 0 & 5 \quad \text{Remainder}
\end{array}
$$

The remainder when $P(x)$ is divided by $x - 2$ is 5.   ∎

2   Notice in the preceding example that $P(2) = 5$ and that the remainder when $P(x)$ is divided by $x - 2$ is 5. This is no accident. This illustrates the **remainder theorem.**

> **Remainder Theorem**
>
> If a polynomial $P(x)$ is divided by $x - c$, then the remainder is $P(c)$.

EXAMPLE 4   Use the remainder theorem and synthetic division to find $P(4)$ if

$$P(x) = 4x^6 - 25x^5 + 35x^4 + 17x^2.$$

Solution: To find $P(4)$ by the remainder theorem, we divide $P(x)$ by $x - 4$. The coefficients of $P(x)$ are 4, -25, 35, 0, 17, 0, and 0. Also, c is 4.

$$
\begin{array}{r|rrrrrrr}
c \searrow 4 & 4 & -25 & 35 & 0 & 17 & 0 & 0 \\
  &   & 16 & -36 & -4 & -16 & 4 & 16 \\
\hline
  & 4 & -9 & -1 & -4 & 1 & 4 & 16
\end{array}
$$

Thus $P(4) = 16$, the remainder.   ∎

EXERCISE SET 4.6

Use synthetic division to find each quotient. See Examples 1 through 3.

1. $\dfrac{x^2 + 3x - 40}{x - 5}$

2. $\dfrac{x^2 - 14x + 24}{x - 2}$

3. $\dfrac{x^2 + 5x - 6}{x + 6}$

4. $\dfrac{x^2 + 12x + 32}{x + 4}$

5. $\dfrac{x^3 - 7x^2 - 13x + 5}{x - 2}$

6. $\dfrac{x^3 + 6x^2 + 4x - 7}{x + 5}$

7. $\dfrac{4x^2 - 9}{x - 2}$

8. $\dfrac{3x^2 - 4}{x - 1}$

For the given polynomial P(x) and the given c, find P(c) by (a) direct substitution and (b) the remainder theorem. See Examples 3 and 4.

9. $P(x) = 3x^2 - 4x - 1; P(2)$

10. $P(x) = x^2 - x + 3; P(5)$

11. $P(x) = 4x^4 + 7x^2 + 9x - 1; P(-2)$

12. $P(x) = 8x^5 + 7x + 4; P(-3)$

13. $P(x) = x^5 + 3x^4 + 3x - 7; P(-1)$

14. $P(x) = 5x^4 - 4x^3 + 2x - 1; P(-1)$

Use synthetic division to find each quotient.

15. $\dfrac{x^3 - 3x^2 + 2}{x - 3}$

16. $\dfrac{x^2 + 12}{x + 2}$

17. $\dfrac{6x^2 + 13x + 8}{x + 1}$

18. $\dfrac{x^3 - 5x^2 + 7x - 4}{x - 3}$

19. $\dfrac{2x^4 - 13x^3 + 16x^2 - 9x + 20}{x - 5}$

20. $\dfrac{3x^4 + 5x^3 - x^2 + x - 2}{x + 2}$

21. $\dfrac{3x^2 - 15}{x + 3}$

22. $\dfrac{3x^2 + 7x - 6}{x + 4}$

23. $\dfrac{3x^3 - 6x^2 + 4x + 5}{x - \dfrac{1}{2}}$

24. $\dfrac{8x^3 - 6x^2 - 5x + 3}{x + \dfrac{3}{4}}$

25. $\dfrac{3x^3 + 2x^2 - 4x + 1}{x - \dfrac{1}{3}}$

26. $\dfrac{9y^3 + 9y^2 - y + 2}{y + \dfrac{2}{3}}$

27. $\dfrac{7x^2 - 4x + 12 + 3x^3}{x + 1}$

28. $\dfrac{x^4 + 4x^3 - x^2 - 16x - 4}{x - 2}$

29. $\dfrac{x^3 - 1}{x - 1}$

30. $\dfrac{y^3 - 8}{y - 2}$

31. $\dfrac{x^2 - 36}{x + 6}$

32. $\dfrac{4x^3 + 12x^2 + x - 12}{x + 3}$

For the given polynomial P(x) and the given c, use the remainder theorem to find P(c).

33. $P(x) = x^3 + 3x^2 - 7x + 4; 1$

34. $P(x) = x^3 + 5x^2 - 4x - 6; 2$

35. $P(x) = 3x^3 - 7x^2 - 2x + 5; -3$

36. $P(x) = 4x^3 + 5x^2 - 6x - 4; -2$

37. $P(x) = 4x^4 + x^2 - 2; -1$

38. $P(x) = x^4 - 3x^2 - 2x + 5; -2$

39. $P(x) = 2x^4 - 3x^2 - 2; \dfrac{1}{3}$

40. $P(x) = 4x^4 - 2x^3 + x^2 - x - 4; \dfrac{1}{2}$

41. $P(x) = x^5 + x^4 - x^3 + 3; \dfrac{1}{2}$

42. $P(x) = x^5 - 2x^3 + 4x^2 - 5x + 6; \dfrac{2}{3}$

Writing in Mathematics

43. Explain the advantage of using the remainder theorem instead of direct substitution.

Skill Review

If $P(x) = 3x^2 - 6x + 5$, $Q(x) = x - 5$, and $R(x) = x^2 - 6x + 3$, find the following. See Section 4.3.

44. $P(2)$ **45.** $Q(0)$ **46.** $R(-1)$ **47.** $P(-2)$

48. $Q(-7)$ **49.** $R(0)$

Find the slope of the line containing each pair of points. See Section 3.3.

50. $(-5, -2), (0, 7)$ **51.** $(3, 6), (-2, 6)$ **52.** $(2, 1), (2, -3)$ **53.** $(4, -1), (5, -2)$

4.7
Algebra of Functions and Composition of Functions

Tape 12

OBJECTIVES

1   Review function notation.

2   Add, subtract, multiply, and divide functions.

3   Compose functions.

1   Recall that, when **y** is a function of **x**, it means that each value of the independent variable **x** corresponds to exactly one value of the dependent variable **y.** The set of all possible x-values is called the domain of the function, and the set of all corresponding y-values is called the range of the function. To denote that y is a function of x, we write

$$y = f(x)$$

Recall that the symbol $f(x)$ is read "f of x". The letter f is not the only one that can be used, any letter is acceptable. For example, $f(x)$, $g(x)$, $G(x)$, $h(x)$, and $p(x)$ all mean "function of x".

HELPFUL HINT

The notation $f(x)$ **does not mean** $f \cdot x$. It is a single function notation symbol.

Since the solutions of a two-variable equation such as $y = -x + 2$ are ordered pairs, some two-variable equations "define" a function. The equation $y = -x + 2$ defines a function, since each x-value corresponds to a single y-value. Thus we are entitled to write

$$y = -x + 2$$

as

$$f(x) = -x + 2$$

Suppose we want to find the **value of the function,** that is, the value of y, when x is 3. This value can be written as $f(3)$.

Since

$$f(x) = -x + 2$$

then

$$f(3) = -3 + 2 = -1$$

Thus $f(3) = -1$ and the ordered pair $(3, -1)$ is a solution of $f(x) = -x + 2$.

EXAMPLE 1   If $g(x) = 2x + 5$, find:

     a. $g(-4)$ **b.** $g(a)$ **c.** $g(x + h)$

Solution: **a.** $g(x) = 2x + 5$ **b.** $g(x) = 2x + 5$

       $g(-4) = 2(-4) + 5$ $g(a) = 2(a) + 5$

               $= -3$ $= 2a + 5$

     c. $g(x) = 2x + 5$

       $g(x + h) = 2(x + h) + 5$

               $= 2x + 2h + 5$ ■

Equations of the form $f(x) = mx + b$ are called **linear functions.** Their graphs are straight lines.

EXAMPLE 2   Graph $H(x) = 3x$.

Solution: If it helps, write $H(x) = 3x$ as $y = 3x$. The function $y = 3x$ or $H(x) = 3x$ is a **linear function** since it can be written in the form $f(x) = mx + b$. This means that its graph is a straight line with slope 3 and y-intercept 0.

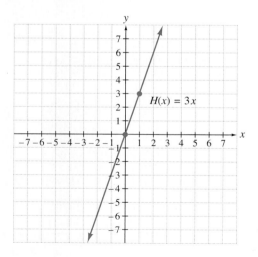

■

2   It is possible to add, subtract, multiply, and divide functions in order to find new functions. This **algebra** of functions is defined next.

Algebra of Functions

Let f and g be functions.
Their **sum,** written $f + g$, is defined by

$$(f + g)(x) = f(x) + g(x)$$

Their **difference,** written as $f - g$, is defined by

$$(f - g)(x) = f(x) - g(x)$$

Their **product,** written as $f \cdot g$, is defined by

$$(f \cdot g)(x) = f(x) \cdot g(x)$$

Their **quotient,** written as $\dfrac{f}{g}$, is defined by

$$\left(\frac{f}{g}\right)(x) = \frac{f(x)}{g(x)}, \qquad g(x) \neq 0$$

EXAMPLE 3   If $f(x) = x - 1$ and $g(x) = 2x - 3$, find the following.
  a. $(f + g)(x)$ **b.** $(f - g)(x)$
  c. $(f \cdot g)(x)$ **d.** $\left(\dfrac{f}{g}\right)(x)$

Solution: Replace $f(x)$ by $x - 1$ and $g(x)$ by $2x - 3$. Then simplify.
  a. $(f + g)(x) = f(x) + g(x)$

$$= (x - 1) + (2x - 3)$$

$$= 3x - 4$$

  b. $(f - g)(x) = f(x) - g(x)$

$$= (x - 1) - (2x - 3)$$

$$= x - 1 - 2x + 3$$

$$= -x + 2$$

  c. $(f \cdot g)(x) = f(x) \cdot g(x)$

$$= (x - 1)(2x - 3)$$

$$= 2x^2 - 5x + 3$$

  d. $\left(\dfrac{f}{g}\right)(x) = \dfrac{f(x)}{g(x)} = \dfrac{x - 1}{2x - 3}, \quad x \neq \dfrac{3}{2}$ ■

There is an interesting, but not surprising relationship between the graphs of functions and the graph of their sum, difference, product, and quotient. For example, the graph of $(f + g)(x)$ can be found by adding the graph of $f(x)$ to the graph of $g(x)$. We add two graphs by adding corresponding y-values.

$g(x) = 2x - 3$

x	y
0	-3
1	-1
2	1
3	3

$f(x) = x - 1$

x	y
0	-1
1	0
2	1
3	2

$(f + g)(x) = 3x - 4$

x	y
0	-4
1	-1
2	2
3	5

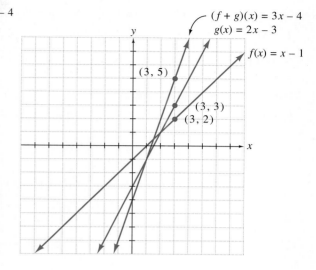

3   We can also find the **composition** of two functions. The notation $f[g(x)]$ means "f composed with g" and can be written as $(\mathbf{f \circ g})(\mathbf{x})$. Also $g[f(x)]$, or $(g \circ f)(x)$, means "g composed with f". If $f(x) = 5x - 2$ and $g(x) = 4x$, then

$$(f \circ g)(x) = f[g(x)]$$

$$= f(4x) \qquad \text{Replace } g(x) \text{ with } 4x.$$

$$= 5(4x) - 2 \qquad \text{Find } f(4x).$$

$$= 20x - 2 \qquad \text{Simplify.}$$

Thus $(f \circ g)(x) = 20x - 2$. Also,

$$(g \circ f)(x) = g[f(x)] = g(5x - 2) = 4(5x - 2) = 20x - 8$$

EXAMPLE 4   If $f(x) = x^2$ and $g(x) = x + 1$, find the following.

a. $(f \circ g)(x)$ **b.** $(g \circ f)(2)$

Solution: **a.** $(f \circ g)(x) = f[g(x)]$

$$= f(x + 1) \qquad \text{Replace } g(x) \text{ with } x + 1.$$

$$= (x + 1)^2 \qquad f(x + 1) = (x + 1)^2.$$

$$= x^2 + 2x + 1$$

b. $(g \circ f)(2) = g[f(2)]$

Next, replace $f(2)$ with 4 since $f(2) = 2^2 = 4$.

$$g[f(2)] = g(4)$$

$$= 4 + 1$$

$$= 5 \qquad \blacksquare$$

EXERCISE SET 4.7

If $f(x) = 3x$, $g(x) = 2x^2 - 5$, and $h(x) = \sqrt{x}$, find the following. See Example 1.

1. $f(5)$

2. $f(-3)$

3. $g(-2)$

4. $g(2)$

5. $h(16)$

6. $h(36)$

7. $h\left(\dfrac{1}{9}\right)$

8. $f(\pi)$

Graph each function. See Example 2.

9. $f(x) = 2x$

10. $g(x) = -5x$

11. $h(x) = x - 3$

12. $p(x) = 4x + 1$

13. $g(x) = -1$

14. $h(x) = 0$

If $f(x) = x^2 + 1$ and $g(x) = 5x$, find the following. See Example 3.

15. $(f + g)(x)$

16. $(g \cdot f)(x)$

17. $(f - g)(x)$

18. $\left(\dfrac{f}{g}\right)(x)$

19. $\left(\dfrac{g}{f}\right)(x)$

20. $(g - f)(x)$

If $f(x) = x^2 + 1$ and $g(x) = 5x$, find the following. See Example 4.

21. $(f \circ g)(x)$

22. $(g \circ f)(x)$

23. $(f \circ f)(1)$

24. $(g \circ g)(-3)$

If $f(x) = -2x$, $g(x) = x^2 + 2$, and $h(x) = 4x + 3$, find the following.

25. $g(7)$

26. $h(-2)$

27. $f(0)$

28. $g(0)$

29. $(f + g)(x)$

30. $(g - f)(x)$

31. $\left(\dfrac{f}{g}\right)(x)$

32. $\left(\dfrac{g}{f}\right)(x)$

33. $(g - h)(x)$

34. $(h \cdot g)(x)$

35. $h(4)$

36. $f(-5)$

37. $(f \circ g)(x)$

38. $(g \circ f)(x)$

39. $(h \circ f)(-3)$

40. $(h \circ g)(4)$

41. $(h + f)(x)$

42. $(h - g)(x)$

43. $(f \circ f)(x)$

44. $(f \circ h)(x)$

45. $f(a + b)$

46. $g(a + b)$

47. $\left(\dfrac{f}{h}\right)x$

48. $\left(\dfrac{h}{g}\right)x$

49. $(g \circ h)(x)$

50. $h(a + b)$

Graph each function.

51. $f(x) = -x$

52. $h(x) = x + 7$

53. $g(x) = 3x + 2$

54. $f(x) = -x + 1$

55. $h(x) = -x - 3$

56. $g(x) = 4x - 3$

57. If $f(x) = 2x + 4$ and $g(x) = \dfrac{1}{2}x - 2$, find $(f \circ g)(x)$ and $(g \circ f)(x)$.

58. If $f(x) = x + 6$ and $g(x) = x - 6$, find $(f \circ g)(x)$ and $(g \circ f)(x)$.

59. Business people are concerned with cost functions, revenue functions, and profit functions. The profit $P(x)$ obtained from x units of a product is equal to the revenue $R(x)$ from selling the x units minus the cost $C(x)$ of manufacturing the x units. Write an equation expressing this relationship among $C(x)$, $R(x)$, and $P(x)$.

60. Suppose the revenue $R(x)$ for x units of a product can be described by $R(x) = 25x$, and the cost $C(x)$ can be described by $C(x) = 50 + x^2 + 4x$. Find the profit $P(x)$ for x units.

61. The area of a circle is a function of the radius of the circle. If area is $A(r)$, radius is r, and $A(r) = \pi r^2$, find the area of a circle with radius 10 cm. That is, find $A(10)$.

62. Maximum heart rate per minute is a function of age. If maximum heart rate is $M(x)$, age is x, and $M(x) = 220 - x$, find the maximum heart rate for a 20-year-old. That is, find $M(20)$.

63. According to some sports enthusiasts, a runner in good shape weighs about 2 pounds per inch of height. (a) Write an equation showing the weight of a runner in good shape as a function of height (in inches). (b) If Samantha is a runner in good shape and is 5 feet 5 inches, determine her approximate weight.

Skill Review

Multiply the following. See Section 4.4.

64. $4x(3x + 2)$

65. $7x(9x - 1)$

66. $-8xy(y^2 + 2)$

67. $-9xy(-9x + 4)$

Solve the following. See Section 2.1.

68. $3(x - 2) + 5x = 1$

69. $-2(x + 7) + 14 = -3x$

70. $\dfrac{x}{3} + \dfrac{1}{2} = 2$

71. $\dfrac{2x}{5} - \dfrac{1}{3} = \dfrac{4}{5}$

CRITICAL THINKING   While learning to use her programmable calculator, Yee programmed the f key so that $f(x) = 2x - 7$. In other words, when Yee enters the number 5, for example, and then presses the f key, the display will show 3 since $f(5) = 2 \cdot 5 - 7 = 3$. Yee feels comfortable with her new calculator and is ready to use it to adjust her final exam scores. Unfortunately, a mistake was overlooked when proofreading the group final exam in beginning algebra and now her students' final exam scores will be adjusted by increasing each score by 4% and then adding 10 additional points.

What function can Yee write and use to program the special g key on her calculator so that she can enter the old score, press the g key, and the new adjusted score will appear in the display? An original final exam score of 800 will be adjusted to what new score? What happens if Yee enters a score and accidently presses the g key twice? How else might an instructor use a programmable calculator to save time?

CHAPTER 4 GLOSSARY

The **numerical coefficient,** or simply the **coefficient,** is the numerical factor of a term.

In the expression a^n, a is called the **base** and n is called the **exponent.**

A **binomial** is a polynomial with two terms.

The **composition** of two functions f and g in x, written as $(f \circ g)(x)$, is the function $f(g(x))$.

The **degree of a term** is the sum of the exponents on the variables contained in the term.

The **degree of a polynomial** is the largest degree of all its terms.

A **monomial** is a polynomial with one term.

A **polynomial** is a finite sum of terms in which all variables have exponents raised to nonnegative integer powers and no variables appear in the denominator. A polynomial that contains only one variable is called a **polynomial in one variable.**

A positive number is written in **scientific notation** if it is written as the product of a number x where $1 \le x < 10$, and a power of 10.

A **trinomial** is a polynomial with three terms.

CHAPTER 4 SUMMARY

SUMMARY OF RULES FOR EXPONENTS   (4.1 and 4.2)

If a and b are real numbers and m and n are integers, then:

Product rule for exponents $a^m \cdot a^n = a^{m+n}$

Zero exponent $a^0 = 1 \qquad (a \ne 0)$

Negative exponent $a^{-n} = \dfrac{1}{a^n}, \qquad (a \ne 0)$

Quotient rule $\dfrac{a^m}{a^n} = a^{m-n} \qquad (a \ne 0)$

Power rules
$$(a^m)^n = a^{m \cdot n}$$
$$(ab)^m = a^m \cdot b^m$$
$$\left(\frac{a}{b}\right)^m = \frac{a^m}{b^m} \quad (b \neq 0)$$

SQUARE OF A BINOMIAL (4.4)
$(a + b)^2 = a^2 + 2ab + b^2$ and $(a - b)^2 = a^2 - 2ab + b^2$

PRODUCT OF THE SUM AND DIFFERENCE OF TWO TERMS (4.4)
$(a + b)(a - b) = a^2 - b^2$

ALGEBRA OF FUNCTIONS (4.7)
$(f + g)(x) = f(x) + g(x), (f - g)(x) = f(x) - g(x)$

$(f \cdot g)(x) = f(x) \cdot g(x), \left(\dfrac{f}{g}\right)(x) = \dfrac{f(x)}{g(x)}, g(x) \neq 0$

COMPOSITION OF FUNCTIONS (4.7)
$(f \circ g)(x) = f[g(x)]$
$(g \circ f)(x) = g[f(x)]$

CHAPTER 4 REVIEW

(4.1) *Evaluate.*

1. $(-2)^2$ **2.** $(-3)^4$ **3.** -2^2 **4.** -3^4
5. 8^0 **6.** -9^0 **7.** -4^{-2} **8.** $(-4)^{-2}$

Simplify each expression. Write using only positive exponents.

9. $-xy^2y^3xy^2z$ **10.** $(-4xy)(-3xy^2b)$ **11.** $a^{-14} \cdot a^5$ **12.** $\dfrac{a^{16}}{a^{17}}$

13. $\dfrac{x^{-7}}{x^4}$ **14.** $\dfrac{9a(a^{-3})}{18a^{15}}$ **15.** $\dfrac{y^{6p-3}}{y^{6p+2}}$

Write in scientific notation.
16. 36,890,000 **17.** -0.000362

Write each number without exponents.
18. 1.678×10^{-6} **19.** 4.1×10^5

(4.2) *Simplify. Write using only positive exponents.*

20. $(8^5)^3$ **21.** $\left(\dfrac{a}{4}\right)^2$ **22.** $(3x)^3$ **23.** $(-4x)^{-2}$

24. $\left(\dfrac{6x}{5}\right)^2$ **25.** $(8^6)^{-3}$ **26.** $\left(\dfrac{4}{3}\right)^{-2}$ **27.** $(-2x^3)^{-3}$

28. $\left(\dfrac{8p^6}{4p^4}\right)^{-2}$

29. $(-3x^{-2}y^2)^3$

30. $\left(\dfrac{x^{-5}y^{-3}}{z^3}\right)^{-5}$

31. $\dfrac{4^{-1}x^3yz}{x^{-2}yx^4}$

32. $(5xyz)^{-4}(x^{-2})^{-3}$

33. $\dfrac{2(3yz)^{-3}}{y^{-3}}$

Simplify each expression.

34. $x^{4a}(3x^{5a})^3$

35. $\dfrac{4y^{3x-3}}{2y^{2x+4}}$

Use scientific notation to find the quotient. Express each quotient in scientific notation.

36. $\dfrac{(0.00012)(144{,}000)}{0.0003}$

37. $\dfrac{(-0.00017)(0.00039)}{3000}$

Simplify. Write using only positive exponents.

38. $\dfrac{27x^{-5}y^5}{18x^{-6}y^2} \cdot \dfrac{x^4y^{-2}}{x^{-2}y^3}$

39. $\dfrac{3x^5}{y^{-4}} \cdot \dfrac{(3xy^{-3})^{-2}}{(z^{-3})^{-4}}$

40. $\dfrac{(x^w)^2}{(x^{w-4})^{-2}}$

(4.3)   *Find the degree of each term.*

41. $-3xy^2z$

42. 7

43. $3x$

Find the degree of each polynomial.

44. $x^2y - 3xy^3z + 5x + 7y$

45. $3x + 2$

Simplify.

46. $4x + 8x - 6x^2$

47. $-8xy^3 + 4xy^3 - 3x^3y$

Add or subtract as indicated.

48. $(3x + 7y) + (4x^2 - 3x + 7) + (y - 1)$

49. $(4x^2 - 6xy + 9y^2) - (8x^2 - 6xy - y^2)$

50. $(3x^2 - 4b + 28) + (9x^2 - 30) - (4x^2 - 6b + 20)$

51. Add $(9xy + 4x^2 + 18)$ and $(7xy - 4x^3 - 9x)$.

52. Subtract $(x - 7)$ from the sum of $(3x^2y - 7xy - 4)$ and $(9x^2y + x)$.

53. $\begin{array}{r} x^2 - 5x + 7 \\ -\ (\ x + 4) \\ \hline \end{array}$

54. $\begin{array}{r} x^3 \qquad\quad + 2xy^2 - y \\ +\ (x - 4xy^2 \qquad\quad - 7) \\ \hline \end{array}$

If $P(x) = 9x^2 - 7x + 8$, find the following.

55. $P(6)$

56. $P(-2)$

57. $P(-3)$

58. $P(-x)$

59. $P(x + h)$

Multiply.

60. $(-4xy^3)(3xy^2t)$

61. $-6xy^3(4xy - 6x + 1)$

62. $-4ab^2(3ab^3 + 7ab + 1)$

63. $(x - 4)(2x + 9)$

64. $(x^2 + 9x + 1)^2$

65. $(-3xa + 4b)^2$

66. Multiply $9x^2 + 4x + 1$ and $4x - 3$ vertically.

67. Multiply $(5x - 9)$ by $(3x + 9)$ using the FOIL method.

68. Multiply $x - \dfrac{1}{3}$ by $x + \dfrac{2}{3}$ using the FOIL method.

Multiply using special products.

69. $(3x - y)^2$

70. $(4x + 9)^2$

71. $(x + 3y)(x - 3y)$

72. $[4 + (3x + y)]^2$

73. $[4 + (3a - b)][4 - (3a - b)]$

74. $[(4y - 2) + x]^2$

75. $[(9y - 3) - y^2]^2$

76. $(x^2 - 9y^3)^2$

77. $(4y^3 + 3x^2)^2$

Multiply. Assume that all variable exponents represent integers.

78. $4a^b(3a^{b+2} - 7)$

79. $(4xy^z - b)^2$

80. $(3x^a - 4)(3x^a + 4)$

(4.5) *Find each quotient. Write using only positive exponents.*

81. $\dfrac{3x^5yb^9}{9xy^7}$

82. Divide $-9xb^4z^3$ by $-4axb^2$.

83. $\dfrac{4xy + 2x^2 - 9}{4xy}$

84. Divide $12xb^2 + 16xb^4$ by $4xb^3$.

Find each quotient.

85. $\dfrac{3x^4 - 25x^2 - 20}{x - 3}$

86. $\dfrac{-x^2 + 2x^4 + 5x - 12}{x - 3}$

87. $\dfrac{2x^4 - x^3 + 2x^2 - 3x + 1}{x - \dfrac{1}{2}}$

88. $\dfrac{x^3 + 3x^2 - 2x + 2}{x - \dfrac{1}{2}}$

89. $\dfrac{3x^4 + 5x^3 + 7x^2 + 3x - 2}{x^2 + x + 2}$

90. $\dfrac{9x^4 - 6x^3 + 3x^2 - 12x - 30}{3x^2 - 2x - 5}$

(4.6) *Use synthetic division to find each quotient.*

91. $\dfrac{3x^3 + 12x - 4}{x - 2}$

92. $\dfrac{3x^3 + 2x^2 - 4x - 1}{x + \dfrac{3}{2}}$

93. $\dfrac{x^5 - 1}{x + 1}$

94. $\dfrac{x^3 - 81}{x - 3}$

95. $\dfrac{x^3 - x^2 + 3x^4 - 2}{x - 4}$

96. $\dfrac{3x^4 - 2x^2 + 10}{x + 2}$

If $P(x) = 3x^5 - 9x + 7$, find the following using the remainder theorem.

97. $P(4)$

98. $P(-5)$

99. $P\left(\dfrac{2}{3}\right)$

100. $P\left(-\dfrac{1}{2}\right)$

(4.7) *If $f(x) = x^2 - 2$, $g(x) = x + 1$ and $h(x) = x^3 - x^2$, find the following.*

101. $(f + g)(x)$

102. $(h - g)(x)$

103. $\left(\dfrac{h}{g}\right)(x)$

104. $(g \cdot f)(x)$

105. $(f \circ g)(x)$

106. $(g \circ f)(x)$

107. $(h \circ g)(2)$

108. $(f \circ f)(x)$

109. $(f \circ g)(-1)$

110. $(h \circ h)(2)$

Graph.

111. $f(x) = x + 1$

112. $g(x) = -3x$

CHAPTER 4 TEST

Evaluate the following.

1. $(-2)^3$

2. 6^{-2}

Simplify. Write using only positive exponents.

3. $-3xy^{-2}(4xy^2)z$

4. $(-9x)^{-2}$

5. $\left(\dfrac{-xy^{-5}z}{xy^3}\right)^{-5}$

6. $\dfrac{144(xy^{-3}z)^3}{12(xy)^{-2}}$

Write in scientific notation.

7. 630,000,000

8. 0.01200

9. Write 5.0×10^{-6} without exponents.

10. Use scientific notation to find the quotient.

$$\frac{(0.0024)(0.00012)}{0.00032}$$

Add or subtract.

11. $(4x^3 - 3x - 4) - (9x^3 + 8x + 5)$

12. $(12x^6 - 6xy^2 + 15) + (2x^6 + 3xy^2 + 6y^2)$

13. If $P(x) = -3x^3 - 4x + 2$, find $P(-3)$.

Multiply or divide.

14. $-3xy(4x + y)$

15. $(3x + 4)(4x - 7)$

16. $(9x^2 + 4y + 2)^2$

17. $[3 - (4x - y)]^2$

18. $\dfrac{4x^2y + 9x + z}{3xz}$

19. $\dfrac{x^6 + 3x^5 - 2x^4 + x^2 - 3x + 2}{x - 2}$

20. Use synthetic division to divide $4x^4 - 3x^3 + 2x^2 - x - 1$ by $x - \dfrac{2}{3}$.

21. If $P(x) = 4x^4 + 7x^2 - 2x - 5$, use the remainder theorem to find $P(-2)$.

If $f(x) = x$, $g(x) = x - 7$, and $h(x) = x^2 - 6x + 5$, find the following.

22. $(h - g)(x)$

23. $(h \cdot f)(x)$

24. $(g \circ f)(x)$

25. $(g \circ h)(x)$

26. Graph $f(x) = 2x - 4$

CHAPTER 4 CUMULATIVE REVIEW

1. Translate each sentence into a mathematical statement.
   a. The sum of 5 and y is greater than or equal to 7.
   b. 11 is not equal to z.
   c. Twenty is less than the difference of 5 and x.

2. Write the additive inverse or the opposite of each.
   a. 8   b. $\dfrac{1}{5}$ c. -9

3. Find each quotient.

a. $\dfrac{20}{-4}$ b. $\dfrac{-9}{-3}$ c. $-\dfrac{3}{8} \div 3$

d. $\dfrac{-40}{10}$ e. $\dfrac{\frac{-1}{10}}{\frac{-2}{5}}$

4. Solve for x: $2x + 5 = 9$.

5. Write the following as mathematical expressions.

a. A piece of board that has length x feet is cut from a 10-foot board. Express the length of the remaining piece of board as an expression in x.

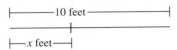

b. Two angles are complementary if the sum of their measures is $90°$. If the measure of one angle is $3x - 10$ degrees, express the measure of its complement in terms of x.

c. If x is the first of two consecutive integers, write their sum as an expression in x.

6. Find three consecutive odd integers such that the sum of the first and the third is eleven more than the second integer.

7. Solve $\left| \dfrac{x}{2} - 1 \right| = 11$.

8. Plot each ordered pair on a Cartesian coordinate system and name the quadrant in which the point is located.

a. $A(2, -1)$ b. $B(0, 5)$ c. $C(-3, 5)$

d. $D(-2, 0)$ e. $\left(-\dfrac{1}{2}, -4 \right)$

9. Find the slope of the line containing the points $(0, 3)$ and $(2, 5)$.

10. Graph the line with slope $-\dfrac{2}{3}$ that contains the point $(-1, 3)$.

11. Find the equation of the line with slope -3 passing through the point $(1, -5)$.

12. Determine whether the relation is also a function.

a. $\{(1, 1), (2, 7), (3, -1), (0, 0)\}$

b. $\{(2, 3), (2, 4), (7, 6)\}$

c. $\{(-2, 5), (-3, 5), (-1, 5)\}$

d. $y = x^2$

e. $y = \dfrac{2x + 3}{6}$

f.

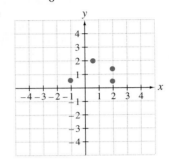

13. Graph the intersection $x \geq 1$ and $y \geq 2x - 1$.

14. Find each product.

a. $(3x^6)(5x)$ b. $(-2x^3p^2)(4xp^{10})$

15. Write the following using only positive exponents. Simplify if possible.

a. 5^{-2} b. $2x^{-3}$ c. $(3x)^{-1}$ d. $\dfrac{m^5}{m^{15}}$

e. $\dfrac{3^3}{3^6}$ f. $2^{-1} + 3^{-2}$ g. $\dfrac{1}{t^{-5}}$

16. Add $11x^3 - 12x^2 + x - 3$ and $x^3 - 10x + 5$.

17. Multiply and simplify the product if possible.

a. $(x + 3)(2x + 5)$

b. $(2x^3 - 3)(5x^2 - 6x + 7)$

18. Divide $10x^2 - 5x + 20$ by 5.

19. Use long division to find the quotient: $\dfrac{6x^2 - 19x + 12}{3x - 5}$.

20. Use synthetic division to divide $2x^3 - x^2 - 13x + 1$ by $x - 3$.

21. If $f(x) = x^2$ and $g(x) = x + 1$, find the following.

a. $(f \circ g)(x)$

b. $(g \circ f)(2)$

CHAPTER **5**

Factoring Polynomials

Many algebraic relationships can be visualized with the use of geometric relationships. (See Critical Thinking, page 242)

INTRODUCTION

When two or more polynomials are multiplied, each polynomial is called a **factor** of the product. "Reversing" this process is called **factoring** a polynomial. It is the process of writing a polynomial as a product of simpler factors. In this chapter, we explore factoring polynomials, solving polynomial equations, and graphing polynomial functions.

5.1
The Greatest Common Factor and Factoring by Grouping

1 Identify the GCF.

2 Factor out the GCF of a polynomial's terms.

3 Factor polynomials by grouping.

Tape 13

1 **Factoring** is the reverse process of multiplying. When an integer is written as the product of two integers, we call each integer a factor. For example, in the product $3 \cdot 5 = 15$, 3 and 5 are called **factors** of 15, and $3 \cdot 5$ is a **factored form** of 15. This is true of polynomials, also. Since

$$(3x - 1)(2x + 5) = 6x^2 + 13x - 5$$

$3x - 1$ and $2x + 5$ are called **factors** of $6x^2 + 13x - 5$, and $(3x - 1)(2x + 5)$ is a **factored form** of $6x^2 + 13x - 5$. The process of writing a polynomial as a product is called **factoring.**

$$6x^2 + 13x - 5 = (3x - 1)(2x + 5)$$
factoring
multiplying

Factoring polynomials is closely related to factoring integers, so we begin by reviewing integer factoring and finding the greatest common factor (GCF) of a list of integers. The GCF of a list of integers is the largest integer that is a factor of each integer. For example, 4 is the GCF of integers 16, 20, and 80, because 4 is the largest factor common to 16, 20, and 80.

To find the GCF, we write each integer as a product of **prime numbers.** A prime number is a natural number greater than 1 whose only natural number factors are 1 and itself. The set of prime numbers is

$$\{2, 3, 5, 7, 11, 13, 17, \ldots\}$$

EXAMPLE 1 Find the GCF of each list of integers.

  a. 40 and 52   **b.** 30, 45, and 75

Solution: **a.** Start by writing each number as a product of primes.

$$40 = 4 \cdot 10 \qquad\qquad 52 = 4 \cdot 13$$
$$= 2 \cdot 2 \cdot 2 \cdot 5 \qquad\qquad = 2 \cdot 2 \cdot 13$$
$$= 2^3 \cdot 5 \qquad\qquad = 2^2 \cdot 13$$

The common factors are two factors of 2, so the GCF is

$$\text{GCF} = 2^2 = 4.$$

b.
$$30 = 3 \cdot 10 \qquad 45 = 9 \cdot 5 \qquad 75 = 3 \cdot 25$$
$$= 3 \cdot 2 \cdot 5 \qquad = 3 \cdot 3 \cdot 5 \qquad = 3 \cdot 5 \cdot 5$$
$$= 2 \cdot 3 \cdot 5 \qquad = 3^2 \cdot 5 \qquad = 3 \cdot 5^2$$

The common factors are one factor of 3 and one factor of 5, so the GCF is

$$\text{GCF} = 3 \cdot 5 = 15.$$ ▪

To find the GCF of a list of variables raised to powers, we use a similar process. The GCF of x^3, x^5, and x^6 is x^3 since each power of x contains at least a factor of x^3.

$$x^3 = x^3$$
$$x^5 = x^3 \cdot x^2$$
$$x^6 = x^3 \cdot x^3$$

In general, the GCF of a list of common variables raised to powers is the common variable raised to an exponent equal to the smallest exponent in the list. To find the GCF of a list of monomials, the following steps can be used.

To Find the GCF of a List of Monomials

Step 1 Find the GCF of the numerical coefficients.
Step 2 Find the GCF of the variable factors.
Step 3 The product of the factors found in *steps 1* and *2* is the GCF of the monomials.

EXAMPLE 2 Find the GCF of $20x^3y$, $10x^2y^2$, and $35x^3$.

Solution: The GCF of the numerical coefficients 20, 10, and 35 is 5. The GCF of the variable factors x^3, x^2, and x^3 is x^2. The variable y is not a common factor because it does not appear in the third monomial. The GCF is thus

$$5 \cdot x^2 \quad \text{or} \quad 5x^2.$$ ▪

2 The goal of this chapter is to present methods of factoring polynomials. In other words, we want to write polynomials as products of simpler polynomials. The first step in factoring polynomials is to use the distributive property and write the polynomial as a product of the GCF of its monomial terms and a simpler polynomial. This is called **factoring out** the GCF.

EXAMPLE 3   Factor.
a. $8x + 4$ **b.** $5y - 2z$ **c.** $6x^2 - 3x^3$

Solution: **a.** The GCF of terms $8x$ and 4 is 4.

$$8x + 4 = 4(2x) + 4(1) \qquad \text{Factor out 4 from each term.}$$
$$= 4(2x + 1) \qquad \text{Apply the distributive property.}$$

The factored form of $8x + 4$ is $4(2x + 1)$. To check, multiply $4(2x + 1)$ to see that the product is $8x + 4$.

b. There is no common factor of the terms $5y$ and $-2z$ other than 1.

c. The greatest common factor of $6x^2$ and $-3x^3$ is $3x^2$. Thus

$$6x^2 - 3x^3 = 3x^2(2) - 3x^2(x)$$
$$= 3x^2(2 - x) \qquad \blacksquare$$

HELPFUL HINT

To check that the GCF has been factored out correctly, multiply the factors together and see that their product is the original polynomial.

EXAMPLE 4   Factor $17x^3y^2 - 34x^4y^2$.

Solution: The GCF of the two terms is $17x^3y^2$, which we factor out of each term.

$$17x^3y^2 - 34x^4y^2 = 17x^3y^2(1) - 17x^3y^2(2x)$$
$$= 17x^3y^2(1 - 2x) \qquad \blacksquare$$

HELPFUL HINT

If the GCF happens to be one of the terms in the polynomial, a factor of 1 will remain for this term when the GCF is factored out. For example, in the polynomial $21x^2 + 7x$ the GCF of $21x^2$ and $7x$ is $7x$, so

$$21x^2 + 7x = 7x(3x) + 7x(1) = 7x(3x + 1)$$

EXAMPLE 5   Factor $-3x^3y + 2x^2y - 5xy$.

Solution: Two possibilities are shown for factoring this polynomial. First, the common factor xy can be factored out.

$$-3x^3y + 2x^2y - 5xy = xy(-3x^2 + 2x - 5)$$

Also, the common factor $-xy$ can be factored out.

$$-3x^3y + 2x^2y - 5xy = -xy(3x^2) + (-xy)(-2x) + (-xy)(5)$$
$$= -xy(3x^2 - 2x + 5)$$

Both of these alternatives are correct. $\blacksquare$

EXAMPLE 6 Factor $2(x - 5) + 3a(x - 5)$.

Solution: The GCF is the binomial $(x - 5)$. Thus

$$2(x - 5) + 3a(x - 5) = (x - 5)(2 + 3a) \quad \blacksquare$$

EXAMPLE 7 Factor $7x(x^2 + 5y) - (x^2 + 5y)$.

Solution: The GCF is the expression $(x^2 + 5y)$. Factor this from each term.

$$7x(x^2 + 5y) - (x^2 + 5y) = 7x(x^2 + 5y) - 1(x^2 + 5y) = (x^2 + 5y)(7x - 1)$$

Notice that we write $-(x^2 + 5y)$ as $-1(x^2 + 5y)$ to aid in factoring. $\blacksquare$

3   Sometimes it is possible to factor a polynomial by grouping the terms of the polynomial and looking for common factors in each group. This method of factoring is called **factoring by grouping.**

EXAMPLE 8 Factor $ab - 6a + 2b - 12$.

Solution: First look for the GCF of all four terms. The GCF for all four terms is 1. Next, group the first two terms and the last two terms and factor out common factors from each group.

$$ab - 6a + 2b - 12 = (ab - 6a) + (2b - 12)$$

Factor a from the first group and 2 from the second group.

$$= a(b - 6) + 2(b - 6)$$

Now we see a GCF of $(b - 6)$. Factor out $(b - 6)$ to get

$$a(b - 6) + 2(b - 6) = (b - 6)(a + 2)$$

This factorization can be checked by multiplying $(b - 6)$ by $(a + 2)$ to verify that the product is the original polynomial. $\blacksquare$

HELPFUL HINT

Notice that the polynomial $5(x - 1) + y(x - 1)$ is **not** in factored form. It is a **sum,** not a **product.** The factored form is $(x - 1)(5 + y)$.

EXAMPLE 9 Factor $m^2n^2 + m^2 - 2n^2 - 2$.

Solution: Once again, the GCF of all four terms is 1. Try grouping the first two terms together and the last two terms together.

$$m^2n^2 + m^2 - 2n^2 - 2 = (m^2n^2 + m^2) + (-2n^2 - 2)$$

Factor m^2 from the first group and 2 from the second group.

$$= m^2(n^2 + 1) + 2(-n^2 - 1)$$

There is no common factor in this resulting polynomial, but notice that $(n^2 + 1)$ and $(-n^2 - 1)$ are opposites. Try grouping the terms differently, as follows:

$$m^2n^2 + m^2 - 2n^2 - 2 = (m^2n^2 + m^2) - (2n^2 + 2) \qquad \text{Watch the signs!}$$

$$= m^2(n^2 + 1) - 2(n^2 + 1) \qquad \begin{array}{l}\text{Factor common factors}\\\text{from the groups of terms.}\end{array}$$

$$= (n^2 + 1)(m^2 - 2) \qquad \text{Factor out a GCF of } (n^2 + 1). \quad \blacksquare$$

MENTAL MATH

Find the GCF of each list of monomials.

1. 6, 12   **2.** 9, 27   **3.** $15x$, 10   **4.** $9x$, 12

5. $13x$, $2x$ **6.** $4y$, $5y$ **7.** $7x$, $14x$ **8.** $8z$, $4z$

EXERCISE SET 5.1

Find the GCF of each list of numbers. See Example 1.

1. 24, 30   **2.** 30, 75   **3.** 84, 140

4. 90, 225   **5.** 20, 36, 60   **6.** 18, 45, 54

7. 30, 48, 72   **8.** 42, 63, 147

Find the GCF of each list of monomials. See Example 2.

9. a^8, a^5, a^3 **10.** b^9, b^2, b^5

11. $x^2y^3z^3$, y^2z^3, xy^2z^2 **12.** xy^2z^3, $x^2y^2z^2$, x^2y^3

13. $6x^3y$, $9x^2y^2$, $12x^2y$ **14.** $4xy^2$, $16xy^3$, $8x^2y^2$

15. $10x^3yz^3$, $20x^2z^5$, $45xz^3$ **16.** $12y^2z^4$, $9xy^3z^4$, $15x^2y^2z^3$

Factor out the GCF in each polynomial. See Examples 3 through 5.

17. $18x - 12$ **18.** $21x + 14$

19. $4y^2 - 16xy^3$ **20.** $3z - 21xz^4$

21. $6x^5 - 8x^4 + 2x^3$ **22.** $9x + 3x^2 - 6x^3$

23. $8a^3b^3 - 4a^2b^2 + 4ab + 16ab^2$ **24.** $12a^3b - 6ab + 18ab^2 - 18a^2b$

Factor out the GCF in each polynomial. See Examples 6 and 7.

25. $6(x + 3) + 5a(x + 3)$ **26.** $2(x - 4) + 3y(x - 4)$

27. $2x(z + 7) + (z + 7)$ **28.** $x(y - 2) + (y - 2)$

29. $3x(x^2 + 5) - 2(x^2 + 5)$ **30.** $4x(2y + 3) - 5(2y + 3)$

Factor each polynomial by grouping. See Examples 8 and 9.

31. $ab + 3a + 2b + 6$ **32.** $ab + 2a + 5b + 10$

33. $ac + 4a - 2c - 8$ **34.** $bc + 8b - 3c - 24$

35. $2xy - 3x - 4y + 6$ **36.** $12xy - 18x - 10y + 15$

37. $12xy - 8x - 3y + 2$ **38.** $20xy - 15 - 4y + 3$

Factor out the GCF in each polynomial.

39. $6x^3 + 9$

40. $6x^2 - 8$

41. $x^3 + 3x^2$

42. $x^4 - 4x^3$

43. $8a^3 - 4a$

44. $12b^4 + 3b^2$

45. $-20x^2y + 16xy^3$

46. $-18xy^3 + 27x^4y$

47. $10a^2b^3 + 5ab^2 - 15ab^3$

48. $10ef - 20e^2f^3 + 30e^3f$

49. $9abc^2 + 6a^2bc - 6ab + 3bc$

50. $4a^2b^2c - 6ab^2c - 4ac + 8a$

51. $4x(y - 2) - 3(y - 2)$

52. $8y(z + 8) - 3(z + 8)$

53. $2m(n - 8) - (n - 8)$

54. $3a(b - 4) - (b - 4)$

55. $15x^3y^2 - 18x^2y^2$

56. $12x^4y^2 - 16x^3y^3$

Factor each polynomial by grouping.

57. $6xy + 10x + 9y + 15$

58. $15xy + 20x + 6y + 8$

59. $xy + 3y - 5x - 15$

60. $xy + 4y - 3x - 12$

61. $6ab - 2a - 9b + 3$

62. $16ab - 8a - 6b + 3$

63. $12xy + 18x + 2y + 3$

64. $20xy + 8x + 5y + 2$

65. $2x^2 + 3xy + 4x + 6y$

66. $3x^2 + 12x + 4xy + 16y$

67. $5x^2 + 5xy - 3x - 3y$

68. $4x^2 + 2xy - 10x - 5y$

69. $x^3 + 3x^2 + 4x + 12$

70. $x^3 + 4x^2 + 3x + 12$

71. $x^3 - x^2 - 2x + 2$

72. $x^3 - 2x^2 - 3x + 6$

73. A manufacturer of tin cans knows that the material needed to construct one tin can is $2\pi r^2 + 2\pi rh$, where radius is r and height is h. Factor the expression.

74. The amount E of current in an electrical circuit is given by the formula $IR_1 + IR_2 = E$. Factor the expression $IR_1 + IR_2$.

75. At the end of T years, the amount of money A in a simple interest savings account from an initial investment of P dollars at rate R is given by the formula

$$A = P + PRT$$

Factor the expression $P + PRT$.

76. An open-topped box has a square base and a height of 10 inches. If each of the bottom edges of the box has length x inches, find the amount of material needed to construct the box. Write the answer in factored form.

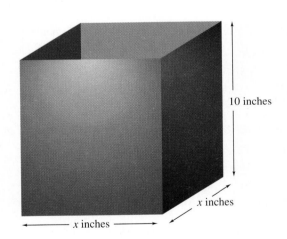

10 inches

x inches

x inches

A Look Ahead

Factor. Assume that variables used as exponents represent positive integers. See the following example.

EXAMPLE   Factor $x^{5a} - x^{3a} + x^{7a}$.

Solution: The variable x is common to all three terms, and the power $3a$ is the smallest of the exponents. So factor out the common factor x^{3a}.

$$x^{5a} - x^{3a} + x^{7a} = x^{3a}(x^{2a}) - x^{3a}(1) + x^{3a}(x^{4a})$$

$$= x^{3a}(x^{2a} - 1 + x^{4a}) \qquad \blacksquare$$

77. $x^{3n} - 2x^{2n} + 5x^n$ **78.** $3y^n + 3y^{2n} + 5y^{8n}$ **79.** $6x^{8a} - 2x^{5a} - 4x^{3a}$ **80.** $3x^{5a} - 6x^{3a} + 9x^{2a}$

Factor. See the following example.

> **EXAMPLE**   Factor $2x^{-1} + 10x^{-3}$.
>
> **Solution:**   The variable x is common to both terms. The smallest power of x is -3, so the GCF is $2x^{-3}$.
>
> $$2x^{-1} + 10x^{-3} = 2x^{-3}(x^2) + 2x^{-3}(5) = 2x^{-3}(x^2 + 5)$$ ■

81. $3x^{-2} + 8x^{-1}$ **82.** $5y^{-2} - 3y^{-1}$
83. $3x^2y^{-3} + 2xy^{-1}$ **84.** $2x^2y^{-2} + 7xy^{-3}$
85. $6x^{-2}y^{-1} - 2x^{-2}y^{-4} + 8xy^{-2}$ **86.** $9x^{-2}y^{-2} - 6x^{-3}y^{-1} - 3x^{-3}y^{-2}$

Writing in Mathematics

87. Define the GCF of a list of integers.
88. When $3x^2 - 9x + 3$ is factored, the result is $3(x^2 - 3x + 1)$. Explain why it is necessary to include the term 1 in this factored form.

89. Consider the following sequence of algebraic steps:
$$x^3 - 6x^2 + 2x - 10 = (x^3 - 6x^2) + (2x - 10)$$
$$= x^2(x - 6) + 2(x - 5)$$

Explain whether the final result is the factored form of the original polynomial.

Skill Review

Simplify the following. See Section 4.1.

90. $(5x^2)(11x^5)$ **91.** $(7y)(-2y^3)$
92. $(5x^2)^3$ **93.** $(-2y^3)^4$

Perform the subtraction. See Section 4.3.

94. Subtract $x^2 - 2x + 1$ from $9x^2 - 6$.   **95.** Subtract $10xy + 3$ from $-6xy + 8$.

5.2
Factoring Trinomials

OBJECTIVES

Tape 13

1   Factor trinomials of the form $x^2 + bx + c$.

2   Factor trinomials of the form $ax^2 + bx + c$.

3   Factor by substitution.

4   Factor trinomials by the AC method.

1   In the previous section, we used factoring by grouping to factor four-term polynomials. In this section, we present techniques for factoring trinomials. Since $(x - 2)(x + 5) = x^2 + 3x - 10$, we say that $(x - 2)(x + 5)$ is the factored form of

$x^2 + 3x - 10$. Taking a close look at how $(x - 2)$ and $(x + 5)$ are multiplied suggests a pattern for factoring trinomials of the form

$$x^2 + bx + c.$$

$$(x - 2)(x + 5) = x^2 + 3x - 10$$

$$x \cdot x = x^2$$
$$-2 + 5 = 3$$
$$-2 \cdot 5 = -10$$

The pattern for factoring is summarized next.

To Factor a Trinomial of the Form $x^2 + bx + c$

Step 1 Find two numbers whose product is c and whose sum is b.
Step 2 The factored form of $x^2 + bx + c$ is

$(x +$ one number$)(x +$ other number$)$

EXAMPLE 1   Factor $x^2 + 10x + 16$.

Solution: Look for two integers whose product is 16 and whose sum is 10. Since our integers must have a positive product and a positive sum, we look at positive factors of 16 only.

Positive Factors of 16	*Sum of Factors*
1, 16	$1 + 16 = 17$
4, 4	$4 + 4 = 8$
2, 8	$2 + 8 = 10$

The correct pair of numbers is 2 and 8 because their product is 16 and their sum is 10. Thus

$$x^2 + 10x + 16 = (x + \text{one number})(x + \text{other number})$$

or

$$x^2 + 10x + 16 = (x + 2)(x + 8)$$

To check, see that $(x + 2)(x + 8) = x^2 + 10x + 16$. ∎

EXAMPLE 2   Factor $x^2 - 12x + 35$.

Solution: Find two integers whose product is 35 and whose sum is -12. Since our integers must have a positive product and a negative sum, we consider negative factors of 35 only. The numbers are -5 and -7.

$$x^2 - 12x + 35 = [x + (-5)][x + (-7)]$$
$$= (x - 5)(x - 7)$$

To check, see that $(x - 5)(x - 7) = x^2 - 12x + 35$. ∎

EXAMPLE 3   Factor $5x^3 - 30x^2 - 35x$.

Solution: First, factor out a GCF of $5x$.

$$5x^3 - 30x^2 - 35x = 5x(x^2 - 6x - 7)$$

Next, factor $x^2 - 6x - 7$ by finding two numbers whose product is -7 and whose sum is -6. The numbers are 1 and -7.

$$5x^3 - 30x^2 - 35x = 5x(x^2 - 6x - 7)$$
$$= 5x(x + 1)(x - 7) \quad \blacksquare$$

HELPFUL HINT

If the polynomial to be factored contains a common factor that is factored out, don't forget to include that common factor in the final factored form of the original polynomial.

EXAMPLE 4   Factor $2n^2 - 38n + 80$.

Solution: The terms of this polynomial have a GCF of 2, which we factor out first.

$$2n^2 - 38n + 80 = 2(n^2 - 19n + 40)$$

Next, factor $n^2 - 19n + 40$ by finding two numbers whose product is 40 and whose sum is -19. Both numbers must be negative since their sum is -19. Possibilities are

$$-1 \text{ and } -40, \qquad -2 \text{ and } -20, \qquad -4 \text{ and } -10, \qquad -5 \text{ and } -8$$

None of the pairs has a sum of -19, so no further factoring with integers is possible. The factored form of $2n^2 - 38n + 80$ is

$$2n^2 - 38n + 80 = 2(n^2 - 19n + 40) \quad \blacksquare$$

We call a polynomial such as $n^2 - 19n + 10$ that cannot be factored with integers a **prime polynomial.**

2   Next, we factor trinomials of the form $ax^2 + bx + c$, where the coefficient a of x^2 is not 1. Don't forget that the first step in factoring any polynomial is to first factor out the GCF of its terms.

EXAMPLE 5   Factor $2x^2 + 11x + 15$.

Solution: Factors of $2x^2$ are $2x$ and x. Try these factors as first terms of the binomials.

$$2x^2 + 11x + 15 = (2x +)(x +)$$

Next, try combinations of factors of 15 until the correct middle term, $11x$, is obtained. We will try positive factors of 15 only since the coefficient of the middle term, 11, is positive. Positive factors of 15 are 1 and 15 and 3 and 5.

$(2x + \underbrace{1)(x + 15)}$
$\underbrace{}_{\displaystyle 1x}$
$30x$
$31x$, incorrect middle term

$(2x + \underbrace{15)(x + 1)}$
$\underbrace{}_{\displaystyle 15x}$
$2x$
$17x$, incorrect middle term

$(2x + \underbrace{3)(x + 5)}$
$\underbrace{}_{\displaystyle 3x}$
$10x$
$13x$, incorrect middle term

$(2x + \underbrace{5)(x + 3)}$
$\underbrace{}_{\displaystyle 5x}$
$6x$
$11x$, **correct** middle term

Thus the factored form of $2x^2 + 11x + 15$ is $(2x + 5)(x + 3)$. ■

To Factor a Trinomial of the Form $ax^2 + bx + c$

Step 1   Write all pairs of factors of ax^2.

Step 2   Write all pairs of factors of c, the constant term.

Step 3   Try various combinations of these factors until the correct middle term bx is found.

Step 4   If no combination exists, the polynomial is **prime.**

EXAMPLE 6   Factor $3x^2 - x - 4$.

Solution: Factors of $3x^2$ are $3x$ and x. Factors of -4 are -1 and 4 or -4 and 1. Try possible combinations of these factors.

$(3x - \underbrace{1)(x + 4)}$
$\underbrace{}_{\displaystyle -1x}$
$12x$
$11x$, incorrect middle term

$(3x + \underbrace{4)(x - 1)}$
$\underbrace{}_{\displaystyle 4x}$
$-3x$
$1x$, incorrect middle term

$(3x - \underbrace{4)(x + 1)}$
$\underbrace{}_{\displaystyle -4x}$
$3x$
$-1x$, **correct** middle term

Thus $3x^2 - x - 4 = (3x - 4)(x + 1)$. ■

EXAMPLE 7   Factor $12x^3y - 22x^2y + 8xy$.

Solution: First factor out the GCF of the terms of this trinomial, $2xy$.

$$12x^3y - 22x^2y + 8xy = 2xy(6x^2 - 11x + 4)$$

Now try to factor the trinomial $6x^2 - 11x + 4$. Factors of $6x^2$ are

$$6x^2 = 2x \cdot 3x, \qquad 6x^2 = 6x \cdot x$$

Try $2x$ and $3x$.

$$6x^2 - 11x + 4 = (2x + \,)(3x + \,)$$

The constant term 4 is positive and the coefficient of the middle term -11 is negative, so factor 4 into negative factors only. Negative factors of 4 are

$$4 = -4(-1), \qquad 4 = -2(-2)$$

Try -4 and -1.

$$2xy(2x \underbrace{- \; 4)(3x}_{-12x} - 1)$$

$$\underbrace{}_{-2x}$$

$$-14x$$

This combination cannot be correct, because one of the factors $(2x - 4)$ has a common factor of 2. This cannot happen if the original polynomial has no common factors. Try -1 and -4.

$$2xy(2x \underbrace{- \; 1)(3x}_{-3x} - 4)$$

$$\underbrace{}_{-8x}$$

$$-11x$$

If this combination had not worked, we would try -2 and -2 and then $6x$ and x as factors of $6x^2$.

$$12x^3y - 22x^2y + 8xy = 2xy(2x - 1)(3x - 4) \qquad \blacksquare$$

HELPFUL HINT

If a trinomial has no common factor (other than 1), then none of its binomial factors will contain a common factor (other than 1).

EXAMPLE 8   Factor $16x^2 + 24xy + 9y^2$.

Solution: No GCF can be factored out of this trinomial. Factors of $16x^2$ are

$$16x^2 = 16x \cdot x, \qquad 16x^2 = 8x \cdot 2x, \qquad 16x^2 = 4x \cdot 4x$$

Factors of $9y^2$ are

$$9y^2 = y \cdot 9y, \qquad 9y^2 = 3y \cdot 3y$$

Try possible combinations until the correct factorization is found.

$$16x^2 + 24xy + 9y^2 = (4x + 3y)(4x + 3y) \quad \text{or} \quad (4x + 3y)^2 \qquad \blacksquare$$

The trinomial $16x^2 + 24xy + 9y^2$ in Example 8 is an example of a **perfect square trinomial** since its factors are two identical binomials. In the next section, we examine a special method for factoring perfect square trinomials.

3   A complicated looking polynomial may be a simpler trinomial "in disguise." Revealing the simpler trinomial is possible by substitution.

EXAMPLE 9   Factor $2(a + 3)^2 - 5(a + 3) - 7$.

Solution: The quantity $(a + 3)$ is in two of the terms of this polynomial. **Substitute** x for $(a + 3)$ and the result is the following simpler trinomial:

$$2(a + 3)^2 - 5(a + 3) - 7 \qquad \text{Original trinomial.}$$
$$\downarrow \qquad\qquad \downarrow$$
$$= \quad 2(x)^2 \quad - \quad 5(x) \quad - 7 \qquad \text{Substitute } x \text{ for } (a + 3).$$

Now factor $2x^2 - 5x - 7$.

$$2x^2 - 5x - 7 = (2x - 7)(x + 1)$$

But the quantity in the original polynomial was $(a + 3)$, not x. Thus we need to reverse the substitution and replace x with $(a + 3)$.

$$(2x - 7) \; (x + 1) \qquad\qquad \text{Factored expression.}$$
$$= [2(a + 3) - 7][(a + 3) + 1] \qquad \text{Substitute } (a + 3) \text{ for } x.$$
$$= (2a + 6 - 7)(a + 3 + 1) \qquad \text{Remove inside parentheses.}$$
$$= (2a - 1)(a + 4) \qquad\qquad \text{Simplify.}$$

Thus $2(a + 3)^2 - 5(a + 3) - 7 = (2a - 1)(a + 4)$.   ■

EXAMPLE 10   Factor $5x^4 + 29x^2 - 42$.

Solution: Again, substitution may help us factor this polynomial more easily. Let $y = x^2$ so that $y^2 = (x^2)^2$ or x^4. Then

$$5x^4 + 29x^2 - 42$$

becomes

$$\downarrow \qquad \downarrow$$
$$5y^2 + 29y - 42$$

which factors as

$$5y^2 + 29y - 42 = (5y - 6)(y + 7)$$

Next, replace y with x^2 to get

$$(5x^2 - 6)(x^2 + 7) \quad ■$$

4   There is another method we can use when factoring trinomials of the form $ax^2 + bx + c$: write the trinomial as a four-term polynomial and then factor by grouping. The method is called the AC method.

To Factor a Trinomial of the Form $ax^2 + bx + c$ by the AC Method

Step 1   Find two numbers whose product is $a \cdot c$ and whose sum is b.
Step 2   Write the term bx as a sum using the factors found in *step 1*.
Step 3   Factor by grouping.

EXAMPLE 11   Factor $6x^2 + 13x + 6$.

Solution: In the trinomial, $a = 6$, $b = 13$, and $c = 6$.

Step 1   Find two numbers whose product is $a \cdot c$ or $6 \cdot 6 = 36$ and whose sum is b, 13.
The two numbers are 4 and 9.

Step 2   Write the middle term $13x$ as the sum $4x + 9x$.

Step 3   Factor $6x^2 + 4x + 9x + 6$ by grouping.

$$(6x^2 + 4x) + (9x + 6) = 2x(3x + 2) + 3(3x + 2)$$
$$= (3x + 2)(2x + 3) \quad \blacksquare$$

MENTAL MATH

1. Find two numbers whose product is 10 and whose sum is 7.
2. Find two numbers whose product is 12 and whose sum is 8.
3. Find two numbers whose product is 24 and whose sum is 11.
4. Find two numbers whose product is 30 and whose sum is 13.

EXERCISE SET 5.2

Factor each trinomial. See Examples 1 and 2.

1. $x^2 + 9x + 18$
2. $x^2 + 9x + 20$
3. $x^2 - 12x + 32$
4. $x^2 - 12x + 27$
5. $x^2 + 10x - 24$
6. $x^2 + 3x - 54$
7. $x^2 - 2x - 24$
8. $x^2 - 9x - 36$

Factor each trinomial completely. See Examples 3 and 4.

9. $3x^2 - 18x + 24$
10. $x^2y^2 + 4xy^2 + 3y^2$
11. $4x^2z + 28xz + 40z$
12. $5x^2 - 45x + 70$
13. $2x^2 + 30x - 108$
14. $3x^2 + 12x - 96$

Factor each trinomial. See Examples 5 through 7.

15. $5x^2 + 16x + 3$
16. $3x^2 + 8x + 4$
17. $2x^2 - 11x + 12$
18. $3x^2 - 19x + 20$
19. $2x^2 + 25x - 20$
20. $6x^2 - 13x - 8$
21. $4x^2 - 12x + 9$
22. $25x^2 - 30x + 9$
23. $12x^2 + 10x - 50$
24. $12y^2 - 48y + 45$
25. $3y^4 - y^3 - 10y^2$
26. $2x^2z + 5xz - 12z$
27. $6x^3 + 8x^2 + 24x$
28. $18y^3 + 12y^2 + 2y$

Factor each trinomial. See Example 8.

29. $x^2 + 8xz + 7z^2$
30. $a^2 - 2ab - 15b^2$

31. $2x^2 - 5xy - 3y^2$

32. $6x^2 + 11xy + 4y^2$

Factor each polynomial completely. See Examples 9 and 10.

33. $x^4 + x^2 - 6$

34. $x^4 - x^2 - 20$

35. $(5x + 1)^2 + 8(5x + 1) + 7$

36. $(3x - 1)^2 + 5(3x - 1) + 6$

37. $x^6 - 7x^3 + 12$

38. $x^6 - 4x^3 - 12$

39. $(a + 5)^2 - 5(a + 5) - 24$

40. $(3c + 6)^2 + 12(3c + 6) - 28$

Use the AC method to factor each trinomial completely. See Example 11.

41. $x^2 - x - 12$

42. $x^2 + 4x - 5$

43. $28y^2 + 22y + 4$

44. $24y^3 - 2y^2 - y$

45. $2x^2 + 15x - 27$

46. $3x^2 + 14x + 15$

Factor each polynomial completely.

47. $x^2 - 24x - 81$

48. $x^2 - 48x - 100$

49. $x^2 - 15x - 54$

50. $x^2 - 15x + 54$

51. $3x^2 - 6x + 3$

52. $8x^2 - 8x + 2$

53. $3x^2 - 5x - 2$

54. $5x^2 - 14x - 3$

55. $8x^2 - 26x + 15$

56. $12x^2 - 17x + 6$

57. $18x^4 + 21x^3 + 6x^2$

58. $20x^5 + 54x^4 + 10x^3$

59. $3a^2 + 12ab + 12b^2$

60. $2x^2 + 16xy + 32y^2$

61. $x^2 + 4x + 5$

62. $x^2 + 6x + 8$

63. $2(x + 4)^2 + 3(x + 4) - 5$

64. $3(x + 3)^2 + 2(x + 3) - 5$

65. $6x^2 - 49x + 30$

66. $4x^2 - 39x + 27$

67. $x^4 - 5x^2 - 6$

68. $x^4 - 5x^2 + 6$

69. $6x^3 - x^2 - x$

70. $12x^3 + x^2 - x$

71. $12a^2 - 29ab + 15b^2$

72. $16y^2 + 6yx - 27x^2$

73. $9x^2 + 30x + 25$

74. $4x^2 + 6x + 9$

75. $3x^2y - 11xy + 8y$

76. $5xy^2 - 9xy + 4x$

77. $2x^2 + 2x - 12$

78. $3x^2 + 6x - 45$

79. $(x - 4)^2 + 3(x - 4) - 18$

80. $(x - 3)^2 - 2(x - 3) - 8$

81. $2x^6 + 3x^3 - 9$

82. $3x^6 - 14x^3 + 8$

83. $72xy^4 - 24xy^2z + 2xz^2$

84. $36xy^2 - 48xyz^2 + 16xz^4$

85. The volume V of a box in terms of its height h is given by the formula $V = 3h^3 - 2h^2 - 8h$. Factor this expression for V.

86. Based on your results from Exercise 85, find the length and width of the box if the height is 5 inches and the dimensions of the box are whole numbers.

A Look Ahead

Factor. Assume that variables used as exponents represent positive integers. See the following example.

EXAMPLE Factor $x^{2n} + 7x^n + 12$.

Solution: Factors of x^{2n} are x^n and x^n, so $x^{2n} + 7x^n + 12 = (x^n + \text{one number})(x^n + \text{other number})$. Factors of 12 whose sum is 7 are 3 and 4. Thus

$$x^{2n} + 7x^n + 12 = (x^n + 4)(x^n + 3) \qquad \blacksquare$$

87. $x^{2n} + 10x^n + 16$

88. $x^{2n} - 7x^n + 12$

89. $x^{2n} - 3x^n - 18$

90. $x^{2n} + 7x^n - 18$

91. $2x^{2n} + 11x^n + 5$

92. $3x^{2n} - 8x^n + 4$

93. $4x^{2n} - 12x^n + 9$

94. $9x^{2n} + 24x^n + 16$

Skill Review

Multiply the following. See Section 4.4.

95. $(x - 2)(x^2 + 2x + 4)$

96. $(y + 1)(y^2 - y + 1)$

If $P(x) = 3x^2 + 2x - 9$, find the following. See Section 4.3.

97. $P(0)$

98. $P(1)$

99. $P(-1)$

100. $P(-2)$

5.3
Factoring by Special Products

OBJECTIVES

Tape 14

1   Factor a perfect square trinomial.

2   Factor the difference of two squares.

3   Factor the sum or difference of two cubes.

---

1   In the last section, we considered a variety of ways to factor trinomials of the form $ax^2 + bx + c$. In one particular example, we factored $16x^2 + 24x + 9$ as

$$16x^2 + 24x + 9 = (4x + 3)^2$$

We call $16x^2 + 24x + 9$ a **perfect square trinomial** because its factors are two identical binomials. A perfect square trinomial can be factored quickly if you recognize the trinomial as a perfect square. The following special formulas can be used.

> **Perfect Square Trinomials**
>
> $$a^2 + 2ab + b^2 = (a + b)^2$$
>
> $$a^2 - 2ab + b^2 = (a - b)^2$$

A trinomial is a perfect square trinomial if it can be written so that its first term is the square of some quantity a, its last term is the square of some quantity b, and its middle term is twice the product of the quantities a and b.

From $a^2 + 2ab + b^2 = (a + b)^2$,   we see that

$$16x^2 + 24x + 9 = (4x)^2 + 2(4x)(3) + 3^2 = (4x + 3)^2$$

EXAMPLE 1 Factor $m^2 + 10m + 25$.

Solution: Notice that the first term $m^2 = (m)^2$, the last term $25 = 5^2$, and $10m = 2 \cdot 5 \cdot m$. Thus,

$$m^2 + 10m + 25 = m^2 + 2(m)(5) + 5^2 = (m + 5)^2 \qquad \blacksquare$$

EXAMPLE 2 Factor $3a^2x - 12abx + 12b^2x$.

Solution: The terms of this trinomial have a GCF of $3x$, which we factor out first.

$$3a^2x - 12abx + 12b^2x = 3x(a^2 - 4ab + 4b^2)$$

Now, the polynomial $a^2 - 4ab + 4b^2$ is a perfect square trinomial. Notice that the first term $a^2 = (a)^2$, the last term $4b^2 = (2b)^2$, and $4ab = 2(a)(2b)$. The factoring can now be completed as

$$3x(a^2 - 4ab + 4b^2) = 3x(a - 2b)^2 \qquad \blacksquare$$

HELPFUL HINT

If you recognize a trinomial as a perfect square trinomial, use the special formulas to factor. However, methods for factoring trinomials from Section 5.2 will also result in the correct factored form.

---

2 We now factor special types of binomials, beginning with the **difference of two squares.** The formula presented in Section 4.4 for the product of a sum and a difference of two terms is used again here. However, the emphasis is now on factoring, rather than on multiplying.

Difference of Two Squares

$$a^2 - b^2 = (a + b)(a - b)$$

EXAMPLE 3 Factor the following:
  a. $x^2 - 9$ **b.** $16y^2 - 9$ **c.** $50 - 8y^2$

Solution: **a.** $x^2 - 9 = x^2 - 3^2$ **b.** $16y^2 - 9 = (4y)^2 - 3^2$

$$= (x + 3)(x - 3) \qquad\qquad = (4y + 3)(4y - 3)$$

  c. $50 - 8y^2 = 2(25 - 4y^2)$

$$= 2(5 + 2y)(5 - 2y) \qquad \blacksquare$$

The binomial $x^2 + 9$ is a **sum of two squares** and cannot be factored using real numbers. **In general, except for factoring out a GCF, the sum of two squares cannot be factored using real numbers.**

HELPFUL HINT

The sum of two squares whose GCF is 1 cannot be factored using real numbers.

EXAMPLE 4   Factor the following:

a. $p^4 - 16$ **b.** $(x + 3)^2 - 36$

Solution: **a.** $p^4 - 16 = (p^2)^2 - 4^2$

$$= (p^2 + 4)(p^2 - 4)$$

The binomial factor $p^2 + 4$ cannot be factored using real numbers, but the binomial factor $p^2 - 4$ is a difference of squares.

$$(p^2 + 4)(p^2 - 4) = (p^2 + 4)(p + 2)(p - 2)$$

b. Factor $(x + 3)^2 - 36$ as the difference of squares.

$$(x + 3)^2 - 36 = (x + 3)^2 - 6^2$$

$$= [(x + 3) + 6][(x + 3) - 6] \qquad \text{Factor.}$$

$$= [x + 3 + 6][x + 3 - 6] \qquad \text{Remove parentheses.}$$

$$= (x + 9)(x - 3) \qquad \text{Simplify.} \quad \blacksquare$$

3   Although the sum of two squares cannot be factored, the sum of two cubes can be factored, as well as the difference of two cubes, as follows.

Sum and Difference of Two Cubes

$$a^3 + b^3 = (a + b)(a^2 - ab + b^2)$$

$$a^3 - b^3 = (a - b)(a^2 + ab + b^2)$$

To check the first formula, find the product of $(a + b)$ and $(a^2 - ab + b^2)$.

$$
\begin{array}{r}
a^2 - ab + b^2 \\
a + b \\
\hline
a^2 b - ab^2 + b^3 \\
a^3 - a^2 b + ab^2 \\
\hline
a^3 + b^3
\end{array}
$$

Then $a^3 + b^3 = (a + b)(a^2 - ab + b^2)$.

The formula $a^3 - b^3 = (a - b)(a^2 + ab + b^2)$ can also be checked by multiplication.

EXAMPLE 5   Factor $x^3 + 8$.

Solution: Use the formula $a^3 + b^3 = (a + b)(a^2 - a \cdot b + b^2)$. Then

$$x^3 + 8 = x^3 + 2^3 = (x + 2)(x^2 - x \cdot 2 + 2^2)$$

Thus, $x^3 + 8 = (x + 2)(x^2 - 2x + 4)$. $\blacksquare$

EXAMPLE 6   Factor $p^3 + 27q^3$.

Solution: $p^3 + 27q^3 = p^3 + (3q)^3 = (p + 3q)[p^2 - (p)(3q) + (3q)^2]$

$$= (p + 3q)(p^2 - 3pq + 9q^2) \quad \blacksquare$$

EXAMPLE 7   Factor $y^3 - 64$.

Solution: This is a difference of cubes since $y^3 - 64 = y^3 - 4^3$.

From $a^3 - b^3 = (a - b)(a^2 + a \cdot b + b^2)$ we have that

$$y^3 - 4^3 = (y - 4)(y^2 + y \cdot 4 + 4^2)$$

$$= (y - 4)(y^2 + 4y + 16) \quad \blacksquare$$

EXAMPLE 8   Factor $125q^2 - n^3q^2$.

Solution: First, factor out a common factor of q^2.

$$125q^2 - n^3q^2 = q^2(125 - n^3)$$

$$= q^2(5^3 - n^3)$$

$$= q^2(5 - n)[5^2 + (5)(n) + (n^2)]$$

$$= q^2(5 - n)(25 + 5n + n^2)$$

Thus $125q^2 - n^3q^2 = q^2(5 - n)(25 + 5n + n^2)$. The trinomial $25 + 5n + n^2$ cannot be factored further. $\blacksquare$

EXAMPLE 9   Factor $x^2 + 4x + 4 - y^2$.

Solution: Factoring by grouping comes to mind since the sum of the first three terms of this polynomial is a perfect square trinomial.

$$x^2 + 4x + 4 - y^2 = (x^2 + 4x + 4) - y^2 \qquad \text{Group the first three terms.}$$

$$= (x + 2)^2 - y^2 \qquad \text{Factor the perfect square trinomial.}$$

This is not factored yet since we have a **difference,** not a **product.** Since $(x + 2)^2 - y^2$ is a difference of squares, we have

$$(x + 2)^2 - y^2 = [(x + 2) + y][(x + 2) - y]$$

$$= (x + 2 + y)(x + 2 - y) \quad \blacksquare$$

EXERCISE SET 5.3

Factor the following. See Examples 1 and 2.

1. $x^2 + 6x + 9$

2. $x^2 - 10x + 25$

3. $4x^2 - 12x + 9$

4. $25x^2 + 10x + 1$

5. $3x^2 - 24x + 48$

6. $x^3 + 14x^2 + 49x$

7. $9y^2x^2 + 12yx^2 + 4x^2$

8. $32x^2 - 16xy + 2y^2$

Factor the following. See Examples 3 and 4.

9. $x^2 - 25$

10. $y^2 - 100$

11. $9 - 4z^2$

12. $16x^2 - y^2$

13. $(y + 2)^2 - 49$

14. $(x - 1)^2 - z^2$

15. $64x^2 - 100$

16. $4x^2 - 36$

Factor the following. See Examples 5 and 6.

17. $x^3 + 27$

18. $y^3 + 1$

19. $m^3 + n^3$

20. $r^3 + 125$

21. $a^3b + 8b^4$

22. $8ab^3 + 27a^4$

Factor the following. See Examples 7 and 8.

23. $z^3 - 1$

24. $x^3 - 8$

25. $x^3y^2 - 27y^2$

26. $64 - p^3$

27. $125y^3 - 8x^3$

28. $54y^3 - 128$

Factor the following. See Example 9.

29. $x^2 + 6x + 9 - y^2$

30. $x^2 + 12x + 36 - y^2$

31. $x^2 - 10x + 25 - y^2$

32. $x^2 - 18x + 81 - y^2$

33. $4x^2 + 4x + 1 - z^2$

34. $9y^2 + 12y + 4 - x^2$

Factor each polynomial completely.

35. $x^2 - 16$

36. $x^2 - 81$

37. $9x^2 - 49$

38. $25x^2 - 4$

39. $x^4 - 81$

40. $x^4 - 256$

41. $x^2 + 8x + 16 - 4y^2$

42. $x^2 + 14x + 49 - 9y^2$

43. $(x + 2y)^2 - 9$

44. $(3x + y)^2 - 25$

45. $x^3 - 1$

46. $x^3 - 8$

47. $x^3 + 125$

48. $x^3 + 216$

49. $4x^2 + 25$

50. $16x^2 + 25$

51. $4a^2 - 81b^2$

52. $49a^2 - 16b^2$

53. $18x^2y - 2y$

54. $12xy^2 - 108x$

55. $(4x - 1)^2 - 100$

56. $(2x - 3)^2 - 64$

57. $x^3 + 64$

58. $x^3 + 343$

59. $x^6 - y^3$

60. $x^3 - y^6$

61. $x^2 + 16x + 64 - x^4$

62. $x^2 + 20x + 100 - x^4$

63. $y^2 - x^2 - 6x - 9$

64. $y^2 - x^2 - 10x - 25$

65. $6x^2 - 24$

66. $28x^2 - 7$

67. $4x^2 + 16$

68. $36x^2 + 16$

69. $x^8 - 1$

70. $x^{16} - 256$

71. $3x^6y^2 + 81y^2$

72. $x^2y^9 + x^2y^3$

73. $(3x - 2)^2 - 16$

74. $(7x + 2)^2 - 9$

75. $(x + y)^3 + 125$

76. $(x + y)^3 + 27$

77. $(x - y)^3 + 1$

78. $(x - y)^3 + 64$

79. $(2x - y)^3 + 27$

80. $(3x - y)^3 + 125$

81. $(2x + 3)^3 - 64$

82. $(4x + 2)^3 - 125$

83. The manufacturer of Antonio's Metal Washers needs to determine the cross-sectional area of each washer. If the outer radius of the washer is R and the radius of the hole is r, express the area of the washer as a polynomial. Factor this polynomial completely.

84. The manufacturer of Tootsie Roll Pops is planning on changing the size of its candy slightly. To compute the new cost, they need a formula for the volume of the candy coating without the Tootsie Roll center. Given the following diagram, express the volume as a polynomial. Factor this polynomial completely.

6 mm

R mm

A Look Ahead

Factor each expression. Assume that variables used as exponents represent positive integers. See the following example.

EXAMPLE   Factor $x^{2n} - 100$.

Solution: This binomial is a difference of squares.

$$x^{2n} - 100 = (x^n)^2 - 10^2$$

$$= (x^n + 10)(x^n - 10) \quad \blacksquare$$

85. $x^{2n} - 25$ **86.** $x^{2n} - 36$ **87.** $x^{2n} - 9$

88. $x^{2n} - 16$ **89.** $36x^{2n} - 49$ **90.** $25x^{2n} - 81$

91. $x^{4n} - 16$ **92.** $x^{4n} - 625$

Writing in Mathematics

93. Knowing that $x - 2$ is a factor of the difference of the fifth powers $x^5 - 32$, explain how to find another factor.

94. Factor $x^6 - 1$ completely, using the following methods from this chapter.
  a. Factor the expression treating it as the difference of two squares, $(x^3)^2 - 1^2$.

  b. Factor the expression treating it as the difference of two cubes, $(x^2)^3 - 1^3$.

  c. Are the answers to parts a and b the same? Why or why not?

Skill Review

Divide by long division. See Section 4.5.

95. $(x^5 - 32) \div (x - 2)$

96. $(x^4 - 3x^3 + 2x - 5) \div (x - 1)$

Simplify the following. Write answers with positive exponents. See Section 4.2.

97. 3^{-3}

98. $x^{-3} \cdot x^7$

99. $\dfrac{y^{-4}}{y^{-9}}$

100. $5 \cdot 4^{-2}$

5.4
Factoring Polynomials Completely

OBJECTIVE   **1**   Practice techniques for factoring polynomials.

Tape 14

1   The key to proficiency in factoring is to practice until you are comfortable with each technique. A summary of the methods of factoring is given next.

To Factor a Polynomial

Step 1   Are there any common factors? If so, factor them out.

Step 2   How many terms are in the polynomial?

    a. If there are **two** terms, decide if one of the following formulas may be applied.

      i. Difference of two squares: $a^2 - b^2 = (a - b)(a + b)$.

      ii. Difference of two cubes: $a^3 - b^3 = (a - b)(a^2 + ab + b^2)$.

      iii. Sum of two cubes: $a^3 + b^3 = (a + b)(a^2 - ab + b^2)$.

    b. If there are **three** terms, try one of the following.

      i. Perfect square trinomial: $a^2 + 2ab + b^2 = (a + b)^2$.
$$a^2 - 2ab + b^2 = (a - b)^2$$

      ii. If not a perfect square trinomial, factor using the methods presented in Section 5.2.

    c. If there are **four** or more terms, try factoring by grouping.

Step 3   See if any factors in the factored polynomial can be factored further.

EXAMPLE 1   Factor each polynomial completely.

    a. $8a^2b - 4ab$ **b.** $36x^2 - 9$ **c.** $2x^2 - 5x - 7$

Solution: **a.** The terms have a common factor of $4ab$, which we factor out.

$$8a^2b - 4ab = 4ab(2a - 1)$$

    b. $36x^2 - 9 = 9(4x^2 - 1)$ Factor out a GCF of 9.

$$= 9(2x + 1)(2x - 1)$$ Factor the difference of squares.

    c. $2x^2 - 5x - 7 = (2x - 7)(x + 1)$ ∎

EXAMPLE 2   Factor each polynomial completely.

    a. $5p^2 + 5 + qp^2 + q$ **b.** $9x^2 + 24x + 16$ **c.** $y^2 + 25$

Solution: **a.** There is no common factor of all terms of $5p^2 + 5 + qp^2 + q$. The polynomial has four terms, so try factoring by grouping.

$$5p^2 + 5 + qp^2 + q = (5p^2 + 5) + (qp^2 + q) \qquad \text{Group the terms.}$$
$$= 5(p^2 + 1) + q(p^2 + 1)$$
$$= (p^2 + 1)(5 + q)$$

b. The trinomial $9x^2 + 24x + 16$ is a perfect square trinomial and $9x^2 + 24x + 16 = (3x + 4)^2$.

c. There is no common factor of $y^2 + 25$ other than 1. This binomial is the sum of two squares and is prime.   ■

EXAMPLE 3   Factor each completely.
a. $27a^3 - b^3$ **b.** $3n^2m^4 - 48m^6$ **c.** $2x^2 - 12x + 18 - 2z^2$
d. $8x^4y^2 + 125xy^2$ **e.** $(x - 5)^2 - 49y^2$

Solution: **a.** This binomial is a difference of two cubes.

$$27a^3 - b^3 = (3a)^3 - b^3$$
$$= (3a - b)[(3a)^2 + (3a)(b) + b^2]$$
$$= (3a - b)(9a^2 + 3ab + b^2)$$

b. $3n^2m^4 - 48m^6 = 3m^4(n^2 - 16m^2)$ Factor out the GCF, $3m^4$.
$$= 3m^4(n + 4m)(n - 4m)$$ Factor the difference of squares.

c. $2x^2 - 12x + 18 - 2z^2 = 2(x^2 - 6x + 9 - z^2)$ The GCF is 2.
$$= 2[(x^2 - 6x + 9) - z^2]$$ Group the first three terms together.
$$= 2[(x - 3)^2 - z^2]$$ Factor the perfect square trinomial.
$$= 2[(x - 3) + z][(x - 3) - z]$$ Factor the difference of squares.
$$= 2(x - 3 + z)(x - 3 - z)$$

d. $8x^4y^2 + 125xy^2 = xy^2(8x^3 + 125)$ The GCF is xy^2.
$$= xy^2[(2x)^3 + 5^3]$$
$$= xy^2(2x + 5)[(2x)^2 - (2x)(5) + 5^2]$$ Factor the sum of cubes.
$$= xy^2(2x + 5)(4x^2 - 10x + 25)$$

e. This binomial is a difference of squares.

$$(x - 5)^2 - 49y^2 = (x - 5)^2 - (7y)^2$$
$$= [(x - 5) + 7y][(x - 5) - 7y]$$
$$= (x - 5 + 7y)(x - 5 - 7y)$$ ■

EXERCISE SET 5.4

Factor completely.

1. $x^2 - 9$
3. $x^2 - 8x - 9$
5. $x^2 + 8$
7. $x^2 - 8x + 16 - y^2$
9. $x^4 - x$
11. $14x^2y - 2xy$
13. $8a^2b - 6ab^2$
15. $x^4 - x^3$
17. $x^4 - 4x^2$
19. $4x^2 - 16$
21. $x - 9x^3$
23. $3x^2 - 8x - 11$
25. $4x^2 + 8x - 12$
27. $4x^2 + 23x + 15$
29. $6x^2 - 8 + 3x^2y - 4y$
31. $2xy + 6y + 8x + 24$
33. $x^3 + 3x^2 - 4x - 12$
35. $4x^2 + 36x + 81$
37. $9x^2 - 30x + 25$
39. $16x^2 + 12x + 9$
41. $48x^2 - 24x + 3$
43. $x^2 + 16$
45. $2x^4 - 2$
47. $a^3 - 8b^3$
49. $125 - x^3$
51. $8x^3 + 27y^3$
53. $2a^3 + 128b^3$
55. $6a^2b^4 - 24a^4$
57. $8x^2y^5 - 2y^3$
59. $x^2 + 6x + 9 - y^2$
61. $2x^2 - 20x + 50 - 2y^2$
63. $3a^2 - 6a + 3 - 3b^2$
65. $27x^4 - xy^3$
67. $16x^3y^4 + 54y$
69. $64x^2y^3 - 8x^2$
71. $(3x + 1)^2 + 3(3x + 1) + 2$
73. $x^4 + 6x^2 - 7$
75. $x^6 + 4x^3 + 3$
77. $(x + y)^2 - 9$
79. $(a - 4)^3 + 1$
81. $(x + 5)^3 + y^3$
83. $(2x + 1)^3 + 27$
85. $27 - (x + 3)^3$

2. $6x - 9$
4. $x^3 + 8$
6. $x^2 - 2xy - 3x + 6y$
8. $12x^2 - 22x - 20$
10. $(2x + 1)^2 - 3(2x + 1) + 2$
12. $24ab^2 - 6ab$
14. $15x^2y^4 - 6x^3y^2$
16. $4y^4 - 2y$
18. $x^6 - x^4$
20. $9x^2 - 81$
22. $y^2 - 25y^4$
24. $5x^2 - 2x - 3$
26. $6x^2 - 6x - 12$
28. $6x^2 + 19x + 10$
30. $10a^2 - 15 + 2a^2b - 3b$
32. $6xy + 30x + 3y + 15$
34. $x^3 + 2x^2 - 9x - 18$
36. $25x^2 + 40x + 16$
38. $4x^2 - 28x + 49$
40. $18x^2 + 12x + 2$
42. $9x^2 - 6x + 4$
44. $x^4 - 16$
46. $3x^4 + 3$
48. $64a^3 - b^3$
50. $27 - y^3$
52. $125x^3 + 8y^3$
54. $5x^3 - 40y^3$
56. $8a^3b^6 - 2a$
58. $6x^3 - 96x^5$
60. $x^2 - 4x + 4 - y^2$
62. $3y^2 - 24y + 48 - 3x^2$
64. $8a^2 + 8a + 2 - 2b^2$
66. $8xy^5 - x^4y^2$
68. $2xy^4 + 250xy$
70. $27x^5y^4 - 216x^2y$
72. $(2x + 3)^2 + 6(2x + 3) + 5$
74. $x^4 - 2x^2 - 8$
76. $x^6 + 6x^3 - 16$
78. $z^2 - (x + 2)^2$
80. $(a + b)^3 + 8$
82. $(y - 1)^3 + 27x^3$
84. $(3x + 2)^3 + 8$
86. $64 - (x + 2)^3$

87. Three inches of matting is placed around a square picture with each side x inches, as shown. Write an expression, in factored form, for the area of the matting.

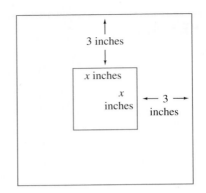

A Look Ahead

See A Look Ahead examples from Sections 5.2 and 5.3.

Factor completely.

88. $x^{2n} - 9$ **89.** $x^{2n} - 25$ **90.** $x^{4n} - 81$

91. $x^{4n} - 16$ **92.** $x^{3p} - 8$ **93.** $y^{3n} - 27$

Writing in Mathematics

94. Explain why a negative constant term of a trinomial such as $x^2 - x - 6$ indicates that the signs of the second terms of its binomial factors are opposite.

Skill Review

Solve the following equations. See Section 2.1.

95. $x - 5 = 0$ **96.** $2x + 7 = 0$ **97.** $3x + 1 = 0$ **98.** $-5x - 15 = 0$

Evaluate each polynomial given the replacement values. See Section 4.3.

99. $x^2 - 5x$; $x = -3$ **100.** $y^3 + 2y - 6$; $y = -1$ **101.** $a + 10$; $a = -7$ **102.** $b^2 - b$; $b = -2$

5.5
Solving Equations by Factoring

OBJECTIVES

1 Solve quadratic equations by factoring.

2 Solve higher-degree equations by factoring.

3 Solve word problems that can be modeled by quadratic equations.

Tape 15

In this section, your efforts to learn factoring start to pay off. We use factoring to solve equations that contain polynomials of degree 2 and higher. The following equations are examples of

Second Degree or Quadratic Equations

$$x^2 = 25 \qquad y^2 + 7y - 8 = 0 \qquad 5n^2 = 2n - 12 \qquad z(z - 3) = 2$$

Quadratic Equation in One Variable

A **quadratic equation** in one variable is one that can be written in the form

$$ax^2 + bx + c = 0$$

where a, b, and c are real numbers and $a \neq 0$.

---

1   The form $ax^2 + bx + c = 0$ is called the **standard form** of a quadratic equation. The following are examples of quadratic equations and the equivalent equations written in standard form.

Quadratic equations:	$x^2 = 25$	$5n^2 = 2n - 12$	$z(z - 3) = 2$
Standard form:	$x^2 - 25 = 0$	$5n^2 - 2n + 12 = 0$	$z^2 - 3z - 2 = 0$

A solution of a quadratic equation in one variable is a value of the variable that makes the equation true. The method presented in this section for solving quadratic equations is called the **factoring method.** This method is based on the **zero-factor property.**

Zero-Factor Property

If a and b are real numbers and $a \cdot b = 0$, then $a = 0$ or $b = 0$.

In other words, if the product of two real numbers is zero, then at least one number must be zero.

EXAMPLE 1   Solve $(x + 2)(x - 6) = 0$.

Solution: By the zero-factor property, $(x + 2)(x - 6) = 0$ only if $x + 2 = 0$ or $x - 6 = 0$.

$$x + 2 = 0 \quad \text{or} \quad x - 6 = 0 \qquad \text{Apply zero-factor property.}$$
$$x = -2 \quad \text{or} \quad x = 6 \qquad \text{Solve each linear equation.}$$

To check, let $x = -2$ and then let $x = 6$ in the original equation.

Let $x = -2$.   Let $x = 6$.

Then $(x + 2)(x - 6) = 0$ Then $(x + 2)(x - 6) = 0$

becomes $(-2 + 2)(-2 - 6) = 0$ becomes $(6 + 2)(6 - 6) = 0$.

$(0)(-8) = 0$ $(8)(0) = 0$

$0 = 0$ True. $0 = 0$ True.

Both -2 and 6 check, and the solution set is $\{-2, 6\}$.   ∎

EXAMPLE 2 Solve $2x^2 + 9x - 5 = 0$.

Solution: To use the zero-factor property, the quadratic expression must be in factored form.

$$2x^2 + 9x - 5 = 0$$

$$(2x - 1)(x + 5) = 0 \qquad \text{Factor.}$$

$$2x - 1 = 0 \quad \text{or} \quad x + 5 = 0 \qquad \text{Set each factor equal to zero.}$$

$$2x = 1$$

$$x = \frac{1}{2} \quad \text{or} \qquad x = -5 \qquad \text{Solve each linear equation.}$$

The solution set is $\left\{-5, \dfrac{1}{2}\right\}$. To check, let $x = \dfrac{1}{2}$ in the original equation; then let $x = -5$ in the original equation. ∎

The general procedure to solve a quadratic equation by factoring is given next.

To Solve Quadratic Equations by Factoring

Step 1 Write the equation in standard form: $ax^2 + bx + c = 0$.
Step 2 Factor the quadratic expression.
Step 3 Set each factor containing a variable equal to 0.
Step 4 Solve the resulting equations.
Step 5 Check each solution in the original equation.

EXAMPLE 3 Solve $3(x^2 + 4) + 5 = -6(x^2 + 2x) + 13$.

Solution: Rewrite the equation so that one side is 0.

$$3(x^2 + 4) + 5 = -6(x^2 + 2x) + 13$$

$$3x^2 + 12 + 5 = -6x^2 - 12x + 13 \qquad \text{Apply the distributive property.}$$

$$9x^2 + 12x + 4 = 0 \qquad \text{Rewrite the equation so that one side is 0.}$$

$$(3x + 2)(3x + 2) = 0 \qquad \text{Factor.}$$

$$3x + 2 = 0 \quad \text{or} \quad 3x + 2 = 0 \qquad \text{Set each factor equal to 0.}$$

$$3x = -2 \quad \text{or} \qquad 3x = -2$$

$$x = -\frac{2}{3} \quad \text{or} \qquad x = -\frac{2}{3} \qquad \text{Solve each equation.}$$

The solution set is $\left\{-\dfrac{2}{3}\right\}$. Check by substituting $-\dfrac{2}{3}$ into the original equation. ∎

If the equation contains fractions, we clear the equation of fractions as a first step.

EXAMPLE 4   Solve $2x^2 = \dfrac{17}{3}x + 1$.

Solution:
$$2x^2 = \frac{17}{3}x + 1$$

$$3(2x^2) = 3\left(\frac{17}{3}x + 1\right)$$ Clear the equation of fractions.

$$6x^2 = 17x + 3$$ Apply the distributive property.

$$6x^2 - 17x - 3 = 0$$ Rewrite the equation in standard form.

$$(6x + 1)(x - 3) = 0$$ Factor.

$$6x + 1 = 0 \quad \text{or} \quad x - 3 = 0$$ Set each factor equal to zero.

$$6x = -1$$

$$x = -\frac{1}{6} \quad \text{or} \quad x = 3$$ Solve each equation.

The solution set is $\left\{-\dfrac{1}{6}, 3\right\}$.   ■

2   Since the zero-factor property can extend to more than two numbers whose product is 0, we can apply it to solving any equation in which one side of the equation is 0 and the other side is a factored expression.

EXAMPLE 5   Solve $x^3 = 4x$.

Solution:
$$x^3 = 4x$$

$$x^3 - 4x = 0$$ Rewrite the equation so that one side is 0.

$$x(x^2 - 4) = 0$$ Factor out the GCF x.

$$x(x + 2)(x - 2) = 0$$ Factor the difference of squares.

$$x = 0 \quad \text{or} \quad x + 2 = 0 \quad \text{or} \quad x - 2 = 0$$ Set each factor equal to 0.

$$x = 0 \quad \text{or} \quad x = -2 \quad \text{or} \quad x = 2$$ Solve each equation.

The solution set is $\{-2, 0, 2\}$. Check by substituting into the original equation.   ■

Notice that the **third**-degree equation of Example 5 yielded **three** solutions.

> **HELPFUL HINT**
>
> In Example 5, it is incorrect to divide both sides of the equation by x. If x is 0, dividing by x is dividing by 0, and this is not allowed.

EXAMPLE 6 Solve $x^3 + 5x^2 = x + 5$.

Solution: First, write the equation so that one side is 0.

$$x^3 + 5x^2 - x - 5 = 0$$

$$(x^3 - x) + (5x^2 - 5) = 0 \qquad \text{Factor by grouping.}$$

$$x(x^2 - 1) + 5(x^2 - 1) = 0$$

$$(x^2 - 1)(x + 5) = 0$$

$$(x + 1)(x - 1)(x + 5) = 0 \qquad \text{Factor the difference of squares.}$$

$$x + 1 = 0 \quad \text{or} \quad x - 1 = 0 \quad \text{or} \quad x + 5 = 0 \qquad \text{Set each factor equal to 0.}$$

$$x = -1 \text{ or} \qquad x = 1 \quad \text{or} \qquad x = -5 \qquad \text{Solve each equation.}$$

The solution set is $\{-5, -1, 1\}$. Check in the original equation. ∎

3 Solutions of word problems may involve the solutions of an equation that is quadratic or has higher degree. We will follow the same procedure for solving word problems that was previously used, making sure all solutions are checked. Many times it is necessary to reject one or more solutions because they don't fit the physical constraints of the problem.

EXAMPLE 7 The length of a rectangular piece of carpet is 2 meters less than 5 times its width. Find the dimensions of the carpet if its area is 16 square meters.

Solution: Let x = width of rectangle; then from the problem we have that $5x - 2$ = length of rectangle.

$(5x - 2)$ meters

x meters

Since the area of a rectangle is the product of width and length, we have the equation $x(5x - 2) = 16$. Solve this equation for x.

$$x(5x - 2) = 16$$

$$5x^2 - 2x = 16$$

$$5x^2 - 2x - 16 = 0$$

$$(5x + 8)(x - 2) = 0$$

$$5x + 8 = 0 \quad \text{or} \quad x - 2 = 0$$

$$5x = -8 \quad \text{or} \qquad x = 2$$

$$x = -\frac{8}{5} \quad \text{or} \qquad x = 2$$

Since x represents the width of a rectangle, the solution $-\frac{8}{5}$ must be rejected and the width of the rectangular piece of carpet is 2 meters. The length of the carpet is $5x - 2$ or $5(2) - 2 = 8$ meters. Notice that the area of the rectangle whose length is 8 meters and whose width is 2 meters is the given 16 square meters.   ■

EXAMPLE 8   The sum of the squares of two consecutive odd negative integers is 130. Find the numbers.

Solution: Let $x = $ first integer; then $x + 2 = $ second integer. The sum of the squares is 130.

$$x^2 + (x + 2)^2 = 130$$

$$x^2 + x^2 + 4x + 4 = 130 \qquad \text{Square the binomial.}$$

$$2x^2 + 4x - 126 = 0 \qquad \text{Rewrite in standard form.}$$

$$2(x^2 + 2x - 63) = 0$$

$$2(x + 9)(x - 7) = 0 \qquad \text{Factor.}$$

$$x + 9 = 0 \quad \text{or} \quad x - 7 = 0 \qquad \text{Set each factor equal to 0.}$$

$$x = -9 \quad \text{or} \qquad x = 7 \qquad \text{Solve.}$$

Since the problem asked for negative integers, reject the solution 7. Thus, $x = -9$ is the first integer, and $x + 2 = -7$ is the second integer. The sum of the squares is 130 since $(-9)^2 + (-7)^2 = 81 + 49 = 130$. The numbers are -9 and -7.   ■

MENTAL MATH

Solve each equation for the variable. See Example 1.

1. $(x - 3)(x + 5) = 0$

2. $(y + 5)(y + 3) = 0$

3. $(z - 3)(z + 7) = 0$

4. $(c - 2)(c - 4) = 0$

5. $x(x - 9) = 0$

6. $w(w + 7) = 0$

EXERCISE SET 5.5

Solve each equation. See Example 1.

1. $(x + 3)(3x - 4) = 0$

2. $(5x + 1)(x - 2) = 0$

3. $3(2x - 5)(4x + 3) = 0$

4. $8(3x - 4)(2x - 7) = 0$

Solve each equation. See Example 2.

5. $x^2 + 11x + 24 = 0$

6. $y^2 - 10y + 24 = 0$

7. $12x^2 + 5x - 2 = 0$

8. $3y^2 - y - 14 = 0$

Solve each equation. See Example 3.

9. $z^2 + 9 = 10z$

10. $n^2 + n = 72$

11. $x(5x + 2) = 3$

12. $n(2n - 3) = 2$

13. $x^2 - 6x = x(8 + x)$

14. $n(3 + n) = n^2 + 4n$

Solve each equation. See Example 4.

15. $\dfrac{z^2}{6} - \dfrac{z}{2} - 3 = 0$

16. $\dfrac{c^2}{20} - \dfrac{c}{4} + \dfrac{1}{5} = 0$

17. $\dfrac{x^2}{2} + \dfrac{x}{20} = \dfrac{1}{10}$

18. $\dfrac{y^2}{30} = \dfrac{y}{15} + \dfrac{1}{2}$

19. $\dfrac{4t^2}{5} = \dfrac{t}{5} + \dfrac{3}{10}$

20. $\dfrac{5x^2}{6} - \dfrac{7x}{2} + \dfrac{2}{3} = 0$

Solve each equation. See Examples 5 and 6.

21. $(x + 2)(x - 7)(3x - 8) = 0$

22. $(4x + 9)(x - 4)(x + 1) = 0$

23. $y^3 = 9y$

24. $n^3 = 16n$

25. $x^3 - x = 2x^2 - 2$

26. $m^3 = m^2 + 12m$

Solve. See Example 7.

27. The area of a square is three times greater than its perimeter. Find the length of the side of the square.

28. The length of a rectangular pool is 5 feet more than its width. Find the dimensions of the pool if its area is 104 square feet.

Solve. See Example 8.

29. The sum of the squares of three consecutive integers is 29. Find the integers.

30. Find two consecutive negative odd integers whose product is 99.

Solve each equation.

31. $(2x + 7)(x - 10) = 0$

32. $(x + 4)(5x - 1) = 0$

33. $3x(x - 5) = 0$

34. $4x(2x + 3) = 0$

35. $x^2 - 2x - 15 = 0$

36. $x^2 + 6x - 7 = 0$

37. $12x^2 + 2x - 2 = 0$

38. $8x^2 + 13x + 5 = 0$

39. $w^2 - 5w = 36$

40. $x^2 + 32 = 12x$

41. $25x^2 - 40x + 16 = 0$

42. $9n^2 + 30n + 25 = 0$

43. $2r^3 + 6r^2 = 20r$

44. $-2t^3 = 108t - 30t^2$

45. $z(5z - 4)(z + 3) = 0$

46. $2r(r + 3)(5r - 4) = 0$

47. $2z(z + 6) = 2z^2 + 12z - 8$

48. $3c^2 - 8c + 2 = c(3c - 8)$

49. $(x - 1)(x + 4) = 24$

50. $(2x - 1)(x + 2) = -3$

51. $\dfrac{x^2}{4} - \dfrac{5}{2}x + 6 = 0$

52. $\dfrac{x^2}{18} + \dfrac{x}{2} + 1 = 0$

53. $y^2 + \dfrac{1}{4} = -y$

54. $\dfrac{x^2}{10} + \dfrac{5}{2} = x$

55. $y^3 + 4y^2 = 9y + 36$

56. $x^3 + 5x^2 = x + 5$

57. $2x^3 = 50x$

58. $m^5 = 36m^3$

59. $x^2 + (x + 1)^2 = 61$

60. $y^2 + (y + 2)^2 = 34$

61. $m^2(3m - 2) = m$

62. $x^2(5x + 3) = 26x$

63. $3x^2 = -x$

64. $y^2 = -5y$

65. $x(x - 3) = x^2 + 5x + 7$

66. $z^2 - 4z + 10 = z(z - 5)$

67. $3(t - 8) + 2t = 7 + t$

68. $7c - 2(3c + 1) = 5(4 - 2c)$

69. $-3(x - 4) + x = 5(3 - x)$

70. $-4(a + 1) - 3a = -7(2a - 3)$

Solve.

71. One number exceeds another by five, and their product is 66. Find the numbers.

72. If the product of two numbers is $\dfrac{15}{4}$ and their sum is 4, find the numbers.

73. An electrician needs to run a cable from the top of a 60-foot tower to a transmitter box located 45 feet away from the base of the tower. Find how long he should cut the cable.

74. Determine whether any three consecutive integers represent the lengths of the sides of a right triangle.

75. The shorter leg of a right triangle is 3 centimeters less than the other leg. Find the length of the two legs if the hypotenuse is 15 centimeters. (Recall the Pythagorean theorem for right triangles: $a^2 + b^2 = c^2$.)

76. A stereo system installer needs to run speaker wire along the two diagonals of a rectangular room whose dimensions are 40 feet by 75 feet. Find how much speaker wire she needs.

77. Marie has a rectangular board 12 inches by 16 inches around which she wants to put a uniform border of shells. If she has enough shells for a border whose area is 128 square inches, determine the width of the border.

78. A gardener has a rose garden that measures 30 feet by 20 feet. He wants to put a uniform border of pine bark around the outside of the garden. Find how wide the border should be if he has enough pine bark to cover 336 square feet.

79. For a TV commercial, a piece of luggage is dropped from a cliff 256 feet above the ground to show the durability of the luggage. The height h of the luggage above the ground after t seconds of falling is given by the equation $h = -16t^2 + 256$. Find how long it takes the luggage to hit the ground.

80. After t seconds, the height h of a model rocket launched from the ground into the air is given by the equation $h = 16t^2 + 80t$. Find how long it takes the rocket to reach a height of 96 feet.

Writing in Mathematics

81. Describe two ways a linear equation differs from a quadratic equation.

82. Explain how solving $2(x - 3)(x - 1) = 0$ differs from solving $2x(x - 3)(x - 1) = 0$.

83. Explain why the zero-factor property works for more than two numbers whose product is 0.

Skill Review

Simplify. See Section 4.1.

84. $\dfrac{10x^2y}{5x^3}$

85. $\dfrac{24xy^3}{6y^4}$

86. $\dfrac{45a^2b}{9b^3}$

87. $\dfrac{36a^3b}{12b^3}$

Determine whether each graph is the graph of a function. See Section 3.5.

88.

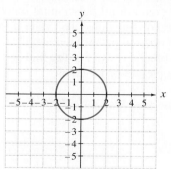

89.

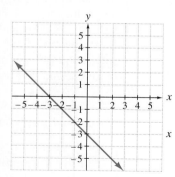

90.

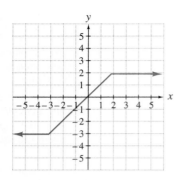

91.

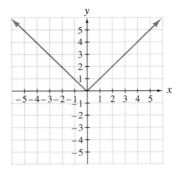

5.6
Graphing Polynomial Functions

OBJECTIVES

Tape 15

1   Define the polynomial function.

2   Graph quadratic functions.

3   Find the vertex of a parabola by using the vertex formula.

4   Graph cubic functions.

---

1   We discussed linear functions defined by equations of the form $f(x) = mx + b$ in Section 4.7. In this section, we discuss two other special cases of **polynomial functions,** that is, **quadratic functions** and **cubic functions.** For example,

$f(x) = 2x - 6$ is a **linear function** since its **degree is one.**
$f(x) = 5x^2 - x + 3$ is a **quadratic function** since its **degree is two.**
$f(x) = 7x^3 + 3x^2 - 1$ is a **cubic function** since its **degree is three.**

All the above functions are also polynomial functions.
   The graph of any polynomial function (linear, quadratic, cubic, and so on) can be sketched by plotting a sufficient number of ordered pairs that satisfy the function and connecting them to form a smooth curve. The graph of all polynomial functions will pass the vertical line test since they are graphs of functions. To graph a linear function defined by $f(x) = mx + b$, recall that two ordered pair solutions will suffice since its graph is a line. To graph other polynomial functions, we need to find and plot more ordered pair solutions to ensure a reasonable picture of its graph.

2   Since we know how to graph linear functions (see Section 4.7), we will now discuss and graph quadratic functions.

> **Quadratic Function**
>
> A quadratic function is a function that can be written in the form
> $f(x) = ax^2 + bx + c$, where a, b, and c are real numbers and $a \neq 0$.

From Sections 3.5 and 4.7, we know that $y = f(x)$ so that an equation of the form $f(x) = ax^2 + bx + c$ may be written as $y = ax^2 + bx + c$. Thus, both $f(x) = ax^2 + bx + c$ and $y = ax^2 + bx + c$ define quadratic functions as long as a is not 0.

Graph the quadratic function defined by $f(x) = x^2$ by plotting points. Choose $-3, -2, -1, 0, 1, 2$, and 3 as x-values and find corresponding $f(x)$ or y-values.

x	$y = f(x)$
-3	9
-2	4
-1	1
0	0
1	1
2	4
3	9

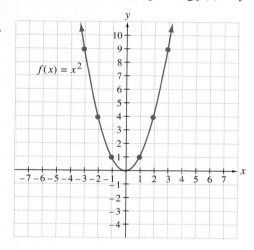

Notice that the graph passes the vertical line test as it should since it is a function. This curve is called a **parabola.** The highest point on a parabola that opens downward or the lowest point on a parabola that opens upward is called the **vertex** of the parabola. The vertex of this parabola is $(0, 0)$, the lowest point on the graph. If we fold the graph along the y-axis, we can see that the two sides of the graph coincide. This means that this curve is symmetric about the y-axis, and the y-axis or the line $x = 0$ is called the **axis of symmetry.** The graph of every quadratic function is a parabola and has an axis of symmetry: the vertical line that passes through the vertex of the parabola.

EXAMPLE 1   Graph the quadratic function $f(x) = -x^2 + 2x - 3$ by plotting points.

Solution: To graph, choose values for x, and find corresponding $f(x)$ or y-values.

x	$y = f(x)$
-2	-11
-1	-6
0	-3
1	-2
2	-3
3	-6

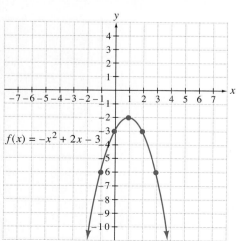

The vertex of this parabola is $(1, -2)$, the highest point on the graph. The vertical line $x = 1$ is the axis of symmetry.   ■

Note that the parabola $f(x) = -x^2 + 2x - 3$ opens downward, while $f(x) = x^2$ opens upward. When the equation of a quadratic function is written in the form $f(x) = ax^2 + bx + c$, the coefficient of the squared variable, a, determines whether the parabola opens downward or upward. If $a > 0$, the parabola opens upward, and if $a < 0$, the parabola opens downward.

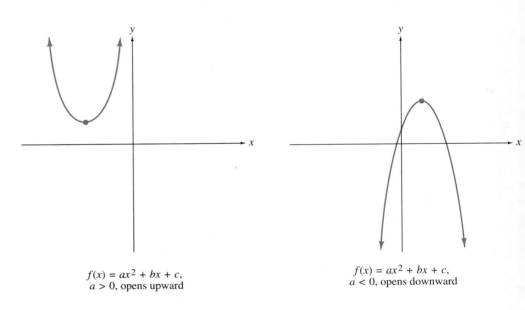

$f(x) = ax^2 + bx + c$,
$a > 0$, opens upward

$f(x) = ax^2 + bx + c$,
$a < 0$, opens downward

3   In both $f(x) = x^2$ and $f(x) = -x^2 + 2x - 3$, the vertex happens to be one of the points we chose to plot. Since this is not always the case, and since plotting the vertex allows us to quickly draw the graph, we need a consistent method for finding the vertex. One method is to use the following formula, which we derive in Section 8.5.

Vertex Formula

The graph of $f(x) = ax^2 + bx + c$, $a \neq 0$, is a parabola with vertex

$$\left(\frac{-b}{2a}, f\left(\frac{-b}{2a} \right) \right)$$

We can also find the x- and y-intercepts of a parabola to aid in graphing. Recall that x-intercepts of the graph of any equation may be found by letting $y = 0$ in the equation and solving for x. Also, y-intercepts may be found by letting $x = 0$ in the equation and solving for y or $f(x)$.

EXAMPLE 2   Graph $f(x) = x^2 + 2x - 3$. Find the vertex and any intercepts.

Solution: To find the vertex, use the vertex formula. For the function $f(x) = x^2 + 2x - 3$, $a = 1$ and $b = 2$. Thus

$$x = \frac{-b}{2a} = \frac{-2}{2(1)} = -1$$

Next, find $f(-1)$.

$$f(-1) = (-1)^2 + 2(-1) - 3$$
$$= 1 - 2 - 3$$
$$= -4$$

The vertex is $(-1, -4)$, and since $a = 1$ is greater than 0, this parabola opens upward. This parabola will have two x-intercepts because its vertex lies below the x-axis and it opens upward. To find the x-intercepts, let y or $f(x) = 0$ and solve for x.

$$f(x) = x^2 + 2x - 3$$
$$0 = x^2 + 2x - 3 \qquad \text{Let } f(x) = 0.$$
$$0 = (x + 3)(x - 1) \qquad \text{Factor.}$$
$$x + 3 = 0 \quad \text{or} \quad x - 1 = 0 \qquad \text{Set each factor equal to 0.}$$
$$x = -3 \quad \text{or} \quad x = 1 \qquad \text{Solve.}$$

The x-intercepts are -3 and 1.

To find the y-intercept, let $x = 0$.

$$f(x) = x^2 + 2x - 3$$
$$f(0) = 0^2 + 2(0) - 3$$
$$f(0) = -3$$

The y-intercept is -3.

Now plot these points and connect them with a smooth curve.

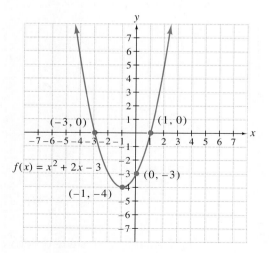

■

HELPFUL HINT

Not all graphs of parabolas have x-intercepts. To see this, first plot the vertex of the parabola and decide whether the parabola opens upward or downward. Then use this information to decide whether the graph of the parabola has x-intercepts.

EXAMPLE 3   Graph $f(x) = 3x^2 - 12x + 13$. Find the vertex and any intercepts.

Solution: To find the vertex, use the vertex formula. For the function $y = 3x^2 - 12x + 13$, $a = 3$ and $b = -12$. Thus

$$x = \frac{-b}{2a} = \frac{-(-12)}{2(3)} = \frac{12}{6} = 2$$

Next, find $f(2)$.

$$f(2) = 3(2)^2 - 12(2) + 13$$
$$= 3(4) - 24 + 13$$
$$= 1$$

The vertex is $(2, 1)$. Also, this parabola opens upward, since $a = 3$ and is greater than 0. Notice that this parabola has no x-intercepts: its vertex lies above the x-axis and it opens upward.

To find the y-intercept, let $x = 0$.

$$f(0) = 3(0)^2 - 12(0) + 13$$
$$= 0 - 0 + 13$$
$$= 13$$

The y-intercept is 13. Use this information along with symmetry of a parabola to sketch the graph of $f(x) = 3x^2 - 12x + 13$.

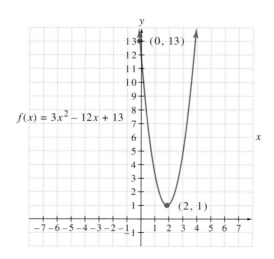

In Section 8.5, we study quadratic functions further.

4   To sketch the graph of a cubic function, we again plot points and then connect the points with a smooth curve.

EXAMPLE 4   Graph $f(x) = x^3 - 4x$. Find any intercepts.

Solution: To find x-intercepts, let y or $f(x) = 0$ and solve for x.

$$f(x) = x^3 - 4x$$

$$0 = x^3 - 4x \qquad \text{Let } f(x) = 0.$$

$$0 = x(x^2 - 4)$$

$$0 = x(x + 2)(x - 2) \qquad \text{Factor.}$$

$$x = 0 \quad \text{or} \quad x + 2 = 0 \quad \text{or} \quad x - 2 = 0 \qquad \text{Set each factor equal to 0.}$$

$$x = 0 \quad \text{or} \quad x = -2 \quad \text{or} \quad x = 2 \qquad \text{Solve.}$$

This graph has three x-intercepts. They are 0, -2 and 2.
To find the y-intercept, let $x = 0$.

$$f(0) = 0^3 - 4(0) = 0$$

Next, select some x-values and find their corresponding $f(x)$ or y-values.

	x	$f(x)$
$f(x) = x^3 - 4x$		
$f(-3) = (-3)^3 - 4(-3) = -27 + 12 = -15$	-3	-15
$f(-1) = (-1)^3 - 4(-1) = -1 + 4 = 3$	-1	3
$f(1) = 1^3 - 4(1) = 1 - 4 = -3$	1	-3
$f(3) = 3^3 - 4(3) = 27 - 12 = 15$	3	15

Plot the intercepts and points and connect them with a smooth curve.

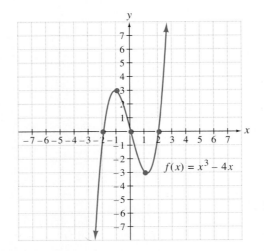

$$f(x) = x^3 - 4x$$

HELPFUL HINT

When a graph has an x-intercept of 0, notice that the y-intercept will also be 0.

> **HELPFUL HINT**
>
> If unsure about the graph of a function, plot more points.

EXAMPLE 5 Graph $f(x) = x^3$. Find any intercepts.

Solution: To find x-intercepts, let y or $f(x) = 0$ and solve for x.

$$f(x) = x^3$$
$$0 = x^3$$
$$0 = x$$

The only x-intercept is 0. This means that the y-intercept is 0 also. Next, choose some x-values and find corresponding y-values.

$$f(x) = x^3$$
$$f(-2) = (-2)^3 = -8$$
$$f(-1) = (-1)^3 = -1$$
$$f(1) = (1)^3 = 1$$
$$f(2) = 2^3 = 8$$

x	$f(x)$
-2	-8
-1	-1
1	1
2	8

Plot the points and sketch the graph of $f(x) = x^3$.

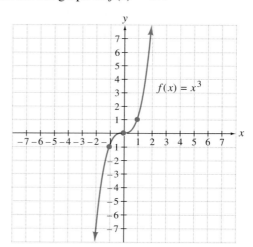

■

MENTAL MATH

State whether the graph of each quadratic function, a parabola, opens upward or downward.

1. $f(x) = 2x^2 + 7x + 10$
2. $f(x) = -3x^2 - 5x$
3. $f(x) = -x^2 + 5$
4. $f(x) = x^2 + 3x + 7$

EXERCISE SET 5.6

Graph each quadratic function by plotting points. See Example 1.

1. $f(x) = 2x^2$

2. $f(x) = -3x^2$

3. $f(x) = x^2 + 1$

4. $f(x) = x^2 - 2$

5. $f(x) = -x^2$

6. $f(x) = \dfrac{1}{2}x^2$

Find the vertex of the graph of each function. See Examples 2 and 3.

7. $f(x) = x^2 + 8x + 7$

8. $f(x) = x^2 + 6x + 5$

9. $f(x) = 3x^2 + 6x + 4$

10. $f(x) = -2x^2 + 2x + 1$

11. $f(x) = -x^2 + 10x + 5$

12. $f(x) = -x^2 - 8x + 2$

Graph each quadratic function. Find and label the vertex and intercepts. See Examples 2 and 3.

13. $f(x) = x^2 + 8x + 7$

14. $f(x) = x^2 + 6x + 5$

15. $f(x) = x^2 - 2x - 24$

16. $f(x) = x^2 - 12x + 35$

17. $f(x) = 2x^2 - 6x$

18. $f(x) = -3x^2 + 6x$

Graph each cubic function. Find any intercepts. See Examples 4 and 5.

19. $f(x) = 4x^3 - 9x$

20. $f(x) = 2x^3 - 5x^2 - 3x$

21. $f(x) = x^3 + 3x^2 - x - 3$ **22.** $f(x) = x^3 + x^2 - 4x - 4$

Graph each function. Find intercepts. If a quadratic function, find the vertex.

23. $f(x) = x^2 + 4x - 5$ **24.** $f(x) = x^2 + 2x - 3$ **25.** $f(x) = (x - 2)(x + 2)(x + 1)$

26. $f(x) = x^3 - 4x^2 + 3x$ **27.** $f(x) = x^2 + 1$ **28.** $f(x) = x^2 + 4$

29. $f(x) = -5x^2 + 5x$ **30.** $f(x) = 3x^2 - 12x$ **31.** $f(x) = x^3 - 9x$

32. $f(x) = x^3 + x^2 - 12x$ **33.** $f(x) = -x^3 - x^2 + 2x$ **34.** $f(x) = x^3 + x^2 - 9x - 9$

35. $f(x) = x^2 - 4x + 4$ **36.** $f(x) = x^2 - 2x + 1$ **37.** $f(x) = -x^3 + x$

38. $f(x) = x^2 + 6x$

39. $f(x) = 2x^2 - x - 3$

40. $f(x) = (x + 2)(x - 2)$

41. $f(x) = -x^3 + 3x^2 + x - 3$

42. $f(x) = -x^3 + 25x$

43. $f(x) = x^2 - 10x + 26$

44. $f(x) = x^2 + 2x + 4$

45. $f(x) = x(x - 4)(x + 2)$

46. $f(x) = 3x(x - 3)(x + 5)$

Writing in Mathematics

47. Explain why the graph of a function never has two or more y-intercepts.

Skill Review

Find the domain and the range of each function graphed. See Section 3.5.

48.

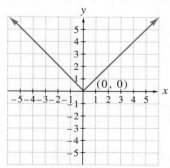

49.

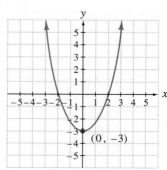

50.

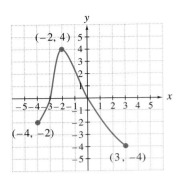

51.

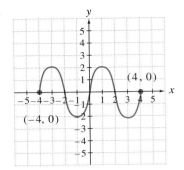

CRITICAL THINKING   A rectangle such as the one below can be used to demonstrate the distributive property by finding its area in two different ways. The area of the rectangle can be represented by the product $a(b + c)$ or by summing the areas of the two smaller rectangles, $ab + ac$. Since the area of the rectangle is the same no matter how one writes it, we have then that $a(b + c) = ab + ac$.

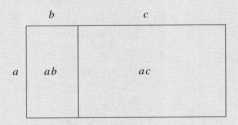

One of the most convincing ways to see that $a^2 + b^2$ does not factor as $(a + b)^2$ or that $(a + b)^2$ is not $a^2 + b^2$ is to sketch a square which is $a + b$ units on each side, and compute the area in two different ways such as above. Does this same idea extend to three dimensions? Sketch a cube that is $a + b$ units on each side and compute the volume in two different ways. What algebraic statement is illustrated?

CHAPTER 5 GLOSSARY

The vertical line that passes through the vertex of a parabola opening upward or downward is called the **axis of symmetry**.

In the product $a \cdot b = c$, a and b are called **factors** of c, and $a \cdot b$ is a **factored form** of c.

The process of writing a polynomial as a product is called **factoring**.

The **greatest common factor (GCF)** of a list of integers is the largest integer that is a factor of each integer.

The graph of a quadratic function is a **parabola** that opens upward or downward.

A **quadratic equation** in one variable is an equation that can be written in the form $ax^2 + bx + c = 0$, where a, b, and c are real numbers and $a \neq 0$.

A **quadratic function** can be written in the form $f(x) = ax^2 + bx + c$, where a, b, and c are real numbers and $a \neq 0$.

The highest point on a parabola that opens downward or the lowest point on a parabola that opens upward is called the **vertex** of the parabola.

CHAPTER 5 SUMMARY

TO FACTOR A POLYNOMIAL (5.4)

Step 1 Are there any common factors? If so, factor them out.

Step 2 How many terms are in the polynomial?

   a. If there are **two** terms, decide if one of the following formulas may be applied.

    i. Difference of two squares: $a^2 - b^2 = (a - b)(a + b)$.

    ii. Difference of two cubes: $a^3 - b^3 = (a - b)(a^2 + ab + b^2)$.

    iii. Sum of two cubes: $a^3 + b^3 = (a + b)(a^2 - ab + b^2)$.

   b. If there are **three** terms, try one of the following.

    i. Perfect square trinomial: $a^2 + 2ab + b^2 = (a + b)^2$.

            $a^2 - 2ab + b^2 = (a - b)^2$

    ii. If not a perfect square trinomial, factor using the methods presented in Section 5.2.

   c. If there are **four** or more terms, try factoring by grouping.

Step 3 See if any factors in the factored polynomial can be factored further.

TO SOLVE QUADRATIC EQUATIONS BY FACTORING (5.5)

Step 1 Write the equation in standard form: $ax^2 + bx + c = 0$.

Step 2 Factor the quadratic expression.

Step 3 Set each factor containing a variable equal to 0.

Step 4 Solve the resulting equations.

Step 5 Check each solution in the original equation.

VERTEX FORMULA (5.6)

The graph of $f(x) = ax^2 + bx + c$, $a \neq 0$, is a parabola with vertex

$$\left(\frac{-b}{2a}, f\left(\frac{-b}{2a}\right)\right)$$

CHAPTER 5 REVIEW

(5.1) *Find the GCF of the monomials in the list.*

1. 24, 60, 84

2. 90, 135, 225

3. x^6, x^8, x^2, x^4

4. y^5, y^3, y^7, y^4

5. $6x^2y^3, 16xy^2, 8x^3$

6. $9x^3y^5, 12y^2, 6x^3y^4$

Factor each polynomial.

7. $16x^3 - 24x^2$

8. $36y - 24y^2$

9. $15x^3y^4z - 3xy^2z^2 + 6x^2y^2z$

10. $20x^4yz^3 - 6x^2y^2z^4 + 4x^3yz^2$

11. $6ab^2 + 8ab - 4a^2b^2$

12. $14a^2b^2 - 21ab^2 + 7ab$

13. $6a(a + 3b) - 5(a + 3b)$

14. $4x(x - 2y) - 5(x - 2y)$

15. $xy - 6y + 3x - 18$

16. $ab - 8b + 4a - 32$

17. $pq - 3p - 5q + 15$

18. $xy - 2x - 5y + 10$

19. $x^3 - x^2 - 2x + 2$

20. $x^3 - x^2 - 5x + 5$

(5.2)   *Completely factor each trinomial.*

21. $x^2 - 14x - 72$

22. $x^2 + 16x - 80$

23. $2x^2 - 18x + 28$

24. $3x^2 + 33x + 54$

25. $6x^3y - 24xy$

26. $4x^4y - 60x^3y - 216x^2y$

27. $2x^2 - 7x - 9$

28. $3x^2 + 2x - 16$

29. $6x^2 + 17x + 10$

30. $15x^2 - 91x + 6$

31. $4x^2 - 12x + 9$

32. $16x^2 + 40x + 25$

33. $25x^2 + 60xy + 36y^2$

34. $9a^2 - 24ab + 16b^2$

35. $4x^2 + 2x - 12$

36. $9x^2 - 12x - 12$

37. $(x + 6)^2 - 5(x + 6) + 6$

38. $(x + 5)^2 + 6(x + 5) + 8$

39. $x^4 - 6x^2 - 16$

40. $x^4 + 8x^2 - 20$

(5.3)   *Factor each polynomial.*

41. $x^2 - 100$

42. $x^2 - 81$

43. $2x^2 - 32$

44. $6x^2 - 54$

45. $81 - x^4$

46. $16 - y^4$

47. $(y + 2)^2 - 25$

48. $(x - 3)^2 - 16$

49. $y^3 - 8$

50. $x^3 - 27$

51. $x^3 + 216$

52. $y^3 + 512$

53. $8 - 27y^3$

54. $1 - 64y^3$

55. $6x^4y + 48xy$

56. $2x^5 + 16x^2y^3$

57. $x^2 - 2x + 1 - y^2$

58. $x^2 - 6x + 9 - 4y^2$

59. $4x^2 + 4x + 1 - 9y^2$

60. $16x^2 + 8x + 1 - y^2$

(5.4)   *Factor each polynomial.*

61. $6xy^2 - 3xy$

62. $2xy^2 - 8x^2y^2$

63. $25x^2 - 100$

64. $16x^2 - 36$

65. $3x^2 - 10x + 8$

66. $2x^2 + 13x + 18$

67. $x^2y + 3x^2 - 4y - 12$

68. $x^2 - 4x - 45$

69. $4x^2 - 14x + 49$

70. $9x^2 + 30x + 25$

71. $4x^2 - 25$

72. $4x^2 + 36$

73. $8x^3 - y^3$

74. $x^3 + 27y^3$

75. $2x^6y + 54x^3y^4$

76. $4xy^2 - 500xy^5$

77. $4a^2 - 24a + 36 - 4b^2$

78. $3x^2 - 36x + 108 - 3y^2$

79. $2x^3y^8 - 128x^3y^5$

80. $54x^7y^4 + 2xy$

(5.5)   *Solve each quadratic or higher-degree equation for the variable.*

81. $(3x - 1)(x + 7) = 0$

82. $3(x + 5)(8x - 3) = 0$

83. $5x(x - 4)(2x - 9) = 0$

84. $6(x + 3)(x - 4)(5x + 1) = 0$

85. $2x^2 = 12x$

86. $4x^3 - 36x = 0$

87. $(1 - x)(3x + 2) = -4x$

88. $2x(x - 12) = -40$

89. $3x^2 + 2x = 12 - 7x$

90. $2x^2 + 3x = 35$

91. $x^3 - 18x = 3x^2$

92. $19x^2 - 42x = -x^3$

93. $12x = 6x^3 + 6x^2$

94. $8x^3 + 10x^2 = 3x$

(5.6) *Graph each polynomial function defined by the equation. Find all intercepts. If the function is a quadratic function, find the vertex.*

95. $f(x) = x^2 + 6x + 9$

96. $f(x) = x^2 - 5x + 4$

97. $f(x) = (x - 1)(x^2 - 2x - 3)$

98. $f(x) = (x + 3)(x^2 - 4x + 3)$

99. $f(x) = 2x^2 - 4x + 5$

100. $f(x) = x^2 - 2x + 3$

101. $f(x) = x^3 - 16x$

102. $f(x) = x^3 + 5x^2 + 6x$

CHAPTER 5 TEST

Factor each polynomial completely. If the polynomial cannot be factored, write prime.

1. $16x^3y - 12x^2y^4$

2. $20a^3b^2 - 35ab^3$

3. $xy - 5x + 3y - 15$

4. $x^3 + 2x^2 + 3x + 6$

5. $x^2 - 11x + 24$

6. $x^2 - 13x - 30$

7. $2x^2 + 17x - 9$

8. $4y^2 + 20y + 25$

9. $6x^2 - 15x - 9$

10. $x^5 + 3x^3 + 2x$

11. $4x^2 - 25$

12. $(3x + 1)^2 - 4$

13. $x^3 + 64$

14. $8x^3 - 125$

15. $3x^2y - 27y^3$

16. $8x^2 - 6x - 9$

17. $6x^2 + 24$

18. $4x^6y^2 - 32y^5$

19. $x^2y - 9y - 3x^2 + 27$

20. $8x^2 - 44x + 20$

Solve the equation for the variable.

21. $3(n - 4)(7n + 8) = 0$

22. $(x - 7)(x + 2) = -20$

23. $3m^3 = 12m$

24. $2x^3 + 5x^2 - 8x - 20 = 0$

25. $\dfrac{3x^2}{5} - \dfrac{2x}{5} = 1$

Graph.

26. $f(x) = x^2 - 4x - 5$

27. $f(x) = x^3 - 1$

CHAPTER 5 CUMULATIVE REVIEW

1. Determine whether the following statements are true or false.

   a. 3 is a real number.

   b. $\dfrac{1}{5}$ is an irrational number.

   c. Every rational number is an integer.

2. Simplify.

   a. $2 + 3 \cdot 2$

   b. $2 \cdot 3^3 + 5^2 - (3 + 4)$

   c. $\dfrac{6^2 - 4}{2 + 2 \cdot 3}$

   d. $\dfrac{|-2|^3}{7^1 - \sqrt{4}}$

3. Find each product.

   a. $(-8)(-1)$

   b. $(-2)\dfrac{1}{6}$

   c. $3(-3)$

   d. $(0)(11)$

   e. $\left(\dfrac{1}{5}\right)\left(-\dfrac{10}{11}\right)$

   f. $(7)(1)(-2)(-3)$

   g. $8(-2)(0)$

4. Solve $\dfrac{x + 5}{2} + \dfrac{1}{2} = 2x - \dfrac{x - 3}{8}$.

5. A square and an equilateral triangle have the same perimeter. Each side of the triangle is 6 centimeters longer than each side of the square. Find the length of each side of the triangle.

6. Graph each set on a number line and then write in interval notation.

   a. $\{x \mid x \geq 2\}$

   b. $\{x \mid x < -1\}$

   c. $\{x \mid 0.5 < x \leq 3\}$

7. Solve $|x| \leq 3$.

8. Find the midpoint of the line segment joining points $P(-3, 3)$ and $Q(1, 0)$.

9. Find the equation of the line through $(4, 0)$ and $(-4, -5)$.

10. Find the equation of the line through $(1, 0)$ and perpendicular to the line $y = 9$.

11. Graph $3x \geq y$.

12. Evaluate the following.

   a. 9^2

   b. 2^3

   c. 7^1

   d. $\left(\dfrac{2}{3}\right)^3$

   e. -2^4

   f. $(-2)^4$

   g. $2 \cdot 4^2$

13. Write each number in scientific notation.
   a. 730,000
   b. 0.00000104

14. Subtract $(12z^5 - 12z^3 + z) - (-3z^4 + z^3 + 12z)$.

15. Multiply $(2x - 7)(3x - 4)$.

16. Divide $2x^3 + 3x^4 - 8x + 6$ by $x^2 - 1$.

17. Factor $ab - 6a + 2b - 12$.

18. Factor $5x^3 - 30x^2 - 35x$.

19. Factor $3a^2x - 12abx + 12b^2x$.

20. Factor $p^3 + 27q^3$.

21. Solve $2x^2 = \dfrac{17}{3}x + 1$.

22. Graph $f(x) = x^2 + 2x - 3$. Find the vertex and any intercepts.

CHAPTER 6

Rational Expressions

Windmills generate electricity. Progressive communities are experimenting with fields of windmills for communal electric needs, examining exactly how windspeed affects the amount of electricity produced. (See Critical Thinking, page 295.)

INTRODUCTION

Dividing one polynomial by another in Sections 4.5 and 4.6, we found quotients of polynomials. When the remainder part of the quotient wasn't 0, the remainder was a fraction, such as $\dfrac{x+1}{x^2}$. This fraction is not a polynomial, since it cannot be written as the sum of whole number powers. Instead, it is called a **rational expression.** In this chapter, we present techniques for operating on these rational expressions. Since they are fractions, operating on rational expressions depends on your ability to work with number fractions and on the factoring techniques of Chapter 5.

6.1
Simplifying Rational Expressions

OBJECTIVES

Tape 16

1 Define a rational expression and identify the values of the variable for which a rational expression is not defined.

2 Write a rational expression in lowest terms.

3 Write a rational expression equivalent to a rational expression with a given denominator.

---

1 Recall that a rational number, or fraction, is the quotient $\dfrac{p}{q}$ of two integers p and q as long as q is not 0. A **rational expression** is the quotient $\dfrac{P}{Q}$ of two polynomials P and Q as long as Q is not 0.

Examples of Rational Expressions

$$\frac{3x+7}{2} \qquad \frac{5x^2-3}{x-1} \qquad \frac{7x-2}{2x^2+7x+6}$$

As with fractions, a rational expression is **undefined** if the denominator is 0. If a variable in a rational expression is replaced with a number that makes the denominator 0, we say that the rational expression is **undefined** for this value of the variable. For example, the rational expression $\dfrac{x^2+2}{x-3}$ is not defined when x is 3, since replacing x with 3 results in a denominator of 0.

EXAMPLE 1 Find the values of x, if any, for which each rational expression is undefined.

a. $\dfrac{8x^3+7x^2+20}{2}$ **b.** $\dfrac{5x^2-3}{x-1}$ **c.** $\dfrac{7x-2}{2x^2+7x+6}$

Solution: A rational expression is undefined when the denominator is 0. To find the values that make the denominator 0, set the denominator equal to 0 and solve for the variable.

a. The denominator of $\dfrac{8x^3 + 7x^2 + 20}{2}$ is never 0, so there is no value of the variable x that makes this expression undefined.

b. Set the denominator $x - 1$ equal to 0 and solve.

$$x - 1 = 0$$
$$x = 1$$

The number 1 is the only value of x for which the rational expression is undefined.

c. Setting the denominator equal to 0 gives a quadratic equation to solve.

$$2x^2 + 7x + 6 = 0 \qquad \text{Set the denominator equal to 0.}$$
$$(2x + 3)(x + 2) = 0 \qquad \text{Factor.}$$
$$2x + 3 = 0 \quad \text{or} \quad x + 2 = 0 \qquad \text{Set each factor equal to 0.}$$
$$x = -\frac{3}{2} \quad \text{or} \qquad x = -2 \qquad \text{Solve.}$$

Thus, $-\dfrac{3}{2}$ and -2 are the values of x for which the rational expression is undefined because these values make the denominator equal to 0. ∎

2 Recall that a fraction is in lowest terms or simplest form if the numerator and denominator have no common factors other than 1 (or -1). For example, $\dfrac{3}{13}$ is in lowest terms since 3 and 13 have no common factors other than 1 (or -1).

To **simplify** a rational expression, or to write it in lowest terms, we use the fundamental principle of rational expressions.

Fundamental Principle of Rational Expressions

For any rational expression $\dfrac{P}{Q}$ and any polynomial R, $R \neq 0$,

$$\frac{P}{Q} = \frac{P \cdot R}{Q \cdot R} \quad \text{and} \quad \frac{P}{Q} = \frac{P \div R}{Q \div R}$$

Thus the fundamental principle says that multiplying or dividing the numerator and denominator of a rational expression by a nonzero polynomial yields an equivalent rational expression.

To simplify a rational expression such as $\dfrac{(x + 2)^2}{x^2 - 4}$, factor the numerator and the denominator and then use the fundamental principle of rational expressions to divide out common factors.

$$\frac{(x + 2)^2}{x^2 - 4} = \frac{(x + 2)(x + 2)}{(x + 2)(x - 2)}$$
$$= \frac{x + 2}{x - 2}$$

In general, the following steps may be used to write rational expressions in lowest terms or simplest form.

> **To Write a Rational Expression in Lowest Terms**
>
> *Step 1* Completely factor the numerator and denominator of the rational expression.
>
> *Step 2* Divide both the numerator and denominator by their GCF.

For now, we assume that variables in a rational expression do not represent values that make the denominator 0.

EXAMPLE 2 Write each rational expression in lowest terms.

a. $\dfrac{24x^6y^5}{8x^7y}$ b. $\dfrac{2x^2}{10x^3 - 2x^2}$

Solution: a. The GCF of the numerator and denominator is $8x^6y$.

$$\frac{24x^6y^5}{8x^7y} = \frac{(8x^6y)\,3y^4}{(8x^6y)\,x} = \frac{3y^4}{x}$$

b. Factor out $2x^2$ from the denominator. Then divide numerator and denominator by their GCF, $2x^2$.

$$\frac{2x^2}{10x^3 - 2x^2} = \frac{2x^2}{2x^2\,(5x - 1)} = \frac{1}{5x - 1} \qquad ∎$$

EXAMPLE 3 Write each rational expression in lowest terms.

a. $\dfrac{2 + x}{x + 2}$ b. $\dfrac{2 - x}{x - 2}$ c. $\dfrac{18 - 2x^2}{x^2 - 2x - 3}$

Solution: a. By the commutative property of addition, $2 + x = x + 2$, so

$$\frac{2 + x}{x + 2} = \frac{x + 2}{x + 2} = 1$$

b. The terms in the numerator of $\dfrac{2 - x}{x - 2}$ differ by sign from the terms of the denominator, so the polynomials are opposites of each other and the expression simplifies to -1. To see this, factor out -1 from the numerator or the denominator. If -1 is factored from the numerator, then

$$\frac{2 - x}{x - 2} = \frac{-1(-2 + x)}{x - 2} = \frac{-1\,(x - 2)}{x - 2} = -1$$

> **HELPFUL HINT**
>
> When the numerator and the denominator of a rational expression are opposites of each other, the expression simplifies to -1.

c. $\dfrac{18 - 2x^2}{x^2 - 2x - 3} = \dfrac{2(9 - x^2)}{(x + 1)(x - 3)}$

$= \dfrac{2(3 + x)(3 - x)}{(x + 1)(x - 3)}$ Factor.

Notice the opposites $3 - x$ and $x - 3$. We write $3 - x$ as $-1(x - 3)$ and simplify.

$$\dfrac{2(3 + x)(3 - x)}{(x + 1)(x - 3)} = \dfrac{2(3 + x) \cdot -1 (x - 3)}{(x + 1) (x - 3)} = -\dfrac{2(3 + x)}{x + 1} \quad \blacksquare$$

HELPFUL HINT

Recall from Section 1.4 that, for a fraction $\dfrac{a}{b}$,

$$\dfrac{a}{-b} = \dfrac{-a}{b} = -\dfrac{a}{b}$$

For example,

$$\dfrac{-(x + 1)}{(x + 2)} = \dfrac{(x + 1)}{-(x + 2)} = -\dfrac{x + 1}{x + 2}$$

EXAMPLE 4   Write each rational expression in lowest terms.

a. $\dfrac{x^3 + 8}{2 + x}$

b. $\dfrac{2y^2 + 2}{y^3 - 5y^2 + y - 5}$

Solution: **a.** $\dfrac{x^3 + 8}{2 + x} = \dfrac{(x + 2)(x^2 - 2x + 4)}{x + 2}$ Factor the sum of two cubes.

$= x^2 - 2x + 4$ Divide out common factors.

b. First, factor the denominator by grouping.

$$y^3 - 5y^2 + y - 5 = (y^3 - 5y^2) + (y - 5)$$
$$= y^2(y - 5) + (y - 5)$$
$$= (y - 5)(y^2 + 1)$$

Then

$$\dfrac{2y^2 + 2}{y^3 - 5y^2 + y - 5} = \dfrac{2(y^2 + 1)}{(y - 5)(y^2 + 1)} = \dfrac{2}{y - 5} \quad \blacksquare$$

3   The fundamental property of fractions also allows us to write a rational expression as an equivalent rational expression with a given denominator. Doing so is necessary to add and subtract rational expressions.

EXAMPLE 5   Write each rational expression as an equivalent rational expression with the given denominator.

a. $\dfrac{3x}{2y}$, denominator $10xy^3$ **b.** $\dfrac{3x+1}{x-5}$, denominator $2x^2 - 11x + 5$

Solution: **a.** $\dfrac{3x}{2y} = \dfrac{?}{10xy^3}$

If the denominator $2y$ is multiplied by $5xy^2$, the result is the given denominator $10xy^3$.

$$2y\,(5xy^2) = 10xy^3$$
$$\quad\uparrow \qquad\qquad \uparrow$$
$$\text{original} \qquad \text{given}$$
$$\text{denominator} \quad \text{denominator}$$

Use the fundamental principle of rational expressions and multiply the numerator and the denominator of the original rational expression by $5xy^2$. Then

$$\dfrac{3x}{2y} = \dfrac{3x\,(5xy^2)}{2y\,(5xy^2)} = \dfrac{15x^2y^2}{10xy^3}$$

b. The factored form of the given denominator, $2x^2 - 11x + 5$, is $(x-5)(2x-1)$.

$$\dfrac{3x+1}{x-5} = \dfrac{?}{(x-5)(2x-1)}$$

Use the fundamental principle of rational expressions and multiply the numerator and denominator of the original rational expression by $2x - 1$.

$$\dfrac{3x+1}{x-5} = \dfrac{(3x+1)(2x-1)}{(x-5)(2x-1)} = \dfrac{6x^2 - x - 1}{(x-5)(2x-1)}$$

To prepare for adding and subtracting rational expressions, we multiply the binomials in the numerator but leave the denominator in factored form. ■

EXERCISE SET 6.1

Find the values of x, if any, for which each rational expression is undefined. See Example 1.

1. $\dfrac{5x-7}{4}$

2. $\dfrac{4-3x}{2}$

3. $-\dfrac{3}{2x}$

4. $\dfrac{6}{5x}$

5. $\dfrac{7}{x-2}$

6. $\dfrac{3}{5+x}$

7. $\dfrac{x+3}{x^2-4}$

8. $\dfrac{7}{x^2-25}$

Write each rational expression in lowest terms. See Example 2.

9. $\dfrac{10x^3}{18x}$

10. $-\dfrac{48a^7}{16a^{10}}$

11. $\dfrac{9x^6y^3}{18x^2y^5}$

12. $\dfrac{10ab^5}{15a^3b^5}$

13. $\dfrac{8q^2}{16q^3 - 16q^2}$

14. $\dfrac{3y}{6y^2 - 30y}$

Write each rational expression in lowest terms. See Example 3.

15. $\dfrac{x + 5}{5 + x}$

16. $\dfrac{x - 5}{5 - x}$

17. $\dfrac{x - 1}{1 - x^2}$

18. $\dfrac{10 + 5x}{x^2 + 2x}$

19. $\dfrac{7 - x}{x^2 - 14x + 49}$

20. $\dfrac{x^2 - 9}{2x^2 - 5x - 3}$

Write each rational expression in lowest terms. See Example 4.

21. $\dfrac{x + 3}{x^3 + 27}$

22. $\dfrac{x^3 - 64}{4 - x}$

23. $\dfrac{2x^3 - 16}{3x - 6}$

24. $\dfrac{x^2 - x + 1}{2x^3 + 2}$

25. $\dfrac{xy - 3y + 2x - 6}{x^2 - 6x + 9}$

26. $\dfrac{2x^3 - 5x^2 + 2x - 5}{3x^2 + 3}$

Write each rational expression as an equivalent rational expression with the given denominator. See Example 5.

27. $\dfrac{5}{2y}$, $4y^3z$

28. $\dfrac{1}{z}$, $5z^5$

29. $\dfrac{3x}{2x - 1}$, $2x^2 + 9x - 5$

30. $\dfrac{5}{3x + 2}$, $3x^2 - 13x - 10$

31. $\dfrac{x - 2}{1}$, $x + 2$

32. $\dfrac{x - 5}{1}$, $x + 1$

Find the values of x, if any, for which each rational expression is undefined.

33. $\dfrac{3x}{x + 11}$

34. $\dfrac{4x}{x + 1}$

35. $\dfrac{x + 3}{4 - 2x}$

36. $\dfrac{4 - x}{21 + 3x}$

37. $\dfrac{4x - 3}{x^2 + 5}$

38. $\dfrac{2 - 3x^2}{3x^2 + 9x}$

39. $\dfrac{5 - 4x}{4x - 2x^2}$

40. $\dfrac{2 - 5x}{1 + 2x^2}$

41. $\dfrac{3 + 2x}{3x^3 + 3x^2 - 6x}$

42. $\dfrac{5 - 3x}{2x^3 - 14x^2 + 20x}$

Write each rational expression in lowest terms.

43. $\dfrac{6x^3}{27x^3}$

44. $\dfrac{18y^7}{30y}$

45. $\dfrac{6a^2b^3}{3a^2b^2}$

46. $\dfrac{27m^4p^2}{3m^4p}$

47. $\dfrac{7 - y}{y - 7}$

48. $\dfrac{17 + x}{x + 17}$

49. $\dfrac{4x - 8}{3x - 6}$

50. $\dfrac{12 - 6x}{30 - 15x}$

51. $\dfrac{2x - 14}{7 - x}$

52. $\dfrac{9 - x}{5x - 45}$

53. $\dfrac{x^2 - 2x - 3}{x^2 - 6x + 9}$

54. $\dfrac{x^2 + 10x + 25}{x^2 + 8x + 15}$

55. $\dfrac{2x^2 + 12x + 18}{x^2 - 9}$

56. $\dfrac{x^2 - 4}{2x^2 + 8x + 8}$

57. $\dfrac{3x + 6}{x^2 + 2x}$

58. $\dfrac{3x + 4}{9x^2 + 4}$

59. $\dfrac{x + 2}{x^2 - 4}$

60. $\dfrac{x^2 - 9}{x - 3}$

61. $\dfrac{2x^2 - x - 3}{2x^3 - 3x^2 + 2x - 3}$

62. $\dfrac{3x^2 - 5x - 2}{6x^3 + 2x^2 + 3x + 1}$

63. $\dfrac{x^4 - 16}{x^2 + 4}$

64. $\dfrac{x^2 + y^2}{x^4 - y^4}$

65. $\dfrac{x^2 + 6x - 40}{10 + x}$

66. $\dfrac{x^2 - 8x + 16}{4 - x}$

67. $\dfrac{2x^2 - 7x - 4}{x^2 - 5x + 4}$

68. $\dfrac{3x^2 - 11x + 10}{x^2 - 7x + 10}$

69. $\dfrac{x^3 - 125}{5 - x}$

70. $\dfrac{4x + 4}{2x^3 + 2}$

71. $\dfrac{8x^3 - 27}{4x - 6}$

72. $\dfrac{9x^2 - 15x + 25}{27x^3 + 125}$

73. $\dfrac{x + 5}{x^2 + 5}$

74. $\dfrac{5x}{5x^2 + 5x}$

Write each rational expression as an equivalent rational expression with the given denominator.

75. $\dfrac{5}{m}$, $6m^3$

76. $\dfrac{3}{x}$, $3x$

77. $\dfrac{7}{m-2}$, $5m-10$

78. $\dfrac{-2}{x+1}$, $10x+10$

79. $\dfrac{y+4}{y-4}$, y^2-16

80. $\dfrac{5}{x-1}$, x^2-1

81. $\dfrac{12x}{x+2}$, x^2+4x+4

82. $\dfrac{x+6}{x-4}$, $x^2-8x+16$

83. $\dfrac{1}{x+2}$, x^3+8

84. $\dfrac{x}{3x-1}$, $27x^3-1$

85. $\dfrac{a}{a+2}$, $ab-3a+2b-6$

86. $\dfrac{5}{x-y}$, $2x^2+5x-2xy-5y$

A Look Ahead

Write each rational expression in lowest terms. Assume that no denominator is 0.

EXAMPLE   Write $\dfrac{x^{2n}+2x^n y^n+y^{2n}}{x^n+y^n}$ in lowest terms.

Solution: Factor the numerator.

$$\frac{x^{2n}+2x^n y^n+y^{2n}}{x^n+y^n} = \frac{(x^n+y^n)(x^n+y^n)}{(x^n+y^n)} = x^n+y^n \quad \blacksquare$$

87. $\dfrac{p^x-4}{4-p^x}$

88. $\dfrac{3+q^n}{q^n+3}$

89. $\dfrac{x^n+4}{x^{2n}-16}$

90. $\dfrac{x^{2k}-9}{3+x^k}$

91. $\dfrac{x^{2k}-4x^k+16}{x^{3k}+64}$

92. $\dfrac{4x^k-12}{x^{2k}+4}$

93. $\dfrac{y^{-2}+2y^{-3}}{y+2}$

94. $\dfrac{x-1}{x^{-1}-x^{-2}}$

95. $\dfrac{2a^{-1}+a^{-2}}{2a^2+a}$

96. $\dfrac{5x^3+3x^2}{5x^{-3}+3x^{-4}}$

Writing in Mathematics

97. The rational expression $\dfrac{x^2-4}{x+2}$ is equivalent to the expression $x-2$ for all real number values of x except -2. In other words, $\dfrac{x^2-4}{x+2} = x-2$ as long as $x \neq -2$. How do you think the graph of $y = \dfrac{x^2-4}{x+2}$ differs from the graph of $y = x-2$?

Skill Review

Perform the indicated operations. See Section 1.4.

98. $\dfrac{6}{35} \cdot \dfrac{28}{9}$

99. $\dfrac{3}{8} \cdot \dfrac{4}{27}$

100. $\dfrac{8}{35} \div \dfrac{4}{5}$

101. $\dfrac{6}{11} \div \dfrac{2}{11}$

Graph the following linear inequalities. See Section 3.6.

102. $x + y \leq 7$

103. $x - y > 2$

6.2
Multiplying and Dividing Rational Expressions

Tape 16

OBJECTIVES

1   Multiply rational expressions.

2   Divide by a rational expression.

Arithmetic operations on rational expressions are performed in the same way as they are on rational numbers.

1

> **Multiplying Rational Expressions**
>
> Let P, Q, R, and S be polynomials. Then
>
> $$\frac{P}{Q} \cdot \frac{R}{S} = \frac{PR}{QS}$$
>
> as long as $Q \neq 0$ and $S \neq 0$.

To multiply rational expressions, the product of their numerators is the numerator of the product and the product of their denominators is the denominator of the product.

EXAMPLE 1   Multiply.

   a. $\dfrac{2x^3}{9y} \cdot \dfrac{y^2}{4x^3}$

   b. $\dfrac{1 + 3n}{2n} \cdot \dfrac{2n - 4}{3n^2 - 2n - 1}$

Solution: **a.** $\dfrac{2x^3}{9y} \cdot \dfrac{y^2}{4x^3} = \dfrac{2x^3 y^2}{36x^3 y}$

To simplify, divide the numerator and the denominator by the common factor $2x^3y$.

$$\frac{2x^3 y^2}{36x^3 y} = \frac{y \,(2x^3 y)}{18 \,(2x^3 y)} = \frac{y}{18}$$

   b. $\dfrac{1 + 3n}{2n} \cdot \dfrac{2n - 4}{3n^2 - 2n - 1} = \dfrac{1 + 3n}{2n} \cdot \dfrac{2(n - 2)}{(3n + 1)(n - 1)}$ Factor.

$$= \frac{(1 + 3n) \cdot 2\,(n - 2)}{2n\,(3n + 1)\,(n - 1)} \qquad \text{Multiply.}$$

$$= \frac{n - 2}{n\,(n - 1)} \qquad \text{Divide out common factors.} \quad \blacksquare$$

The following steps may be used to multiply rational expressions.

> **To Multiply Rational Expressions**
>
> *Step 1* Completely factor the numerators and denominators.
> *Step 2* Multiply the numerators and multiply the denominators.
> *Step 3* Write the product in lowest terms by dividing both the numerator and the denominator by their GCF.

EXAMPLE 2 Multiply.

a. $\dfrac{2x^2 + 3x - 2}{-4x - 8} \cdot \dfrac{16x^2}{4x^2 - 1}$

b. $(ac - ad + bc - bd) \cdot \dfrac{a + b}{d - c}$

Solution: a. $\dfrac{2x^2 + 3x - 2}{-4x - 8} \cdot \dfrac{16x^2}{4x^2 - 1} = \dfrac{(2x - 1)(x + 2)}{-4(x + 2)} \cdot \dfrac{16x^2}{(2x + 1)(2x - 1)}$ Factor.

$$= \dfrac{4 \cdot 4x^2\,(2x - 1)(x + 2)}{-1 \cdot 4(x + 2)\,(2x + 1)\,(2x - 1)}$$ Multiply.

$$= -\dfrac{4x^2}{2x + 1}$$ Divide out common factors.

b. First, factor $ac - ad + bc - bd$ by grouping.

$$ac - ad + bc - bd = (ac - ad) + (bc - bd)$$ Group terms.
$$= a(c - d) + b(c - d)$$
$$= (c - d)(a + b)$$

To multiply, write $(c - d)(a + b)$ as a fraction whose denominator is 1.

$$(ac - ad + bc - bd) \cdot \dfrac{a + b}{d - c} = \dfrac{(c - d)(a + b)}{1} \cdot \dfrac{a + b}{d - c}$$ Factor.

$$= \dfrac{(c - d)(a + b)(a + b)}{d - c}$$ Multiply.

Write $(c - d)$ as $-1(d - c)$ and simplify.

$$\dfrac{-1\,(d - c)\,(a + b)(a + b)}{d - c} = -(a + b)^2 \quad \blacksquare$$

2   To divide by a rational expression, multiply by its reciprocal. Recall that two numbers are reciprocals of each other if their product is 1. Similarly, if $\dfrac{P}{Q}$ is a rational expression, then $\dfrac{Q}{P}$ is its **reciprocal,** since

$$\dfrac{P}{Q} \cdot \dfrac{Q}{P} = \dfrac{P \cdot Q}{Q \cdot P} = 1$$

The following are examples of expressions and their reciprocals.

Expression	Reciprocal
$\dfrac{3}{x}$	$\dfrac{x}{3}$
$\dfrac{2 + x^2}{4x - 3}$	$\dfrac{4x - 3}{2 + x^2}$
x^3	$\dfrac{1}{x^3}$
0	no reciprocal

Division by rational expressions is defined as follows.

Dividing Rational Expressions

Let P, Q, R, and S be polynomials. Then

$$\frac{P}{Q} \div \frac{R}{S} = \frac{P}{Q} \cdot \frac{S}{R} = \frac{PS}{QR}$$

as long as $Q \neq 0$, $S \neq 0$, and $R \neq 0$.

Notice that division of rational expressions is the same as for rational numbers.

EXAMPLE 3 Divide.

a. $\dfrac{3x}{5y} \div \dfrac{9y}{x^5}$

b. $\dfrac{8m^2}{3m^2 - 12} \div \dfrac{40}{2 - m}$

Solution:

a. $\dfrac{3x}{5y} \div \dfrac{9y}{x^5} = \dfrac{3x}{5y} \cdot \dfrac{x^5}{9y}$ Multiply by the reciprocal of the divisor.

$\qquad\qquad = \dfrac{x^6}{15y^2}$ Simplify.

b. $\dfrac{8m^2}{3m^2 - 12} \div \dfrac{40}{2 - m} = \dfrac{8m^2}{3m^2 - 12} \cdot \dfrac{2 - m}{40}$ Multiply by the reciprocal of the divisor.

$\qquad\qquad = \dfrac{8m^2(2 - m)}{3(m + 2)(m - 2) \cdot 40}$ Factor and multiply.

$\qquad\qquad = \dfrac{8 \, m^2 \cdot \, -1 \, (m - 2)}{3(m + 2) \, (m - 2) \cdot 8 \cdot 5}$ Write $(2 - m)$ as $-1(m - 2)$.

$\qquad\qquad = -\dfrac{m^2}{15(m + 2)}$ Simplify. ∎

> **HELPFUL HINT**
>
> When dividing rational expressions, do not divide out common factors until the division problem is rewritten as a multiplication problem.

EXAMPLE 4   Divide $\dfrac{8x^3 + 125}{x^4 + 5x^2 + 4} \div \dfrac{2x + 5}{2x^2 + 8}$.

Solution: $\dfrac{8x^3 + 125}{x^4 + 5x^2 + 4} \div \dfrac{2x + 5}{2x^2 + 8} = \dfrac{8x^3 + 125}{x^4 + 5x^2 + 4} \cdot \dfrac{2x^2 + 8}{2x + 5}$

$$= \dfrac{(2x + 5)(4x^2 - 10x + 25) \cdot 2(x^2 + 4)}{(x^2 + 1)(x^2 + 4) \cdot (2x + 5)}$$

$$= \dfrac{2(4x^2 - 10x + 25)}{x^2 + 1} \qquad \blacksquare$$

EXERCISE SET 6.2

Multiply as indicated. Write all answers in lowest terms. See Examples 1 and 2.

1. $\dfrac{4}{x} \cdot \dfrac{x^2}{8}$

2. $\dfrac{x}{3} \cdot \dfrac{9}{x^3}$

3. $\dfrac{2a^2b}{6ac} \cdot \dfrac{3c^2}{4ab}$

4. $\dfrac{5ab^4}{6abc} \cdot \dfrac{2bc^2}{10ab^2}$

5. $\dfrac{2x}{5} \cdot \dfrac{5x + 10}{6(x + 2)}$

6. $\dfrac{3x}{7} \cdot \dfrac{14 - 7x}{9(2 - x)}$

7. $\dfrac{2x - 4}{15} \cdot \dfrac{6}{2 - x}$

8. $\dfrac{10 - 2x}{7} \cdot \dfrac{14}{5x - 25}$

9. $\dfrac{18a - 12a^2}{4a^2 + 4a + 1} \cdot \dfrac{4a^2 + 8a + 3}{4a^2 - 9}$

10. $\dfrac{a - 5b}{a^2 + ab} \cdot \dfrac{b^2 - a^2}{10b - 2a}$

11. $\dfrac{x^2 - 6x - 16}{2x^2 - 128} \cdot \dfrac{x^2 + 16x + 64}{3x^2 + 30x + 48}$

12. $\dfrac{2x^2 + 12x - 32}{x^2 + 16x + 64} \cdot \dfrac{x^2 + 10x + 16}{x^2 - 3x - 10}$

13. $\dfrac{4x + 8}{x + 1} \cdot \dfrac{2 - x}{3x - 15} \cdot \dfrac{2x^2 - 8x - 10}{x^2 - 4}$

14. $\dfrac{3x - 15}{2 - x} \cdot \dfrac{x + 1}{4x + 8} \cdot \dfrac{x^2 - 4}{2x^2 - 8x - 10}$

Divide as indicated. Write all answers in lowest terms. See Examples 3 and 4.

15. $\dfrac{4}{x} \div \dfrac{8}{x^2}$

16. $\dfrac{x}{3} \div \dfrac{x^3}{9}$

17. $\dfrac{4ab}{3c^2} \div \dfrac{2a^2b}{6ac}$

18. $\dfrac{6abc}{5ab^4} \div \dfrac{2bc^2}{10ab^2}$

19. $\dfrac{2x}{5} \div \dfrac{6x + 12}{5x + 10}$

20. $\dfrac{7}{3x} \div \dfrac{14 - 7x}{18 - 9x}$

21. $\dfrac{2(x + y)}{5} \div \dfrac{6(x + y)}{25}$

22. $\dfrac{4}{x - 2y} \div \dfrac{9}{3x - 6y}$

23. $\dfrac{x^2 - 6x + 9}{x^2 - x - 6} \div \dfrac{x^2 - 9}{4}$

24. $\dfrac{x^2 - 4}{3x + 6} \div \dfrac{2x^2 - 8x + 8}{x^2 + 4x + 4}$

25. $\dfrac{x^2 - 6x - 16}{2x^2 - 128} \div \dfrac{x^2 + 10x + 16}{x^2 + 16x + 64}$

26. $\dfrac{a^2 - a - 6}{a^2 - 81} \div \dfrac{a^2 - 7a - 18}{4a + 36}$

27. $\dfrac{14x^4}{y^5} \div \dfrac{2x^2}{y^7} \div \dfrac{2x^4}{7y^5}$

28. $\dfrac{x^2}{7y^2} \div \left(\dfrac{2x^2}{y^7} \div \dfrac{2x^2}{7y^5} \right)$

Perform the indicated operation. Write all answers in lowest terms.

29. $\dfrac{3xy^3}{4x^3y^2} \cdot \dfrac{-8x^3y^4}{9x^4y^7}$

30. $-\dfrac{2xyz^3}{5x^2z^2} \cdot \dfrac{10xy}{x^3}$

31. $\dfrac{8a}{3a^4b^2} \div \dfrac{4b^5}{6a^2b}$

32. $\dfrac{3y^3}{14x^4} \div \dfrac{8y^3}{7x}$

33. $\dfrac{a^2b}{a^2-b^2} \cdot \dfrac{a+b}{4a^3b}$

34. $\dfrac{3ab^2}{a^2-4} \cdot \dfrac{a-2}{6a^2b^2}$

35. $\dfrac{x^2-9}{4} \div \dfrac{x^2-6x+9}{x^2-x-6}$

36. $\dfrac{a-5b}{a^2+ab} \div \dfrac{15b-3a}{b^2-a^2}$

37. $\dfrac{9x+9}{4x+8} \cdot \dfrac{2x+4}{3x^2-3}$

38. $\dfrac{x^2-1}{10x+30} \cdot \dfrac{12x+36}{3x-3}$

39. $\dfrac{a+b}{ab} \div \dfrac{a^2-b^2}{4a^3b}$

40. $\dfrac{6a^2b^2}{a^2-4} \div \dfrac{3ab^2}{a-2}$

41. $\dfrac{2x^2-4x-30}{5x^2-40x-75} \div \dfrac{x^2-8x+15}{x^2-6x+9}$

42. $\dfrac{4a+36}{a^2-7a-18} \div \dfrac{a^2-a-6}{a^2-81}$

43. $\dfrac{2x^3-16}{6x^2+6x-36} \cdot \dfrac{9x+18}{3x^2+6x+12}$

44. $\dfrac{x^2-3x+9}{5x^2-20x-105} \cdot \dfrac{x^2-49}{x^3+27}$

45. $\dfrac{15b-3a}{b^2-a^2} \div \dfrac{a-5b}{ab+b^2}$

46. $\dfrac{4x+4}{x-1} \div \dfrac{x^2-4x-5}{x^2-1}$

47. $\dfrac{a^3+a^2b+a+b}{a^3+a} \cdot \dfrac{6a^2}{2a^2-2b^2}$

48. $\dfrac{a^2-2a}{ab-2b+3a-6} \cdot \dfrac{8b+24}{3a+6}$

49. $\dfrac{5a}{12} \cdot \dfrac{2}{25a^2} \cdot \dfrac{15a}{2}$

50. $\dfrac{4a}{7} \div \dfrac{a^2}{14} \cdot \dfrac{3}{a}$

51. $\dfrac{3x-x^2}{x^3-27} \div \dfrac{x}{x^2+3x+9}$

52. $\dfrac{x^2-3x}{x^3-27} \div \dfrac{2x}{2x^2+6x+18}$

53. $\dfrac{4a}{7} \div \left(\dfrac{a^2}{14} \cdot \dfrac{3}{a}\right)$

54. $\dfrac{a^2}{14} \cdot \dfrac{3}{a} \div \dfrac{4a}{7}$

55. $\dfrac{8b+24}{3a+6} \div \dfrac{ab-2b+3a-6}{a^2-4a+4}$

56. $\dfrac{2a^2-2b^2}{a^3+a^2b+a+b} \div \dfrac{6a^2}{a^3+a}$

57. $\dfrac{4}{x} \div \dfrac{3xy}{x^2} \cdot \dfrac{6x^2}{x^4}$

58. $\dfrac{4}{x} \cdot \dfrac{3xy}{x^2} \div \dfrac{6x^2}{x^4}$

59. $\dfrac{3x^2-5x-2}{y^2+y-2} \cdot \dfrac{y^2+4y-5}{12x^2+7x+1} \div \dfrac{5x^2-9x-2}{8x^2-2x-1}$

60. $\dfrac{x^2+x-2}{3y^2-5y-2} \cdot \dfrac{12y^2+y-1}{x^2+4x-5} \div \dfrac{8y^2-6y+1}{5y^2-9y-2}$

61. $\dfrac{5a^2-20}{3a^2-12a} \div \dfrac{a^3+2a^2}{2a^2-8a} \cdot \dfrac{9a^3+6a^2}{2a^2-4a}$

62. $\dfrac{5a^2-20}{3a^2-12a} \div \left(\dfrac{a^3+2a^2}{2a^2-8a} \cdot \dfrac{9a^3+6a^2}{2a^2-4a}\right)$

63. $\dfrac{5x^4+3x^2-2}{x-1} \cdot \dfrac{x+1}{x^4-1}$

64. $\dfrac{3x^4-10x^2-8}{x-2} \cdot \dfrac{3x+6}{15x^2+10}$

65. Find the area of the rectangle.

$$\dfrac{x+2}{x} \text{ meters}$$

$$\dfrac{5x}{x^2-4} \text{ meters}$$

66. Find the area of the triangle.

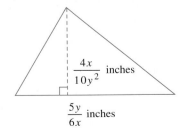

$$\dfrac{4x}{10y^2} \text{ inches}$$

$$\dfrac{5y}{6x} \text{ inches}$$

67. The parallelogram has area $\dfrac{x^2 + x - 2}{x^3}$ square feet and height $\dfrac{x^2}{x - 1}$ feet. Express the length of its base as a rational expression in x ($A = b \cdot h$).

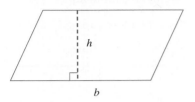

68. A lottery prize of $\dfrac{15x^3}{y^2}$ dollars is to be divided among $5x$ people. Express the amount of money each person is to receive as a rational expression in x and y.

A Look Ahead

Perform the indicated operation. Write all answers in lowest terms. See the following example.

EXAMPLE   Perform the following operation.

$$\frac{x^{2n} - 3x^n - 18}{x^{2n} - 9} \cdot \frac{3x^n + 9}{x^{2n}}$$

Solution:

$$\frac{x^{2n} - 3x^n - 18}{x^{2n} - 9} \cdot \frac{3x^n + 9}{x^{2n}} = \frac{(x^n + 3)(x^n - 6) \cdot 3(x^n + 3)}{(x^n + 3)(x^n - 3) \cdot x^{2n}}$$

$$= \frac{3(x^n - 6)(x^n + 3)}{x^{2n}(x^n - 3)} \quad ∎$$

69. $\dfrac{x^{2n} - 4}{7x} \cdot \dfrac{14x^3}{x^n - 2}$

70. $\dfrac{x^{2n} + 4x^n + 4}{4x - 3} \cdot \dfrac{8x^2 - 6x}{x^n + 2}$

71. $\dfrac{y^{2n} + 9}{10y} \cdot \dfrac{y^n - 3}{y^{4n} - 81}$

72. $\dfrac{y^{4n} - 16}{y^{2n} + 4} \cdot \dfrac{6y}{y^n + 2}$

73. $\dfrac{y^{2n} - y^n - 2}{2y^n - 4} \div \dfrac{y^{2n} - 1}{1 + y^n}$

74. $\dfrac{y^{2n} + 7y^n + 10}{10} \div \dfrac{y^{2n} + 4y^n + 4}{5y^n + 25}$

Skill Review

Perform the indicated operation. See Section 1.4.

75. $\dfrac{4}{5} + \dfrac{3}{5}$

76. $\dfrac{4}{10} - \dfrac{7}{10}$

77. $\dfrac{5}{28} - \dfrac{2}{21}$

78. $\dfrac{5}{13} + \dfrac{2}{7}$

Use synthetic division to divide the following. See Section 4.6.

79. $(x^3 - 6x^2 + 3x - 4) \div (x - 1)$

80. $(5x^4 - 3x^2 + 2) \div (x + 2)$

6.3
Adding and Subtracting Rational Expressions

Tape 17

OBJECTIVES

1 Add or subtract rational expressions with common denominators.

2 Identify the least common denominator of two or more rational expressions.

3 Add and subtract rational expressions with unlike denominators.

---

1 Rational expressions, like rational numbers, may be added or subtracted. We define the sum or difference of rational expressions similar to the sum or difference of rational numbers.

> **Adding or Subtracting Rational Expressions with Common Denominators**
>
> If $\dfrac{P}{Q}$ and $\dfrac{R}{Q}$ are rational expressions, then
>
> $$\frac{P}{Q} + \frac{R}{Q} = \frac{P+R}{Q} \quad \text{and} \quad \frac{P}{Q} - \frac{R}{Q} = \frac{P-R}{Q}$$

To add or subtract rational expressions with common denominators, add or subtract the numerators, and write the sum or difference over the common denominator.

EXAMPLE 1   Add or subtract.

    a. $\dfrac{5}{7} + \dfrac{x}{7}$ **b.** $\dfrac{x}{4} + \dfrac{5x}{4}$

    c. $\dfrac{x^2}{x+7} - \dfrac{49}{x+7}$ **d.** $\dfrac{x}{3y^2} - \dfrac{x+1}{3y^2}$

Solutions: We have common denominators, so add or subtract the numerators and place the sum or difference over the common denominator.

    a. $\dfrac{5}{7} + \dfrac{x}{7} = \dfrac{5+x}{7}$

    b. $\dfrac{x}{4} + \dfrac{5x}{4} = \dfrac{x+5x}{4} = \dfrac{6x}{4} = \dfrac{3x}{2}$

    c. $\dfrac{x^2}{x+7} - \dfrac{49}{x+7} = \dfrac{x^2-49}{x+7}$

    Next, write this rational expression in lowest terms.

$$\frac{x^2-49}{x+7} = \frac{(x+7)(x-7)}{x+7} = x-7$$

d. $\dfrac{x}{3y^2} - \dfrac{x+1}{3y^2} = \dfrac{x - (x+1)}{3y^2}$ Subtract numerators.

$\qquad\qquad\quad = \dfrac{x - x - 1}{3y^2}$ Apply the distributive property.

$\qquad\qquad\quad = -\dfrac{1}{3y^2}$ Simplify.   ∎

2   To add or subtract rational expressions with unlike denominators, first write the rational expressions as equivalent rational expressions with common denominators.

The **least common denominator (LCD)** is usually the easiest common denominator to work with. The LCD of a list of rational expressions is a polynomial of least degree whose factors include all the denominator factors in the list.

Use the following steps to find the LCD.

To Find the Least Common Denominator (LCD)

Step 1   Factor each denominator completely.

Step 2   The least common denominator (LCD) is the product of all unique factors formed in *step 1*, each raised to a power equal to the greatest number of times that the factor appears in any one factored denominator.

EXAMPLE 2   Find the LCD of the rational expressions in each list.

a. $\dfrac{2}{15x^5y^2}, \dfrac{3z}{5xy^3}$

b. $\dfrac{7}{z+1}, \dfrac{z}{z-1}$

c. $\dfrac{m-1}{m^2-25}, \dfrac{2m}{2m^2-9m-5}, \dfrac{7}{m^2-10m+25}$

Solution: **a.** Factor each denominator.

$$15x^5y^2 = 3 \cdot 5 \cdot x^5 \cdot y^2$$
$$5xy^3 = 5 \cdot x \cdot y^3$$

The unique factors are 3, 5, x, and y.
The greatest number of times that 3 appears in one denominator is 1.
The greatest number of times that 5 appears in one denominator is 1.
The greatest number of times that x appears in one denominator is 5.
The greatest number of times that y appears in one denominator is 3.
The LCD is the product of $3^1 \cdot 5^1 \cdot x^5 \cdot y^3 = 15x^5y^3$.

b. The denominators $z + 1$ and $z - 1$ do not factor further. Each factor appears once, so the

$$\text{LCD} = (z+1)(z-1)$$

c. First, factor each denominator.

$$m^2 - 25 = (m + 5)(m - 5)$$

$$2m^2 - 9m - 5 = (2m + 1)(m - 5)$$

$$m^2 - 10m + 25 = (m - 5)(m - 5)$$

The LCD $= (m + 5)(2m + 1)(m - 5)^2$, which is the product of each unique factor raised to a power equal to the greatest number of times it appears in one factored denominator. ∎

3 To add or subtract rational expressions with unlike denominators, follow the steps shown.

To Add or Subtract Rational Expressions with Unlike Denominators

Step 1 Find the LCD of the rational expressions.

Step 2 Write each rational expression as an equivalent rational expression whose denominator is the LCD found in *step 1*.

Step 3 Add or subtract numerators, and write the sum or difference over the common denominator.

Step 4 Write the answer in lowest terms.

EXAMPLE 3 Perform the indicated operation.

a. $\dfrac{2}{x^2 y} + \dfrac{5}{3x^3 y}$ **b.** $\dfrac{3z}{z + 2} + \dfrac{2z}{z - 2}$ **c.** $\dfrac{5k}{k^2 - 4} - \dfrac{2}{k^2 + k - 2}$

Solution: **a.** The LCD is $3x^3 y$. Write each fraction as an equivalent fraction with denominator $3x^3 y$.

$$\frac{2}{x^2 y} + \frac{5}{3x^3 y} = \frac{2 \cdot 3x}{x^2 y \cdot 3x} + \frac{5}{3x^3 y}$$

$$= \frac{6x}{3x^3 y} + \frac{5}{3x^3 y}$$

$$= \frac{6x + 5}{3x^3 y} \qquad \text{Add the numerators.}$$

b. The LCD is the product of the two denominators: $(z + 2)(z - 2)$.

$$\frac{3z}{z + 2} + \frac{2z}{z - 2} = \frac{3z \cdot (z - 2)}{(z + 2) \cdot (z - 2)} + \frac{2z \cdot (z + 2)}{(z - 2) \cdot (z + 2)} \qquad \begin{array}{l}\text{Write equivalent} \\ \text{rational expressions.}\end{array}$$

$$= \frac{3z(z - 2) + 2z(z + 2)}{(z + 2)(z - 2)} \qquad \text{Add.}$$

$$= \frac{3z^2 - 6z + 2z^2 + 4z}{(z + 2)(z - 2)}$$

$$= \frac{5z^2 - 2z}{(z + 2)(z - 2)} \qquad \begin{array}{l}\text{Simplify the} \\ \text{numerator.}\end{array}$$

c. $\dfrac{5k}{k^2 - 4} - \dfrac{2}{k^2 + k - 2} = \dfrac{5k}{(k + 2)(k - 2)} - \dfrac{2}{(k + 2)(k - 1)}$　Factor to find the LCD.

The LCD is $(k + 2)(k - 2)(k - 1)$. Write equivalent rational expressions with the LCD as denominators.

$$\dfrac{5k}{(k + 2)(k - 2)} - \dfrac{2}{(k + 2)(k - 1)} = \dfrac{5k(k - 1)}{(k + 2)(k - 2)(k - 1)} - \dfrac{2(k - 2)}{(k + 2)(k - 1)(k - 2)}$$

$$= \dfrac{5k(k - 1) - 2(k - 2)}{(k + 2)(k - 2)(k - 1)}$$　Subtract.

$$= \dfrac{5k^2 - 5k - 2k + 4}{(k + 2)(k - 2)(k - 1)}$$

$$= \dfrac{5k^2 - 7k + 4}{(k + 2)(k - 2)(k - 1)}$$　Simplify the numerator.

Since the numerator polynomial is prime, this rational expression is in lowest terms.　∎

EXAMPLE 4　Perform the indicated operation.

a. $\dfrac{x}{x - 3} - 5$　**b.** $\dfrac{7}{x - y} + \dfrac{3}{y - x}$

Solution:　**a.** $\dfrac{x}{x - 3} - 5 = \dfrac{x}{x - 3} - \dfrac{5}{1}$　Write 5 as $\dfrac{5}{1}$.

The LCD is $x - 3$.

$$= \dfrac{x}{x - 3} - \dfrac{5 \cdot (x - 3)}{1 \cdot (x - 3)}$$

$$= \dfrac{x - 5(x - 3)}{x - 3}$$　Subtract.

$$= \dfrac{x - 5x + 15}{x - 3}$$

$$= \dfrac{-4x + 15}{x - 3}$$　Simplify.

b. Notice that the denominators $x - y$ and $y - x$ are opposites of one another. To write equivalent rational expressions with the LCD, write one denominator such as $y - x$ as $-1(x - y)$.

$$\dfrac{7}{x - y} + \dfrac{3}{y - x} = \dfrac{7}{x - y} + \dfrac{3}{-1(x - y)}$$

$$= \dfrac{7}{x - y} + \dfrac{-3}{x - y}$$

$$= \dfrac{4}{x - y}$$　Add the numerators.　∎

EXAMPLE 5 Add.

$$\frac{2x - 1}{2x^2 - 9x - 5} + \frac{x + 3}{6x^2 - x - 2}$$

Solution:

$$\frac{2x - 1}{2x^2 - 9x - 5} + \frac{x + 3}{6x^2 - x - 2} = \frac{2x - 1}{(2x + 1)(x - 5)} + \frac{x + 3}{(2x + 1)(3x - 2)} \qquad \text{Factor the denominators.}$$

The LCD is $(2x + 1)(x - 5)(3x - 2)$.

$$= \frac{(2x - 1) \cdot (3x - 2)}{(2x + 1)(x - 5) \cdot (3x - 2)} + \frac{(x + 3) \cdot (x - 5)}{(2x + 1)(3x - 2) \cdot (x - 5)}$$

$$= \frac{6x^2 - 7x + 2}{(2x + 1)(x - 5)(3x - 2)} + \frac{x^2 - 2x - 15}{(2x + 1)(x - 5)(3x - 2)}$$

$$= \frac{6x^2 - 7x + 2 + x^2 - 2x - 15}{(2x + 1)(x - 5)(3x - 2)} \qquad \text{Add.}$$

$$= \frac{7x^2 - 9x - 13}{(2x + 1)(x - 5)(3x - 2)} \qquad \text{Simplify the numerator.}$$

The numerator polynomial is prime and the rational expression is in lowest terms. ■

EXAMPLE 6 Perform the indicated operations.

$$\frac{7}{x - 1} + \frac{2(x + 1)}{(x - 1)^2} - \frac{2}{x^2}$$

Solution: The LCD is $x^2(x - 1)^2$.

$$\frac{7}{x - 1} + \frac{2(x + 1)}{(x - 1)^2} - \frac{2}{x^2} = \frac{7 \cdot x^2(x - 1)}{(x - 1) \cdot x^2(x - 1)} + \frac{2(x + 1) \cdot x^2}{(x - 1)^2 \cdot x^2} - \frac{2 \cdot (x - 1)^2}{x^2 \cdot (x - 1)^2}$$

$$= \frac{7x^3 - 7x^2}{x^2(x - 1)^2} + \frac{2x^3 + 2x^2}{x^2(x - 1)^2} - \frac{2x^2 - 4x + 2}{x^2(x - 1)^2}$$

$$= \frac{7x^3 - 7x^2 + 2x^3 + 2x^2 - 2x^2 + 4x - 2}{x^2(x - 1)^2} \qquad \text{Watch your signs.}$$

$$= \frac{9x^3 - 7x^2 + 4x - 2}{x^2(x - 1)^2} \qquad \begin{array}{l}\text{Add or subtract like terms} \\ \text{in the numerator.}\end{array}$$

Since the numerator and denominator have no common factors, this rational expression is in lowest terms. ■

EXERCISE SET 6.3

Perform the indicated operation. Write answers in lowest terms. See Example 1.

1. $\dfrac{2}{x} - \dfrac{5}{x}$

2. $\dfrac{4}{x^2} + \dfrac{2}{x^2}$

3. $\dfrac{2}{x - 2} + \dfrac{x}{x - 2}$

4. $\dfrac{x}{5-x} + \dfrac{2}{5-x}$

5. $\dfrac{x^2}{x+2} - \dfrac{4}{x+2}$

6. $\dfrac{4}{x-2} - \dfrac{x^2}{x-2}$

7. $\dfrac{2x-6}{x^2+x-6} + \dfrac{3-3x}{x^2+x-6}$

8. $\dfrac{5x+2}{x^2+2x-8} + \dfrac{2-4x}{x^2+2x-8}$

Find the LCD of the rational expressions in the list. See Example 2.

9. $\dfrac{2}{7}, \dfrac{3}{5x}$

10. $\dfrac{4}{5y}, \dfrac{3}{4y^2}$

11. $\dfrac{3}{x}, \dfrac{2}{x+1}$

12. $\dfrac{5}{2x}, \dfrac{7}{2+x}$

13. $\dfrac{12}{x+7}, \dfrac{8}{x-7}$

14. $\dfrac{1}{2x-1}, \dfrac{x}{2x+1}$

15. $\dfrac{5}{3x+6}, \dfrac{2x}{2x-4}$

16. $\dfrac{2}{3a+9}, \dfrac{5}{5a-15}$

17. $\dfrac{2a}{a^2-b^2}, \dfrac{1}{a^2-2ab+b^2}$

18. $\dfrac{2a}{a^2+8a+16}, \dfrac{7a}{a^2+a-12}$

Perform the indicated operation. Write answers in lowest terms. See Example 3.

19. $\dfrac{4}{3x} + \dfrac{3}{2x}$

20. $\dfrac{10}{7x} - \dfrac{5}{2x}$

21. $\dfrac{5}{2y^2} - \dfrac{2}{7y}$

22. $\dfrac{4}{11x^4y} - \dfrac{1}{4x^2y^3}$

23. $\dfrac{x-3}{x+4} - \dfrac{x+2}{x-4}$

24. $\dfrac{x-1}{x-5} - \dfrac{x+2}{x+5}$

25. $\dfrac{1}{x-5} + \dfrac{x}{x^2-x-20}$

26. $\dfrac{x+1}{x^2-x-20} - \dfrac{2}{x+4}$

Perform the indicated operation. Write answers in lowest terms. See Example 4.

27. $\dfrac{1}{a-b} + \dfrac{1}{b-a}$

28. $\dfrac{1}{a-3} - \dfrac{1}{3-a}$

29. $x+1+\dfrac{1}{x-1}$

30. $5 - \dfrac{1}{x-1}$

31. $\dfrac{5}{x-2} + \dfrac{x+4}{2-x}$

32. $\dfrac{3}{5-x} + \dfrac{x+2}{x-5}$

Perform the indicated operation. Write all answers in lowest terms. See Example 5.

33. $\dfrac{y+1}{y^2-6y+8} - \dfrac{3}{y^2-16}$

34. $\dfrac{x+2}{x^2-36} - \dfrac{x}{x^2+9x+18}$

35. $\dfrac{x+4}{3x^2+11x+6} + \dfrac{x}{2x^2+x-15}$

36. $\dfrac{x+3}{5x^2+12x+4} + \dfrac{6}{x^2-x-6}$

37. $\dfrac{7}{x^2-x-2} + \dfrac{x}{x^2+4x+3}$

38. $\dfrac{a}{a^2+10a+25} + \dfrac{4}{a^2+6a+5}$

Perform the indicated operation. Write all answers in lowest terms. See Example 6.

39. $\dfrac{2}{x+1} - \dfrac{3x}{3x+3} + \dfrac{1}{2x+2}$

40. $\dfrac{5}{3x-6} - \dfrac{x}{x-2} + \dfrac{3+2x}{5x-10}$

41. $\dfrac{3}{x+3} + \dfrac{5}{x^2+6x+9} - \dfrac{x}{x^2-9}$

42. $\dfrac{x+2}{x^2-2x-3} + \dfrac{x}{x-3} - \dfrac{4}{x+1}$

Add or subtract as indicated. Write answers in lowest terms.

43. $\dfrac{4}{3x^2y^3} + \dfrac{5}{3x^2y^3}$

44. $\dfrac{7}{2xy^4} + \dfrac{1}{2xy^4}$

45. $\dfrac{x-5}{2x} - \dfrac{x+5}{2x}$

46. $\dfrac{x+4}{4x} - \dfrac{x-4}{4x}$

47. $\dfrac{3}{2x+10} + \dfrac{8}{3x+15}$

48. $\dfrac{10}{3x-3} + \dfrac{1}{7x-7}$

49. $\dfrac{-2}{x^2-3x} - \dfrac{1}{x^3-3x^2}$

50. $\dfrac{-3}{2a+8} - \dfrac{8}{a^2+4a}$

51. $\dfrac{ab}{a^2-b^2} + \dfrac{b}{a+b}$

52. $\dfrac{x}{25-x^2} + \dfrac{2}{3x-15}$

53. $\dfrac{5}{x^2-4} - \dfrac{3}{x^2+4x+4}$

54. $\dfrac{3z}{z^2-9} - \dfrac{2}{3-z}$

55. $\dfrac{2}{a^2+2a+1} + \dfrac{3}{a^2-1}$

56. $\dfrac{9x+2}{3x^2-2x-8} + \dfrac{7}{3x^2+x-4}$

Perform the indicated operation. Write answers in simplest form.

57. $\left(\dfrac{2}{3} - \dfrac{1}{x}\right) \cdot \left(\dfrac{3}{x} + \dfrac{1}{2}\right)$

58. $\left(\dfrac{2}{3} - \dfrac{1}{x}\right) \div \left(\dfrac{3}{x} + \dfrac{1}{2}\right)$

59. $\left(\dfrac{1}{x} + \dfrac{2}{3}\right) - \left(\dfrac{1}{x} - \dfrac{2}{3}\right)$

60. $\left(\dfrac{1}{2} + \dfrac{2}{x}\right) - \left(\dfrac{1}{2} - \dfrac{1}{x}\right)$

61. $\left(\dfrac{2a}{3}\right)^2 \div \left(\dfrac{a^2}{a+1} - \dfrac{1}{a+1}\right)$

62. $\left(\dfrac{x+2}{2x} - \dfrac{x-2}{2x}\right) \cdot \left(\dfrac{5x}{4}\right)^2$

63. $\left(\dfrac{2x}{3}\right)^2 \div \left(\dfrac{x}{3}\right)^2$

64. $\left(\dfrac{2x}{3}\right)^2 \cdot \left(\dfrac{3}{x}\right)^2$

65. $\dfrac{x}{x^2-9} + \dfrac{3}{x^2-6x+9} - \dfrac{1}{x+3}$

66. $\dfrac{3}{x^2-9} - \dfrac{x}{x^2-6x+9} + \dfrac{1}{x+3}$

67. $\left(\dfrac{x}{x+1} - \dfrac{x}{x-1}\right) \div \dfrac{x}{2x+2}$

68. $\dfrac{x}{2x+2} \div \left(\dfrac{x}{x+1} + \dfrac{x}{x-1}\right)$

69. $\dfrac{4}{x} \cdot \left(\dfrac{2}{x+2} - \dfrac{2}{x-2}\right)$

70. $\dfrac{1}{x+1} \cdot \left(\dfrac{5}{x} + \dfrac{2}{x-3}\right)$

71. Find the perimeter of the square.

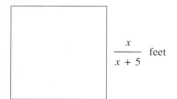

$\dfrac{x}{x+5}$ feet

72. Find the perimeter of the quadrilateral.

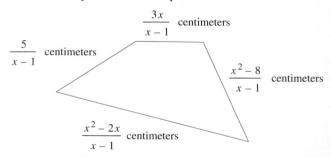

$\dfrac{3x}{x-1}$ centimeters

$\dfrac{5}{x-1}$ centimeters

$\dfrac{x^2-8}{x-1}$ centimeters

$\dfrac{x^2-2x}{x-1}$ centimeters

A Look Ahead

Perform the indicated operation. See the following example.

EXAMPLE   Add $x^{-1} + 3x^{-2}$.

Solution: $x^{-1} + 3x^{-2} = \dfrac{1}{x} + \dfrac{3}{x^2}$

$$= \dfrac{1 \cdot x}{x \cdot x} + \dfrac{3}{x^2}$$

$$= \dfrac{x}{x^2} + \dfrac{3}{x^2}$$

$$= \dfrac{x + 3}{x^2} \quad \blacksquare$$

73. $x^{-1} + (2x)^{-1}$ **74.** $3y^{-1} + (4y)^{-1}$ **75.** $4x^{-2} - 3x^{-1}$

76. $(4x)^{-2} - (3x)^{-1}$ **77.** $x^{-3}(2x + 1) - 5x^{-2}$ **78.** $4x^{-3} + x^{-4}(5x + 7)$

Skill Review

Use the distributive property to multiply the following. See Sections 1.4 and 6.2.

79. $12\left(\dfrac{2}{3} + \dfrac{1}{6}\right)$ **80.** $14\left(\dfrac{1}{7} + \dfrac{3}{14}\right)$

81. $x^2\left(\dfrac{4}{x^2} + 1\right)$ **82.** $5y^2\left(\dfrac{1}{y^2} - \dfrac{1}{5}\right)$

Find each root. See Section 1.3.

83. $\sqrt{100}$ **84.** $\sqrt{25}$

6.4
Simplifying Complex Fractions

OBJECTIVES

Tape 18

1   Identify complex fractions.

2   Simplify complex fractions by simplifying the numerator and denominator and then dividing.

3   Simplify complex fractions by multiplying by a common denominator.

4   Simplify expressions with negative exponents.

1   A fraction whose numerator, denominator, or both contain one or more rational expressions is called a **complex fraction.**

Examples of Complex Fractions

$$\frac{\dfrac{1}{a}}{\dfrac{b}{2}} \qquad \frac{\dfrac{x}{2y^2}}{\dfrac{6x-2}{9y}} \qquad \frac{x+\dfrac{1}{y}}{y+1}$$

2   Two methods for simplifying complex fractions are introduced. The first method evolves from the definition of a fraction as a quotient.

> **To Simplify a Complex Fraction: Method I**
>
> *Step 1*   Simplify the numerator and the denominator of the complex fraction so that each is a single fraction.
>
> *Step 2*   Perform the indicated division by multiplying the numerator of the complex fraction by the reciprocal of the denominator of the complex fraction.
>
> *Step 3*   Simplify if possible.

EXAMPLE 1   Simplify each complex fraction.

$$\textbf{a. } \frac{\dfrac{2x}{27y^2}}{\dfrac{6x^2}{9}} \qquad \textbf{b. } \frac{\dfrac{5x}{x+2}}{\dfrac{10}{x-2}} \qquad \textbf{c. } \frac{x+\dfrac{1}{y}}{y+\dfrac{1}{x}}$$

Solution: **a.** The numerator of the complex fraction is already a single fraction and so is the denominator. Perform the indicated division by multiplying the numerator $\dfrac{2x}{27y^2}$ by the reciprocal of the denominator $\dfrac{6x^2}{9}$. Then simplify.

$$\frac{\dfrac{2x}{27y^2}}{\dfrac{6x^2}{9}} = \frac{2x}{27y^2} \div \frac{6x^2}{9}$$

$$= \frac{2x}{27y^2} \cdot \frac{9}{6x^2} \qquad \text{Multiply by the reciprocal of } \frac{6x^2}{9}.$$

$$= \frac{2x \cdot 9}{27y^2 \cdot 6x^2}$$

$$= \frac{1}{9xy^2}$$

b. $\dfrac{\dfrac{5x}{x+2}}{\dfrac{10}{x-2}} = \dfrac{5x}{x+2} \cdot \dfrac{x-2}{10}$ Multiply by the reciprocal of $\dfrac{10}{x-2}$.

$= \dfrac{5x(x-2)}{2 \cdot 5(x+2)}$

$= \dfrac{x(x-2)}{2(x+2)}$ Simplify.

c. First, simplify the numerator and the denominator of the complex fraction separately.

$\dfrac{x+\dfrac{1}{y}}{y+\dfrac{1}{x}} = \dfrac{\dfrac{x \cdot y}{1 \cdot y}+\dfrac{1}{y}}{\dfrac{y \cdot x}{1 \cdot x}+\dfrac{1}{x}}$ The LCD is y.

The LCD is x.

$= \dfrac{\dfrac{xy+1}{y}}{\dfrac{yx+1}{x}}$ Add.

Add.

$= \dfrac{xy+1}{y} \cdot \dfrac{x}{xy+1}$ Multiply by the reciprocal of $\dfrac{yx+1}{x}$.

$= \dfrac{x\,(xy+1)}{y\,(xy+1)}$

$= \dfrac{x}{y}$ ■

3   Next, we look at another method for simplifying complex fractions. With this method we multiply the numerator and the denominator of the complex fraction by the LCD of all fractions in the complex fraction.

To Simplify a Complex Fraction: Method II

Step 1   Multiply the numerator and the denominator of the complex fraction by the LCD of the fractions in both the numerator and the denominator.

Step 2   Simplify.

EXAMPLE 2   Simplify each complex fraction.

a. $\dfrac{\dfrac{5x}{x+2}}{\dfrac{10}{x-2}}$ **b.** $\dfrac{x+\dfrac{1}{y}}{y+\dfrac{1}{x}}$

Solution: **a.** The least common denominator of $\dfrac{5x}{x+2}$ and $\dfrac{10}{x-2}$ is $(x+2)(x-2)$. Multiply both the numerator $\dfrac{5x}{x+2}$ and the denominator $\dfrac{10}{x-2}$ by the LCD.

$$\dfrac{\dfrac{5x}{x+2}}{\dfrac{10}{x-2}} = \dfrac{\left(\dfrac{5x}{x+2}\right)\cdot(x+2)(x-2)}{\left(\dfrac{10}{x-2}\right)\cdot(x+2)(x-2)} \qquad \text{Simplify.}$$

$$= \dfrac{5\,x\cdot(x-2)}{2\cdot 5\cdot(x+2)} \qquad \text{Simplify.}$$

$$= \dfrac{x(x-2)}{2(x+2)}$$

b. The least common denominator of $\dfrac{1}{y}$ and $\dfrac{1}{x}$ is xy.

$$\dfrac{x+\dfrac{1}{y}}{y+\dfrac{1}{x}} = \dfrac{\left(x+\dfrac{1}{y}\right)\cdot xy}{\left(y+\dfrac{1}{x}\right)\cdot xy}$$

$$= \dfrac{x\cdot xy+\dfrac{1}{y}\cdot x\,y}{y\cdot xy+\dfrac{1}{x}\cdot x\,y} \qquad \text{Apply the distributive property.}$$

$$= \dfrac{x^2y+x}{xy^2+y} \qquad \text{Simplify.}$$

$$= \dfrac{x\,(xy+1)}{y\,(xy+1)} \qquad \text{Factor.}$$

$$= \dfrac{x}{y} \qquad \text{Simplify.} \qquad \blacksquare$$

4

EXAMPLE 3 Simplify.

$$\dfrac{x^{-1}+2xy^{-1}}{x^{-2}-x^{-2}y^{-1}}$$

Solution: This fraction does not appear to be a complex fraction. However, if we write it using only positive exponents, we see that it is a complex fraction.

$$\dfrac{x^{-1}+2xy^{-1}}{x^{-2}-x^{-2}y^{-1}} = \dfrac{\dfrac{1}{x}+\dfrac{2x}{y}}{\dfrac{1}{x^2}-\dfrac{1}{x^2y}}$$

The LCD of $\dfrac{1}{x}, \dfrac{2x}{y}, \dfrac{1}{x^2}$, and $\dfrac{1}{x^2 y}$ is x^2y. Multiply both the numerator and the denominator by x^2y.

$$= \dfrac{\left(\dfrac{1}{x} + \dfrac{2x}{y}\right) \cdot x^2 y}{\left(\dfrac{1}{x^2} - \dfrac{1}{x^2 y}\right) \cdot x^2 y}$$

$$= \dfrac{\dfrac{1}{x} \cdot x^2 y + \dfrac{2x}{y} \cdot x^2 y}{\dfrac{1}{x^2} \cdot x^2 y - \left(\dfrac{1}{x^2 y}\right) \cdot x^2 y} \qquad \text{Apply the distributive property.}$$

$$= \dfrac{xy + 2x^3}{y - 1} \qquad \text{Simplify.} \qquad \blacksquare$$

EXERCISE SET 6.4

Simplify each complex fraction. See Examples 1 and 2.

1. $\dfrac{\dfrac{1}{3}}{\dfrac{2}{5}}$

2. $\dfrac{\dfrac{3}{5}}{\dfrac{4}{5}}$

3. $\dfrac{\dfrac{4}{x}}{\dfrac{5}{2x}}$

4. $\dfrac{\dfrac{5}{2x}}{\dfrac{4}{x}}$

5. $\dfrac{\dfrac{10}{3x}}{\dfrac{5}{6x}}$

6. $\dfrac{\dfrac{15}{2x}}{\dfrac{5}{6x}}$

7. $\dfrac{1 + \dfrac{2}{5}}{2 + \dfrac{3}{5}}$

8. $\dfrac{2 + \dfrac{1}{7}}{3 - \dfrac{4}{7}}$

9. $\dfrac{\dfrac{4}{x-1}}{\dfrac{x}{x-1}}$

10. $\dfrac{\dfrac{x}{x+2}}{\dfrac{2}{x+2}}$

11. $\dfrac{1 - \dfrac{2}{x}}{x - \dfrac{4}{9x}}$

12. $\dfrac{5 - \dfrac{3}{x}}{x + \dfrac{2}{3x}}$

13. $\dfrac{\dfrac{1}{x+1} - 1}{\dfrac{1}{x-1} + 1}$

14. $\dfrac{1 + \dfrac{1}{x-1}}{1 - \dfrac{1}{x+1}}$

Simplify. See Example 3.

15. $\dfrac{x^{-1}}{x^{-2} + y^{-2}}$

16. $\dfrac{a^{-3} + b^{-1}}{a^{-2}}$

17. $\dfrac{2a^{-1} + 3b^{-2}}{a^{-1} - b^{-1}}$

18. $\dfrac{x^{-1} + y^{-1}}{3x^{-2} + 5y^{-2}}$

19. $\dfrac{1}{x - x^{-1}}$

20. $\dfrac{x^{-2}}{x + 3x^{-1}}$

Simplify.

21. $\dfrac{\dfrac{x+1}{7}}{\dfrac{x+2}{7}}$

22. $\dfrac{\dfrac{y}{10}}{\dfrac{x+1}{10}}$

23. $\dfrac{\dfrac{1}{2}-\dfrac{1}{3}}{\dfrac{3}{4}+\dfrac{2}{5}}$

24. $\dfrac{\dfrac{5}{6}-\dfrac{1}{2}}{\dfrac{1}{3}+\dfrac{1}{8}}$

25. $\dfrac{\dfrac{x+1}{3}}{\dfrac{2x-1}{6}}$

26. $\dfrac{\dfrac{x+3}{12}}{\dfrac{4x-5}{15}}$

27. $\dfrac{\dfrac{x}{3}}{\dfrac{2}{x+1}}$

28. $\dfrac{\dfrac{x-1}{5}}{\dfrac{3}{x}}$

29. $\dfrac{\dfrac{2}{x}+3}{\dfrac{4}{x^2}-9}$

30. $\dfrac{2+\dfrac{1}{x}}{4x-\dfrac{1}{x}}$

31. $\dfrac{1-\dfrac{x}{y}}{\dfrac{x^2}{y^2}-1}$

32. $\dfrac{1-\dfrac{2}{x}}{x-\dfrac{4}{x}}$

33. $\dfrac{\dfrac{-2x}{x-y}}{\dfrac{y}{x^2}}$

34. $\dfrac{\dfrac{7y}{x^2+xy}}{\dfrac{y^2}{x^2}}$

35. $\dfrac{\dfrac{2}{x}+\dfrac{1}{x^2}}{\dfrac{y}{x^2}}$

36. $\dfrac{\dfrac{5}{x^2}-\dfrac{2}{x}}{\dfrac{1}{x}+2}$

37. $\dfrac{\dfrac{x}{9}-\dfrac{1}{x}}{1+\dfrac{3}{x}}$

38. $\dfrac{\dfrac{x}{4}-\dfrac{4}{x}}{1-\dfrac{4}{x}}$

39. $\dfrac{\dfrac{x-1}{x^2-4}}{1+\dfrac{1}{x-2}}$

40. $\dfrac{\dfrac{2}{x+5}+\dfrac{4}{x+3}}{\dfrac{3x+13}{x^2+8x+15}}$

41. $\dfrac{\dfrac{4}{5-x}+\dfrac{5}{x-5}}{\dfrac{2}{x}+\dfrac{3}{x-5}}$

42. $\dfrac{\dfrac{3}{x-4}-\dfrac{2}{4-x}}{\dfrac{2}{x-4}-\dfrac{2}{x}}$

43. $\dfrac{\dfrac{x+2}{x}-\dfrac{2}{x-1}}{\dfrac{x+1}{x}+\dfrac{x+1}{x-1}}$

44. $\dfrac{\dfrac{5}{a+2}-\dfrac{1}{a-2}}{\dfrac{3}{2+a}+\dfrac{6}{2-a}}$

45. $\dfrac{\dfrac{x-2}{x+2}+\dfrac{x+2}{x-2}}{\dfrac{x-2}{x+2}-\dfrac{x+2}{x-2}}$

46. $\dfrac{\dfrac{x-1}{x+1}-\dfrac{x+1}{x-1}}{\dfrac{x-1}{x+1}+\dfrac{x+1}{x-1}}$

47. $\dfrac{\dfrac{2}{y^2}-\dfrac{5}{xy}-\dfrac{3}{x^2}}{\dfrac{2}{y^2}+\dfrac{7}{xy}+\dfrac{3}{x^2}}$

48. $\dfrac{\dfrac{2}{x^2}-\dfrac{1}{xy}-\dfrac{1}{y^2}}{\dfrac{1}{x^2}-\dfrac{3}{xy}+\dfrac{2}{y^2}}$

49. $\dfrac{a^{-1}+1}{a^{-1}-1}$

50. $\dfrac{a^{-1}-4}{4+a^{-1}}$

51. $\dfrac{3x^{-1}+(2y)^{-1}}{x^{-2}}$

52. $\dfrac{5x^{-2}-3y^{-1}}{x^{-1}+y^{-1}}$

53. $\dfrac{2a^{-1}+(2a)^{-1}}{a^{-1}+2a^{-2}}$

54. $\dfrac{a^{-1}+2a^{-2}}{2a^{-1}+(2a)^{-1}}$

55. $\dfrac{5x^{-1}+2y^{-1}}{x^{-2}y^{-2}}$

56. $\dfrac{x^{-2}y^{-2}}{5x^{-1}+2y^{-1}}$

57. $\dfrac{5x^{-1}-2y^{-1}}{25x^{-2}-4y^{-2}}$

58. $\dfrac{3x^{-1}+3y^{-1}}{4x^{-2}-9y^{-2}}$

59. $(x^{-1}+y^{-1})^{-1}$

60. $\dfrac{xy}{x^{-1}+y^{-1}}$

61. $\dfrac{x}{1 - \dfrac{1}{1 + \dfrac{1}{x}}}$

62. $\dfrac{x}{1 - \dfrac{1}{1 - \dfrac{1}{x}}}$

A Look Ahead

Simplify. See the following example.

EXAMPLE   Simplify.

$$\frac{2(a + b)^{-1} - 5(a - b)^{-1}}{4(a^2 - b^2)^{-1}}$$

Solution: $\dfrac{2(a + b)^{-1} - 5(a - b)^{-1}}{4(a^2 - b^2)^{-1}} = \dfrac{\dfrac{2}{a + b} - \dfrac{5}{a - b}}{\dfrac{4}{a^2 - b^2}}$

$$= \frac{\left(\dfrac{2}{a + b} - \dfrac{5}{a - b}\right) \cdot (a + b)(a - b)}{\left[\dfrac{4}{(a + b)(a - b)}\right] \cdot (a + b)(a - b)}$$

$$= \frac{\dfrac{2}{a + b} \cdot (a + b)(a - b) - \dfrac{5}{a - b} \cdot (a + b)(a - b)}{\dfrac{4(a + b)(a - b)}{(a + b)(a - b)}}$$

$$= \frac{2(a - b) - 5(a + b)}{4}$$

$$= \frac{-3a - 7b}{4} \quad \text{or} \quad -\frac{3a + 7b}{4} \quad \blacksquare$$

63. $\dfrac{1}{1 - (1 - x)^{-1}}$

64. $\dfrac{1}{1 + (1 + x)^{-1}}$

65. $\dfrac{(x + 2)^{-1} + (x - 2)^{-1}}{(x^2 - 4)^{-1}}$

66. $\dfrac{(y - 1)^{-1} - (y + 4)^{-1}}{(y^2 + 3y - 4)^{-1}}$

67. $\dfrac{3(a + 1)^{-1} + 4a^{-2}}{(a^3 + a^2)^{-1}}$

68. $\dfrac{9x^{-1} - 5(x - y)^{-1}}{4(x - y)^{-1}}$

Skill Review

Solve each equation for x. See Sections 2.1 and 5.5.

69. $7x + 2 = x - 3$

70. $4 - 2x = 17 - 5x$

71. $x^2 = 4x - 4$

72. $5x^2 + 10x = 15$

73. $\dfrac{x}{3} - 5 = 13$

74. $\dfrac{2x}{9} + 1 = \dfrac{7}{9}$

6.5
Solving Equations with Rational Expressions

OBJECTIVE | **1** | Solve equations containing rational expressions.

Tape 18

1 To solve equations containing rational expressions, we first clear the equation of fractions by multiplying both sides of the equation by the LCD of all rational expressions.

EXAMPLE 1 Solve $\dfrac{8x}{5} + \dfrac{3}{2} = \dfrac{3x}{5}$.

Solution: The LCD of $\dfrac{8x}{5}, \dfrac{3}{2}$, and $\dfrac{3x}{5}$ is 10. Multiply both sides of the equation by 10.

$$\frac{8x}{5} + \frac{3}{2} = \frac{3x}{5}$$

$$10\left(\frac{8x}{5} + \frac{3}{2}\right) = 10\left(\frac{3x}{5}\right) \qquad \text{Multiply by the LCD.}$$

$$10 \cdot \frac{8x}{5} + 10 \cdot \frac{3}{2} = 10 \cdot \frac{3x}{5} \qquad \text{Apply the distributive property.}$$

$$16x + 15 = 6x \qquad \text{Simplify.}$$

$$15 = -10x$$

$$-\frac{15}{10} = x \quad \text{or} \quad x = -\frac{3}{2} \qquad \text{Solve.}$$

Verify this solution by substituting $-\dfrac{3}{2}$ for x in the original equation. The solution set is $\left\{-\dfrac{3}{2}\right\}$. ∎

The important difference in the equations in this section is that the denominator of a rational expression may contain a variable. Recall that a rational expression is undefined for values of the variable that make the denominator 0. Thus special precautions must be taken when an equation contains rational expressions with variables in the denominator. If a proposed solution makes the denominator 0, then it must be rejected as a solution. Such proposed solutions are called **extraneous solutions.**

EXAMPLE 2 Solve $\dfrac{3}{x} - \dfrac{x + 21}{3x} = \dfrac{5}{3}$.

Solution: The LCD of denominators x, $3x$, and 3 is $3x$. Multiply both sides by $3x$.

$$\frac{3}{x} - \frac{x + 21}{3x} = \frac{5}{3}$$

$$3x\left(\frac{3}{x} - \frac{x + 21}{3x}\right) = 3x\left(\frac{5}{3}\right)$$

$$3x\left(\frac{3}{x}\right) - 3x\left(\frac{x + 21}{3x}\right) = 3x\left(\frac{5}{3}\right) \qquad \text{Apply the distributive property.}$$

$$9 - (x + 21) = 5x$$

$$9 - x - 21 = 5x$$

$$-12 = 6x$$

$$-2 = x \qquad \qquad \text{Solve.}$$

The proposed solution is -2. Check the solution in the original equation.

$$\frac{3}{-2} - \frac{-2 + 21}{3(-2)} = \frac{5}{3}$$

$$-\frac{9}{6} + \frac{19}{6} = \frac{5}{3}$$

$$\frac{10}{6} = \frac{5}{3} \qquad \text{True.}$$

The solution set is $\{-2\}$.   ■

To Solve an Equation Containing Rational Expressions

Step 1 Multiply both sides of the equation by the LCD of all rational expressions in the equation.

Step 2 Use the distributive property to remove any grouping symbols such as parentheses.

Step 3 Determine whether the equation is linear or quadratic and solve accordingly.

Step 4 Check the solution in the original equation.

EXAMPLE 3   Solve $\dfrac{x + 6}{x - 2} = \dfrac{2(x + 2)}{x - 2}$.

Solution: First, multiply both sides of the equation by the LCD, $x - 2$.

$$\frac{x + 6}{x - 2} = \frac{2(x + 2)}{x - 2}$$

$$(x - 2)\left(\frac{x + 6}{x - 2}\right) = (x - 2)\left[\frac{2(x + 2)}{x - 2}\right] \qquad \text{Multiply both sides by } x - 2.$$

$$x + 6 = 2(x + 2) \qquad\qquad \text{Simplify.}$$

$$x + 6 = 2x + 4 \qquad\qquad \text{Use the distributive property.}$$

$$x = 2 \qquad\qquad\qquad \text{Solve.}$$

Now check the proposed solution 2 in the **original equation.**

$$\frac{2 + 6}{2 - 2} = \frac{2(2 + 2)}{2 - 2}$$

$$\frac{8}{0} = \frac{2(4)}{0}$$

Since the denominators are 0, 2 is an extraneous solution. There is no solution to the original equation. The solution set is $\varnothing$ or $\{\ \}$.   ■

EXAMPLE 4   Solve $\dfrac{z}{2z^2 + 3z - 2} - \dfrac{1}{2z} = \dfrac{3}{z^2 + 2z}$.

Solution: Factor the denominators to find that the LCD is $2z(z + 2)(2z - 1)$. Multiply both sides by the LCD.

$$\frac{z}{2z^2 + 3z - 2} - \frac{1}{2z} = \frac{3}{z^2 + 2z}$$

$$\frac{z}{(2z - 1)(z + 2)} - \frac{1}{2z} = \frac{3}{z(z + 2)}$$

$$2z(z + 2)(2z - 1)\left[\frac{z}{(2z - 1)(z + 2)} - \frac{1}{2z}\right]$$

$$= 2z(z + 2)(2z - 1)\left[\frac{3}{z(z + 2)}\right]$$

$$2z(z + 2)(2z - 1)\left[\frac{z}{(2z - 1)(z + 2)}\right] - 2z(z + 2)(2z - 1)\left(\frac{1}{2z}\right)$$

$$= 2z(z + 2)(2z - 1)\left[\frac{3}{z(z + 2)}\right] \qquad \text{Apply the distributive property.}$$

$$2z(z) - (z + 2)(2z - 1) = 3 \cdot 2(2z - 1) \qquad\qquad \text{Simplify.}$$

$$2z^2 - (2z^2 + 3z - 2) = 12z - 6$$

$$2z^2 - 2z^2 - 3z + 2 = 12z - 6$$

$$-3z + 2 = 12z - 6$$

$$-15z = -8$$

$$z = \frac{8}{15} \qquad\qquad\qquad\qquad \text{Solve.}$$

The proposed solution $\dfrac{8}{15}$ does not make any denominator 0; the solution set is $\left\{\dfrac{8}{15}\right\}$.   ■

EXAMPLE 5   Solve $\dfrac{2x}{x - 3} + \dfrac{6 - 2x}{x^2 - 9} = \dfrac{x}{x + 3}$.

Solution: Factor the second denominator to find that the LCD is $(x + 3)(x - 3)$. Multiply both sides of the equation by $(x + 3)(x - 3)$. By the distributive property, this is the same as multiplying each term by $(x + 3)(x - 3)$.

$$\frac{2x}{x - 3} + \frac{6 - 2x}{x^2 - 9} = \frac{x}{x + 3}$$

$$(x + 3)(x - 3)\left(\frac{2x}{x - 3}\right) + (x + 3)(x - 3)\left[\frac{6 - 2x}{(x + 3)(x - 3)}\right]$$

$$= (x + 3)(x - 3)\left(\frac{x}{x + 3}\right)$$

$$2x(x + 3) + (6 - 2x) = x(x - 3) \qquad \text{Simplify.}$$

$$2x^2 + 6x + 6 - 2x = x^2 - 3x \qquad \text{Apply the}$$
$$\qquad\qquad\qquad\qquad\qquad\qquad \text{distributive property.}$$

$$x^2 + 7x + 6 = 0$$

Next, we solve this quadratic equation by the factoring method.

$$(x + 6)(x + 1) = 0 \qquad \text{Factor.}$$

$$x = -6 \quad \text{or} \quad x = -1 \qquad \text{Set each factor equal to 0.}$$

Neither -6 nor -1 makes any denominator 0. The solution set is $\{-6, -1\}$. ∎

EXERCISE SET 6.5

Solve each equation. See Example 1.

1. $\dfrac{x}{2} - \dfrac{x}{3} = 12$

2. $x = \dfrac{x}{2} - 4$

3. $\dfrac{x}{3} = \dfrac{1}{6} + \dfrac{x}{4}$

4. $\dfrac{x}{2} = \dfrac{21}{10} - \dfrac{x}{5}$

Solve each equation. See Example 2.

5. $\dfrac{2}{x} + \dfrac{1}{2} = \dfrac{5}{x}$

6. $\dfrac{5}{3x} + 1 = \dfrac{7}{6}$

7. $\dfrac{x + 3}{x} = \dfrac{5}{x}$

8. $\dfrac{4 - 3x}{2x} = -\dfrac{8}{2x}$

Solve each equation. See Example 3.

9. $\dfrac{x + 5}{x + 3} = \dfrac{8}{x + 3}$

10. $\dfrac{5}{x - 2} - \dfrac{2}{x + 4} = -\dfrac{4}{x^2 + 2x - 8}$

11. $\dfrac{1}{x - 1} + \dfrac{1}{x + 1} = \dfrac{2}{x^2 - 1}$

Solve each equation. See Example 4.

12. $\dfrac{1}{x - 1} = \dfrac{2}{x + 1}$

13. $\dfrac{6}{x + 3} = \dfrac{4}{x - 3}$

14. $\dfrac{1}{x - 4} - \dfrac{3x}{x^2 - 16} = \dfrac{2}{x + 4}$

15. $\dfrac{3}{2x + 3} - \dfrac{1}{2x - 3} = \dfrac{4}{4x^2 - 9}$

Solve each equation. See Example 5.

16. $\dfrac{1}{x - 4} = \dfrac{8}{x^2 - 16}$

17. $\dfrac{2}{x^2 - 4} = \dfrac{1}{2x - 4}$

18. $\dfrac{1}{x - 2} - \dfrac{2}{x^2 - 2x} = 1$

19. $\dfrac{12}{3x^2 + 12x} = 1 - \dfrac{1}{x + 4}$

Solve each equation.

20. $\dfrac{5}{x} = \dfrac{20}{12}$

21. $\dfrac{2}{x} = \dfrac{10}{5}$

22. $1 - \dfrac{4}{a} = 5$

23. $7 + \dfrac{6}{a} = 5$

24. $\dfrac{1}{2x} - \dfrac{1}{x + 1} = \dfrac{1}{3x^2 + 3x}$

25. $\dfrac{2}{x - 5} + \dfrac{1}{2x} = \dfrac{5}{3x^2 - 15x}$

26. $\dfrac{1}{x} - \dfrac{x}{25} = 0$

27. $\dfrac{x}{4} + \dfrac{5}{x} = 3$

28. $5 - \dfrac{2}{2y - 5} = \dfrac{3}{2y - 5}$

29. $1 - \dfrac{5}{y + 7} = \dfrac{4}{y + 7}$

30. $\dfrac{x - 1}{x + 2} = \dfrac{2}{3}$

31. $\dfrac{6x + 7}{2x + 9} = \dfrac{5}{3}$

32. $\dfrac{x + 3}{x + 2} = \dfrac{1}{x + 2}$

33. $\dfrac{2x + 1}{4 - x} = \dfrac{9}{4 - x}$

34. $\dfrac{1}{a - 3} + \dfrac{2}{a + 3} = \dfrac{1}{a^2 - 9}$

35. $\dfrac{12}{9 - a^2} + \dfrac{3}{3 + a} = \dfrac{2}{3 - a}$

36. $\dfrac{64}{x^2 - 16} + 1 = \dfrac{2x}{x - 4}$

37. $2 + \dfrac{3}{x} = \dfrac{2x}{x + 3}$

38. $\dfrac{-15}{4y + 1} + 4 = y$

39. $\dfrac{36}{x^2 - 9} + 1 = \dfrac{2x}{x + 3}$

40. $\dfrac{28}{x^2 - 9} + \dfrac{2x}{x - 3} + \dfrac{6}{x + 3} = 0$

41. $\dfrac{x^2 - 20}{x^2 - 7x + 12} = \dfrac{3}{x - 3} + \dfrac{5}{x - 4}$

42. $\dfrac{x + 2}{x^2 + 7x + 10} = \dfrac{1}{3x + 6} - \dfrac{1}{x + 5}$

43. $\dfrac{3}{2x - 5} + \dfrac{2}{2x + 3} = 0$

*Perform the indicated operation and simplify **or** solve the equation for the variable.*

44. $\dfrac{2}{x^2 - 4} = \dfrac{1}{x + 2} - \dfrac{3}{x - 2}$

45. $\dfrac{3}{x^2 - 25} = \dfrac{1}{x + 5} + \dfrac{2}{x - 5}$

46. $\dfrac{5}{x^2 - 3x} + \dfrac{4}{2x - 6}$

47. $\dfrac{5}{x^2 - 3x} \div \dfrac{4}{2x - 6}$

48. $\dfrac{x - 1}{x + 1} + \dfrac{x + 7}{x - 1} = \dfrac{4}{x^2 - 1}$

49. $\left(1 - \dfrac{y}{x}\right) \div \left(1 - \dfrac{x}{y}\right)$

50. $\dfrac{a^2 - 9}{a - 6} \cdot \dfrac{a^2 - 5a - 6}{a^2 - a - 6}$

51. $\dfrac{2}{a - 6} + \dfrac{3a}{a^2 - 5a - 6} - \dfrac{a}{5a + 5}$

52. $\dfrac{2x + 3}{3x - 2} = \dfrac{4x + 1}{6x + 1}$

53. $\dfrac{5x - 3}{2x} = \dfrac{10x + 3}{4x + 1}$

54. $\dfrac{a}{9a^2 - 1} + \dfrac{2}{6a - 2}$

55. $\dfrac{3}{4a - 8} - \dfrac{a + 2}{a^2 - 2a}$

56. $\dfrac{-3}{x^2} - \dfrac{1}{x} + 2 = 0$

57. $\dfrac{x}{2x + 6} + \dfrac{5}{x^2 - 9}$

58. $\dfrac{x - 8}{x^2 - x - 2} + \dfrac{2}{x - 2}$

59. $\dfrac{x - 8}{x^2 - x - 2} + \dfrac{2}{x - 2} = \dfrac{3}{x + 1}$

60. $\dfrac{3}{a} - 5 = \dfrac{7}{a} - 1$

61. $\dfrac{7}{3z - 9} + \dfrac{5}{z}$

A Look Ahead

Solve each equation. See the following example.

> **EXAMPLE**   Solve $3 + 5x^{-1} = 2x^{-2}$.
>
> **Solution:** $3 + 5x^{-1} = 2x^{-2}$
>
> $$3 + \dfrac{5}{x} = \dfrac{2}{x^2}$$

$$x^2 \cdot \left(3 + \frac{5}{x}\right) = x^2 \cdot \frac{2}{x^2}$$

$$x^2 \cdot 3 + x^2 \cdot \frac{5}{x} = x^2 \cdot \frac{2}{x^2}$$

$$3x^2 + 5x = 2$$

$$3x^2 + 5x - 2 = 0$$

$$(3x - 1)(x + 2) = 0$$

$$x = \frac{1}{3} \quad \text{or} \quad x = -2$$

Neither proposed solution makes the denominator 0. The solution set is $\left\{\frac{1}{3}, -2\right\}$. ■

62. $x^{-2} - 19x^{-1} + 48 = 0$ **63.** $x^{-2} - 5x^{-1} - 36 = 0$ **64.** $p^{-2} + 4p^{-1} - 5 = 0$ **65.** $6p^{-2} - 5p^{-1} + 1 = 0$

Solve each equation. See the following example.

EXAMPLE   Solve $\left(\frac{x}{x+1}\right)^2 - 7\left(\frac{x}{x+1}\right) + 10 = 0$.

Solution: Let $u = \frac{x}{x+1}$ and solve for u. Then substitute back and solve for x.

$$\left(\frac{x}{x+1}\right)^2 - 7\left(\frac{x}{x+1}\right) + 10 = 0$$

$$u^2 - 7u + 10 = 0 \qquad \text{Let } u = \frac{x}{x+1}.$$

$$(u - 5)(u - 2) = 0 \qquad \text{Factor.}$$

$$u = 5 \quad \text{or} \quad u = 2 \qquad \text{Solve.}$$

Since $u = \frac{x}{x+1}$, we have that $5 = \frac{x}{x+1}$ and $2 = \frac{x}{x+1}$. Thus there are two rational equations to solve.

1. $\quad 5 = \frac{x}{x+1}$ **2.** $\quad 2 = \frac{x}{x+1}$

$$5 \cdot (x + 1) = x \qquad\qquad 2 \cdot (x + 1) = x$$

$$5x + 5 = x \qquad\qquad 2x + 2 = x$$

$$5 = -4x \qquad\qquad 2 = -x$$

$$x = -\frac{5}{4} \qquad\qquad x = -2$$

Since neither $-\frac{5}{4}$ nor -2 makes the denominator 0, the solution set is $\left\{-\frac{5}{4}, -2\right\}$. ■

66. $(x - 1)^2 + 3(x - 1) + 2 = 0$ **67.** $(4 - x)^2 - 5(4 - x) + 6 = 0$

68. $\left(\dfrac{3}{x-1}\right)^2 + 2\left(\dfrac{3}{x-1}\right) + 1 = 0$

69. $\left(\dfrac{5}{2+x}\right)^2 + \left(\dfrac{5}{2+x}\right) - 20 = 0$

Skill Review

Write each sentence as an equation and solve. See Section 2.3.

70. Four more than 3 times a number is 19.

71. The sum of two consecutive integers is 147.

72. The length of a rectangle is 5 inches more than the width. Its perimeter is 50 inches.

73. The sum of a number and its reciprocal is $\dfrac{5}{2}$.

Simplify the following. See Sections 1.3 and 1.4.

74. $-|-6| - (-5)$

75. $\sqrt{49} - (10 - 6)^2$

76. $|4 - 8| + (4 - 8)$

77. $(-4)^2 - 5^2$

6.6
Applications of Rational Expressions

Tape 19

OBJECTIVES

1   Solve an equation containing rational expressions for a specified variable.

2   Solve word problems by writing equations containing rational expressions.

1   In Section 2.2, we solved equations for a specified variable. In this section, we continue practicing this skill by solving equations containing rational expressions for a specified variable. The steps given in Section 2.2 for solving these equations are repeated here.

To Solve Literal Equations

Step 1   Clear the equation of fractions or rational expressions by multiplying each side of the equation by the least common denominator (LCD) of all denominators in the equation.

Step 2   Use the distributive property to remove grouping symbols such as parentheses.

Step 3   Add or subtract like terms on the same side of the equation.

Step 4   Use the addition property of equality to rewrite the equation as an equivalent equation with terms containing the specified variable on one side and all other terms on the other side.

Step 5   Use the distributive property and the multiplication property of equality to isolate the specified variable.

EXAMPLE 1   Solve $\dfrac{1}{x} + \dfrac{1}{y} = \dfrac{1}{z}$ for x.

Solution: To clear this equation of fractions, multiply both sides of the equation by xyz, the LCD of $\frac{1}{x}$, $\frac{1}{y}$, and $\frac{1}{z}$.

$$\frac{1}{x} + \frac{1}{y} = \frac{1}{z}$$

$$xyz\left(\frac{1}{x} + \frac{1}{y}\right) = xyz\left(\frac{1}{z}\right)$$

$$xyz\left(\frac{1}{x}\right) + xyz\left(\frac{1}{y}\right) = xyz\left(\frac{1}{z}\right)$$

$$yz + xz = xy$$

Next, subtract xz from both sides so that all terms containing the specified variable x are on one side of the equation and all other terms are on the other side.

$$yz = xy - xz$$

Use the distributive property to factor x from $xy - xz$ and then the multiplication property of equality to solve for x.

$$yz = x(y - z)$$

$$\frac{yz}{y - z} = x \quad \text{or} \quad x = \frac{yz}{y - z} \qquad \text{Divide both sides by } y - z. \qquad \blacksquare$$

2   Word problems can sometimes describe situations that can be modeled by equations containing rational expressions. Examples 2 through 5 give you practice in solving such problems.

EXAMPLE 2   Find the number that, when subtracted from the numerator and added to the denominator of $\frac{9}{19}$, changes $\frac{9}{19}$ into a fraction equivalent to $\frac{1}{3}$.

Solution: Let n be the number to be subtracted from the numerator and added to the denominator. The equation is

$$\frac{9 - n}{19 + n} = \frac{1}{3}$$

To solve for n, begin by multiplying both sides by the LCD, $3(19 + n)$.

$$3(19 + n) \cdot \left(\frac{9 - n}{19 + n}\right) = 3(19 + n)\left(\frac{1}{3}\right) \qquad \text{Multiply by the LCD.}$$

$$3(9 - n) = 19 + n \qquad \text{Simplify.}$$

$$27 - 3n = 19 + n$$

$$8 = 4n$$

$$2 = n \qquad \text{Solve.}$$

Check in the stated problem. If we subtract 2 from the numerator and add 2 to the denominator of $\frac{9}{19}$, we have $\frac{9 - 2}{19 + 2} = \frac{7}{21} = \frac{1}{3}$. The problem checks, so the number is 2. $\blacksquare$

EXAMPLE 3 The intensity I (in foot-candles) of light d feet from its source is given by the equation

$$I = \frac{320}{d^2}$$

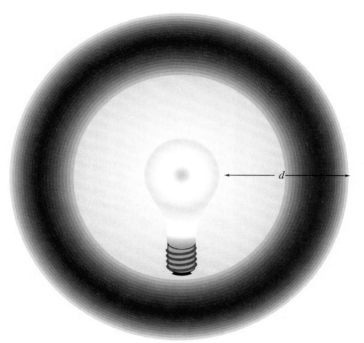

How far away is the source if the intensity of light is 5 foot-candles?

Solution: The formula describing the relationship between I and d is given, as well as a value of I. Replace I by 5 in the formula and solve for d.

$$I = \frac{320}{d^2}$$

$$5 = \frac{320}{d^2} \qquad \text{Let } I = 5.$$

$$d^2(5) = d^2\left(\frac{320}{d^2}\right) \qquad \text{Multiply both sides by } d^2.$$

$$5d^2 = 320 \qquad \text{Simplify.}$$

$$d^2 = 64 \qquad \text{Divide by 5.}$$

Then, since $8^2 = 64$ and also $(-8)^2 = 64$, we have that

$$d = 8 \quad \text{or} \quad d = -8$$

Since d represents distance and distance cannot be negative, the proposed solution -8 must be rejected. Check the solution 8 feet in the given formula. ∎

The following work example leads to an equation containing rational expressions.

EXAMPLE 4   Melissa can clean the house in 4 hours while her husband Zack can do the same job in 5 hours. They have agreed to clean together so that they can finish in time to watch a movie on TV that starts in 2 hours. How long will it take them to clean the house together? Can they finish before the movie starts?

Solution: The key idea here is the relationship between the **time** (hours) it takes to complete the job and the **part of the job** completed in 1 unit of time (hour). For example, if the **time** it takes Melissa to complete the job is 4 hours, the **part of the job** she can complete in 1 hour is $\frac{1}{4}$. Similarly, Zack can complete $\frac{1}{5}$ of the job in 1 hour.

Let t represent the **time** in hours it takes Melissa and Zack to clean the house together. Then $\frac{1}{t}$ represents the **part of the job** they complete in 1 hour. This information is summarized as follows:

	Hours to Complete	**Part of Job** Completed in 1 Hour
Melissa	4	$\frac{1}{4}$
Zack	5	$\frac{1}{5}$
Together	t	$\frac{1}{t}$

Adding the part of the job Melissa completes in 1 hour and the part of the job Zack completes in 1 hour, we find the part of the job they do together in 1 hour.

$$\frac{1}{4} + \frac{1}{5} = \frac{1}{t}$$

$$20t\left(\frac{1}{4} + \frac{1}{5}\right) = 20t\left(\frac{1}{t}\right) \qquad \text{Multiply both sides by the LCD } 20t.$$

$$5t + 4t = 20$$

$$9t = 20$$

$$t = \frac{20}{9} \quad \text{or} \quad 2\frac{2}{9} \qquad \text{Solve.}$$

The proposed solution is $2\frac{2}{9}$. That is, Melissa and Zack take $2\frac{2}{9}$ hours to clean the house together. This proposed solution is reasonable since $2\frac{2}{9}$ hours is more than half of Melissa's time and less than half of Zack's time. Check this solution in the originally stated problem. Can they finish before the movie starts? If they can clean the house together in $2\frac{2}{9}$ hours, they could not complete the job before the movie starts. ∎

EXAMPLE 5   In his boat, Steve Deitmer takes $1\frac{1}{2}$ times as long to go 72 miles upstream as he does to return. If the boat cruises at 30 mph in still water, what is the speed of the current?

Solution: Since we are asked to find the speed of the current, **let x represent the current's speed.** The speed of the boat traveling downstream is made faster by the current and is represented by $30 + x$. The speed of the boat traveling upstream is made slower by the current and is represented by $30 - x$. Steve travels 72 miles upstream and the same distance downstream. This information is summarized in the following chart.

	distance	rate	time $\left(\dfrac{d}{r}\right)$
Upstream	72	$30 - x$	$\dfrac{72}{30 - x}$ $\left(\dfrac{d}{r}\right)$
Downstream	72	$30 + x$	$\dfrac{72}{30 + x}$ $\left(\dfrac{d}{r}\right)$

Since the time spent traveling upstream is $1\frac{1}{2}$ times the time spent traveling downstream, we have

time upstream	is	$1\frac{1}{2}$	times	time downstream
$\dfrac{72}{30 - x}$	$=$	$\dfrac{3}{2}$		$\dfrac{72}{30 + x}$

Multiply both sides by the LCD $2(30 + x)(30 - x)$.

$$2(30 + x)(30 - x)\left(\frac{72}{30 - x}\right) = 2(30 + x)(30 - x)\left(\frac{3}{2} \cdot \frac{72}{30 + x}\right)$$

$$72 \cdot 2(30 + x) = 3 \cdot 72 \cdot (30 - x) \qquad \text{Simplify.}$$
$$2(30 + x) = 3(30 - x) \qquad \text{Divide by 72.}$$
$$60 + 2x = 90 - 3x \qquad \text{Simplify.}$$
$$5x = 30$$
$$x = 6 \qquad \text{Solve.}$$

Thus the current's speed is 6 mph. Check this solution in the originally stated problem. ∎

EXERCISE SET 6.6

Solve each equation for the specified variable. See Example 1.

1. $F = \frac{9}{5}C + 32$; C

2. $V = \frac{1}{3}\pi r^2 h$; h

3. $\frac{1}{R} = \frac{1}{R_1} + \frac{1}{R_2}$; R

4. $\frac{1}{R} = \frac{1}{R_1} + \frac{1}{R_2}$; R_1

5. $S = \frac{n(a + L)}{2}$; n

6. $S = \frac{n(a + L)}{2}$; a

Solve. See Example 2.

7. The sum of a number and 5 times its reciprocal is 6. Find the number(s).

8. The quotient of a number and 9 times its reciprocal is 1. Find the number(s).

9. If a number is added to the numerator of $\frac{12}{41}$ and twice the number is added to the denominator of $\frac{12}{41}$, the resulting fraction is equivalent to $\frac{1}{3}$. Find the number.

10. If a number is subtracted from the numerator of $\frac{13}{8}$ and added to the denominator of $\frac{13}{8}$, the resulting fraction is equivalent to $\frac{2}{5}$. Find the number.

In electronics, the relationship among the combined resistance r of two resistors r_1 and r_2 wired in a parallel circuit is described by the formula $\frac{1}{r} = \frac{1}{r_1} + \frac{1}{r_2}$. Use this formula to solve Exercises 11 through 13. See Example 3.

11. If the combined resistance is 2 ohms and one of the two resistances is 3 ohms, find the other resistance.

12. Find the combined resistance of two resistors of 12 ohms each when they are wired in a parallel circuit.

13. The relationship among resistance and two resistors wired in a parallel circuit may be extended to three resistors, r_1, r_2, and r_3. Write an equation you believe may describe the relationship, and use it to find the combined resistance if r_1 is 5, r_2 is 6, and r_3 is 2.

Solve. See Example 4.

14. Alan can type a research paper in 6 hours. With Steve's help, the paper can be typed in 4 hours. Find how long it takes Steve to type the paper alone.

15. An experienced roofer can roof a house in 26 hours. A beginning roofer needs 39 hours for the same job. Find how long it takes for the two to do the job working together.

16. A new printing press can print newspapers twice as fast as the old one. The old one can print the afternoon edition in 4 hours. Find how long it takes to print the afternoon edition if both printers are operating.

17. Three computers can do a sorting task in 20 minutes, 30 minutes, and 60 minutes, respectively. Find how long it takes to do the sorting task if all three computers run together.

Solve. See Example 5.

18. A small plane and a car leave the same town at sunrise and head for a town 450 miles away. The speed of the plane is three times the speed of the car, and the plane arrives 6 hours ahead of the car. Find the speed of the car.

19. Mattie drove 150 miles in the same amount of time that it took a plane to travel 600 miles. The speed of the plane was 150 mph faster than the speed of the car. Find the speed of the plane.

20. The speed of a boat in still water is 24 mph. If the boat travels 54 miles upstream in the same time that it takes to travel 90 miles downstream, find the speed of the current.

21. The speed of Lazy River's current is 5 mph. If a boat travels 20 miles downstream in the same time that it takes to travel 10 miles upstream, find the speed of the boat in still water.

Solve each equation for the specified variable.

22. $A = \dfrac{h(a + b)}{2}$; b

23. $A = \dfrac{h(a + b)}{2}$; h

24. $\dfrac{P_1 V_1}{T_1} = \dfrac{P_2 V_2}{T_2}$; T_2

25. $H = \dfrac{kA(T_1 - T_2)}{L}$; T_2

26. $f = \dfrac{f_1 f_2}{f_1 + f_2}$; f_2

27. $I = \dfrac{E}{R + r}$; r

28. $\lambda = \dfrac{2L}{n}$; L

29. $S = \dfrac{a_1 - a_n r}{1 - r}$; a_1

30. $\dfrac{\theta}{\omega} = \dfrac{2L}{c}$; c

31. $F = \dfrac{-GMm}{r^2}$; M

Solve.

32. The sum of the reciprocals of two consecutive odd integers is $\dfrac{20}{99}$. Find the two integers.

33. The sum of the reciprocals of two consecutive integers is $-\dfrac{15}{56}$. Find the two integers.

34. If Sarah can do a job in 5 hours and Dick and Sarah working together can do the same job in 2 hours, find how long it takes Dick to do the job alone.

35. One hose can fill a goldfish pond in 45 minutes, and two hoses can fill the same pond in 20 minutes. Find how long it takes the second hose alone to fill the pond.

36. The speed of a bicyclist is 10 mph faster than the speed of a walker. If the bicyclist travels 26 miles in the same amount of time that the walker travels 6 miles, find the speed of the bicyclist.

37. Two trains leave at the same time going in opposite directions. One train travels 15 mph faster than the other. In 6 hours the trains are 630 miles apart. Find the speed of each.

38. The numerator of a fraction is 4 less than the denominator. If both the numerator and the denominator are increased by 2, the resulting fraction is equivalent to $\dfrac{2}{3}$. Find the fraction.

39. The denominator of a fraction is 1 more than the numerator. If both the numerator and the denominator are decreased by 3, the resulting fraction is equivalent to $\dfrac{4}{5}$. Find the fraction.

40. Moo Dairy has 3 machines to fill half-gallon milk cartons. The machines can fill the daily quota in 5 hours, 6 hours, and 7.5 hours, respectively. Find how long it takes to fill the daily quota if all three machines are running.

41. The inlet pipe of an oil tank can fill the tank in 1 hour 30 minutes. The outlet pipe can empty the tank in 1 hour. Find how long it takes to empty a full tank if both pipes are open.

42. A plane flies 465 miles with the wind and 345 miles against the wind in the same length of time. If the speed of the wind is 20 mph, find the speed of the plane in still air.

43. Two rockets are launched. The first travels at 9000 mph. Fifteen minutes later the second is launched at 10,000 mph. Find the distance at which both rockets are an equal distance from the earth.

44. Two joggers, one averaging 8 mph and one averaging 6 mph, start from a designated initial point. The slower jogger arrives at the end of the run a half-hour after the other jogger. Find the distance of the run.

Skill Review

Solve the equation for x. See Section 6.5.

45. $\dfrac{x}{5} = \dfrac{x+2}{3}$

46. $\dfrac{x}{4} = \dfrac{x+3}{6}$

47. $\dfrac{x-3}{2} = \dfrac{x-5}{6}$

48. $\dfrac{x-6}{4} = \dfrac{x-2}{5}$

Factor the following. See Section 5.4.

49. $2x^3 - 9x^2 - 18x$

50. $5yx^2 - 45y$

51. $x^3 + 8$

52. $y^3 - 27$

6.7
Proportion and Variation

Tape 19

OBJECTIVES

1 Solve a proportion for a variable.

2 Write an equation expressing direct variation.

3 Write an equation expressing inverse variation.

4 Write an equation expressing joint variation.

1   A quotient or fraction can be called a **ratio.** For example, the fraction $\frac{3}{4}$ can be interpreted and read as **the ratio of 3 to 4.** A **proportion** is a statement that two ratios are equal. The statement

$$\frac{a}{b} = \frac{c}{d}$$

is a proportion, as long as neither b nor d is 0. In this proportion, b and c are called the **means,** and a and d are called the **extremes.** If both sides of the proportion are multiplied by the LCD bd, we have

$$\frac{a}{b} = \frac{c}{d} \qquad \text{Original proportion.}$$

$$bd\left(\frac{a}{b}\right) = bd\left(\frac{c}{d}\right) \qquad \text{Multiply both sides by the LCD } bd.$$

$$da = bc \qquad \text{Simplify.}$$

In other words, in a proportion the product of the means equals the product of the extremes. Equating the product of the means to the product of the extremes is called **cross multiplying** and each product is called a **cross product.**

EXAMPLE 1   Solve the proportion $\dfrac{x+2}{3} = \dfrac{x}{5}$ for x.

Solution: Start by setting the product of the means equal to the product of the extremes or cross multiplying.

$$\frac{x+2}{3} = \frac{x}{5} \qquad \text{Original proportion.}$$

$$5(x+2) = 3(x) \qquad \text{Cross multiply.}$$

$$5x + 10 = 3x$$

$$10 = -2x \qquad \text{Subtract } 5x \text{ from both sides.}$$

$$-5 = x \qquad \text{Divide by } -2.$$

The solution set is $\{-5\}$.   ∎

2   The concepts of proportion are closely related to the concepts of variation. In this section, we are interested in **direct, inverse,** and **joint** variations. A very familiar example of direct variation is the relationship of the circumference C of a circle to its radius r. The formula $C = 2\pi r$ expresses the fact that the circumference is always 2π times the radius. In other words, C is always a constant multiple (2π) of r. Because it is, we say that **C varies directly as r** or that **C is directly proportional to r.**

Direct Variation

y varies directly as x or **y is directly proportional to x** if there is a nonzero constant k such that

$$y = kx$$

k is called the **constant of variation** or the **constant of proportionality.**

EXAMPLE 2   Suppose that y varies directly as x. If y is 5 when x is 30, find the constant of variation. Also, find y when $x = 90$.

Solution: Since y varies directly as x, we write $y = kx$. If $y = 5$ when $x = 30$, we have that

$$y = kx$$

$$5 = k(30) \qquad \text{Replace } y \text{ with 5 and } x \text{ with 30.}$$

$$\frac{1}{6} = k \qquad \text{Solve for } k.$$

The constant of variation is $\frac{1}{6}$.

After finding the constant of variation k, the direct variation equation can be written as $y = \frac{1}{6}x$. Next, find y when x is 90.

$$y = \frac{1}{6}x$$

$$= \frac{1}{6}(90) \qquad \text{Let } x = 90.$$

$$= 15$$

When $x = 90$, y must be 15. ■

Notice that the direct variation equation $y = kx$ is not only a linear equation, but y is also a function of x.

EXAMPLE 3   Hooke's law states that the distance a spring stretches is directly proportional to the weight attached to the spring. If a 40-pound weight attached to the spring stretches the spring 5 inches, find the distance that a 65-pound weight attached to the spring stretches the spring.

Solution: The distance the spring stretches is directly proportional to the weight attached. Let d represent the distance stretched, and let w represent the weight attached. The constant of variation is represented by k. Because d is directly proportional to w, we write

$$d = kw$$

When 40 pounds is attached, the spring stretches 5 inches. That is, when $w = 40$, $d = 5$.

$$5 = k(40) \qquad \text{Replace by the known values.}$$

$$\frac{1}{8} = k \qquad \text{Solve for } k.$$

Thus we have

$$d = \frac{1}{8}w$$

To find the stretch when a weight of 65 pounds is attached, replace w with 65 and solve for d.

$$d = \frac{1}{8}(65)$$

$$= \frac{65}{8} = 8\frac{1}{8} \quad \text{or} \quad 8.125$$

Thus, the spring stretches 8.125 inches when a 65-pound weight is attached. ∎

3 When y is proportional to the **reciprocal** of another variable x, we say that y **varies inversely as x,** or that **y is inversely proportional to x.** An example of the inverse variation relationship is the relationship between the pressure that a gas exerts and the volume of its container. As the volume of a container decreases, the pressure of the gas it contains increases.

Inverse Variation

y varies inversely as x or, **y is inversely proportional to x** if there is a nonzero constant k such that

$$y = \frac{k}{x}$$

k is called the **constant of variation** or the **constant of proportionality.**

EXAMPLE 4 Suppose that u varies inversely as w. If u is 3 when w is 5, find u when w is 30.

Solution: Since u varies inversely as w, we have $u = \dfrac{k}{w}$. Let $u = 3$, $w = 5$, and solve for k.

$$u = \frac{k}{w}$$

$$3 = \frac{k}{5} \qquad \text{Let } u = 3 \text{ and } w = 5.$$

$$15 = k \qquad \text{Multiply both sides by 5 or cross multiply.}$$

The constant of variation k is 15. This gives the inverse variation equation:

$$u = \frac{15}{w}$$

Now find u when $w = 30$.

$$u = \frac{15}{30} \qquad \text{Let } w = 30.$$

$$= \frac{1}{2}$$

Thus, when $w = 30$, $u = \dfrac{1}{2}$. ∎

EXAMPLE 5   Boyle's law says that, if the temperature stays the same, the pressure P of a gas is inversely proportional to the volume V. If a cylinder in a steam engine has a pressure of 960 kilopascals when the volume is 1.4 cubic meters, find the pressure when the volume increases to 2.5 cubic meters.

Solution: An inverse relationship exists between pressure P and volume V.

$$P = \frac{k}{V} \qquad P \text{ varies inversely as } V.$$

$$960 = \frac{k}{1.4} \qquad \text{Let } P = 960 \text{ and } V = 1.4.$$

$$1344 = k \qquad \text{Cross multiply.}$$

Thus, the value of k is 1344. Replace k by 1344 in the variation equation:

$$P = \frac{1344}{V}$$

Next, find P when V is 2.5 cubic meters.

$$P = \frac{1344}{2.5} \qquad \text{Let } V = 2.5.$$

$$P = 537.6$$

When the volume is 2.5 cubic meters, the pressure is 537.6 kilopascals.   ■

4   It can also occur that the ratio of a variable to the product of many other variables is constant. For example, the ratio of distance traveled to the product of speed and time traveled is constantly 1.

$$\frac{d}{rt} = 1 \quad \text{or} \quad d = rt$$

Such a relationship is called **joint variation.**

Joint Variation

If the ratio of a variable y to the product of 2 or more variables is constant, then **y varies jointly as** or is **jointly proportional to** the other variables. If

$$y = kxz$$

k is the constant of variation or the constant of proportionality.

EXAMPLE 6   The lateral surface area of a cylinder varies jointly as its radius and height. Express surface area S in terms of radius r and height h.

Solution: Because the surface area varies jointly as the radius r and the height h, we equate S to the constant multiple, of r and h:

$$S = krh \qquad ■$$

EXERCISE SET 6.7

Solve each proportion for x. See Example 1.

1. $\dfrac{x-3}{8} = \dfrac{x}{9}$

2. $\dfrac{x+3}{6} = \dfrac{x}{5}$

3. $\dfrac{x}{3} = \dfrac{x-4}{5}$

4. $\dfrac{x}{4} = \dfrac{x-12}{7}$

5. $\dfrac{x+6}{16} = \dfrac{3x-2}{8}$

6. $\dfrac{x+11}{10} = \dfrac{3x-9}{2}$

7. $\dfrac{2x-1}{6} = \dfrac{3x}{10}$

8. $\dfrac{5x-5}{6} = \dfrac{5x}{9}$

Write each statement as an equation. See Examples 2 through 6.

9. *A* is directly proportional to *B*.

10. *C* varies inversely as *D*.

11. *X* is inversely proportional to *Z*.

12. *G* varies directly with *M*.

13. *N* varies directly with the square of *P*.

14. *A* varies jointly with *D* and *E*.

15. *T* is inversely proportional to *R*.

16. *G* is inversely proportional to *H*.

17. *P* varies directly with *R*.

18. *T* is directly proportional to *S*.

Solve. See Examples 2 and 3.

19. *A* varies directly as *B*. If *A* is 60 when *B* is 12, find *A* when *B* is 9.

20. *C* varies directly as *D*. If *C* is 42 when *D* is 14, find *C* when *D* is 6.

21. Charles's law states that, if the pressure *P* stays the same, the volume *V* of a gas is directly proportional to its temperature *T*. If a balloon is filled with 20 cubic meters of a gas at a temperature of 300 K, find the new volume if the temperature rises to 360 K while the pressure stays the same.

22. The amount *P* of pollution varies directly with the population *N* of people. Kansas City has a population of 450,000 and produces 260,000 tons of pollutants. Find how many tons of pollution we should expect St. Louis to produce, if we know that its population is 980,000.

Solve. See Examples 4 and 5.

23. *H* is inversely proportional to *J*. If *H* is 4 when *J* is 5, find *H* when *J* is 2.

24. *D* varies inversely as *A*. If *D* is 16 when *A* is 2, find *D* when *A* is 8.

25. If the voltage *V* in an electric circuit is held constant, the current *I* is inversely proportional to the resistance *R*. If the current is 40 amperes when the resistance is 270 ohms, find the current when the resistance is 150 ohms.

26. Pairs of markings a set distance apart are made on highways so that police can detect drivers exceeding the speed limit. Over a fixed distance, the speed *R* varies inversely with the time *T*. In one particular pair of markings, *R* is 45 mph when *T* is 6 seconds. Find the speed of a car that travels the given distance in 5 seconds.

Write each statement as an equation. See Example 6.

27. *x* varies jointly as *y* and *z*.

28. *P* varies jointly as *R* and the square of *S*.

29. *r* varies jointly as *s* and the cube of *t*.

30. *a* varies jointly as *b* and *c*.

Solve each proportion.

31. $\dfrac{x}{5} = \dfrac{x+2}{4}$

32. $\dfrac{2x-1}{3} = \dfrac{x}{2}$

33. $\dfrac{8x+1}{15} = \dfrac{12x+6}{9}$

34. $\dfrac{3x+7}{10} = \dfrac{9x+10}{8}$

35. $\dfrac{2x-3}{20} = \dfrac{x+5}{5}$

36. $\dfrac{6x-5}{4} = \dfrac{x+3}{9}$

37. $\dfrac{3x-1}{6} = \dfrac{2x-5}{14}$

38. $\dfrac{5x-1}{9} = \dfrac{3x-2}{12}$

39. Q is directly proportional to R. If Q is 4 when R is 20, find Q when R is 35.

40. S is directly proportional to T. If S is 4 when T is 16, find S when T is 40.

41. M varies directly with P. If M is 8 when P is 20, find M when P is 24.

42. F is directly proportional to G. If F is 18 when G is 10, find F when G is 16.

43. B is inversely proportional to C. If B is 12 when C is 3, find B when C is 18.

44. U varies inversely with V. If U is 14 when V is 4, find U when V is 7.

45. W varies inversely with X. If W is 18 when X is 6, find W when X is 40.

46. Z is inversely proportional to Y. If Z is 9 when Y is 6, find Z when Y is 24.

47. The weight of a synthetic ball varies directly with the cube of its radius. A ball with a radius of 2 inches weighs 1.20 pounds. Find the weight of a ball of the same material with a 3-inch radius.

48. At sea, the distance to the horizon is directly proportional to the square root of the elevation of the observer. If a person who is 36 feet above the water can see 7.4 miles, find how far a person 64 feet above the water can see.

49. Because it is more efficient to produce larger numbers of items, the cost of producing Dysan computer disks is inversely proportional to the number produced. If 4000 can be produced at a cost of $1.20 each, find the cost per disk when 6000 are produced.

50. The weight of an object on or above the surface of the earth varies inversely as the square of the distance between the object and the center of the earth. If a person weighs 160 pounds on the surface of the earth, find the individual's weight if he moves 200 miles above the earth. (Assume that the radius of the earth is 4000 miles.)

51. The number of cars manufactured on an assembly line at a General Motors plant varies jointly as the number of workers and the time they work. If 200 workers can produce 60 cars in 2 hours, find how many cars 240 workers should be able to make in 3 hours.

52. The volume of a cone varies jointly as the square of its radius and its height. If the volume of a cone is 32π when the radius is 4 inches and the height is 6 inches, find the volume of a cone when the radius is 3 inches and the height is 5 inches.

53. When a wind blows perpendicularly against a flat surface, its force is jointly proportional to the surface area and the velocity of the wind. A sail whose surface area is 12 square feet experiences a 20-pound force when the wind speed is 10 miles per hour. Find the force on an 8-square-foot sail if the wind speed is 12 miles per hour.

54. The horsepower that can be safely transmitted to a shaft varies jointly as its angular speed of rotation (in revolutions per minute) and the cube of its diameter. A 2-inch shaft making 120 revolutions per minute safely transmits 40 horsepower. Find how much horsepower can be safely transmitted by a 3-inch shaft making 80 revolutions per minute.

55. A circular column has a *safe load* that is directly proportional to the fourth power of its diameter and inversely proportional to the square of its length. An 8-inch pillar 10 feet long can safely support a 16-ton load. Find the load that a 6-inch pillar made of the same material can support, if it is 8 feet long.

56. The maximum safe load for a rectangular beam varies jointly as its width and the square of its height and inversely as its length. If a beam 6 inches wide, 4 inches high, and 10 feet long supports 12 tons, find how much a similar beam can support if the beam is 8 inches wide, 5 inches high, and 16 feet long.

57. The area of a circle is directly proportional to the square of its radius. If the radius is tripled, determine how the area changes.

58. The horsepower to drive a boat varies directly as the cube of the speed of the boat. If the speed of the boat is to double, determine the corresponding increase in horsepower required.

59. The intensity of light varies inversely as the square of the distance from the light source. If a person doubles her distance from a light source, determine what happens to the intensity of light at her new location.

60. The volume of a cylinder varies jointly as the height and the square of the radius. If the height is halved and the radius is doubled, determine what happens to the volume.

Writing in Mathematics

61. It has been said that, in the long run, lifetime income is directly proportional to the number of years of education an individual has. Explain what is meant by this statement.

62. The number of careless errors found in a production run is inversely proportional to the amount of time needed to perform the job. Explain what is meant by this statement.

Skill Review

Find the circumference and area of each circle. The center of each circle is represented by zero.

63.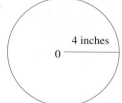
4 inches
0

64.
6 centimeters
0

65.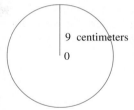
9 centimeters
0

66.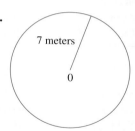
7 meters
0

CRITICAL THINKING   A progressive community in California is experimenting with electricity generated by windmills. City engineers are analyzing data they have gathered about their field of windmills. The engineers are familiar with other research demonstrating that the amount of electricity a windmill generates hourly is directly proportional to the cube of the miles-per-hour windspeed. For example, a windmill that generates 15 watts per hour when the windspeed is 10 miles per hour will generate 120 watts per hour when the windspeed is 20 miles per hour, 405 watts per hour when the windspeed is 30 miles per hour, and so on.

The engineers' data shows that for several days the windspeed was more or less steady at 20 miles per hour, and sure enough the windmills generated the expected 120 watts per hour. The weather forecast for the coming few days is that windspeed will fluctuate wildly, but still the average windspeed is expected to be 20 miles per hour. Should the engineers expect the windmills to still generate 120 watts per hour?

During one three-day period, each windmill generated 100 watts. The weather person predicts the windspeed for the coming few days will drop by a third. How many watts per hour should the engineers now expect each windmill to generate? Generally, if one windspeed yields k watts per hour, how many watts per hour does a third of the windspeed yield?

CHAPTER 6 GLOSSARY

A fraction whose numerator, denominator, or both contains one or more rational expressions is called a **complex fraction.**

Equating the product of the means of a proportion to the product of the extremes is called **cross multiplying,** and each product is called a **cross product.**

A proposed solution that makes the denominator of a rational expression 0 is called an **extraneous solution.**

In the proportion $\dfrac{a}{b} = \dfrac{c}{d}$, a and d are called the **extremes** and b and c are called the **means.**

A **proportion** is a statement that 2 ratios are equal.

A **rational expression** is the quotient $\dfrac{P}{Q}$ of two polynomials P and Q as long as Q is not 0.

A rational expression is **undefined** for a value of the variable that makes the denominator 0.

We say that y **varies directly as** x or y **is directly proportional to** x if there is a nonzero constant k such that $y = kx$. The constant k is called the **constant of variation** or the **constant of proportionality.**

The variable y **varies inversely as** x or y **is inversely proportional to** x if there is a nonzero constant k such that $y = k/x$.

If the ratio of a variable y to the product of two or more variables is constant, then y **varies jointly as** or is **jointly proportional to the other variables.**

CHAPTER 6 SUMMARY

Let $\dfrac{P}{Q}$ and $\dfrac{R}{S}$ be rational expressions.

FUNDAMENTAL PRINCIPLE OF RATIONAL EXPRESSIONS (6.1)

For any rational expression $\dfrac{P}{Q}$ and any polynomial R, $R \neq 0$,

$$\frac{P}{Q} = \frac{P \cdot R}{Q \cdot R} \quad \text{and} \quad \frac{P}{Q} = \frac{P \div R}{Q \div R}$$

MULTIPLYING RATIONAL EXPRESSIONS (6.2)

$$\frac{P}{Q} \cdot \frac{R}{S} = \frac{PR}{QS}$$

DIVIDING RATIONAL EXPRESSIONS (6.2)

$$\frac{P}{Q} \div \frac{R}{S} = \frac{P}{Q} \cdot \frac{S}{R} = \frac{PS}{QR}, \quad \text{as long as } R \text{ is not } 0.$$

ADDING OR SUBTRACTING RATIONAL EXPRESSIONS WITH COMMON DENOMINATORS (6.3)

If $\dfrac{P}{Q}$ and $\dfrac{R}{Q}$ are rational expressions, then

$$\frac{P}{Q} + \frac{R}{Q} = \frac{P + R}{Q} \quad \text{and} \quad \frac{P}{Q} - \frac{R}{Q} = \frac{P - R}{Q}$$

TO ADD OR SUBTRACT RATIONAL EXPRESSIONS WITH UNLIKE DENOMINATORS (6.3)

Step 1 Find the LCD of the rational expressions.

Step 2 Write each rational expression as an equivalent rational expression whose denominator is the LCD found in *step 1*.

Step 3 Add or subtract numerators, and write the sum or difference over the common denominator.

Step 4 Write the answer in lowest terms.

TO SOLVE AN EQUATION CONTAINING RATIONAL EXPRESSIONS (6.5)

Step 1 Multiply both sides of the equation by the LCD of all rational expressions in the equation.

Step 2   Use the distributive property to remove any grouping symbols such as parentheses.

Step 3   Determine whether the equation is linear or quadratic and solve accordingly.

Step 4   Check the solution in the original equation.

CHAPTER 6 REVIEW

(6.1)   *Find the values of x, if any, for which each rational expression is undefined.*

1. $\dfrac{3 - 5x}{7}$

2. $\dfrac{4}{x - 5}$

3. $\dfrac{4x}{3x - 12}$

4. $\dfrac{x + 3}{7x - 4}$

5. $\dfrac{4 - x}{4x^3 - 4x^2 - 24x}$

Write each rational expression in lowest terms.

6. $\dfrac{15x^4}{45x^2}$

7. $\dfrac{x + 2}{2 + x}$

8. $\dfrac{18m^6p^2}{10m^4p}$

9. $\dfrac{x - 12}{12 - x}$

10. $\dfrac{5x - 15}{25x - 75}$

11. $\dfrac{22x + 8}{11x + 4}$

12. $\dfrac{2x}{2x^2 - 2x}$

13. $\dfrac{x + 7}{x^2 - 49}$

14. $\dfrac{2x^2 + 4x - 30}{x^2 + x - 20}$

15. $\dfrac{xy - 3x + 2y - 6}{x^2 + 4x + 4}$

(6.2)   *Perform the indicated operation. Write answers in lowest terms.*

16. $\dfrac{5}{x^3} \cdot \dfrac{x^2}{15}$

17. $\dfrac{3x^4yz^3}{15x^2y^2} \cdot \dfrac{10xy}{z^6}$

18. $\dfrac{4 - x}{5} \cdot \dfrac{15}{2x - 8}$

19. $\dfrac{x^2 - 6x - 9}{2x^2 - 18} \cdot \dfrac{4x + 12}{5x - 15}$

20. $\dfrac{a - 4b}{a^2 + ab} \cdot \dfrac{b^2 - a^2}{8b - 2a}$

21. $\dfrac{x^2 - x - 12}{2x^2 - 32} \cdot \dfrac{x^2 + 8x + 16}{3x^2 + 21x + 36}$

22. $\dfrac{2x^3 + 54}{5x^2 + 5x - 30} \cdot \dfrac{6x + 12}{3x^2 - 9x + 27}$

23. $\dfrac{3}{4x} \div \dfrac{8}{2x^2}$

24. $\dfrac{4x + 8y}{3} \div \dfrac{5x + 10y}{9}$

25. $\dfrac{5ab}{14c^3} \div \dfrac{10a^4b^2}{6ac^5}$

26. $\dfrac{2}{5x} \div \dfrac{4 - 18x}{6 - 27x}$

27. $\dfrac{x^2 - 25}{3} \div \dfrac{x^2 - 10x + 25}{x^2 - x - 20}$

28. $\dfrac{a - 4b}{a^2 + ab} \div \dfrac{20b - 5a}{b^2 - a^2}$

29. $\dfrac{7x + 28}{2x + 4} \div \dfrac{x^2 + 2x - 8}{x^2 - 2x - 8}$

30. $\dfrac{3x + 3}{x - 1} \div \dfrac{x^2 - 6x - 7}{x^2 - 1}$

31. $\dfrac{2x - x^2}{x^3 - 8} \div \dfrac{x^2}{x^2 + 2x + 4}$

32. $\dfrac{5a^2 - 20}{a^3 + 2a^2 + a + 2} \div \dfrac{7a}{a^3 + a}$

33. $\dfrac{2a}{21} \div \dfrac{3a^2}{7} \cdot \dfrac{4}{a}$

34. $\dfrac{5x - 15}{3 - x} \cdot \dfrac{x + 2}{10x + 20} \cdot \dfrac{x^2 - 9}{x^2 - x - 6}$

35. $\dfrac{4a + 8}{5a^2 - 20} \cdot \dfrac{3a^2 - 6a}{a + 3} \div \dfrac{2a^2}{5a + 15}$

(6.3)   *Find the LCD of the rational expressions in the list.*

36. $\dfrac{4}{9}, \dfrac{5}{2}$

37. $\dfrac{5}{4x^2y^5}, \dfrac{3}{10x^2y^4}, \dfrac{x}{6y^4}$

38. $\dfrac{5}{2x}, \dfrac{7}{x - 2}$

39. $\dfrac{3}{5x}, \dfrac{2}{x - 5}$

40. $\dfrac{1}{5x^3}, \dfrac{4}{x^2 + 3x - 28}, \dfrac{11}{10x^2 - 30x}$

Perform the indicated operation. Write answers in lowest terms.

41. $\dfrac{2}{15} + \dfrac{4}{15}$

42. $\dfrac{4}{x-4} + \dfrac{x}{x-4}$

43. $\dfrac{4}{3x^2} + \dfrac{2}{3x^2}$

44. $\dfrac{1}{x-2} - \dfrac{1}{4-2x}$

45. $\dfrac{2x+1}{x^2+x-6} + \dfrac{2-x}{x^2+x-6}$

46. $\dfrac{7}{2x} + \dfrac{5}{6x}$

47. $\dfrac{1}{3x^2y^3} - \dfrac{1}{5x^4y}$

48. $\dfrac{1}{10-x} + \dfrac{x-1}{x-10}$

49. $\dfrac{x-2}{x+1} - \dfrac{x-3}{x-1}$

50. $\dfrac{x}{9-x^2} - \dfrac{2}{5x-15}$

51. $2x+1 - \dfrac{1}{x-3}$

52. $\dfrac{2}{a^2-2a+1} + \dfrac{3}{a^2-1}$

53. $\dfrac{x}{9x^2+12x+16} - \dfrac{3x+4}{27x^3-64}$

Perform the indicated operation. Write answers in lowest terms.

54. $\dfrac{2}{x-1} - \dfrac{3x}{3x-3} + \dfrac{1}{2x-2}$

55. $\dfrac{3}{2x} \cdot \left(\dfrac{2}{x+1} - \dfrac{2}{x-3} \right)$

56. $\left(\dfrac{3x}{4} \right)^2 \cdot \left(\dfrac{2}{x} \right)^3$

57. $\left(\dfrac{2}{x} - \dfrac{1}{5} \right) \cdot \left(\dfrac{2}{x} + \dfrac{1}{3} \right)$

58. $\dfrac{2}{x^2-16} - \dfrac{3x}{x^2+8x+16} + \dfrac{3}{x+4}$

59. $\left(\dfrac{x}{x+5} \right)^2 - 1$

60. $\dfrac{x}{x^2-6x+9} + \dfrac{3}{x-3} \cdot \dfrac{2x^2-18}{4x}$

(6.4) *Simplify each complex fraction.*

61. $\dfrac{\frac{2}{5}}{\frac{3}{5}}$

62. $\dfrac{1 - \frac{3}{4}}{2 + \frac{1}{4}}$

63. $\dfrac{\frac{1}{x} - \frac{2}{3x}}{\frac{5}{2x} - \frac{1}{3}}$

64. $\dfrac{\frac{x^2}{15}}{\frac{x+1}{5x}}$

65. $\dfrac{\frac{3}{y^2}}{\frac{6}{y^3}}$

66. $\dfrac{\frac{x+2}{3}}{\frac{5}{x-2}}$

67. $\dfrac{2 - \frac{3}{2x}}{x - \frac{2}{5x}}$

68. $\dfrac{1 + \frac{x}{y}}{\frac{x^2}{y^2} - 1}$

69. $\dfrac{\frac{5}{x} + \frac{1}{xy}}{\frac{3}{x^2}}$

70. $\dfrac{\frac{x}{3} - \frac{3}{x}}{1 + \frac{3}{x}}$

71. $\dfrac{\frac{1}{x-1} + 1}{\frac{1}{x+1} - 1}$

72. $\dfrac{2}{1 - \frac{2}{x}}$

73. $\dfrac{1}{1 + \dfrac{2}{1 - \frac{1}{x}}}$

74. $\dfrac{\frac{x^2+5x-6}{4x+3}}{\frac{(x+6)^2}{8x+6}}$

75. $\dfrac{\frac{x-3}{x+3} + \frac{x+3}{x-3}}{\frac{x-3}{x+3} - \frac{x+3}{x-3}}$

76. $\dfrac{\frac{3}{x-1} - \frac{2}{1-x}}{\frac{2}{x-1} - \frac{2}{x}}$

(6.5) *Solve each equation for x.*

77. $\dfrac{2}{5} = \dfrac{x}{15}$

78. $\dfrac{3}{x} + \dfrac{1}{3} = \dfrac{5}{x}$

79. $4 + \dfrac{8}{x} = 8$

80. $\dfrac{2x+3}{5x-9} = \dfrac{3}{2}$

81. $\dfrac{1}{x-2} - \dfrac{3x}{x^2-4} = \dfrac{2}{x+2}$

82. $\dfrac{7}{x} - \dfrac{x}{7} = 0$

83. $\dfrac{x-2}{x^2-7x+10} = \dfrac{1}{5x-10} - \dfrac{1}{x-5}$

Solve the equations for x or perform the indicated operation. Simplify.

84. $\dfrac{5}{x^2 - 7x} + \dfrac{4}{2x - 14}$

85. $3 - \dfrac{5}{x} - \dfrac{2}{x^2} = 0$

86. $\dfrac{4}{3 - x} - \dfrac{7}{2x - 6} + \dfrac{5}{x}$

(6.6) *Solve the equation for the specified variable.*

87. $A = \dfrac{h(a + b)}{2}, a$

88. $\dfrac{1}{R} = \dfrac{1}{R_1} + \dfrac{1}{R_2}, R_2$

89. $I = \dfrac{E}{R + r}, R$

90. $A = P + Prt, r$

91. $H = \dfrac{kA(T_1 - T_2)}{L}, A$

Solve.

92. The sum of a number and twice its reciprocal is 3. Find the number(s).

93. If a number is added to the numerator of $\dfrac{3}{7}$, and twice that number is added to the denominator of $\dfrac{3}{7}$, the result is equivalent to $\dfrac{10}{21}$. Find the number.

94. Mary is three-fourths as old as her friend Mark. If the sum of their ages is 42, find their ages.

95. The denominator of a fraction is 2 more than the numerator. If the numerator is decreased by 3 and the denominator is increased by 5, the resulting fraction is equivalent to $\dfrac{2}{3}$. Find the fraction.

96. The sum of the reciprocals of two consecutive integers is $\dfrac{29}{210}$. Find the two integers.

97. The sum of the reciprocals of two consecutive even integers is $-\dfrac{9}{40}$. Find the two integers.

98. Three boys can paint a fence in 4 hours, 5 hours, and 6 hours, respectively. Find how long it will take all three boys to paint the fence.

99. If Sue can type mailing labels in 6 hours and Tom and Sue working together can type mailing labels in 4 hours, find how long it takes Tom to type mailing labels alone.

100. The inlet pipe of a water tank can fill the tank in 2 hours and 30 minutes. The outlet pipe can empty the tank in 2 hours. Find how long it takes to empty a full tank if both pipes are open.

101. Timmy drove 210 miles in the same amount of time that it took a plane to travel 1715 miles. The speed of the plane was 430 mph faster than the speed of the car. Find the speed of the plane.

102. The combined resistance R of two resistors, r_1 and r_2, in parallel is given by the formula $\dfrac{1}{R} = \dfrac{1}{r_1} + \dfrac{1}{r_2}$. If the combined resistance is $\dfrac{30}{11}$ ohms and one of the two resistors is 5 ohms, find the other resistor.

103. The speed of a boat in still water is 32 mph. If the boat travels 72 miles upstream in the same time that it takes to travel 120 miles downstream, find the current of the stream.

104. A plane flies 445 miles with the wind and 355 miles against the wind in the same length of time. If the speed of the plane in still air is 400 mph, find the speed of the wind.

105. The speed of a jogger is 3 mph faster than the speed of a walker. If the jogger travels 14 miles in the same amount of time that the walker travels 8 miles, find the speed of the walker.

106. Two trains leave Tucson at the same time traveling on parallel tracks. In 6 hours the faster train is 382 miles from Tucson and the trains are 112 miles apart. Find how fast each train is traveling.

(6.7) *Solve each proportion for x.*

107. $\dfrac{3x - 5}{14} = \dfrac{x}{4}$

108. $\dfrac{2 - 5x}{12} = \dfrac{x}{8}$

Solve each of the following variation problems.

109. A is directly proportional to B. If $A = 6$ when $B = 14$, find A when $B = 21$.

110. C is inversely proportional to D. If $C = 12$ when $D = 8$, find C when $D = 24$.

111. According to Boyle's law, the pressure exerted by a gas is inversely proportional to the volume, as long as the temperature stays the same. If a gas exerts a pressure of 1250 pounds per square inch when the volume is 2 cubic feet, find the volume when the pressure is 800 pounds per square inch.

112. The surface area of a sphere varies directly as the square of its radius. If the surface area is 36 square inches when the radius is 3 inches, find the surface area when the radius is 4 inches.

CHAPTER 6 TEST

Find the values of x, if any, for which each rational expression is undefined.

1. $\dfrac{2x}{15}$

2. $\dfrac{9}{2x^2 - 7x - 15}$

Write each rational expression in lowest terms.

3. $\dfrac{5x^7}{3x^4}$

4. $\dfrac{7x - 21}{24 - 8x}$

5. $\dfrac{2x + 6}{x^2 - 9}$

6. $\dfrac{x^2 - 4x}{x^2 + 5x - 36}$

Perform the indicated operation. Write answers in lowest terms.

7. $\dfrac{x}{x - 2} \cdot \dfrac{x^2 - 4}{5x}$

8. $\dfrac{2x^2 - x - 3}{5x + 10} \cdot \dfrac{3x + 6}{4x - 6}$

9. $\dfrac{2x^3 + 16}{6x^2 + 12x} \cdot \dfrac{5}{x^2 - 2x + 4}$

10. $\dfrac{26ab}{7c} \div \dfrac{13a^2c^5}{14a^4b^3}$

11. $\dfrac{3x^2 - 12}{x^2 + 2x - 8} \div \dfrac{6x + 18}{x + 4}$

12. $\dfrac{4x - 12}{2x - 9} \div \dfrac{3 - x}{4x^2 - 81} \cdot \dfrac{x + 3}{5x + 15}$

Find the LCD of the rational expressions in the list.

13. $\dfrac{2}{25x}, \dfrac{3}{15x^3}$

Perform the indicated operation.

14. $\dfrac{5}{4x^3} + \dfrac{7}{4x^3}$

15. $\dfrac{3 + 2x}{10 - x} + \dfrac{13 + x}{x - 10}$

16. $\dfrac{3}{x^2 - x - 6} + \dfrac{2}{x^2 - 5x + 6}$

17. $\dfrac{x}{4x^2 - 2x + 1} - \dfrac{2x + 1}{8x^3 + 1}$

18. $\dfrac{5}{x - 7} - \dfrac{2x}{3x - 21} + \dfrac{x}{2x - 14}$

19. $\dfrac{3x}{5} \cdot \left(\dfrac{5}{x} - \dfrac{5}{2x}\right)$

20. $\dfrac{5x + 3y}{x^2 - y^2} \cdot \left(\dfrac{x}{y} - \dfrac{y}{x}\right)$

21. $\dfrac{3}{x + y} + \dfrac{2x^2 - 2xy + 4x - 4y}{4x + 8} \div \dfrac{y^2 - x^2}{2y}$

Simplify each complex fraction.

22. $\dfrac{\dfrac{4x}{13}}{\dfrac{20x}{13}}$

23. $\dfrac{\dfrac{5}{x} - \dfrac{7}{3x}}{\dfrac{9}{8x} - \dfrac{1}{x}}$

24. $\dfrac{\dfrac{7}{2x} + \dfrac{2}{xy}}{\dfrac{3}{xy^2}}$

25. $\dfrac{\dfrac{x^2 - 5x + 6}{x + 3}}{\dfrac{x^2 - 4x + 4}{x^2 - 9}}$

Solve each equation for x.

26. $\dfrac{5x + 3}{3x - 7} = \dfrac{19}{7}$

27. $\dfrac{5}{x - 5} + \dfrac{x}{x + 5} = -\dfrac{29}{21}$

28. $\dfrac{x}{x - 4} = 3 - \dfrac{4}{x - 4}$

29. Solve for x: $\dfrac{x+b}{a} = \dfrac{4x-7a}{b}$

30. The product of one more than a number and twice the reciprocal of the number is $\dfrac{12}{5}$. Find the number.

31. If Jan can weed the garden in 2 hours and her husband in 1 hour and 30 minutes, find how long it takes them to weed the garden together.

32. Suppose that W is inversely proportional to V. If $W = 20$ when $V = 12$, find W when $V = 15$.

33. Suppose that Q is jointly proportional to R and the square of S. If $Q = 24$ when $R = 3$ and $S = 4$, find Q when $R = 2$ and $S = 3$.

34. When an anvil is dropped into a gorge, the speed with which it strikes the ground is directly proportional to the square root of the distance it falls. An anvil that falls 400 feet hits the ground at a speed of 160 feet per second. Find the height of a cliff over the gorge if a dropped anvil hits the ground at a speed of 128 feet per second.

CHAPTER 6 CUMULATIVE REVIEW

1. Determine whether each statement is true or false.
  a. $3 \in \{1, 2, 3\}$ **b.** $7 \notin \{1, 2, 3\}$
  c. $\{2\} \in \{1, 2, 3\}$

2. Simplify the following by combining like terms.
  a. $3x - 5x$ **b.** $7y + y$
  c. $4z + 6$

3. Solve $3y - 2x = 7$ for y.

4. Complete the ordered pair solutions for the equation $3x - y = 12$.
  a. $(\ , 0)$ **b.** $(1, \ )$ **c.** $(0, \ )$

5. Find the equation of the line through $(4, 4)$ and parallel to the line $2x + 3y = -6$.

6. Find each quotient.
  a. $\dfrac{x^7}{x^4}$ **b.** $\dfrac{5^8}{5^2}$ **c.** $\dfrac{20x^6}{4x^5}$ **d.** $\dfrac{12y^{10}z^7}{14y^8z^7}$

7. If $P(x) = 7x^2 - 3x + 1$ and $Q(x) = 3x - 2$, find the following.
  a. $P(1)$ **b.** $Q(1)$ **c.** $P(-2)$ **d.** $Q(0)$

8. Multiply $(x + 3)$ by $(2x + 5)$.

9. Graph $H(x) = 3x$.

10. Factor.
  a. $8x + 4$ **b.** $5y - 2z$ **c.** $6x^2 - 3x^3$

11. Factor $x^2 + 10x + 16$.

12. Factor $2(a + 3)^2 - 5(a + 3) - 7$.

13. Factor $m^2 + 10m + 25$.

14. Factor $x^3 + 8$.

15. Solve $2x^2 + 9x - 5 = 0$.

16. The length of a rectangular piece of carpet is 2 meters less than 5 times its width. Find the dimensions of the carpet if its area is 16 square meters.

17. Write each rational expression in lowest terms.
  a. $\dfrac{24x^6y^5}{8x^7y}$ **b.** $\dfrac{2x^2}{10x^3 - 2x^2}$

18. Multiply.
  a. $\dfrac{2x^3}{9y} \cdot \dfrac{y^2}{4x^3}$ **b.** $\dfrac{1 + 3n}{2n} \cdot \dfrac{2n - 4}{3n^2 - 2n - 1}$

19. Add or subtract.
  a. $\dfrac{5}{7} + \dfrac{x}{7}$ **b.** $\dfrac{x}{4} + \dfrac{5x}{4}$ **c.** $\dfrac{x^2}{x + 7} - \dfrac{49}{x + 7}$
  d. $\dfrac{x}{3y^2} - \dfrac{x + 1}{3y^2}$

20. Solve $\dfrac{3}{x} - \dfrac{x + 21}{3x} = \dfrac{5}{3}$.

21. Solve the proportion $\dfrac{x + 2}{3} = \dfrac{x}{5}$ for x.

CHAPTER 7

Rational Exponents, Radicals, and Complex Numbers

To start a new business, Peter borrowed money from his uncle, who stipulated an annually compounded interest. Now, 4 years and 6 months later, Peter is prepared to pay back his Uncle, and he must give his uncle all the interest due. (See Critical Thinking, page 339.)

INTRODUCTION

In this chapter we introduce **rational exponents.** As the name implies, rational exponents are exponents that are rational numbers. We present an interpretation of rational exponents that is consistent with the meaning and rules already established for integer exponents, and we present two forms of notation for rational exponents: the exponent notation and the radical notation. We conclude this chapter with **complex numbers,** a natural extension of the real number system that gives meaning to rational exponents applied to negative real numbers.

7.1
Rational Exponents

OBJECTIVES

1 Define a raised to the $\frac{1}{n}$th power.

2 Define a raised to the $\frac{m}{n}$th power.

3 Simplify expressions containing rational exponents.

Tape 20

1 In this section we give meaning to exponential expressions with rational exponents such as

$$4^{1/2} \qquad 16^{3/4} \qquad (-27)^{-2/3}$$

First, let's define what we mean by the symbol $a^{1/n}$.

> **Definition of $a^{1/n}$ if a Is Positive**
>
> If n is a positive integer and $a \geq 0$, then $a^{1/n}$ is the nonnegative real number whose nth power is a. In symbols, this is
>
> $$(a^{1/n})^n = a$$

We call $a^{1/n}$ the nth root of a.

EXAMPLE 1 Evaluate the following.
a. $4^{1/2}$ **b.** $64^{1/3}$ **c.** $81^{1/4}$ **d.** $0^{1/6}$ **e.** $-9^{1/2}$

Solution: **a.** $4^{1/2} = 2$ since $2^2 = 4$ and 2 is positive.
b. $64^{1/3} = 4$ since $4^3 = 64$ and 4 is positive.
c. $81^{1/4} = 3$ since $3^4 = 81$.
d. $0^{1/6} = 0$ since $0^6 = 0$.
e. $-9^{1/2} = -3$ since $9^{1/2} = 3$. ■

Next we define $a^{1/n}$ if a is negative.

Definition of $a^{1/n}$ if a Is Negative

If n is an odd positive integer and a is a negative number, then $a^{1/n}$ is the real number whose nth power is a. In symbols, this is

$$(a^{1/n})^n = a$$

If n is an even positive integer and a is a negative number, then $a^{1/n}$ is not a real number.

EXAMPLE 2   Evaluate the following.

    a. $(-8)^{1/3}$ **b.** $(-32)^{1/5}$ **c.** $(-16)^{1/4}$ **d.** $\left(-\dfrac{1}{64}\right)^{1/3}$

Solution: **a.** $(-8)^{1/3} = -2$ since $(-2)^3 = -8$. **b.** $(-32)^{1/5} = -2$ since $(-2)^5 = -32$.

    c. $(-16)^{1/4}$ is not a real number because there is no real number whose fourth power is -16.

    d. $\left(-\dfrac{1}{64}\right)^{1/3} = -\dfrac{1}{4}$ since $\left(-\dfrac{1}{4}\right)^3 = -\dfrac{1}{64}$.

We summarize some important conclusions about $a^{1/n}$ next.

Let a be a real number and n be a positive integer.

			Examples

If a is positive, then $a^{1/n}$ is positive. $27^{1/3} = 3$

If $a = 0$, then $a^{1/n} = 0$. $0^{1/4} = 0$

If a is negative and $\begin{cases} n \text{ is odd, then } a^{1/n} \text{ is negative.} \\ n \text{ is even, then } a^{1/n} \text{ is not a real number.} \end{cases}$ $(-8)^{1/3} = -2$

    $(-4)^{1/2}$ is not a real number.

The definition of $a^{1/n}$ can be extended to include rational exponents with numerators other than 1. For example, if rules for integer exponents are to be true for rational exponents, then $8^{2/3} = (8^{1/3})^2$ or $8^{2/3} = (8^2)^{1/3}$. This suggests the following definition.

    2

Definition of $a^{m/n}$

If m and n are positive integers with m/n in lowest terms, then

$$a^{m/n} = (a^{1/n})^m = (a^m)^{1/n}$$

as long as all roots are real numbers.

From this definition, we can evaluate $8^{2/3}$ in two ways:

$$8^{2/3} = (8^{1/3})^2 = 2^2 = 4$$

or

$$8^{2/3} = (8^2)^{1/3} = 64^{1/3} = 4$$

Both methods give us the same correct result. It is usually easier to find $a^{m/n}$ by finding $a^{1/n}$ first and then raising this result to the mth power.

EXAMPLE 3   Evaluate the following.

   a. $4^{3/2}$ **b.** $-16^{3/4}$ **c.** $(-16)^{3/4}$ **d.** $(-27)^{2/3}$ **e.** $\left(\dfrac{1}{9}\right)^{3/2}$

Solution: **a.** $4^{3/2} = (4^{1/2})^3 = 2^3 = 8$

   b. $-16^{3/4} = -(16^{3/4}) = -(16^{1/4})^3 = -(2)^3 = -8$

   c. $(-16)^{3/4}$ is not a real number since $(-16)^{1/4}$ is not a real number.

   d. $(-27)^{2/3} = [(-27)^{1/3}]^2 = (-3)^2 = 9$

   e. $\left(\dfrac{1}{9}\right)^{3/2} = \left[\left(\dfrac{1}{9}\right)^{1/2}\right]^3 = \left(\dfrac{1}{3}\right)^3 = \dfrac{1}{27}$ ■

The rational exponents we have given meaning to exclude negative rational numbers. To complete the set of definitions, we define $a^{-m/n}$.

Definition of $a^{-m/n}$

$$a^{-m/n} = \dfrac{1}{a^{m/n}}$$

as long as $a^{m/n}$ is a nonzero real number.

EXAMPLE 4   Evaluate the following.

   a. $16^{-3/4}$ **b.** $(-27)^{-2/3}$

Solution: **a.** $16^{-3/4} = \dfrac{1}{16^{3/4}} = \dfrac{1}{(16^{1/4})^3} = \dfrac{1}{2^3} = \dfrac{1}{8}$

   b. $(-27)^{-2/3} = \dfrac{1}{(-27)^{2/3}} = \dfrac{1}{[(-27)^{1/3}]^2} = \dfrac{1}{(-3)^2} = \dfrac{1}{9}$ ■

HELPFUL HINT

Notice that the sign of the base is not affected by the sign of its exponent. For example,

$$9^{-3/2} = \dfrac{1}{9^{3/2}} = \dfrac{1}{27}$$

Also

$$(-27)^{-1/3} = \dfrac{1}{(-27)^{1/3}} = -\dfrac{1}{3}$$

---

3   We repeat now the rules for exponents that were developed in Chapter 3. These rules apply not just to integer exponents, but to all rational exponents. Knowing, for example, how to simplify $2^3 2^5$, we may use the same procedure to simplify $2^{2/3} 2^{3/5}$. These rules are stated next.

Rules for Exponents

If m and n are rational numbers and a and b are real numbers such that the following exponential expressions are defined, then

Product rule	$a^m a^n = a^{m+n}$
Quotient rule	$\dfrac{a^m}{a^n} = a^{m-n},\ a \neq 0$
Power rules	$(a^m)^n = a^{mn}$
	$(ab)^m = a^m b^m$
	$\left(\dfrac{a}{b}\right)^m = \dfrac{a^m}{b^m},\ b \neq 0$

EXAMPLE 5 Simplify the following expressions. Assume all variables represent positive numbers. Write your answers using only positive exponents.

a. $x^{1/2} x^{1/3}$ **b.** $\dfrac{7^{1/3}}{7^{4/3}}$ **c.** $\dfrac{(2x^{2/5} y^{-1/3})^5}{x^2 y}$

Solution: **a.** $x^{1/2} x^{1/3} = x^{(1/2 + 1/3)} = x^{3/6 + 2/6} = x^{5/6}$ **b.** $\dfrac{7^{1/3}}{7^{4/3}} = 7^{1/3 - 4/3} = 7^{-3/3} = 7^{-1} = \dfrac{1}{7}$

c. We begin by using the power rule $(ab)^m = a^m b^m$ to simplify the numerator.

$$\frac{(2x^{2/5} y^{-1/3})^5}{x^2 y} = \frac{2^5 (x^{2/5})^5 (y^{-1/3})^5}{x^2 y} = \frac{32 x^2 y^{-5/3}}{x^2 y}$$

$$= 32 x^{2-2} y^{-5/3 - 3/3} \qquad \text{Apply the quotient rule.}$$

$$= 32 x^0 y^{-8/3}$$

$$= \frac{32}{y^{8/3}} \quad \blacksquare$$

EXAMPLE 6 Multiply. Assume all variables represent positive numbers.
a. $z^{2/3}(z^{1/3} - z^5)$ **b.** $(x^{1/3} - 5)(x^{1/3} + 2)$

Solution: **a.** $z^{2/3}(z^{1/3} - z^5) = z^{2/3} z^{1/3} - z^{2/3} z^5$ Apply the distributive property.

$$= z^{(2/3 + 1/3)} - z^{(2/3 + 5)} \qquad \text{Use the product rule.}$$

$$= z^{3/3} - z^{(2/3 + 15/3)}$$

$$= z - z^{17/3}$$

b. $(x^{1/3} - 5)(x^{1/3} + 2) = x^{2/3} + 2x^{1/3} - 5x^{1/3} - 10$

$$= x^{2/3} - 3x^{1/3} - 10 \quad \blacksquare$$

EXAMPLE 7 Factor $x^{-1/2}$ from the expression $3x^{-1/2} - 7x^{5/2}$. Assume that all variables represent positive numbers.

Solution:
$$3x^{-1/2} - 7x^{5/2} = (x^{-1/2})(3) - (x^{-1/2})(7x^{6/2})$$

$$= x^{-1/2}(3 - 7x^3)$$

To check, multiply $x^{-1/2}(3 - 7x^3)$ to see that the product is $3x^{-1/2} - 7x^{5/2}$. $\blacksquare$

EXERCISE SET 7.1

Evaluate. See Example 1.

1. $49^{1/2}$ **2.** $64^{1/3}$ **3.** $27^{1/3}$ **4.** $8^{1/3}$

5. $\left(\dfrac{1}{16}\right)^{1/4}$ **6.** $\left(\dfrac{1}{64}\right)^{1/2}$ **7.** $169^{1/2}$ **8.** $81^{1/4}$

Evaluate. See Example 2.

9. $(-27)^{1/3}$ **10.** $(-100)^{1/2}$ **11.** $(-625)^{1/4}$ **12.** $(-64)^{1/3}$

13. $-64^{1/2}$ **14.** $-16^{1/4}$ **15.** $(-1)^{1/5}$ **16.** $(-32)^{1/5}$

Evaluate. See Example 3.

17. $16^{3/4}$ **18.** $4^{5/2}$ **19.** $(-64)^{2/3}$

20. $(-8)^{4/3}$ **21.** $(-16)^{3/4}$ **22.** $(-9)^{3/2}$

Evaluate. See Example 4.

23. $8^{-4/3}$ **24.** $64^{-2/3}$ **25.** $(-64)^{-2/3}$

26. $(-8)^{-4/3}$ **27.** $(-4)^{-3/2}$ **28.** $(-16)^{-5/4}$

Simplify each expression. Write using positive exponents. Assume that all variables represent positive numbers. See Example 5.

29. $a^{2/3}a^{5/3}$ **30.** $b^{9/5}b^{8/5}$ **31.** $(4u^2v^{-6})^{3/2}$

32. $(32^{1/5}x^{2/3}y^{1/3})^3$ **33.** $\dfrac{b^{1/2}b^{3/4}}{-b^{1/4}}$ **34.** $\dfrac{a^{1/4}a^{-1/2}}{a^{2/3}}$

Multiply. Assume that all variables represent positive numbers. See Example 6.

35. $y^{1/2}(y^{1/2} - y^{2/3})$ **36.** $x^{1/2}(x^{1/2} + x^{3/2})$ **37.** $x^{2/3}(2x - 2)$

38. $3x^{1/2}(x + y)$ **39.** $(2x^{1/3} + 3)(2x^{1/3} - 3)$ **40.** $(y^{1/2} + 5)(y^{1/2} + 5)$

Factor the common factor from the given expression. Assume that all variables represent positive numbers. See Example 7.

41. $x^{8/3}$; $x^{8/3} + x^{10/3}$ **42.** $x^{3/2}$; $x^{5/2} - x^{3/2}$ **43.** $x^{1/5}$; $x^{2/5} - 3x^{1/5}$

44. $x^{2/7}$; $x^{3/7} - 2x^{2/7}$ **45.** $x^{-1/3}$; $5x^{-1/3} + x^{2/3}$ **46.** $x^{-3/4}$; $x^{-3/4} + 3x^{1/4}$

Evaluate.

47. $25^{1/2}$ **48.** $100^{1/2}$ **49.** $64^{1/3}$ **50.** $125^{1/3}$

51. $-9^{1/2}$ **52.** $-36^{1/2}$ **53.** $(-64)^{1/2}$ **54.** $(-25)^{1/2}$

55. $\left(\dfrac{1}{49}\right)^{1/2}$ **56.** $\left(\dfrac{4}{9}\right)^{1/2}$ **57.** $16^{3/4}$ **58.** $27^{2/3}$

59. $8^{-2/3}$ **60.** $81^{-3/4}$ **61.** $\left(-\dfrac{1}{9}\right)^{1/2}$ **62.** $\left(\dfrac{1}{256}\right)^{1/4}$

63. $\left(-\dfrac{1}{125}\right)^{2/3}$ **64.** $\left(\dfrac{27}{8}\right)^{2/3}$ **65.** $\left(-\dfrac{1}{256}\right)^{1/4}$ **66.** $-64^{-2/3}$

67. $\left(\dfrac{16}{9}\right)^{1/2}$ **68.** $\left(\dfrac{16}{9}\right)^{3/2}$ **69.** $\left(\dfrac{49}{25}\right)^{3/2}$ **70.** $64^{-4/3}$

Simplify each expression. Write using positive exponents. Assume that all variables represent positive numbers.

71. $x^{-5/3}x^{1/2}$

72. $x^{-1/3}x^{1/5}$

73. $\dfrac{w^{3/4}}{w^{1/4}}$

74. $\dfrac{y^{1/2}}{y^{2/3}}$

75. $(x^{-1/5}y^{1/12})^3$

76. $(a^{10}b^{15})^{-1/5}$

77. $(a^3b^9)^{-2/3}a^{2/3}$

78. $(x^4y^2z^6)^{-3/4}x^{1/2}$

79. $(16x^4y^2)^{1/2}$

80. $(9a^4b^8)^{1/4}$

81. $\left[\dfrac{y^{2/3}y^{-5/6}}{y^{1/9}}\right]^9$

82. $\left[\dfrac{a^{1/3}a^{-2/3}}{a^{1/2}}\right]^{-4}$

83. $\left[\dfrac{b^2a^{-3/4}}{b^{1/2}}\right]^{-1/2}$

84. $\dfrac{(x^{-5/6}x^3)^{-2}}{x^{4/3}a^{-2}}$

Multiply. Assume that all variables represent positive numbers.

85. $4x^{2/3}(6x^{5/2} - y^{2/3})$

86. $8x^{2/5}(6x^{3/2} - y^{1/2})$

87. $(x^{1/2} + 2)(x^{1/2} - 3)$

88. $(2a^{2/3} + b^{1/2})(3a^{2/3} + 2)$

89. $(x^{1/2} - 2x)^2$

90. $(3y^{1/4} + 2)^2$

Factor the common factor from the given expression. Assume that all variables represent positive numbers.

91. $x^{3/4}; 9x^{3/4} - x^{7/4}$

92. $y^{2/3}; 7y^{5/3} - 3y^{2/3}$

93. $3x^{5/3}; 9x^{5/3} + 15x^{7/3}$

94. $4x^{1/2}; 8x^{1/2} - 16x^{3/2}$

95. $y^{-1/5}; 3y^{-1/5} - 5y^{6/5}$

96. $x^{-2/7}; 4x^{5/7} - 13x^{-2/7}$

A Look Ahead

Simplify each expression. Assume that a and b are rational numbers and that x and y are positive numbers. See the following example.

EXAMPLE   Simplify $\dfrac{(x^2y^{-3/4})^a}{x^a}$. Assume that a is a rational number and x and y are positive numbers.

Solution: $\dfrac{(x^2y^{-3/4})^a}{x^a} = \dfrac{(x^2)^a(y^{-3/4})^a}{x^a}$

$= \dfrac{x^{2a}y^{-3a/4}}{x^a}$

$= x^{(2a-a)}y^{-3a/4}$

$= \dfrac{x^a}{y^{3a/4}}$ ■

97. $x^{a/5} \cdot x^{a/3}$

98. $y^{-b/2} \cdot y^{b/3}$

99. $\dfrac{y^{2a/3}}{y^{5a/6}}$

100. $\dfrac{x^{3a/7}}{x^{9/14}}$

101. $\dfrac{(x^by^{-1/2})^2}{x^{5b}}$

102. $\dfrac{(x^{a/3}y^{1/6})^3}{(x^{-a}y^a)^2}$

Writing in Mathematics

103. Explain why $(-64)^{1/2}$ is not a real number, while $(-64)^{1/3}$ is a real number.

104. Explain the steps necessary to factor $x^{1/2}$ from $3x^{5/2} - yx^{1/2}$.

Skill Review

Factor completely. See Section 5.4.

105. $x^3 - 6x^2 + 8x$

106. $2x^3 - 2x^2 - 4x$

107. $x^3 + 27y^3$

108. $8x^3 - 125y^3$

Simplify the following. See Section 6.1.

109. $\dfrac{x^2 - 9}{3 - x}$

110. $\dfrac{x^2 + 4x + 4}{2 + x}$

7.2
Simplifying Radicals

OBJECTIVES

Tape 20

1	Use radical notation.
2	Write radical expressions as expressions containing rational exponents.
3	Write exponential expressions as expressions containing radicals.
4	Use the product rule for radicals to simplify radical expressions.
5	Use the quotient rule for radicals to simplify radical expressions.

1   At the beginning of this chapter, we said that we have two notations for rational exponents. The first, exponential notation, we have developed. Now we give the second notation, radical notation.

Radical Notation

If n is a positive integer and $a^{1/n}$ is a real number, then
$$a^{1/n} = \sqrt[n]{a}$$

The expression $\sqrt[n]{a}$ is called a **radical expression.** The symbol $\sqrt{}$ is called the **radical sign.** The expression under the radical sign is called the **radicand,** and the positive integer n is called the **index.**

$$\text{index} \longrightarrow \sqrt[n]{a} \longleftarrow \text{radicand}$$
$$\text{radical sign} \longrightarrow$$

When no index is written, the index is assumed to be 2. For example, $\sqrt{25}$ is $\sqrt[2]{25}$. When we simplify a radical expression, we say we **take the root** or **find the root.**

EXAMPLE 1   Simplify. Assume that all variables represent positive numbers.

a. $\sqrt{25}$ **b.** $\sqrt[3]{-8}$ **c.** $\sqrt{0}$ **d.** $\sqrt[4]{-16}$ **e.** $\sqrt{\dfrac{1}{36}}$ **f.** $\sqrt[3]{-\dfrac{1}{8}}$
g. $\sqrt{x^6}$ **h.** $\sqrt[3]{x^{15}}$

Solution: **a.** $\sqrt{25} = 5$ because $5^2 = 25$. Read the expression $\sqrt{25}$ as "the square root of 25."

b. $\sqrt[3]{-8} = -2$ because $(-2)^3 = -8$. Read the expression $\sqrt[3]{-8}$ as "the cube root of -8."

c. $\sqrt{0} = 0$ because $0^2 = 0$.　**d.** $\sqrt[4]{-16}$ is not a real number.

e. $\sqrt{\dfrac{1}{36}} = \dfrac{1}{6}$ because $\left(\dfrac{1}{6}\right)^2 = \dfrac{1}{36}$.　**f.** $\sqrt[3]{-\dfrac{1}{8}} = -\dfrac{1}{2}$ because $\left(-\dfrac{1}{2}\right)^3 = -\dfrac{1}{8}$.

g. $\sqrt{x^6} = x^3$ because $(x^3)^2 = x^6$.　**h.** $\sqrt[3]{x^{15}} = x^5$ because $(x^5)^3 = x^{15}$.　∎

2　To emphasize that $\sqrt[n]{a}$ and $a^{1/n}$ are different notations that represent identical numbers, we repeat the conclusions about the sign of $a^{1/n}$ from the previous section, this time using radical notation.

Let a be a real number and n be a positive integer.

Examples

If a is positive, then $\sqrt[n]{a}$ is positive.　　　　$\sqrt[3]{27} = 3$
If $a = 0$, then $\sqrt[n]{a} = 0$.　　　　$\sqrt[4]{0} = 0$

If a is negative and $\begin{cases} n \text{ is odd, then } \sqrt[n]{a} \text{ is negative.} \\ n \text{ is even, then } \sqrt[n]{a} \text{ is not a real number.} \end{cases}$　　$\sqrt[3]{-8} = -2$
$\sqrt{-4}$ is not a real number.

EXAMPLE 2　Write using rational exponents. Assume that all variables represent positive real numbers.

a. $\sqrt[5]{x}$　**b.** $\sqrt[3]{17x^2y^5}$　**c.** $\sqrt{x - 5a}$　**d.** $3\sqrt{2p} - 5\sqrt[3]{p^2}$

Solution:　**a.** $\sqrt[5]{x} = x^{1/5}$

b. $\sqrt[3]{17x^2y^5} = (17x^2y^5)^{1/3}$

We can further simplify this expression using a power rule for exponents.

$$(17x^2y^5)^{1/3} = 17^{1/3}x^{2/3}y^{5/3}$$

c. $\sqrt{x - 5a} = (x - 5a)^{1/2}$

d. $3\sqrt{2p} - 5\sqrt[3]{p^2} = 3(2p)^{1/2} - 5(p^2)^{1/3} = 3(2p)^{1/2} - 5p^{2/3}$　∎

Next we define $a^{m/n}$ in terms of radicals.

3　In the previous section, we said that $a^{m/n} = (a^m)^{1/n} = (a^{1/n})^m$. Since $a^{1/n} = \sqrt[n]{a}$, we have the following in radical notation:

Radical Notation for $a^{m/n}$

$a^{m/n} = (\sqrt[n]{a})^m$ or $a^{m/n} = \sqrt[n]{a^m}$ as long as these radicals are defined.

EXAMPLE 3　Write the following using radical notation. Assume that all variables represent positive real numbers.

a. $x^{1/5}$　**b.** $5x^{2/3}$　**c.** $3(p + q)^{1/2}$

Solution: **a.** $x^{1/5} = \sqrt[5]{x}$ **b.** $5x^{2/3} = 5\sqrt[3]{x^2}$ or $5x^{2/3} = 5(\sqrt[3]{x})^2$
      c. $3(p + q)^{1/2} = 3\sqrt{p + q}$ ∎

HELPFUL HINT

The **denominator** of a rational exponent is the index of the corresponding radical. For example, $x^{1/5} = \sqrt[5]{x}$ and $z^{2/3} = \sqrt[3]{z^2}$ or $z^{2/3} = (\sqrt[3]{z})^2$.

Notice that every positive number has two square roots. For example, the square roots of 4 are 2 and -2 since $2^2 = 4$ and $(-2)^2 = 4$. We use the notation $\sqrt{4}$ to mean the **positive** or **principal square root** of 4 and $-\sqrt{4}$ to mean the negative square root of 4. Thus

$$\sqrt{4} = 2 \quad \text{and} \quad -\sqrt{4} = -2$$

In general, a^2 has two square roots, a and $-a$, because $(a)^2 = a^2$ and $(-a)^2 = a^2$. Since the notation $\sqrt{a^2}$ indicates the positive square root of a^2 only, we write

$$\sqrt{a^2} = |a|$$

to ensure that the square root is positive. This is true when the index is any even positive integer.

If n is an even positive integer, $\sqrt[n]{a^n} = |a|$.
If n is an odd positive integer, $\sqrt[n]{a^n} = a$.

EXAMPLE 4   Simplify.

      a. $\sqrt{(-3)^2}$ **b.** $\sqrt{x^2}$ **c.** $\sqrt[4]{(x - 2)^4}$ **d.** $\sqrt[3]{(-5)^3}$

Solution: **a.** $\sqrt{(-3)^2} = |-3| = 3$

      b. $\sqrt{x^2} = |x|$

      c. $\sqrt[4]{(x - 2)^4} = |x - 2|$

      d. $\sqrt[3]{(-5)^3} = -5$ ∎

---

4   Next, we state the product rule for radicals, which is the relationship $(ab)^{1/n} = a^{1/n}b^{1/n}$ stated in radical notation.

Product Rule for Radicals

If $\sqrt[n]{a}$ and $\sqrt[n]{b}$ are real numbers, then

$$\sqrt[n]{ab} = \sqrt[n]{a} \cdot \sqrt[n]{b}$$

This rule can be used to simplify radical expressions. We say a radical expression is in simplest form when the following are true.

> **A Radical Is in Simplest Form When:**
>
> **1.** No radicand has a prime factor that appears the same or more times than the index of the radical.
> **2.** No radicand contains fractions.
> **3.** No denominator contains a radical.
> **4.** No factors other than 1 are common to an index of a radical and an exponent contained in the radicand.

Before simplifying radical expressions, you should be familiar with the following terms.

Term	Definition	Examples
Perfect square	Square of a whole number	$2^2 = \mathbf{4}, 5^2 = \mathbf{25}$
Perfect cube	Cube of a whole number	$2^3 = \mathbf{8}, 4^3 = \mathbf{64}$
Perfect fourth power	Whole number raised to the fourth power	$2^4 = \mathbf{16}, 3^4 = \mathbf{81}$

and so on.

EXAMPLE 5 Simplify the following.

　　a. $\sqrt{50}$　**b.** $\sqrt[3]{24}$　**c.** $\sqrt{26}$　**d.** $\sqrt[4]{32}$

Solution: **a.** Factor 50 so that one factor is the largest perfect square that divides 50. The largest perfect square factor of 50 is 25, so we write 50 as $25 \cdot 2$ and use the product rule for radicals to simplify.

$$\sqrt{50} = \sqrt{25 \cdot 2} = \sqrt{25} \cdot \sqrt{2} = 5\sqrt{2}$$

b. Since the index is 3, we find the largest perfect cube factor of 24. We write 24 as $8 \cdot 3$ since 8 is the largest perfect cube factor of 24.

$$\sqrt[3]{24} = \sqrt[3]{8 \cdot 3} = \sqrt[3]{8} \cdot \sqrt[3]{3} = 2\sqrt[3]{3}$$

c. The largest perfect square factor of 26 is 1, so $\sqrt{26}$ cannot be simplified further.

d. $\sqrt[4]{32} = \sqrt[4]{16 \cdot 2} = \sqrt[4]{16} \cdot \sqrt[4]{2} = 2\sqrt[4]{2}$ ■

EXAMPLE 6 Use the product rule to simplify. Assume that variables represent positive numbers.

　　a. $\sqrt{25x^3}$

　　b. $\sqrt[3]{54x^6y^8}$

　　c. $\sqrt[4]{81z^{11}}$

Solution: **a.** $\sqrt{25x^3} = \sqrt{25x^2 \cdot x} = \sqrt{25x^2} \cdot \sqrt{x} = 5x\sqrt{x}$

b. $\sqrt[3]{54x^6y^8} = \sqrt[3]{27x^6y^6 \cdot 2y^2} = \sqrt[3]{27x^6y^6} \cdot \sqrt[3]{2y^2} = 3x^2y^2\sqrt[3]{2y^2}$

c. $\sqrt[4]{81z^{11}} = \sqrt[4]{81z^8 \cdot z^3} = \sqrt[4]{81z^8} \cdot \sqrt[4]{z^3} = 3z^2\sqrt[4]{z^3}$ ■

5   We can also use the quotient rule for radicals to simplify radical expressions. The quotient rule for radicals is the rational exponent rule

$$\left(\frac{a}{b}\right)^{1/n} = \frac{a^{1/n}}{b^{1/n}}$$

stated in radical notation.

Quotient Rule for Radicals

If $\sqrt[n]{a}$ and $\sqrt[n]{b}$ are real numbers and $\sqrt[n]{b}$ is not zero, then

$$\sqrt[n]{\frac{a}{b}} = \frac{\sqrt[n]{a}}{\sqrt[n]{b}}$$

EXAMPLE 7   Use the quotient rule to simplify. Assume that all variables represent positive numbers.

a. $\sqrt{\dfrac{5}{49}}$ **b.** $\sqrt[3]{-\dfrac{8}{27}}$ **c.** $\sqrt{\dfrac{36x^4}{25y^2}}$ **d.** $\sqrt[3]{\dfrac{16x^3}{z^6}}$

Solution: **a.** $\sqrt{\dfrac{5}{49}} = \dfrac{\sqrt{5}}{\sqrt{49}} = \dfrac{\sqrt{5}}{7}$ **b.** $\sqrt[3]{-\dfrac{8}{27}} = \dfrac{\sqrt[3]{-8}}{\sqrt[3]{27}} = \dfrac{-2}{3}$ or $-\dfrac{2}{3}$

c. $\sqrt{\dfrac{36x^4}{25y^2}} = \dfrac{\sqrt{36x^4}}{\sqrt{25y^2}} = \dfrac{6x^2}{5y}$

d. $\sqrt[3]{\dfrac{16x^3}{z^6}} = \dfrac{\sqrt[3]{16x^3}}{\sqrt[3]{z^6}} = \dfrac{\sqrt[3]{8x^3 \cdot 2}}{z^2} = \dfrac{\sqrt[3]{8x^3} \cdot \sqrt[3]{2}}{z^2} = \dfrac{2x\sqrt[3]{2}}{z^2}$ ∎

   CALCULATOR BOX

Approximate Numbers Raised to Fractional Powers:

A calculator can be used to approximate numbers raised to fractional powers. For example, to approximate $34^{1/2}$, recall that $34^{1/2} = \sqrt{34}$. The square root key can then be used. Press $\boxed{34}$ $\boxed{\sqrt{}}$. The display should read $\boxed{5.8309519}$, which is an approximation for $\sqrt{34}$. Then $\sqrt{34} \approx 5.83$ rounded to 2 decimal places.

To approximate $18^{3/4}$, write $\dfrac{3}{4}$ as a decimal and then use the $\boxed{y^x}$ key. Press $\boxed{18}$ $\boxed{y^x}$ $\boxed{(}$ $\boxed{3}$ $\boxed{\div}$ $\boxed{4}$ $\boxed{)}$ $\boxed{=}$. The display will read $\boxed{8.7388519}$. Then $18^{3/4} \approx 8.74$ rounded to 2 decimal places.

Use a calculator to approximate the following. Round your approximations to two decimal places.

1. $\sqrt{4348}$ **2.** $\sqrt[3]{-14}$
3. $60^{1/3}$ **4.** $25^{1/5}$
5. $256^{2/3}$ **6.** $(2.5477)^{7/8}$

EXERCISE SET 7.2

Simplify. Assume that all variables represent positive real numbers. See Example 1.

1. $\sqrt{36}$ **2.** $\sqrt{81}$ **3.** $\sqrt[3]{27}$ **4.** $\sqrt[3]{64}$

5. $\sqrt{49x^6}$ **6.** $\sqrt{4y^2}$ **7.** $\sqrt[3]{-125x^3}$ **8.** $\sqrt[5]{-32y^{10}}$

Write using rational exponents. Assume that all variables represent positive real numbers. See Example 2.

9. $\sqrt{3}$ **10.** $\sqrt[3]{y}$ **11.** $\sqrt[3]{y^5}$ **12.** $\sqrt[4]{x^3}$

13. $\sqrt[5]{4y^7}$ **14.** $\sqrt[4]{11x^5}$ **15.** $\sqrt{(y+1)^3}$ **16.** $\sqrt{(3+y^2)^5}$

17. $2\sqrt{x} - 3\sqrt{y}$ **18.** $4\sqrt{2x} + \sqrt{xy}$

Write using radical notation. Assume that all variables represent positive real numbers. See Example 3.

19. $a^{3/7}$ **20.** $b^{2/17}$ **21.** $2x^{5/3}$ **22.** $(2x)^{5/3}$

23. $(4t)^{-5/6}$ **24.** $4t^{-5/6}$ **25.** $(4x-1)^{3/5}$ **26.** $(3x+2y^2)^{9/5}$

Simplify. Use absolute value bars where necessary. See Example 4.

27. $\sqrt{(-4)^2}$ **28.** $\sqrt[4]{(-7)^4}$ **29.** $\sqrt[5]{x^5}$ **30.** $\sqrt[4]{b^4}$

31. $\sqrt{a^2}$ **32.** $\sqrt[3]{y^3}$ **33.** $\sqrt[6]{(x-4)^6}$ **34.** $\sqrt[4]{(x^2-1)^4}$

Simplify each radical. See Example 5.

35. $\sqrt{32}$ **36.** $\sqrt{27}$ **37.** $\sqrt[3]{192}$ **38.** $\sqrt[3]{108}$

39. $5\sqrt{75}$ **40.** $3\sqrt{8}$

Simplify each radical. Assume that all variables represent positive real numbers. See Example 6.

41. $\sqrt{100x^5}$ **42.** $\sqrt{64y^9}$ **43.** $\sqrt[3]{16y^7}$ **44.** $\sqrt[3]{64y^9}$

45. $\sqrt[4]{a^8b^7}$ **46.** $\sqrt[5]{32z^{12}}$

Simplify each radical. Assume that all variables represent positive real numbers. See Example 7.

47. $\sqrt{\dfrac{3}{25}}$ **48.** $\sqrt{\dfrac{10}{9}}$ **49.** $\sqrt{\dfrac{49}{4x^2}}$ **50.** $\sqrt{\dfrac{81y^{12}}{z^4}}$

51. $\sqrt[3]{\dfrac{y^7}{8x^6}}$ **52.** $\sqrt[3]{\dfrac{-27}{x^9}}$

Simplify each radical. Assume that all variables represent positive real numbers.

53. $\sqrt{121}$ **54.** $\sqrt[3]{125}$ **55.** $\sqrt[3]{8x^3}$ **56.** $\sqrt{16x^8}$

57. $\sqrt{y^5}$ **58.** $\sqrt[3]{y^5}$ **59.** $\sqrt{20}$ **60.** $\sqrt{24}$

61. $\sqrt{25a^2b^3}$ **62.** $\sqrt{9x^5y^7}$ **63.** $\sqrt[3]{-27x^9}$ **64.** $\sqrt[3]{-8a^{21}b^6}$

65. $\sqrt[4]{a^{16}b^4}$ **66.** $\sqrt[4]{x^8y^{12}}$ **67.** $\sqrt[3]{50x^{14}}$ **68.** $\sqrt[3]{40y^{10}}$

69. $\sqrt[5]{-32x^{10}y}$ **70.** $\sqrt[5]{-243z^9}$ **71.** $-\sqrt{32a^8b^7}$ **72.** $-\sqrt{20ab^6}$

73. $\sqrt{\dfrac{6}{49}}$ **74.** $\sqrt{\dfrac{8}{81}}$ **75.** $\sqrt{\dfrac{5x^2}{4y^2}}$ **76.** $\sqrt{\dfrac{y^{10}}{9x^6}}$

77. $-\sqrt[3]{\dfrac{z^7}{27x^3}}$ **78.** $-\sqrt[3]{\dfrac{64a}{b^9}}$ **79.** $\sqrt[4]{\dfrac{x^7}{16}}$ **80.** $\sqrt[4]{\dfrac{y}{81x^4}}$

Writing in Mathematics

81. Explain why absolute value bars are necessary to simplify $\sqrt[4]{a^4}$ as $|a|$.

82. Explain why absolute value bars are unnecessary to simplify $\sqrt[3]{a^3}$ as a.

Skill Review

Evaluate. See Sections 4.1 and 4.2.

83. 5^{-1}

84. 7^{-2}

85. -2^{-3}

86. -3^{-4}

Simplify. Write answers with positive exponents. Assume that all variables represent nonzero real numbers.

87. $x^{-2}x^{-3}$

88. $\dfrac{a^{-2}}{a^{-3}}$

7.3
Adding and Subtracting Radical Expressions

OBJECTIVE   **1**   Add or subtract radical expressions.

Tape 21

1   We have learned that only like terms can be added or subtracted. To add or subtract like terms, we use the distributive property. For example,

$$2x + 3x = (2 + 3)x = 5x \quad \text{and} \quad 7x^2y - 4x^2y = (7 - 4)x^2y = 3x^2y$$

The distributive property can also be used to add **like radicals.**

> **Like Radicals**
>
> Radicals with the same index and the same radicand are like radicals.

For example, $2\sqrt{7} + 3\sqrt{7} = (2 + 3)\sqrt{7} = 5\sqrt{7}$. Also

$$5\sqrt{3x} - 7\sqrt{3x} = (5 - 7)\sqrt{3x} = -2\sqrt{3x}$$

The expression $2\sqrt{7} + 2\sqrt[3]{7}$ cannot be simplifed since $2\sqrt{7}$ and $2\sqrt[3]{7}$ are not like radicals.

EXAMPLE 1   Add or subtract. Assume that variables represent positive real numbers.

a. $\sqrt{20} + 2\sqrt{45}$ **b.** $\sqrt[3]{54} - 5\sqrt[3]{16} + \sqrt[3]{2}$ **c.** $\sqrt{27x} - 2\sqrt{9x} + \sqrt{72x}$

d. $\sqrt[3]{98} + \sqrt{98}$ **e.** $\sqrt[3]{48y^4} + \sqrt[3]{6y^4}$

Solution: First, simplify each radical. Then add or subtract any like radicals.

a. $\sqrt{20} + 2\sqrt{45} = \sqrt{4} \cdot \sqrt{5} + 2 \cdot \sqrt{9} \cdot \sqrt{5}$

$$= 2\sqrt{5} + 2 \cdot 3 \cdot \sqrt{5}$$

$$= 2\sqrt{5} + 6\sqrt{5} = 8\sqrt{5}$$

b. $\sqrt[3]{54} - 5\sqrt[3]{16} + \sqrt[3]{2} = \sqrt[3]{27} \cdot \sqrt[3]{2} - 5 \cdot \sqrt[3]{8} \cdot \sqrt[3]{2} + \sqrt[3]{2}$

$$= 3\sqrt[3]{2} - 5 \cdot 2\sqrt[3]{2} + \sqrt[3]{2}$$

$$= 3\sqrt[3]{2} - 10\sqrt[3]{2} + \sqrt[3]{2}$$

$$= -6\sqrt[3]{2}$$

c. $\sqrt{27x} - 2\sqrt{9x} + \sqrt{72x} = \sqrt{9} \cdot \sqrt{3x} - 2 \cdot \sqrt{9} \cdot \sqrt{x} + \sqrt{36} \cdot \sqrt{2x}$

$$= 3\sqrt{3x} - 2 \cdot 3 \cdot \sqrt{x} + 6\sqrt{2x}$$

$$= 3\sqrt{3x} - 6\sqrt{x} + 6\sqrt{2x}$$

d. We can simplify $\sqrt{98}$, but since the indexes are different, these radicals cannot be added.

$$\sqrt[3]{98} + \sqrt{98} = \sqrt[3]{98} + \sqrt{49} \cdot \sqrt{2} = \sqrt[3]{98} + 7\sqrt{2}$$

e. $\sqrt[3]{48y^4} + \sqrt[3]{6y^4} = \sqrt[3]{8y^3} \cdot \sqrt[3]{6y} + \sqrt[3]{y^3} \cdot \sqrt[3]{6y}$

$$= 2y\sqrt[3]{6y} + y\sqrt[3]{6y}$$

$$= 3y\sqrt[3]{6y} \quad \blacksquare$$

The following summarizes how to simplify radical expressions.

> **To Simplify Radical Expressions**
>
> **1.** Write each radical term in simplest form.
> **2.** Add or subtract any like radicals.

EXAMPLE 2   Simplify.

a. $\dfrac{\sqrt{45}}{4} - \dfrac{\sqrt{5}}{3}$ **b.** $\sqrt[3]{\dfrac{7x}{8}} + 2\sqrt[3]{7x}$

Solution: **a.** $\dfrac{\sqrt{45}}{4} - \dfrac{\sqrt{5}}{3} = \dfrac{3\sqrt{5}}{4} - \dfrac{\sqrt{5}}{3} = \dfrac{3\sqrt{5} \cdot 3}{4 \cdot 3} - \dfrac{\sqrt{5} \cdot 4}{3 \cdot 4} = \dfrac{9\sqrt{5}}{12} - \dfrac{4\sqrt{5}}{12} = \dfrac{5\sqrt{5}}{12}$

b. $\sqrt[3]{\dfrac{7x}{8}} + 2\sqrt[3]{7x} = \dfrac{\sqrt[3]{7x}}{\sqrt[3]{8}} + 2\sqrt[3]{7x} = \dfrac{\sqrt[3]{7x}}{2} + 2\sqrt[3]{7x}$

$$= \dfrac{\sqrt[3]{7x}}{2} + \dfrac{2\sqrt[3]{7x} \cdot 2}{2} = \dfrac{\sqrt[3]{7x}}{2} + \dfrac{4\sqrt[3]{7x}}{2} = \dfrac{5\sqrt[3]{7x}}{2} \quad \blacksquare$$

MENTAL MATH

Simplify. Assume that all variables represent positive real numbers.

1. $2\sqrt{3} + 4\sqrt{3}$ **2.** $5\sqrt{7} + 3\sqrt{7}$ **3.** $8\sqrt{x} - 5\sqrt{x}$

4. $3\sqrt{y} + 10\sqrt{y}$ **5.** $7\sqrt[3]{x} + 5\sqrt[3]{x}$ **6.** $8\sqrt[3]{z} - 2\sqrt[3]{z}$

EXERCISE SET 7.3

Simplify. Assume that all variables represent positive real numbers. See Example 1.

1. $\sqrt{8} - \sqrt{32}$

2. $\sqrt{27} - \sqrt{75}$

3. $2\sqrt{2x^3} + 4x\sqrt{8x}$

4. $3\sqrt{45x^3} + x\sqrt{5x}$

5. $2\sqrt{50} - 3\sqrt{125} + \sqrt{98}$

6. $4\sqrt{32} - \sqrt{18} + 2\sqrt{128}$

7. $\sqrt[3]{16x} - \sqrt[3]{54x}$

8. $2\sqrt[3]{3a^4} - 3a\sqrt[3]{81a}$

9. $\sqrt{9b^3} - \sqrt{25b^3} + \sqrt{49b^3}$

10. $\sqrt{4x^7} + 9x^2\sqrt{x^3} - 5x\sqrt{x^5}$

Simplify. Assume that all variables represent positive rea

11. $\dfrac{5\sqrt{2}}{3} + \dfrac{2\sqrt{2}}{5}$

12. $\dfrac{\sqrt{3}}{2} + \dfrac{4\sqrt{3}}{3}$

13. $\sqrt[3]{\dfrac{11}{8}} - \dfrac{\sqrt[3]{11}}{6}$

14. $\dfrac{2\sqrt[3]{4}}{7} - \dfrac{\sqrt[3]{4}}{14}$

15. $\dfrac{\sqrt{20x}}{9} + \sqrt{\dfrac{5x}{9}}$

16. $\dfrac{3x\sqrt{7}}{5} + \sqrt{\dfrac{7x^2}{100}}$

Simplify. Assume that all variables represent positive real numbers.

17. $7\sqrt{9} - 7 + \sqrt{3}$

18. $\sqrt{16} - 5\sqrt{10} + 7$

19. $2 + 3\sqrt{y^2} - 6\sqrt{y^2} + 5$

20. $3\sqrt{7} - \sqrt[3]{x} + 4\sqrt{7} - 3\sqrt[3]{x}$

21. $3\sqrt{108} - 2\sqrt{18} - 3\sqrt{48}$

22. $-\sqrt{75} + \sqrt{12} - 3\sqrt{3}$

23. $-5\sqrt[3]{625} + \sqrt[3]{40}$

24. $-2\sqrt[3]{108} - \sqrt[3]{32}$

25. $\sqrt{9b^3} - \sqrt{25b^3} + \sqrt{16b^3}$

26. $\sqrt{4x^7y^5} + 9x^2\sqrt{x^3y^5} - 5xy\sqrt{x^5y^5}$

27. $5y\sqrt{8y} + 2\sqrt{50y^3}$

28. $3\sqrt{8x^2y^3} - 2x\sqrt{32y^3}$

29. $\sqrt[3]{54xy^3} - 5\sqrt[3]{2xy^3} + y\sqrt[3]{128x}$

30. $2\sqrt[3]{24x^3y^4} + 4x\sqrt[3]{81y^4}$

31. $6\sqrt[3]{11} + 8\sqrt{11} - 12\sqrt{11}$

32. $3\sqrt[3]{5} + 4\sqrt{5}$

33. $-2\sqrt[4]{x^7} + 3\sqrt[4]{16x^7}$

34. $6\sqrt[3]{24x^3} - 2\sqrt[3]{81x^3} - x\sqrt[3]{3}$

35. $\dfrac{4\sqrt{3}}{3} - \dfrac{\sqrt{12}}{3}$

36. $\dfrac{\sqrt{45}}{10} + \dfrac{7\sqrt{5}}{10}$

37. $\dfrac{\sqrt[3]{8x^4}}{7} + \dfrac{3x\sqrt[3]{x}}{7}$

38. $\dfrac{\sqrt[4]{48}}{5x} - \dfrac{2\sqrt[4]{3}}{10x}$

39. $\sqrt{\dfrac{28}{x^2}} + \sqrt{\dfrac{7}{4x^2}}$

40. $\dfrac{\sqrt{99}}{5x} - \sqrt{\dfrac{44}{x^2}}$

41. $\sqrt[3]{\dfrac{16}{27}} - \dfrac{\sqrt[3]{54}}{6}$

42. $\dfrac{\sqrt[3]{3}}{10} + \sqrt[3]{\dfrac{24}{125}}$

43. $-\dfrac{\sqrt[3]{2x^4}}{9} + \sqrt[3]{\dfrac{250x^4}{27}}$

44. $\dfrac{\sqrt[3]{y^5}}{8} + \dfrac{5y\sqrt[3]{y^2}}{4}$

45. Find the perimeter of the trapezoid.

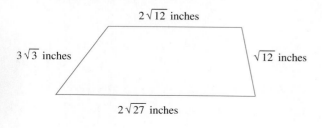

$2\sqrt{12}$ inches

$3\sqrt{3}$ inches $\sqrt{12}$ inches

$2\sqrt{27}$ inches

46. Find the perimeter of the triangle.

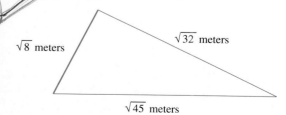

$\sqrt{8}$ meters $\sqrt{32}$ meters

$\sqrt{45}$ meters

47. Baseboard needs to be installed around the perimeter of a rectangular room. Find how much baseboard should be ordered by finding the perimeter of the room.

$3\sqrt{20}$ feet

$\sqrt{125}$ feet

48. A border of wallpaper is to be used around the perimeter of the odd-shaped room shown. Find how much wallpaper border is needed by finding the perimeter of the room.

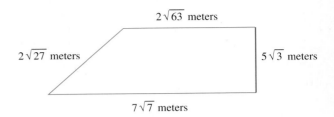

$2\sqrt{63}$ meters

$2\sqrt{27}$ meters

$5\sqrt{3}$ meters

$7\sqrt{7}$ meters

Skill Review

Multiply. See Section 4.4.

49. $(x + 4)(x - 4)$

50. $(3x + 2)(3x - 2)$

51. $(2x + 5)^2$

52. $(5xy - 4)(xy - 11)$

Solve. See Section 6.5.

53. $\dfrac{1}{x - 2} + \dfrac{5}{x + 2} = \dfrac{4}{x^2 - 4}$

54. $\dfrac{8}{x} + \dfrac{5}{x^2 + 3x} = \dfrac{7}{x + 3}$

7.4
Multiplying and Dividing Radical Expressions

Tape 21

OBJECTIVES

1   Multiply radical expressions.

2   Rationalize denominators.

1   In this section we multiply and divide radical expressions. Radical expressions are multiplied using many of the same properties used to multiply polynomial expressions. For instance, to multiply $\sqrt{2}(\sqrt{6} - 3\sqrt{2})$, we use the distributive property and multiply $\sqrt{2}$ by each term inside the parentheses.

$$\sqrt{2}(\sqrt{6} - 3\sqrt{2}) = \sqrt{2}(\sqrt{6}) - \sqrt{2}(3\sqrt{2})$$
$$= \sqrt{12} - 3\sqrt{2 \cdot 2}$$
$$= \sqrt{4 \cdot 3} - 3 \cdot 2$$
$$= 2\sqrt{3} - 6$$

EXAMPLE 1   Multiply.

a. $\sqrt{3}(5 + \sqrt{30})$ **b.** $(\sqrt{5} - \sqrt{6})(\sqrt{7} + 1)$ **c.** $(7\sqrt{x} + 5)(3\sqrt{x} - \sqrt{5})$

d. $(4\sqrt{3} - 1)^2$ **e.** $(\sqrt{2x} - 5)(\sqrt{2x} + 5)$

Solution: **a.** $\sqrt{3}(5 + \sqrt{30}) = \sqrt{3}(5) + \sqrt{3}(\sqrt{30})$

$$= 5\sqrt{3} + \sqrt{3 \cdot 30}$$

$$= 5\sqrt{3} + \sqrt{3 \cdot 3 \cdot 10}$$

$$= 5\sqrt{3} + 3\sqrt{10}$$

b. To multiply, we can use the FOIL method.

$$\overset{\text{First}\qquad\text{Outside}\qquad\text{Inside}\qquad\text{Last}}{(\sqrt{5} - \sqrt{6})(\sqrt{7} + 1) = \sqrt{5} \cdot \sqrt{7} + \sqrt{5} \cdot 1 - \sqrt{6} \cdot \sqrt{7} - \sqrt{6} \cdot 1}$$

$$= \sqrt{35} + \sqrt{5} - \sqrt{42} - \sqrt{6}$$

c. $(7\sqrt{x} + 5)(3\sqrt{x} - \sqrt{5}) = 7\sqrt{x}(3\sqrt{x}) - 7\sqrt{x}(\sqrt{5}) + 5(3\sqrt{x}) - 5(\sqrt{5})$

$$= 21x - 7\sqrt{5x} + 15\sqrt{x} - 5\sqrt{5}$$

d. $(4\sqrt{3} - 1)^2 = (4\sqrt{3} - 1)(4\sqrt{3} - 1)$

$$= 4\sqrt{3}(4\sqrt{3}) - 4\sqrt{3}(1) - 1(4\sqrt{3}) - 1(-1)$$

$$= 16 \cdot 3 - 4\sqrt{3} - 4\sqrt{3} + 1$$

$$= 48 - 8\sqrt{3} + 1$$

$$= 49 - 8\sqrt{3}$$

e. $(\sqrt{2x} - 5)(\sqrt{2x} + 5) = \sqrt{2x} \cdot \sqrt{2x} + 5\sqrt{2x} - 5\sqrt{2x} - 5 \cdot 5$

$$= 2x - 25 \quad\blacksquare$$

2   The radical expression $\dfrac{\sqrt{3}}{\sqrt{2}}$ is not simplified because the denominator contains a radical. The process of removing a radical from the denominator of a radical expression is called **rationalizing the denominator.** To rationalize the denominator of $\dfrac{\sqrt{3}}{\sqrt{2}}$, we use the fundamental principle of fractions and multiply the numerator and the denominator by $\sqrt{2}$.

$$\frac{\sqrt{3}}{\sqrt{2}} = \frac{\sqrt{3} \cdot \sqrt{2}}{\sqrt{2} \cdot \sqrt{2}} = \frac{\sqrt{6}}{\sqrt{4}} = \frac{\sqrt{6}}{2}$$

EXAMPLE 2   Simplify the following. Assume that all variables represent positive integers.

a. $\dfrac{\sqrt{27}}{\sqrt{5}}$ **b.** $\dfrac{2\sqrt{16}}{\sqrt{9x}}$ **c.** $\sqrt[3]{\dfrac{1}{2}}$

Solution: **a.** First, we simplify $\sqrt{27}$; then we rationalize the denominator.

$$\frac{\sqrt{27}}{\sqrt{5}} = \frac{\sqrt{9 \cdot 3}}{\sqrt{5}} = \frac{3\sqrt{3}}{\sqrt{5}}$$

To rationalize the denominator, multiply the numerator and denominator by $\sqrt{5}$.

$$\frac{3\sqrt{3}}{\sqrt{5}} = \frac{3\sqrt{3} \cdot \sqrt{5}}{\sqrt{5} \cdot \sqrt{5}} = \frac{3\sqrt{15}}{5}$$

b. First, we simplify the radicals and then rationalize the denominator.

$$\frac{2\sqrt{16}}{\sqrt{9x}} = \frac{2(4)}{3\sqrt{x}} = \frac{8}{3\sqrt{x}}$$

To rationalize the denominator, multiply the numerator and denominator by $\sqrt{x}$. Then

$$\frac{8}{3\sqrt{x}} = \frac{8 \cdot \sqrt{x}}{3\sqrt{x} \cdot \sqrt{x}} = \frac{8\sqrt{x}}{3x}$$

c. $\sqrt[3]{\frac{1}{2}} = \frac{\sqrt[3]{1}}{\sqrt[3]{2}} = \frac{1}{\sqrt[3]{2}}$. Now we rationalize the denominator. Since this is a cube root, we want to multiply the denominator $\sqrt[3]{2}$ by a value that will make the radicand a perfect cube. If we multiply $\sqrt[3]{2}$ by $\sqrt[3]{2^2}$, we get $\sqrt[3]{2^3} = \sqrt[3]{8} = 2$.

$$= \frac{1 \cdot \sqrt[3]{2^2}}{\sqrt[3]{2} \cdot \sqrt[3]{2^2}} = \frac{\sqrt[3]{4}}{\sqrt[3]{8}} = \frac{\sqrt[3]{4}}{2} \qquad \blacksquare$$

A different method is needed to rationalize a denominator that is a sum or difference of two terms. For example, let's rationalize the denominator of an expression like $\dfrac{5}{\sqrt{3} - 2}$. To eliminate the radical from this denominator, multiply both the numerator and denominator by $\sqrt{3} + 2$, the **conjugate** of $\sqrt{3} - 2$. In general, the conjugate of $a + b$ is $a - b$. Then

$$\frac{5}{\sqrt{3} - 2} = \frac{5\,(\sqrt{3} + 2)}{(\sqrt{3} - 2)\,(\sqrt{3} + 2)}$$

$$= \frac{5(\sqrt{3} + 2)}{\sqrt{3} \cdot \sqrt{3} + 2\sqrt{3} - 2\sqrt{3} - 4}$$

$$= \frac{5(\sqrt{3} + 2)}{3 - 4}$$

$$= \frac{5(\sqrt{3} + 2)}{-1}$$

$$= -5(\sqrt{3} + 2) \quad \text{or} \quad -5\sqrt{3} - 10$$

EXAMPLE 3 Rationalize each denominator.

a. $\dfrac{2}{3\sqrt{2} + 5}$ **b.** $\dfrac{\sqrt{6} + 2}{\sqrt{5} - \sqrt{3}}$ **c.** $\dfrac{7\sqrt{y}}{\sqrt{12x}}$ **d.** $\dfrac{2\sqrt{m}}{3\sqrt{x} + \sqrt{m}}$

Solution: **a.** Multiply the numerator and denominator by the conjugate of $3\sqrt{2} + 5$.

$$\frac{2}{3\sqrt{2} + 5} = \frac{2\,(3\sqrt{2} - 5)}{(3\sqrt{2} + 5)\,(3\sqrt{2} - 5)}$$

$$= \frac{2(3\sqrt{2} - 5)}{(3\sqrt{2})^2 - 5^2}$$

Use the pattern discussed above to multiply $(3\sqrt{2} + 5)$ and $(3\sqrt{2} - 5)$.

$$= \frac{2(3\sqrt{2} - 5)}{18 - 25}$$

$$= \frac{2(3\sqrt{2} - 5)}{-7} \quad \text{or} \quad -\frac{2(3\sqrt{2} - 5)}{7} \quad \text{or} \quad \frac{10 - 6\sqrt{2}}{7}$$

b. Multiply the numerator and denominator by the conjugate of $\sqrt{5} - \sqrt{3}$.

$$\frac{\sqrt{6} + 2}{\sqrt{5} - \sqrt{3}} = \frac{(\sqrt{6} + 2)(\sqrt{5} + \sqrt{3})}{(\sqrt{5} - \sqrt{3})(\sqrt{5} + \sqrt{3})}$$

$$= \frac{\sqrt{6}\sqrt{5} + \sqrt{6}\sqrt{3} + 2\sqrt{5} + 2\sqrt{3}}{(\sqrt{5})^2 - (\sqrt{3})^2}$$

$$= \frac{\sqrt{30} + \sqrt{18} + 2\sqrt{5} + 2\sqrt{3}}{5 - 3}$$

$$= \frac{\sqrt{30} + 3\sqrt{2} + 2\sqrt{5} + 2\sqrt{3}}{2}$$

c. Notice that the denominator of this example is **not the sum or difference of two terms.** For this reason, we simplify the radical expression and then multiply the numerator and denominator by a factor so that the resulting denominator is a rational expression.

$$\frac{7\sqrt{y}}{\sqrt{12x}} = \frac{7\sqrt{y}}{\sqrt{4}\sqrt{3x}} = \frac{7\sqrt{y} \cdot \sqrt{3x}}{2\sqrt{3x} \cdot \sqrt{3x}} = \frac{7\sqrt{3xy}}{2 \cdot 3x} = \frac{7\sqrt{3xy}}{6x}$$

d. Multiply by the conjugate of $3\sqrt{x} + \sqrt{m}$ to eliminate the radicals from the denominator.

$$\frac{2\sqrt{m}}{3\sqrt{x} + \sqrt{m}} = \frac{2\sqrt{m}(3\sqrt{x} - \sqrt{m})}{(3\sqrt{x} + \sqrt{m})(3\sqrt{x} - \sqrt{m})} = \frac{6\sqrt{mx} - 2m}{(3\sqrt{x})^2 - (\sqrt{m})^2}$$

$$= \frac{6\sqrt{mx} - 2m}{9x - m} \quad ∎$$

EXERCISE SET 7.4

Multiply, and then simplify if possible. Assume that all variables represent positive real numbers. See Example 1.

1. $\sqrt{7}(\sqrt{5} + \sqrt{3})$ **2.** $\sqrt{5}(\sqrt{15} - \sqrt{35})$ **3.** $(\sqrt{5} - \sqrt{2})^2$ **4.** $(3x - \sqrt{2})(3x - \sqrt{2})$

5. $\sqrt{3x}(\sqrt{3} - \sqrt{x})$ **6.** $\sqrt{5y}(\sqrt{y} + \sqrt{5})$ **7.** $(2\sqrt{x} - 5)(3\sqrt{x} + 1)$ **8.** $(8\sqrt{y} + z)(4\sqrt{y} - 1)$

9. $(\sqrt[3]{a} - 4)(\sqrt[3]{a} + 5)$ **10.** $(\sqrt[3]{a} + 2)(\sqrt[3]{a} + 7)$

Simplify by rationalizing the denominator. Assume that all variables represent positive numbers. See Example 2.

11. $\frac{1}{\sqrt{3}}$ **12.** $\frac{\sqrt{2}}{\sqrt{6}}$ **13.** $\sqrt{\frac{1}{5}}$ **14.** $\sqrt{\frac{1}{2}}$

15. $\frac{4}{\sqrt[3]{3}}$ **16.** $\frac{6}{\sqrt[3]{9}}$ **17.** $\frac{3}{\sqrt{8x}}$ **18.** $\frac{5}{\sqrt{27a}}$

19. $\dfrac{3}{\sqrt[3]{4x^2}}$ **20.** $\dfrac{5}{\sqrt[3]{3y}}$

Simplify by rationalizing the denominator. Assume that all variables represent positive numbers. See Example 3.

21. $\dfrac{5}{2 - \sqrt{7}}$ **22.** $\dfrac{3}{\sqrt{5} - 2}$

23. $\dfrac{-7}{\sqrt{x} - 3}$ **24.** $\dfrac{-8}{\sqrt{y} + 4}$

25. $\dfrac{\sqrt{2} - \sqrt{3}}{\sqrt{2} + \sqrt{3}}$ **26.** $\dfrac{\sqrt{3} + \sqrt{4}}{\sqrt{2} + \sqrt{3}}$

27. $\dfrac{\sqrt{a} + 1}{2\sqrt{a} - \sqrt{b}}$ **28.** $\dfrac{2\sqrt{a} - 3}{2\sqrt{a} - \sqrt{b}}$

Multiply, and then simplify if possible. Assume that all variables represent positive numbers.

29. $6(\sqrt{2} - 2)$ **30.** $\sqrt{5}(6 - \sqrt{5})$

31. $\sqrt{2}(\sqrt{2} + x\sqrt{6})$ **32.** $\sqrt{3}(\sqrt{3} - 2\sqrt{5x})$

33. $(2\sqrt{7} + 3\sqrt{5})(\sqrt{7} - 2\sqrt{5})$ **34.** $(\sqrt{6} - 4\sqrt{2})(3\sqrt{6} + 1)$

35. $(\sqrt{x} - y)(\sqrt{x} + y)$ **36.** $(3\sqrt{x} + 2)(\sqrt{3x} - 2)$

37. $(\sqrt{3} + x)^2$ **38.** $(\sqrt{y} - 3x)^2$

39. $(\sqrt{5x} - 3\sqrt{2})(\sqrt{5x} - 3\sqrt{3})$ **40.** $(5\sqrt{3x} - \sqrt{y})(4\sqrt{x} + 1)$

41. $(\sqrt[3]{4} + 2)(\sqrt[3]{2} - 1)$ **42.** $(\sqrt[3]{3} + \sqrt[3]{2})(\sqrt[3]{9} - \sqrt[3]{4})$

43. $(\sqrt[3]{x} + 1)(\sqrt[3]{x} - 4\sqrt{x} + 7)$ **44.** $(\sqrt[3]{3x} + 3)(\sqrt[3]{2x} - 3x - 1)$

Simplify by rationalizing the denominator. Assume that all variables represent positive numbers.

45. $\sqrt{\dfrac{2}{5}}$ **46.** $\sqrt{\dfrac{3}{7}}$ **47.** $\dfrac{9}{\sqrt{3a}}$ **48.** $\dfrac{x}{\sqrt{5}}$

49. $\dfrac{3}{\sqrt[3]{2}}$ **50.** $\dfrac{5}{\sqrt[3]{9}}$ **51.** $\dfrac{2\sqrt{3}}{\sqrt{7}}$ **52.** $\dfrac{-5\sqrt{2}}{\sqrt{11}}$

53. $\dfrac{8}{1 + \sqrt{10}}$ **54.** $\dfrac{-3}{\sqrt{6} - 2}$ **55.** $\dfrac{-4}{\sqrt{12x}}$ **56.** $\dfrac{5}{\sqrt{75y}}$

57. $\dfrac{x}{\sqrt[3]{16x^2}}$ **58.** $\dfrac{3a}{\sqrt[3]{2a}}$ **59.** $\sqrt{\dfrac{18x^4y^6}{3z}}$ **60.** $\sqrt{\dfrac{8x^5y}{2z}}$

61. $\dfrac{\sqrt{2} - 1}{\sqrt{2} + 1}$ **62.** $\dfrac{\sqrt{8} - \sqrt{3}}{\sqrt{2} + \sqrt{3}}$ **63.** $\dfrac{\sqrt{x} + 1}{\sqrt{x} - 1}$ **64.** $\dfrac{\sqrt{x} + \sqrt{y}}{\sqrt{x} - \sqrt{y}}$

65. $\dfrac{4\sqrt{x} - y}{\sqrt{x} + \sqrt{y}}$ **66.** $\dfrac{2\sqrt{x} + \sqrt{y}}{\sqrt{2x} - \sqrt{y}}$

67. $\dfrac{\sqrt{x}}{\sqrt{x} + \sqrt{y}}$ **68.** $\dfrac{2\sqrt{a}}{2\sqrt{x} - \sqrt{y}}$

69. $\dfrac{3 + \sqrt{y}}{y - \sqrt{y}}$ **70.** $\dfrac{\sqrt{2} + \sqrt{x}}{\sqrt{2} - \sqrt{x}}$

71. $\dfrac{2\sqrt{27} - \sqrt{3}}{\sqrt{3}}$

72. $\dfrac{3\sqrt{45} + 8\sqrt{20}}{\sqrt{5}}$

73. $\dfrac{4}{\sqrt{x} - 2\sqrt{y}}$

74. $\dfrac{\sqrt{8x}}{x + \sqrt{y}}$

75. $\dfrac{4 - \sqrt{3x}}{\sqrt{7} - \sqrt{x}}$

76. $\dfrac{\sqrt{5} - \sqrt{5m}}{\sqrt{5} - \sqrt{3m}}$

77. Find the volume of the box.

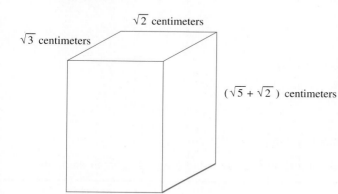

$\sqrt{2}$ centimeters

$\sqrt{3}$ centimeters

$(\sqrt{5} + \sqrt{2})$ centimeters

78. Find the area of the triangle.

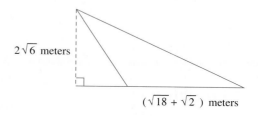

$2\sqrt{6}$ meters

$(\sqrt{18} + \sqrt{2})$ meters

A Look Ahead

Rationalize each numerator. See the following example. Assume that all variables represent positive real numbers.

EXAMPLE   Rationalize the numerator of $\dfrac{\sqrt{x} + 2}{\sqrt{2x}}$.

Solution: Since we wish to rationalize the numerator, we multiply both the numerator and denominator by the conjugate of the numerator, $\sqrt{x} + 2$.

$$\dfrac{\sqrt{x} + 2}{\sqrt{2x}} = \dfrac{(\sqrt{x} + 2)(\sqrt{x} - 2)}{\sqrt{2x}(\sqrt{x} - 2)} = \dfrac{(\sqrt{x})^2 - 2^2}{x\sqrt{2} - 2\sqrt{2x}} = \dfrac{x - 4}{x\sqrt{2} - 2\sqrt{2x}}$$

79. $\dfrac{\sqrt{2}}{5}$

80. $\dfrac{\sqrt{x}}{7}$

81. $\dfrac{\sqrt{5} + 2}{\sqrt{2}}$

82. $\dfrac{2 - \sqrt{7}}{-5}$

83. $\dfrac{\sqrt{x} + 3}{\sqrt{x}}$

84. $\dfrac{5 + \sqrt{2}}{\sqrt{2x}}$

Writing in Mathematics

85. Explain why rationalizing the denominator does not change the value of the original expression.

Skill Review

Solve each equation. See Section 5.5.

86. $(x - 6)(2x + 1) = 0$

87. $(y + 2)(5y + 4) = 0$

88. $x^2 - 8x = -12$

89. $x^3 = x$

Simplify. See Section 6.4.

90. $\dfrac{\dfrac{x}{6}}{\dfrac{2x}{3} + \dfrac{1}{2}}$

91. $\dfrac{\dfrac{1}{y} + \dfrac{4}{5}}{\dfrac{-3}{20}}$

7.5
Radical Equations

Tape 22

OBJECTIVES

1 Solve equations containing radical expressions.

2 Find an unknown side of a right triangle using the Pythagorean theorem.

1 In this section, we present techniques to solve equations containing radical expressions such as

$$\sqrt{2x - 3} = 9$$

We use the power rule to help us solve these radical equations.

> **Power Rule**
>
> If both sides of an equation are raised to the same power, **all** solutions of the original equation are **among** the solutions of the new equation.

This rule **does not** say that raising both sides of an equation to a power yields an equivalent equation. A solution of the new equation **may or may not** be a solution of the original equation. Thus **each solution of the new equation must be checked** to make sure it is a solution of the original equation.

EXAMPLE 1 Solve $\sqrt{2x - 3} = 9$ for x.

Solution: Use the power rule to square both sides of the equation to eliminate the radical.

$$\sqrt{2x - 3} = 9$$
$$(\sqrt{2x - 3})^2 = 9^2$$
$$2x - 3 = 81$$
$$2x = 84$$
$$x = 42$$

Now check the solution in the original equation.

$$\sqrt{2x - 3} = 9$$

$$\sqrt{2(42) - 3} = 9 \qquad \text{Let } x = 42.$$

$$\sqrt{84 - 3} = 9$$

$$\sqrt{81} = 9$$

$$9 = 9 \qquad \text{True.}$$

The solution checks so we conclude that the solution set is {42}. ∎

EXAMPLE 2   Solve $\sqrt{-10x - 1} + 3x = 0$ for x.

Solution: First, isolate the radical on one side of the equation. To do this, we subtract $3x$ from both sides.

$$\sqrt{-10x - 1} + 3x = 0$$

$$\sqrt{-10x - 1} + 3x - 3x = 0 - 3x$$

$$\sqrt{-10x - 1} = -3x$$

Next, use the power rule to eliminate the radical.

$$(\sqrt{-10x - 1})^2 = (-3x)^2$$

$$-10x - 1 = 9x^2$$

Since this is a quadratic equation, set the equation equal to 0 and try to solve by factoring.

$$9x^2 + 10x + 1 = 0$$

$$(9x + 1)(x + 1) = 0 \qquad \text{Factor.}$$

$$9x + 1 = 0 \quad \text{or} \quad x + 1 = 0 \qquad \text{Set each factor equal to 0.}$$

$$x = -\frac{1}{9} \quad \text{or} \quad x = -1$$

Possible solutions are $-\frac{1}{9}$ and -1. Now we check our work in the original equation.

$$\text{Let } x = -\frac{1}{9} \qquad\qquad\qquad\qquad \text{Let } x = -1$$

$$\sqrt{-10x - 1} + 3x = 0 \qquad\qquad \sqrt{-10x - 1} + 3x = 0$$

$$\sqrt{-10\left(-\frac{1}{9}\right) - 1} + 3\left(-\frac{1}{9}\right) = 0 \qquad \sqrt{-10(-1) - 1} + 3(-1) = 0$$

$$\sqrt{\frac{10}{9} - \frac{9}{9}} - \frac{3}{9} = 0 \qquad\qquad \sqrt{10 - 1} - 3 = 0$$

$$\sqrt{\frac{1}{9}} - \frac{1}{3} = 0 \qquad\qquad\qquad \sqrt{9} - 3 = 0$$

$$\frac{1}{3} - \frac{1}{3} = 0 \quad \text{True} \qquad\qquad 3 - 3 = 0 \quad \text{True}$$

Both solutions check. The solution set is $\left\{-\frac{1}{9}, -1\right\}$. ∎

To Solve a Radical Equation

Step 1 Write the equation so that one radical with variables is by itself on one side of the equation.

Step 2 Raise each side of the equation to a power equal to the index of the radical.

Step 3 Add or subtract any like terms.

Step 4 If the equation still contains a radical term, repeat *steps 1* through *3*.

Step 5 Solve the equation.

Step 6 Check all proposed solutions in the original equation for extraneous solutions.

EXAMPLE 3   Solve for x: $\sqrt[3]{x + 1} + 5 = 3$.

Solution: First, we isolate the radical by subtracting 5 from both sides of the equation.

$$\sqrt[3]{x + 1} + 5 = 3$$
$$\sqrt[3]{x + 1} = -2$$

Next, raise both sides of the equation to the third power to eliminate the radical.

$$(\sqrt[3]{x + 1})^3 = (-2)^3$$
$$x + 1 = -8$$
$$x = -9$$

The solution checks in the original equation, so the solution set is $\{-9\}$. ∎

EXAMPLE 4   Solve $\sqrt{4 - x} = x - 2$ for x.

Solution:
$$\sqrt{4 - x} = x - 2$$
$$(\sqrt{4 - x})^2 = (x - 2)^2$$
$$4 - x = x^2 - 4x + 4$$
$$x^2 - 3x = 0 \quad \text{Write the quadratic equation in standard form.}$$
$$x(x - 3) = 0 \quad \text{Factor.}$$
$$x = 0 \quad \text{or} \quad x - 3 = 0$$
$$x = 3$$

Check the possible solutions.

$\sqrt{4 - x} = x - 2$	$\sqrt{4 - x} = x - 2$
$\sqrt{4 - 0} = 0 - 2$ Let $x = 0$.	$\sqrt{4 - 3} = 3 - 2$ Let $x = 3$.
$2 = -2$ False.	$1 = 1$ True.

The proposed solution 3 checks, but 0 does not. When a proposed solution does not check, it is an **extraneous root or solution.** Since 0 is an extraneous solution, the solution set is $\{3\}$. ∎

> **HELPFUL HINT**
>
> In Example 4, notice that $(x - 2)^2 = x^2 - 4x + 4$. Make sure binomials are squared correctly.

EXAMPLE 5   Solve $\sqrt{2x + 5} + \sqrt{2x} = 3$.

Solution: Isolate a radical by subtracting $\sqrt{2x}$ from both sides.

$$\sqrt{2x + 5} + \sqrt{2x} = 3$$

$$\sqrt{2x + 5} = 3 - \sqrt{2x}$$

Use the power rule to begin eliminating the radicals. Square both sides.

$$(\sqrt{2x + 5})^2 = (3 - \sqrt{2x})^2$$

$$2x + 5 = 9 - 6\sqrt{2x} + 2x \qquad \text{Multiply: } (3 - \sqrt{2x})(3 - \sqrt{2x}).$$

There is still a radical in the equation, so isolate the radical again. Then square both sides.

$$2x + 5 = 9 - 6\sqrt{2x} + 2x$$

$$6\sqrt{2x} = 4 \qquad\qquad\qquad \text{Isolate the radical.}$$

$$(6\sqrt{2x})^2 = 4^2 \qquad\qquad \text{Square both sides of the equation}$$
$$\qquad\qquad\qquad\qquad\qquad \text{to eliminate the radical.}$$

$$36(2x) = 16$$

$$72x = 16$$

$$x = \frac{16}{72} \qquad\qquad\qquad \text{Solve.}$$

$$x = \frac{2}{9}$$

The proposed solution, $\frac{2}{9}$, does check in the original equation, so the solution set is $\left\{\frac{2}{9}\right\}$. ∎

> **HELPFUL HINT**
>
> Make sure expressions are squared correctly. In Example 5, we squared $(3 - \sqrt{2x})$.
>
> $$(3 - \sqrt{2x})^2 = (3 - \sqrt{2x})(3 - \sqrt{2x})$$
> $$= 3 \cdot 3 - 3\sqrt{2x} - 3\sqrt{2x} + \sqrt{2x} \cdot \sqrt{2x}$$
> $$= 9 - 6\sqrt{2x} + 2x$$

2 Recall that the Pythagorean theorem states that in a right triangle the length of the hypotenuse squared equals the sum of each of the lengths of the legs squared.

Pythagorean Theorem

If a and b are the lengths of the legs of a right triangle and c is the length of the hypotenuse, then $a^2 + b^2 = c^2$.

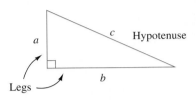

EXAMPLE 6 Find the length of the unknown leg of the following right triangle.

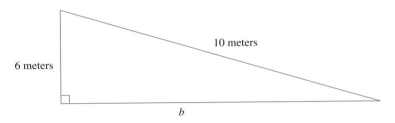

Solution: In the formula $a^2 + b^2 = c^2$, c is the hypotenuse. Let $c = 10$, the length of the hypotenuse, let $a = 6$ and solve for b. Then $a^2 + b^2 = c^2$ becomes

$$6^2 + b^2 = 10^2$$
$$36 + b^2 = 100$$
$$b^2 = 64 \qquad \text{Subtract 36 from both sides.}$$
$$b = 8 \quad \text{and} \quad b = -8 \qquad \text{Because } b^2 = 64.$$

Since we are solving for a length, we will list the positive solution only. The other leg of the triangle is 8 meters long. ■

EXERCISE SET 7.5

Solve. See Example 1.

1. $\sqrt{2x} = 4$

2. $\sqrt{3x} = 3$

3. $\sqrt{x - 3} = 2$

4. $\sqrt{x + 1} = 5$

5. $\sqrt{2x} = -4$

6. $\sqrt{5x} = -5$

Solve. See Example 2.

7. $\sqrt{4x - 3} - 5 = 0$

8. $\sqrt{x - 3} - 1 = 0$

9. $\sqrt{2x - 3} - 2 = 1$

10. $\sqrt{3x + 3} - 4 = 8$

Solve. See Example 3.

11. $\sqrt[3]{6x} = -3$ **12.** $\sqrt[3]{4x} = -2$ **13.** $\sqrt[3]{x-2} - 3 = 0$ **14.** $\sqrt[3]{2x-6} - 4 = 0$

Solve. See Example 4.

15. $\sqrt{13-x} = x - 1$ **16.** $\sqrt{2x-3} = 3 - x$

17. $x - \sqrt{4-3x} = -8$ **18.** $2x + \sqrt{x+1} = 8$

Solve. See Example 5.

19. $\sqrt{y+5} = 2 - \sqrt{y-4}$ **20.** $\sqrt{x+3} + \sqrt{x-5} = 3$

21. $\sqrt{x-3} + \sqrt{x+2} = 5$ **22.** $\sqrt{2x-4} - \sqrt{3x+4} = -2$

Find the length of the unknown side in each triangle. See Example 6.

23.

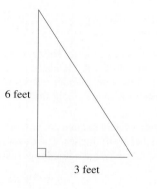

6 feet

3 feet

24.

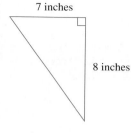

7 inches

8 inches

25.

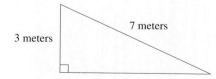

3 meters

7 meters

26.

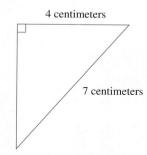

4 centimeters

7 centimeters

Solve.

27. $\sqrt{3x-2} = 5$ **28.** $\sqrt{5x-4} = 9$ **29.** $-\sqrt{2x} + 4 = -6$

30. $-\sqrt{3x+9} = -12$ **31.** $\sqrt{3x+1} + 2 = 0$ **32.** $\sqrt{3x+1} - 2 = 0$

33. $\sqrt[4]{4x+1} - 2 = 0$ **34.** $\sqrt[4]{2x-9} - 3 = 0$ **35.** $\sqrt{4x-3} = 5$

36. $\sqrt{3x+9} = 12$ **37.** $\sqrt[3]{6x-3} - 3 = 0$ **38.** $\sqrt[3]{3x} + 4 = 7$

39. $\sqrt[3]{2x-3} - 2 = -5$ **40.** $\sqrt[3]{x-4} - 5 = -7$ **41.** $\sqrt{x+4} = \sqrt{2x-5}$

42. $\sqrt{3y+6} = \sqrt{7y-6}$ **43.** $x - \sqrt{1-x} = -5$ **44.** $x - \sqrt{x-2} = 4$

45. $\sqrt[3]{-6x-1} = \sqrt[3]{-2x-5}$ **46.** $x + \sqrt{x+5} = 7$ **47.** $\sqrt{5x-1} - \sqrt{x} + 2 = 3$

48. $\sqrt{2x-1} - 4 = -\sqrt{x-4}$ **49.** $\sqrt{2x-1} = \sqrt{1-2x}$ **50.** $\sqrt{3x+4} - 1 = \sqrt{2x+1}$

Find the length of the unknown side of each triangle.

51.

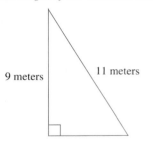

9 meters   11 meters

52.

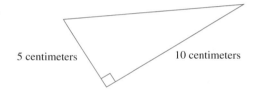

5 centimeters   10 centimeters

53.

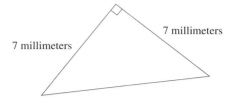

7 millimeters   7 millimeters

54.

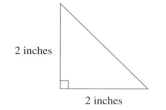

2 inches

2 inches

55. A wire is needed to support a vertical pole 15 feet high. The cable will be anchored to a stake 8 feet from the base of the pole. How much cable is needed?

15 feet

8 feet

56. A spotlight is mounted on the eaves of a house 12 feet

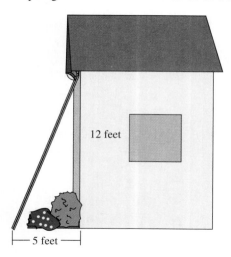

12 feet

5 feet

above the ground. A flower bed runs between the house and the sidewalk, so the closest the ladder can be placed to the house is 5 feet. How long a ladder is needed so that an electrician can reach the place where the light is to be mounted?

57. A furniture upholsterer wished to cut a cover from a piece of fabric that is 45 inches by 45 inches. The cover must be cut on the bias of the fabric. What is the longest strip that can be cut?

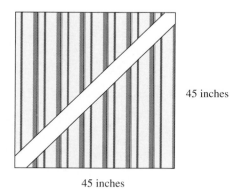

45 inches

45 inches

58. A wire is to be attached to support a telephone pole. Because of surrounding buildings, sidewalks, and roadway, the wire must be anchored exactly 15 feet from the base of the pole. Telephone company workers find they have only 30 feet of cable, and 2 feet of that must be used to attach the cable to the pole and to the stake on the ground. How high from the base of the pole can the wire be attached?

15 feet

59. The cost C in dollars per day to operate a small delivery service is given by $C = 80\sqrt[3]{n} + 500$, where n is the number of deliveries per day. In July, the manager decides that it is necessary to keep delivery costs below $1620.00. Find the greatest number of deliveries this company can make per day and still keep overhead below $1620.00.

60. The formula $v = \sqrt{2gh}$ relates the velocity v, in feet per second, of an object after it falls h feet accelerated

by gravity g feet per second squared. If g is approximately 32 feet per second squared, find how far an object has fallen if its velocity is 80 feet per second.

61. Two tractors are pulling a tree stump from a field. If two forces A and B pull at right angles (90°) to each other, the size of the resulting force R is given by the formula

$$R = \sqrt{A^2 + B^2}$$

If tractor A is exerting 600 pounds of force and the resulting force is 850 pounds, find how much force tractor B is exerting.

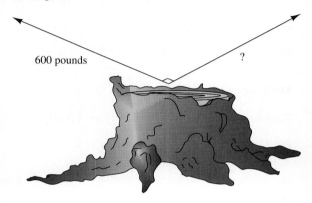

600 pounds   ?

A Look Ahead

Solve. See the following example.

EXAMPLE   Solve $(t^2 - 3t) - 2\sqrt{t^2 - 3t} = 0$.

Solution: Substitution can be used to make this problem somewhat simpler. Let $x = t^2 - 3t$.

$$(t^2 - 3t) - 2\sqrt{t^2 - 3t} = 0$$
$$x - 2\sqrt{x} = 0 \qquad \text{Let } x = t^2 - 3t.$$
$$x = 2\sqrt{x}$$
$$x^2 = (2\sqrt{x})^2$$
$$x^2 = 4x$$
$$x^2 - 4x = 0$$
$$x(x - 4) = 0$$
$$x = 0 \quad \text{or} \quad x - 4 = 0$$
$$x = 4$$

Now we "undo" the substitution.

$x = 0$ Replace x with $t^2 - 3t$.	$x = 4$ Replace x with $t^2 - 3t$.

$$x = 0 \qquad \text{Replace } x \text{ with } t^2 - 3t.$$
$$t^2 - 3t = 0$$
$$t(t - 3) = 0$$
$$t = 0 \quad \text{or} \quad t - 3 = 0$$
$$t = 3$$

$$x = 4 \qquad \text{Replace } x \text{ with } t^2 - 3t.$$
$$t^2 - 3t = 4$$
$$t^2 - 3t - 4 = 0$$
$$(t - 4)(t + 1) = 0$$
$$t - 4 = 0 \quad \text{or} \quad t + 1 = 0$$
$$t = 4 \quad \text{or} \qquad t = -1$$

In this problem we have four possible solutions: 0, 3, 4, and -1. All four solutions check in the original equation, so the solution set is $\{-1, 0, 3, 4\}$.   ∎

62. $3\sqrt{x^2 - 8x} = x^2 - 8x$

63. $\sqrt{(x^2 - x) + 7} = 2(x^2 - x) - 1$

64. $7 - (x^2 - 3x) = \sqrt{(x^2 - 3x) + 5}$

65. $x^2 + 6x = 4\sqrt{x^2 + 6x}$

Skill Review

Find the slope of the line containing each pair of points. See Section 3.3.

66. $(-1, 5), (0, 2)$

67. $(6, 9), (-3, 2)$

68. $(4, 7), (-1, 5)$

69. $(8, -3), (5, 0)$

Use the vertical line test to determine whether each graph represents the graph of a function. See Section 3.5.

70.

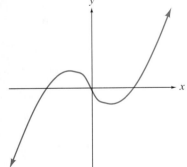

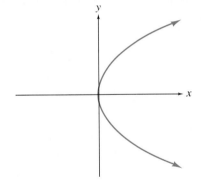

7.6
Complex Numbers

1	Define imaginary and complex numbers.
2	Add or subtract complex numbers.
3	Multiply complex numbers.
4	Divide complex numbers.
5	Raise i to powers.

1   Our work with radical expressions has excluded expressions like $\sqrt{-16}$ because $\sqrt{-16}$ is not a real number: there is no real number whose square is -16. In this section we discuss a number system that includes roots of negative numbers. This number system is the **complex number system** and it includes real numbers. The complex number system allows us to solve equations such as $x^2 + 1 = 0$ that have no real number solutions. The set of complex numbers includes the **imaginary unit.**

Imaginary Unit

The imaginary unit, written i, is the number whose square is -1. That is,

$$i^2 = -1 \quad \text{and} \quad i = \sqrt{-1}.$$

Using i, we can write $\sqrt{-16}$ as

$$\sqrt{-16} = \sqrt{16(-1)} = \sqrt{16}\,\sqrt{-1} = 4i.$$

EXAMPLE 1   Write using i notation.
   a. $\sqrt{-36}$
   b. $\sqrt{-5}$

Solution: **a.** $\sqrt{-36} = \sqrt{36}\,\sqrt{-1} = 6i.$
   b. $\sqrt{-5} = \sqrt{-1(5)} = \sqrt{-1}\,\sqrt{5} = i\sqrt{5}$. Since $\sqrt{5}\,i$ can be easily confused with $\sqrt{5i}$, we write $\sqrt{5}\,i$ as $i\sqrt{5}$.   ∎

Now that we have practiced working with the imaginary unit, complex numbers will be defined.

Complex Numbers

A complex number is a number that can be written in the form $a + bi$, where a and b are real numbers.

The number a is the **real part** of $a + bi$, and the number b is the **imaginary part** of $a + bi$. Notice that the real numbers are a subset of the complex numbers since any real number can be written in the form of a complex number. For example,

$$16 = 16 + 0i$$

In general, a complex number $a + bi$ is a real number if $b = 0$. Also, a complex number is called a **pure imaginary number** if $a = 0$. For example,

$$3i = 0 + 3i \quad \text{and} \quad i\sqrt{7} = 0 + i\sqrt{7}$$

are imaginary numbers.

The following diagram shows the relationship between complex numbers and their subsets. The shaded region represents the set of real numbers.

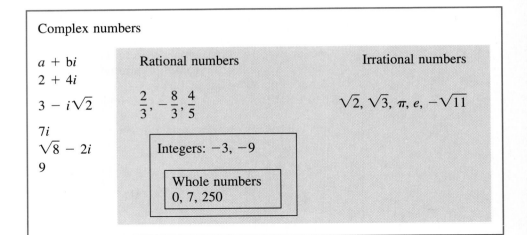

Complex numbers

$a + bi$
$2 + 4i$
$3 - i\sqrt{2}$
$7i$
$\sqrt{8} - 2i$
9

Rational numbers

$\dfrac{2}{3}, -\dfrac{8}{3}, \dfrac{4}{5}$

Integers: $-3, -9$

Whole numbers
$0, 7, 250$

Irrational numbers

$\sqrt{2}, \sqrt{3}, \pi, e, -\sqrt{11}$

2   Two complex numbers $a + bi$ and $c + di$ are equal if and only if $a = c$ and $b = d$. Complex numbers can be added or subtracted by adding or subtracting their real parts and then adding or subtracting their imaginary parts.

Sum or Difference of Complex Numbers

If $a + bi$ and $c + di$ are complex numbers, then their sum is
$$(a + bi) + (c + di) = (a + c) + (b + d)i$$
Their difference is
$$(a + bi) - (c + di) = a + bi - c - di = (a - c) + (b - d)i$$

EXAMPLE 2   Add or subtract the complex numbers. Write the sum or difference in the form $a + bi$.
   a. $(2 + 3i) + (-3 + 2i)$ **b.** $(5i) - (1 - i)$ **c.** $(-3 - 7i) - (-6)$

Solution: **a.** $(2 + 3i) + (-3 + 2i) = (2 - 3) + (3 + 2)i = -1 + 5i$

b. $5i - (1 - i) = 5i - 1 + i$
$$= -1 + (5 + 1)i$$
$$= -1 + 6i$$

c. $(-3 - 7i) - (-6) = -3 - 7i + 6$
$$= (-3 + 6) - 7i$$
$$= 3 - 7i \quad \blacksquare$$

3   We will use the relationship $i^2 = -1$ to simplify when multiplying two complex numbers.

EXAMPLE 3   Multiply the complex numbers. Write the product in the form $a + bi$.
   a. $(2 - 5i)(4 + i)$ **b.** $(2 - i)^2$ **c.** $(7 + 3i)(7 - 3i)$

Solution: Multiply complex numbers as though they were binomials.

a. $(2 - 5i)(4 + i) = 2(4) + 2(i) - 5i(4) - 5i(i)$

$$= 8 + 2i - 20i - 5i^2$$

$$= 8 - 18i - 5(-1) \qquad \text{Let } i^2 = -1.$$

$$= 8 - 18i + 5$$

$$= 13 - 18i$$

b. $(2 - i)^2 = (2 - i)(2 - i)$

$$= 2(2) - 2(i) - 2(i) + i^2$$

$$= 4 - 4i + (-1) \qquad \text{Let } i^2 = -1.$$

$$= 3 - 4i$$

c. $(7 + 3i)(7 - 3i) = 7(7) - 7(3i) + 3i(7) - 3i(3i)$

$$= 49 - 21i + 21i - 9i^2$$

$$= 49 - 9(-1) \qquad \text{Let } i^2 = -1.$$

$$= 49 + 9$$

$$= 58 \quad \blacksquare$$

From Example 3c, notice that the product of $7 + 3i$ and $7 - 3i$ is a real number. These two complex numbers are called **complex conjugates** of one another. In general, we have the following:

Complex Conjugates

The complex numbers $(a + bi)$ and $(a - bi)$ are called **complex conjugates** of each other, and $(a + bi)(a - bi) = a^2 + b^2$.

To see that the product of a complex number $a + bi$ and its conjugate $a - bi$ is the real number $a^2 + b^2$, we multiply.

$$(a + bi)(a - bi) = a^2 - abi + abi - b^2i^2$$

$$= a^2 - b^2(-1)$$

$$= a^2 + b^2$$

4   We use complex conjugates to divide by a complex number.

EXAMPLE 4   Find the quotient. Write in the form $a + bi$.

a. $\dfrac{2 + i}{1 - i}$

b. $\dfrac{7}{3i}$

Solution: **a.** Multiply the numerator and denominator by the complex conjugate of $1 - i$ to eliminate the imaginary number in the denominator.

$$\frac{2 + i}{1 - i} = \frac{(2 + i)(1 + i)}{(1 - i)(1 + i)}$$

$$= \frac{2(1) + 2(i) + 1(i) + i^2}{1^2 - i^2}$$

$$= \frac{2 + 3i - 1}{1 + 1}$$

$$= \frac{1 + 3i}{2} \quad \text{or} \quad \frac{1}{2} + \frac{3}{2}i$$

To check that $\frac{2 + i}{1 - i} = \frac{1}{2} + \frac{3}{2}i$, multiply $\left(\frac{1}{2} + \frac{3}{2}i\right)(1 - i)$ to verify that the product is $2 + i$.

b. Multiply the numerator and denominator by the conjugate of $3i$. Note that $3i = 0 + 3i$, so its conjugate is $0 - 3i$ or $-3i$.

$$\frac{7}{3i} = \frac{7(-3i)}{(3i)(-3i)} = \frac{-21i}{-9i^2} = \frac{-21i}{-9(-1)} = \frac{-21i}{9} = \frac{-7i}{3} \quad \text{or} \quad 0 - \frac{7}{3}i \quad \blacksquare$$

The product rule for radicals does not necessarily hold true for imaginary numbers. **To multiply imaginary numbers, each must be in complex form bi.** For example, to multiply $\sqrt{-4}$ and $\sqrt{-9}$, first write each number in the form bi.

$$\sqrt{-4}\,\sqrt{-9} = 2i(3i) = 6i^2 = -6$$

EXAMPLE 5 Multiply or divide the following as indicated.

a. $\sqrt{-3}\,\sqrt{-5}$ **b.** $\sqrt{-36}\,\sqrt{-1}$ **c.** $\sqrt{8}\,\sqrt{-2}$ **d.** $\dfrac{\sqrt{-125}}{\sqrt{5}}$

Solution: Write each imaginary number in the form bi first.

a. $\sqrt{-3}\,\sqrt{-5} = i\sqrt{3}\,(i\sqrt{5}) = i^2\sqrt{15} = -\sqrt{15}$

b. $\sqrt{-36}\,\sqrt{-1} = 6i(i) = 6i^2 = 6(-1) = -6$

c. $\sqrt{8}\,\sqrt{-2} = 2\sqrt{2}\,(i\sqrt{2}) = 2i(\sqrt{2}\,\sqrt{2}) = 2i(2) = 4i$

d. $\dfrac{\sqrt{-125}}{\sqrt{5}} = \dfrac{i\sqrt{125}}{\sqrt{5}} = i\sqrt{25} = 5i$ $\blacksquare$

5 We can use the fact that $i^2 = 1$ to find higher powers of i. To find i^3, we rewrite it as the product of i^2 and i.

$$i^3 = i^2 \cdot i = (-1)i = -i$$

Continue this process and find higher powers of i.

$$i^4 = i^2 \cdot i^2 = (-1)(-1) = 1$$

$$i^5 = i^4 \cdot i = 1 \cdot i = i$$

If we continue finding powers of i, we generate a pattern.

$$i^1 = i \qquad i^5 = i \qquad i^9 = i$$
$$i^2 = -1 \qquad i^6 = -1 \qquad i^{10} = -1$$
$$i^3 = -i \qquad i^7 = -i \qquad i^{11} = -i$$
$$i^4 = 1 \qquad i^8 = 1 \qquad i^{12} = 1$$

The values i, -1, $-i$, and 1 repeat as i is raised to higher and higher powers. This pattern allows us to find other powers of i.

EXAMPLE 6 Find the following powers of i.
a. i^7 b. i^{20} c. i^{46} d. i^{-12}

Solution: a. $i^7 = i^4 \cdot i^3 = 1(-i) = -i$ b. $i^{20} = (i^4)^5 = 1^5 = 1$
c. $i^{46} = i^{44} \cdot i^2 = (i^4)^{11} \cdot i^2 = 1^{11}(-1) = -1$
d. $i^{-12} = \dfrac{1}{i^{12}} = \dfrac{1}{(i^4)^3} = \dfrac{1}{(1)^3} = \dfrac{1}{1} = 1$ ∎

MENTAL MATH

Simplify. See Example 1.

1. $\sqrt{-81}$ **2.** $\sqrt{-49}$ **3.** $\sqrt{-7}$ **4.** $\sqrt{-3}$
5. $-\sqrt{16}$ **6.** $-\sqrt{4}$ **7.** $\sqrt{-64}$ **8.** $\sqrt{-100}$

EXERCISE SET 7.6

Simplify. See Example 1.

1. $\sqrt{-24}$ **2.** $\sqrt{-32}$ **3.** $-\sqrt{-36}$ **4.** $-\sqrt{-121}$
5. $8\sqrt{-63}$ **6.** $4\sqrt{-20}$ **7.** $-\sqrt{54}$ **8.** $\sqrt{-63}$

Add or subtract. Write the sum or difference in the form $a + bi$. See Example 2.

9. $(4 - 7i) + (2 + 3i)$ **10.** $(2 - 4i) - (2 - i)$ **11.** $(6 + 5i) - (8 - i)$
12. $(8 - 3i) + (-8 + 3i)$ **13.** $6 - (8 + 4i)$ **14.** $(9 - 4i) - 9$

Multiply. Write the product in the form $a + bi$. See Example 3.

15. $6i(2 - 3i)$ **16.** $5i(4 - 7i)$ **17.** $(\sqrt{3} + 2i)(\sqrt{3} - 2i)$
18. $(\sqrt{5} - 5i)(\sqrt{5} + 5i)$ **19.** $(4 - 2i)^2$ **20.** $(6 - 3i)^2$

Write each quotient in the form $a + bi$. See Example 4.

21. $\dfrac{4}{i}$ **22.** $\dfrac{5}{6i}$ **23.** $\dfrac{7}{4 + 3i}$

24. $\dfrac{9}{1 - 2i}$

25. $\dfrac{3 + 5i}{1 + i}$

26. $\dfrac{6 + 2i}{4 - 3i}$

Multiply or divide. See Example 5.

27. $\sqrt{-2} \cdot \sqrt{-7}$

28. $\sqrt{-11} \cdot \sqrt{-3}$

29. $\sqrt{-5} \cdot \sqrt{-10}$

30. $\sqrt{-2} \cdot \sqrt{-6}$

31. $\sqrt{16} \cdot \sqrt{-1}$

32. $\sqrt{3} \cdot \sqrt{-27}$

33. $\dfrac{\sqrt{-9}}{\sqrt{3}}$

34. $\dfrac{\sqrt{49}}{\sqrt{-10}}$

35. $\dfrac{\sqrt{-80}}{\sqrt{-10}}$

36. $\dfrac{\sqrt{-40}}{\sqrt{-8}}$

Find each power of i. See Example 6.

37. i^8

38. i^{10}

39. i^{21}

40. i^{15}

41. i^{11}

42. i^{40}

43. i^{-6}

44. i^{-9}

Perform the indicated operation. Write the result in the form $a + bi$.

45. $(7i)(-9i)$

46. $(-6i)(-4i)$

47. $(6 - 3i) - (4 - 2i)$

48. $(-2 - 4i) - (6 - 8i)$

49. $(6 - 2i)(3 + i)$

50. $(2 - 4i)(2 - i)$

51. $(8 - \sqrt{-3}) - (2 + \sqrt{-12})$

52. $(8 - \sqrt{-4}) - (2 + \sqrt{-16})$

53. $(1 - i)(1 + i)$

54. $(6 + 2i)(6 - 2i)$

55. $\dfrac{16 + 15i}{-3i}$

56. $\dfrac{2 - 3i}{-7i}$

57. $(9 + 8i)^2$

58. $(4 - 7i)^2$

59. $\dfrac{2}{3 + i}$

60. $\dfrac{5}{3 - 2i}$

61. $(5 - 6i) - 4i$

62. $(6 - 2i) + 7i$

63. $\dfrac{2 - 3i}{2 + i}$

64. $\dfrac{6 + 5i}{6 - 5i}$

65. $(2 + 4i) + (6 - 5i)$

66. $(5 - 3i) + (7 - 8i)$

Writing in Mathematics

67. Describe how to find the conjugate of a complex number.

68. Explain why the product of a complex number and its complex conjugate is a real number.

Skill Review

The sum of the measures of the angles of a triangle is 180°.

Find the unknown angle in each triangle.

69.

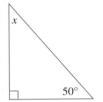

70.

71.

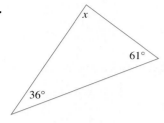

72.

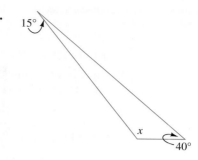

Solve each compound inequality. See Section 2.6.

73. $-9 \leq 2x + 3 \leq 7$

74. $-2 < \dfrac{x}{2} - 1 < 1$

> **CRITICAL THINKING**   Peter's uncle Dennis demonstrated great faith in his nephew's abilities when he loaned him $10,000 to help finance a new business. Now, four years and 6 months later, that faith has been justified: the business is a success and Peter is prepared to pay his uncle back.
>
> Dennis loaned the money stipulating that Peter must pay 8% interest compounded annually. Thus, for each year the loan was outstanding, Peter pays 8%, not only on the original $10,000, but also on the interest due from the preceding years the loan was outstanding. Peter wonders, though, what to do about the extra 6 months. He figures his uncle is due some amount of interest money for the extra 6 months, but how much is not clear.
>
> What reasonable ways can you suggest for evaluating the interest Peter owes for the extra 6 months?

CHAPTER 7 GLOSSARY

The **conjugate** of $a + b$ is $a - b$.

The complex numbers $(a + bi)$ and $(a - bi)$ are called **complex conjugates** of each other, and their product $(a + bi)(a - bi) = a^2 + b^2$.

A **complex number** is a number that can be written in the form $a + bi$, where a and b are real numbers.

The **imaginary unit**, i, is the number whose square is -1. That is, $i^2 = -1$ and $i = \sqrt{-1}$.

The positive integer n in $\sqrt[n]{}$ is called the **index**.

Radicals with the same index and the same radicand are **like radicals**.

The expression $\sqrt[n]{a}$ is called a **radical expression**.

The symbol $\sqrt{}$ is called the **radical sign**.

The expression under the radical sign is called the **radicand**.

CHAPTER 7 SUMMARY

DEFINITION OF $a^{1/n}$ IF a IS POSITIVE   (7.1)

If n is a positive integer and $a \geq 0$, then $a^{1/n}$ is the nonnegative real number whose nth power is a. In symbols, this is

$$(a^{1/n})^n = a$$

DEFINITION OF $a^{1/n}$ IF a IS NEGATIVE   (7.1)

If n is an odd positive integer and a is a negative number, then $a^{1/n}$ is the real number whose nth power is a. In symbols, this is

$$(a^{1/n})^n = a$$

If n is an even positive integer and a is a negative number, then $a^{1/n}$ is not a real number.

DEFINITION OF $a^{m/n}$ (7.1)

If m and n are positive integers with m/n in lowest terms, then

$$a^{m/n} = (a^{1/n})^m = (a^m)^{1/n}$$

as long as all roots are real numbers.

DEFINITION OF $a^{-m/n}$ (7.1)

$$a^{-m/n} = \frac{1}{a^{m/n}}$$

as long as $a^{m/n}$ is a nonzero real number.

RADICAL NOTATION   (7.2)

If n is a positive integer and $a^{1/n}$ is a real number, then

$$a^{1/n} = \sqrt[n]{a}$$

RADICAL NOTATION FOR $a^{m/n}$ (7.2)

$a^{m/n} = (\sqrt[n]{a})^m$ or $a^{m/n} = \sqrt[n]{a^m}$ as long as these radicals are defined.

PRODUCT RULE FOR RADICALS   (7.2)

If $\sqrt[n]{a}$ and $\sqrt[n]{b}$ are real numbers, then

$$\sqrt[n]{ab} = \sqrt[n]{a} \cdot \sqrt[n]{b}$$

QUOTIENT RULE FOR RADICALS

If $\sqrt[n]{a}$ and $\sqrt[n]{b}$ are real numbers and $\sqrt[n]{b}$ is not zero, then

$$\sqrt[n]{\frac{a}{b}} = \frac{\sqrt[n]{a}}{\sqrt[n]{b}}$$

POWER RULE   (7.5)

If both sides of an equation are raised to the same power, **all** solutions of the original equation are **among** the solutions of the new equation.

CHAPTER 7 REVIEW

(7.1) *Evaluate the following.*

1. $\left(\dfrac{1}{81}\right)^{1/4}$

2. $\left(-\dfrac{1}{27}\right)^{1/3}$

3. $(-27)^{-1/3}$

4. $(-64)^{-1/3}$

5. $-9^{3/2}$

6. $64^{-1/3}$

7. $(-25)^{5/2}$

8. $\left(\dfrac{25}{49}\right)^{-3/2}$

9. $\left(\dfrac{8}{27}\right)^{-2/3}$

10. $\left(-\dfrac{1}{36}\right)^{-1/4}$

Simplify each expression. Assume that all variables represent positive numbers. Write using only positive exponents.

11. $a^{1/3}a^{4/3}a^{1/2}$

12. $\dfrac{b^{1/3}}{b^{4/3}}$

13. $(a^{-1/2})^{-2}$

14. $(x^{1/2}x^{3/4})^{-2}$

15. $(a^{1/2}a^{-2})^3$

16. $(a^8b^4c^4)^{3/4}$

17. $(x^{-3}y^6)^{1/3}$

18. $\left(\dfrac{b^{3/4}}{a^{-1/2}}\right)^8$

19. $\dfrac{x^{1/4}x^{-1/2}}{x^{2/3}}$

20. $\left(\dfrac{49c^{5/3}}{a^{-1/4}b^{5/6}}\right)^{-3/2}$

21. $\dfrac{(x^{-2}y^4)^{1/2}}{(x^{1/2})^4}$

22. $a^{-1/4}(a^{5/4} - a^{9/4})$

23. $x^{4/3}(x^{2/3} + x^{-1/3})$

(7.2) *Take the root. Assume that all variables represent positive numbers.*

24. $\sqrt{81}$

25. $\sqrt[4]{81}$

26. $\sqrt[3]{-8}$

27. $\sqrt[4]{-16}$

28. $-\sqrt{\dfrac{1}{49}}$

29. $\sqrt{x^{64}}$

30. $-\sqrt{36}$

31. $\sqrt[3]{64}$

Write using rational exponents.

32. $4x\sqrt[3]{(3x)^5}$

33. $\sqrt[5]{5x^2y^3}$

Write using radical notation.

34. $7x(3^{1/3}y^{2/3})$

35. $5(xy^2z^5)^{1/3}$

Simplify. Use absolute value bars when necessary.

36. $\sqrt{(-x)^2}$

37. $\sqrt[4]{(x^2 - 4)^4}$

38. $\sqrt[3]{(-27)^3}$

39. $\sqrt[5]{(-5)^5}$

40. $-\sqrt[5]{x^5}$

41. $\sqrt[4]{-x^4}$

42. $\sqrt[8]{-y^8}$

43. $\sqrt[5]{-y^5}$

44. $\sqrt[9]{-x^9}$

Simplify. Assume that all variables represent positive numbers.

45. $\sqrt[3]{-a^6b^9}$

46. $\sqrt{16a^4b^{12}}$

47. $\sqrt[5]{32a^5b^{10}}$

48. $\sqrt[5]{-32x^{15}y^{20}}$

(7.3) *Perform the indicated operation. Assume that all variables represent positive numbers.*

49. $x\sqrt{75xy} - \sqrt{27x^3y}$

50. $2\sqrt{32x^2y^3} - xy\sqrt{98y}$

51. $7ab\sqrt{a^3b^3} + 2ab\sqrt{a^3b^3}$

52. $\sqrt[3]{128} + \sqrt[3]{250}$

53. $2a\sqrt[4]{16ab^5} + 3b\sqrt[4]{256a^5b}$

54. $3\sqrt[4]{32a^5} - a\sqrt[4]{162a}$

55. $2\sqrt{50} - 3\sqrt{125} + \sqrt{98}$

56. $3\sqrt{108} - 2\sqrt{18} - 3\sqrt{48}$

57. $2a\sqrt[4]{32b^5} - 3b\sqrt[4]{162a^4b} + \sqrt[4]{2a^4b^5}$

58. $6y\sqrt[4]{48x^5} - 2x\sqrt[4]{243xy^4} - 4\sqrt[4]{3x^5y^4}$

(7.4) *Multiply. Assume that all variables represent positive numbers.*

59. $\sqrt{3}(\sqrt{27} - \sqrt{3})$

60. $\sqrt{10}(\sqrt{10} - \sqrt{5})$

61. $(\sqrt{x} - 3)^2$

62. $(\sqrt{2x} + 4)^2$

63. $(\sqrt{2} - 3)(\sqrt{2} + 4)$

64. $(\sqrt{5} - 5)(2\sqrt{5} + 2)$

65. $(2\sqrt{x} - 3\sqrt{y})(2\sqrt{x} + 3\sqrt{y})$

66. $(2\sqrt{3x} - \sqrt{y})(2\sqrt{3x} + \sqrt{y})$

67. $(\sqrt{a} - 2)(\sqrt{a} - 3)$

68. $(\sqrt{x} + 4)(\sqrt{x} - 7)$

69. $(\sqrt[3]{a} + 2)^2$

70. $(\sqrt[3]{x} - 4)^2$

71. $(2\sqrt{x} - \sqrt{y})(3\sqrt{x} + \sqrt{y})$

72. $(\sqrt{3x} - 2\sqrt{y})(3\sqrt{3x} - \sqrt{y})$

Simplify by rationalizing the denominator. Assume that all variables represent positive numbers.

73. $\sqrt{\dfrac{x}{5}}$

74. $\sqrt{\dfrac{y}{2}}$

75. $\dfrac{\sqrt{40x^3y^2}}{\sqrt{80x^2y^3}}$

76. $\dfrac{\sqrt{15a^2b^5}}{\sqrt{30a^5b^3}}$

77. $\sqrt[3]{\dfrac{15x^6y^7}{z^2}}$

78. $\sqrt[3]{\dfrac{32y^{12}z^{10}}{2x}}$

79. $\dfrac{5}{2 - \sqrt{7}}$

80. $\dfrac{3}{\sqrt{y} - 2}$

81. $\dfrac{\sqrt{2} - \sqrt{3}}{\sqrt{2} + \sqrt{3}}$

82. $\dfrac{2\sqrt{27} - \sqrt{3}}{\sqrt{3}}$

(7.5) *Solve each equation for the variable.*

83. $\sqrt[3]{4x} = -2$

84. $\sqrt{5x} = -5$

85. $\sqrt[3]{x - 2} = 3$

86. $\sqrt[3]{2x - 6} = 4$

87. $\sqrt{2x + 1} = -4$

88. $\sqrt[4]{2x - 9} = 3$

89. $\sqrt[3]{3x - 9} - \sqrt[3]{2x + 12} = 0$

90. $\sqrt[3]{x - 12} - \sqrt[3]{5x + 16} = 0$

91. $\sqrt{x + 6} = \sqrt{x + 2}$

92. $2x - 5\sqrt{x} = 3$

Find each unknown length.

93.

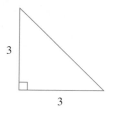

3

3

94.

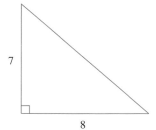

7

8

95.
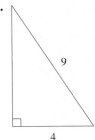

96.
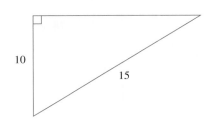

(7.6)   *Perform the indicated operation and simplify. Write the result in the form a + bi.*

97. $\sqrt{-8}$

98. $-\sqrt{-6}$

99. $\sqrt{-4} + \sqrt{-16}$

100. $\sqrt{-25} - 2i$

101. $(12 - 6i) + (3 + 2i)$

102. $(-8 - 7i) - (5 - 4i)$

103. $(\sqrt{3} + \sqrt{2}) + (3\sqrt{2} - \sqrt{-8})$

104. $(3 - \sqrt{-72}) + (4 - \sqrt{-32})$

105. $2i(2 - 5i)$

106. $-3i(6 - 4i)$

107. $(3 + 2i)(1 + i)$

108. $(2 - 3i)(4 + \sqrt{-4})$

109. $\dfrac{2 + 3i}{2i}$

110. $\dfrac{1 + i}{-3i}$

CHAPTER 7 TEST

Raise to the power or take the root. Assume that all variables represent positive numbers. Write using only positive exponents.

1. $\sqrt[3]{-216}$

2. $-\sqrt[4]{x^{64}}$

3. $\left(\dfrac{1}{125}\right)^{1/3}$

4. $\left(\dfrac{1}{125}\right)^{-1/3}$

5. $\left(\dfrac{8x^3}{27}\right)^{2/3}$

6. $\sqrt[3]{-a^{18}b^9}$

7. $\left(\dfrac{64c^{4/3}}{a^{-2/3}b^{5/6}}\right)^{1/2}$

8. $a^{-2/3}(a^{5/4} - a^3)$

Take the root. Use absolute value bars when necessary.

9. $\sqrt[4]{(4xy)^4}$

10. $\sqrt[3]{(-27)^3}$

Simplify by rationalizing the denominator. Assume that all variables represent positive numbers.

11. $\sqrt{\dfrac{9}{y}}$

12. $\dfrac{\sqrt{15x^2y}}{\sqrt{30xy^2}}$

13. $\dfrac{4 - \sqrt{x}}{4 + 2\sqrt{x}}$

14. $\dfrac{\sqrt[3]{ab}}{\sqrt[3]{ab^2}}$

Perform the indicated operations. Assume that all variables represent positive numbers.

15. $\sqrt{125x^3} - 3\sqrt{20x^3}$

16. $a\sqrt[4]{32ab^2} - a\sqrt[4]{2ab^2} + a\sqrt[4]{162ab^2}$

17. $\sqrt{3}(\sqrt{16} - \sqrt{2})$

18. $(\sqrt{x} + 1)^2$

19. $(\sqrt{2} - 4)(\sqrt{3} + 1)$

20. $(\sqrt{5} + 5)(\sqrt{5} - 5)$

Solve.

21. $x = \sqrt{x - 2} + 2$

22. $\sqrt{x^2 - 7} + 3 = 0$

23. $\sqrt{x + 5} = \sqrt{2x - 1}$

Perform the indicated operation and simplify. Write the result in the form a + bi.

24. $\sqrt{-2}$

25. $-\sqrt{-8}$

26. $(12 - 6i) - (12 - 3i)$

27. $(6 - 2i)(6 + 2i)$

28. $(4 + 3i)^2$

29. $\dfrac{1 + 4i}{1 - i}$

30. Find x.

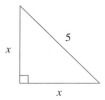

31. Find y.

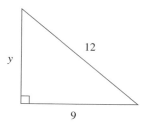

CHAPTER 7 CUMULATIVE REVIEW

1. Solve for x: $3x + 5 = 3(x + 2)$.

2. Solve $|3x + 2| = |5x - 8|$.

3. Solve for x: $|5x + 1| + 1 \leq 10$.

4. If L_1 is a line through $(3, 2)$ and $(8, 5)$ and L_2 is a line through $(5, 7)$ and $(0, 4)$, determine whether L_1 and L_2 are parallel.

5. Graph the union: $x + \dfrac{1}{2}y \geq -4$ or $y \leq -2$.

6. Find each product:
  a. $2^2 \cdot 2^5$
  b. $x^7 x^3$
  c. $y \cdot y^2 \cdot y^4$

7. Simplify the following. Write answers using positive exponents:
  a. $(5x^2)^3$
  b. $\left(\dfrac{2}{3}\right)^3$
  c. $\left(\dfrac{3p^4}{q^5}\right)^2$
  d. $\left(\dfrac{2^{-3}}{y}\right)^{-2}$
  e. $(x^{-5}y^2z^{-1})^7$

8. Simplify the expression. Assume that a and b are nonzero integers and that x and y are not 0.
  a. $x^{-b}(2x^b)^2$
  b. $\dfrac{(y^{3a})^2}{y^{a-6}}$

9. Find the following products:
  a. $2x(5x - 4)$
  b. $-3x^2(4x^2 - 6x + 1)$
  c. $-xy(7x^2y + 3xy - 11)$

10. Find the quotient: $\dfrac{3x^5y^7 - 15x^3y - 6x}{6x^2y^3}$.

11. Find the GCF of $20x^3y$, $10x^2y^2$, and $35x^3$.

12. Factor $2x^2 + 11x + 15$.

13. Factor $y^3 - 64$.

14. Write each rational expression in lowest terms:
  a. $\dfrac{2 + x}{x + 2}$
  b. $\dfrac{2 - x}{x - 2}$
  c. $\dfrac{18 - 2x^2}{x^2 - 2x - 3}$

15. Divide:
  a. $\dfrac{3x}{5y} \div \dfrac{9y}{x^2}$
  b. $\dfrac{8m^2}{3m^2 - 12} \div \dfrac{40}{2 - m}$

16. Simplify: $\dfrac{x^{-1} + 2xy^{-1}}{x^{-2} - x^{-2}y^{-1}}$:

17. Solve $\dfrac{2x}{x - 3} + \dfrac{6 - 2x}{x^2 - 9} = \dfrac{x}{x + 3}$.

18. Evaluate the following:
  a. $4^{3/2}$ **b.** $-16^{3/4}$ **c.** $(-16)^{3/4}$
  d. $(-27)^{2/3}$ **e.** $\left(\dfrac{1}{9}\right)^{3/2}$

19. Write the following using radical notation. Assume that all variables represent positive real numbers.

   a. $x^{1/5}$ **b.** $5x^{2/3}$ **c.** $3(p + q)^{1/2}$

20. Simplify the following. Assume that all variables represent positive real numbers.

   a. $\sqrt{20} + 2\sqrt{45}$

   b. $\sqrt[3]{54} - 5\sqrt[3]{16} + \sqrt[3]{2}$

   c. $\sqrt{27x} - 2\sqrt{9x} + \sqrt{72x}$

   d. $\sqrt[3]{98}$

   e. $\sqrt[3]{48}$

21. Solve

22. Write us

   a. $\sqrt{-36}$

23. Find the

   a. i^7

CHAPTER **8**

Quadratic Equations and Inequalities and Conic Sections

Sidewalks are constructed from separate concrete blocks rather than one continuous concrete slab, because concrete expands in the heat of the sun. Engineers must account for this expansion when planning the dimensions of the blocks. (See Critical Thinking, page 401.)

INTRODUCTION

We continue in this chapter the work begun in Section 5.5, when we solved quadratic equations in one variable by factoring. Two additional methods for solving quadratic equations are analyzed, as well as methods for solving nonlinear inequalities in one variable. We devote the final three sections of this chapter to quadratic equations in two variables and their graphs: the parabola, circle, ellipse, and hyperbola.

8.1
Solving Quadratic Equations by Completing the Square

OBJECTIVES

1 Use the square root property to solve quadratic equations.

2 Write perfect square trinomials.

3 Solve quadratic equations by completing the square.

Tape 23

1 In Chapter 5, we solved quadratic equations by factoring. Recall that a **quadratic** or **second-degree equation** is an equation that can be written in the form $ax^2 + bx + c = 0$ where a, b, and c are real numbers and a is not 0. To solve a quadratic equation such as $x^2 = 9$ by factoring, we use the zero-factor theorem. To use the zero-factor theorem the equation must be written in standard form, $ax^2 + bx + c = 0$.

$$x^2 = 9$$

$$x^2 - 9 = 0 \qquad \text{Subtract 9 from both sides.}$$

$$(x + 3)(x - 3) = 0 \qquad \text{Factor.}$$

$$x + 3 = 0 \quad \text{or} \quad x - 3 = 0 \qquad \text{Set each factor equal to 0.}$$

$$x = -3 \qquad\quad x = 3 \qquad \text{Solve.}$$

The solution set is $\{-3, 3\}$, the positive and negative **square roots** of 9. Not all quadratic equations can be solved by factoring. Thus we need to explore other methods of solving quadratic equations. Notice that the solutions of the equation $x^2 = 9$ are two numbers whose square is 9.

$$3^2 = 9 \quad \text{and} \quad (-3)^2 = 9$$

Thus we can solve the equation $x^2 = 9$ by taking the square root of both sides. Be sure to include both $\sqrt{9}$ and $-\sqrt{9}$ as solutions.

$$x^2 = 9$$

$$x = \pm\sqrt{9}$$

$$x = \pm 3 \qquad \text{The notation } \pm 3 \text{ (read as plus or minus 3)}$$
$$\text{indicates the pair of numbers } +3 \text{ and } -3.$$

This illustrates the square root property.

> **Square Root Property**
>
> If b is a real number and if $a^2 = b$, then $a = \pm\sqrt{b}$.

EXAMPLE 1 Use the square root property to solve $x^2 = 50$.

Solution:
$$x^2 = 50$$
$$x = \pm\sqrt{50}$$
$$x = \pm 5\sqrt{2}$$

The solution set is $\{5\sqrt{2}, -5\sqrt{2}\}$. ∎

EXAMPLE 2 Use the square root property to solve $(x + 1)^2 = 12$ for x.

Solution: By the square root property, we have

$$(x + 1)^2 = 12$$
$$x + 1 = \pm\sqrt{12} \qquad \text{Apply the square root property.}$$
$$x + 1 = \pm 2\sqrt{3} \qquad \text{Simplify the radical.}$$
$$x = -1 \pm 2\sqrt{3} \qquad \text{Subtract 1 from both sides.}$$

The solution set is $\{-1 + 2\sqrt{3}, -1 - 2\sqrt{3}\}$. ∎

EXAMPLE 3 Solve $(2x - 5)^2 = -16$.

Solution:
$$(2x - 5)^2 = -16$$
$$2x - 5 = \pm\sqrt{-16} \qquad \text{Apply the square root property.}$$
$$2x - 5 = \pm 4i \qquad \text{Simplify the radical.}$$
$$2x = 5 \pm 4i \qquad \text{Add 5 to both sides.}$$
$$x = \frac{5 \pm 4i}{2} \qquad \text{Divide both sides by 2.}$$

The solution set is $\left\{\dfrac{5 + 4i}{2}, \dfrac{5 - 4i}{2}\right\}$. ∎

2   Notice from Examples 2 and 3 that, if we write a quadratic equation so that one side is a binomial squared, we can solve using the square root property. To write the square of a binomial, we write perfect square trinomials. Recall that a perfect square trinomial is a trinomial that can be factored into two identical binomial factors.

Perfect Square Trinomials	*Factored Form*
$x^2 + 8x + 16$	$(x + 4)^2$
$x^2 - 6x + 9$	$(x - 3)^2$

Notice in each perfect square trinomial that **the constant term of the trinomial is the square of half the coefficient of the x-term.** For example,

$$x^2 + 8x + 16 \qquad\qquad x^2 - 6x + 9$$
$$\frac{1}{2}(8) = 4 \text{ and } 4^2 = 16 \qquad \frac{1}{2}(-6) = -3 \text{ and } (-3)^2 = 9$$

3   The process of writing a quadratic equation so that one side is a perfect square trinomial is called **completing the square.**

EXAMPLE 4   Solve $p^2 + 2p = 4$ by completing the square.

Solution: First, add the square of half the coefficient of p to both sides so that the resulting trinomial will be a perfect square trinomial. The coefficient of p is 2.

$$\frac{1}{2}(2) = 1 \quad \text{and} \quad 1^2 = 1$$

Add 1 to both sides of the original equation.

$$p^2 + 2p = 4$$

$$p^2 + 2p + 1 = 4 + 1 \qquad \text{Add 1 to both sides.}$$

$$(p + 1)^2 = 5 \qquad \text{Factor.}$$

We may now apply the square root property and solve for p.

$$p + 1 = \pm\sqrt{5} \qquad \text{Use the square root property.}$$

$$p = -1 \pm \sqrt{5} \qquad \text{Subtract 1 from both sides.}$$

Notice that there are two solutions: $-1 + \sqrt{5}$ and $-1 - \sqrt{5}$. The solution set is $\{-1 + \sqrt{5}, -1 - \sqrt{5}\}$. ∎

EXAMPLE 5   Solve $m^2 - 7m - 1 = 0$ for m by completing the square.

Solution: First, add 1 to both sides of the equation so that the left side has no constant term.

$$m^2 - 7m - 1 = 0$$

$$m^2 - 7m = 1$$

Now find the proper constant term that makes the left side a perfect square trinomial by squaring half the coefficient of m. Add this result to both sides of the equation.

$$\frac{1}{2}(-7) = -\frac{7}{2} \quad \text{and} \quad \left(-\frac{7}{2}\right)^2 = \frac{49}{4}$$

$$m^2 - 7m + \frac{49}{4} = 1 + \frac{49}{4} \qquad \text{Add } \frac{49}{4} \text{ to both sides of the equation.}$$

$$\left(m - \frac{7}{2}\right)^2 = \frac{53}{4} \qquad \text{Factor the perfect square trinomial.}$$

$$m - \frac{7}{2} = \pm\sqrt{\frac{53}{4}} \qquad \text{Apply the square root property.}$$

$$m = \frac{7}{2} \pm \frac{\sqrt{53}}{2} \qquad \text{Add } \frac{7}{2} \text{ to both sides and simplify } \sqrt{\frac{53}{4}}.$$

$$m = \frac{7 \pm \sqrt{53}}{2} \qquad \text{Simplify.}$$

The solution set is $\left\{\dfrac{7 + \sqrt{53}}{2}, \dfrac{7 - \sqrt{53}}{2}\right\}$. ∎

EXAMPLE 6 Solve $2x^2 - 8x + 3 = 0$.

Solution: Our procedure for finding the constant term to complete the square works only if the coefficient of the squared variable term is 1. Therefore, to solve this equation, the first step is to divide both sides by 2, the coefficient of x^2.

$$2x^2 - 8x + 3 = 0$$

$$x^2 - 4x + \frac{3}{2} = 0 \qquad \text{Divide both sides by 2.}$$

$$x^2 - 4x = -\frac{3}{2} \qquad \text{Subtract } \frac{3}{2} \text{ from both sides.}$$

Next, find the square of half of -4.

$$\frac{1}{2}(-4) = -2 \quad \text{and} \quad (-2)^2 = 4$$

Add 4 to both sides of the equation.

$$x^2 - 4x + 4 = -\frac{3}{2} + 4$$

$$(x - 2)^2 = \frac{5}{2} \qquad \text{Factor the perfect square.}$$

$$x - 2 = \pm\sqrt{\frac{5}{2}} \qquad \text{Apply the square root property.}$$

$$x - 2 = \pm\frac{\sqrt{10}}{2} \qquad \text{Rationalize the denominator.}$$

$$x = 2 \pm \frac{\sqrt{10}}{2} \qquad \text{Add 2 to both sides.}$$

$$= \frac{4}{2} \pm \frac{\sqrt{10}}{2} \qquad \text{Find the common denominator.}$$

$$= \frac{4 \pm \sqrt{10}}{2} \qquad \text{Simplify.}$$

The solution set is $\left\{ \dfrac{4 + \sqrt{10}}{2}, \dfrac{4 - \sqrt{10}}{2} \right\}$. ■

The following steps may be used to solve a quadratic equation such as $ax^2 + bx + c = 0$ by completing the square.

To Solve a Quadratic Equation in x by Completing the Square

Step 1 If the coefficient of x^2 is 1, go to *step 2*. Otherwise, divide both sides of the equation by the coefficient of x^2.

Step 2 Isolate all variable terms on one side of the equation.

Step 3 Complete the square for the resulting binomial by adding the square of half of x to both sides of the equation.

Step 4 Factor the resulting perfect square trinomial.

Step 5 Apply the square root property to solve for x.

EXAMPLE 7   Solve $3x^2 - 9x + 8 = 0$ by completing the square.

Solution:

$$3x^2 - 9x + 8 = 0$$

$$x^2 - 3x + \frac{8}{3} = 0 \qquad \text{Divide both sides of the equation by 3.}$$

$$x^2 - 3x = -\frac{8}{3} \qquad \text{Subtract } \frac{8}{3} \text{ from both sides.}$$

Since $\frac{1}{2}(-3) = -\frac{3}{2}$ and $\left(-\frac{3}{2}\right)^2 = \frac{9}{4}$, we add $\frac{9}{4}$ to both sides of the equation.

$$x^2 - 3x + \frac{9}{4} = -\frac{8}{3} + \frac{9}{4}$$

$$\left(x - \frac{3}{2}\right)^2 = -\frac{5}{12} \qquad \text{Factor the perfect square trinomial.}$$

$$x - \frac{3}{2} = \pm\sqrt{-\frac{5}{12}} \qquad \text{Apply the square root property.}$$

$$x - \frac{3}{2} = \pm\frac{i\sqrt{5}}{2\sqrt{3}} \qquad \text{Simplify the radical.}$$

$$x - \frac{3}{2} = \pm\frac{i\sqrt{15}}{6} \qquad \text{Rationalize the denominator.}$$

$$x = \frac{3}{2} \pm \frac{i\sqrt{15}}{6} \qquad \text{Add } \frac{3}{2} \text{ to both sides.}$$

$$= \frac{9}{6} \pm \frac{i\sqrt{15}}{6} \qquad \text{Find a common denominator.}$$

$$= \frac{9 \pm i\sqrt{15}}{6} \qquad \text{Simplify.}$$

The solution set is $\left\{ \dfrac{9 + i\sqrt{15}}{6}, \dfrac{9 - i\sqrt{15}}{6} \right\}$.   ■

EXERCISE SET 8.1

Use the square root property to solve each equation. These equations have real number solutions. See Example 1.

1. $x^2 = 16$ **2.** $x^2 = 49$ **3.** $x^2 - 7 = 0$ **4.** $x^2 - 11 = 0$

5. $x^2 = 18$ **6.** $y^2 = 20$ **7.** $3z^2 - 30 = 0$ **8.** $2x^2 = 4$

Use the square root property to solve each equation. See Example 1.

9. $x^2 + 9 = 0$ **10.** $x^2 + 4 = 0$ **11.** $x^2 - 6 = 0$

12. $y^2 - 10 = 0$ **13.** $2z^2 + 16 = 0$ **14.** $3p^2 + 36 = 0$

Use the square root property to solve each equation. These equations have real number solutions. See Example 2.

15. $(x + 5)^2 = 9$ **16.** $(y - 3)^2 = 4$

17. $(z - 6)^2 = 18$ **18.** $(y + 4)^2 = 27$

19. $(2x - 3)^2 = 8$ **20.** $(4x + 9)^2 = 6$

Use the square root property to solve each equation. See Example 3.

21. $(x - 1)^2 = -16$ **22.** $(y + 2)^2 = -25$
23. $(z + 7)^2 = 5$ **24.** $(x + 10)^2 = 11$
25. $(x + 3)^2 = -8$ **26.** $(y - 4)^2 = -18$

Add the proper constant to each binomial so that the resulting trinomial is a perfect square trinomial. Then factor the trinomial.

27. $x^2 + 16x$ **28.** $y^2 + 2y$
29. $z^2 - 12z$ **30.** $x^2 - 8x$

31. $p^2 + 9p$ **32.** $n^2 + 5n$
33. $x^2 + x$ **34.** $y^2 - y$

Solve each equation by completing the square. These equations have real number solutions. See Examples 4 and 5.

35. $x^2 + 8x = -15$ **36.** $y^2 + 6y = -8$
37. $x^2 + 6x + 2 = 0$ **38.** $x^2 - 2x - 2 = 0$
39. $x^2 + x - 1 = 0$ **40.** $x^2 + 3x - 2 = 0$

Solve each equation by completing the square. See Examples 4 and 5.

41. $y^2 + 2y + 2 = 0$ **42.** $x^2 + 4x + 6 = 0$

43. $x^2 - 6x + 3 = 0$ **44.** $x^2 - 7x - 1 = 0$

Solve each equation by completing the square. These equations have real number solutions. See Examples 6 and 7.

45. $x^2 + x - 1 = 0$ **46.** $y^2 + y - 7 = 0$

47. $3p^2 - 12p + 2 = 0$ **48.** $2x^2 + 14x - 1 = 0$

49. $4y^2 - 12y - 2 = 0$ **50.** $6x^2 - 3 = 6x$

51. $2x^2 + 7x = 4$ **52.** $3x^2 - 4x = 4$

Solve each equation by completing the square. See Examples 6 and 7.

53. $2a^2 + 8a = -12$ **54.** $3x^2 + 12x = -14$

55. $5x^2 + 15x - 1 = 0$ **56.** $16y^2 + 16y - 1 = 0$

57. $2x^2 - x + 6 = 0$ **58.** $4x^2 - 2x + 5 = 0$

Solve each equation by completing the square. These equations have real number solutions.

59. $x^2 - 4x - 5 = 0$ **60.** $y^2 + 6y - 8 = 0$
61. $x^2 + 8x + 1 = 0$ **62.** $x^2 - 10x + 2 = 0$

63. $3y^2 + 6y - 4 = 0$

64. $2y^2 + 12y + 3 = 0$

65. $2x^2 - 3x - 5 = 0$

66. $5x^2 + 3x - 2 = 0$

Solve each equation by completing the square.

67. $x^2 + 10x + 28 = 0$

68. $y^2 + 8y + 18 = 0$

69. $z^2 + 3z - 4 = 0$

70. $y^2 + y - 2 = 0$

71. $2x^2 - 4x + 3 = 0$

72. $9x^2 - 36x = -40$

73. $3x^2 + 3x = 5$

74. $5y^2 - 15y = 1$

Writing in Mathematics

75. When solving a quadratic equation by factoring, one side of the equation must be 0. Explain why.

Skill Review

Evaluate each expression if $x = 2$, $y = -1$, and $z = 3$. See Section 1.4.

76. $x^3 - 2yz$

77. $x^3 + y^3 + z^3$

Evaluate each expression. See Section 7.1.

78. $8^{2/3}$

79. $9^{3/2}$

Simplify each expression. See Section 7.1.

80. $y^{1/2} \cdot y^{3/4}$

81. $\dfrac{y^{1/2}}{y^{3/4}}$

8.2
Solving Quadratic Equations by the Quadratic Formula

OBJECTIVES

Tape 23

1   Solve quadratic equations using the quadratic formula.

2   Determine the number and type of solutions of a quadratic equation using the discriminant.

1   Any quadratic equation can be solved by completing the square. Since the same sequence of steps is repeated each time we complete the square, let's complete the square for a general quadratic equation, $ax^2 + bx + c = 0$. By doing so, we find a pattern for the solutions of a quadratic equation known as the **quadratic formula.**

Recall that to complete the square for an equation such as $ax^2 + bx + c = 0$ we first divide both sides by the coefficient of x.

$$ax^2 + bx + c = 0$$

$$x^2 + \frac{b}{a}x + \frac{c}{a} = 0 \qquad \text{Divide both sides by } a, \text{ the coefficient of } x^2.$$

$$x^2 + \frac{b}{a}x = -\frac{c}{a} \qquad \text{Subtract the constant } \frac{c}{a} \text{ from both sides.}$$

Find the square of half the coefficient of x, $\frac{b}{a}$.

$$\frac{1}{2}\left(\frac{b}{a}\right) = \frac{b}{2a} \quad \text{and} \quad \left(\frac{b}{2a}\right)^2 = \frac{b^2}{4a^2}$$

Add this result to both sides of the equation.

$$x^2 + \frac{b}{a}x + \frac{b^2}{4a^2} = -\frac{c}{a} + \frac{b^2}{4a^2} \qquad \text{Add } \frac{b^2}{4a^2} \text{ to both sides.}$$

$$x^2 + \frac{b}{a}x + \frac{b^2}{4a^2} = \frac{-c \cdot 4a}{a \cdot 4a} + \frac{b^2}{4a^2} \qquad \begin{array}{l}\text{Find a common denominator} \\ \text{on the right side.}\end{array}$$

$$x^2 + \frac{b}{a}x + \frac{b^2}{4a^2} = \frac{b^2 - 4ac}{4a^2} \qquad \text{Simplify the right side.}$$

$$\left(x + \frac{b}{2a}\right)^2 = \frac{b^2 - 4ac}{4a^2} \qquad \begin{array}{l}\text{Factor the perfect square trinomial} \\ \text{on the left side.}\end{array}$$

$$x + \frac{b}{2a} = \pm\sqrt{\frac{b^2 - 4ac}{4a^2}} \qquad \text{Apply the square root property.}$$

$$x + \frac{b}{2a} = \pm\frac{\sqrt{b^2 - 4ac}}{2a} \qquad \text{Simplify the radical.}$$

$$x = -\frac{b}{2a} \pm \frac{\sqrt{b^2 - 4ac}}{2a} \qquad \text{Subtract } \frac{b}{2a} \text{ from both sides.}$$

$$x = \frac{-b \pm \sqrt{b^2 - 4ac}}{2a} \qquad \text{Simplify.}$$

This equation identifies the solutions of the general quadratic equation in standard form and is called the quadratic formula. It can be used to solve any equation written in standard form $ax^2 + bx + c = 0$ as long as a is not 0.

Quadratic Formula

A quadratic equation written in the form $ax^2 + bx + c = 0$ has the solutions

$$x = \frac{-b \pm \sqrt{b^2 - 4ac}}{2a}$$

EXAMPLE 1   Solve $3x^2 + 16x + 5 = 0$ for x.

Solution: This equation is in standard form, so $a = 3$, $b = 16$, and $c = 5$. Substitute these values into the quadratic formula.

$$x = \frac{-b \pm \sqrt{b^2 - 4ac}}{2a} \qquad \text{Quadratic formula.}$$

$$= \frac{-16 \pm \sqrt{16^2 - 4(3)(5)}}{2(3)} \qquad \text{Let } a = 3, b = 16, \text{ and } c = 5.$$

$$= \frac{-16 \pm \sqrt{256 - 60}}{6}$$

$$= \frac{-16 \pm \sqrt{196}}{6} = \frac{-16 \pm 14}{6}$$

$$x = \frac{-16 + 14}{6} = -\frac{1}{3} \quad \text{or} \quad x = \frac{-16 - 14}{6} = -\frac{30}{6} = -5$$

The solution set is $\left\{-\frac{1}{3}, -5\right\}$. ∎

HELPFUL HINT

A quadratic equation **should** be written in standard form $ax^2 + bx + c = 0$ in order to correctly identify a, b, and c in the quadratic formula.

EXAMPLE 2   Solve $2x^2 - 4x = 3$.

Solution: First, write the equation in standard form by subtracting 3 from both sides.

$$2x^2 - 4x - 3 = 0$$

Now, $a = 2$, $b = -4$, and $c = -3$. Substitute these values into the quadratic formula.

$$x = \frac{-b \pm \sqrt{b^2 - 4ac}}{2a}$$

$$= \frac{-(-4) \pm \sqrt{(-4)^2 - 4(2)(-3)}}{2(2)}$$

$$= \frac{4 \pm \sqrt{16 + 24}}{4}$$

$$= \frac{4 \pm \sqrt{40}}{4} = \frac{4 \pm 2\sqrt{10}}{4}$$

$$= \frac{2\,(2 \pm \sqrt{10})}{2 \cdot 2} = \frac{2 \pm \sqrt{10}}{2}$$

The solution set is $\left\{\dfrac{2 + \sqrt{10}}{2}, \dfrac{2 - \sqrt{10}}{2}\right\}$. ∎

HELPFUL HINT

Consider the expression $\dfrac{4 \pm 2\sqrt{10}}{4}$ in the preceding example. Note that 2 **must** be factored out of both terms of the numerator **before** simplifying. Then

$$\frac{4 \pm 2\sqrt{10}}{4} = \frac{2\,(2 \pm \sqrt{10})}{2 \cdot 2} = \frac{2 \pm \sqrt{10}}{2}$$

EXAMPLE 3 Solve $\dfrac{1}{4}m^2 - m + \dfrac{1}{2} = 0$.

Solution: First, we multiply both sides of the equation by 4 to clear fractions.

$$4\left(\frac{1}{4}m^2 - m + \frac{1}{2}\right) = 4 \cdot 0$$

$$m^2 - 4m + 2 = 0 \qquad \text{Simplify.}$$

Substitute $a = 1$, $b = -4$, and $c = 2$ into the quadratic formula and simplify.

$$m = \frac{-(-4) \pm \sqrt{(-4)^2 - 4(1)(2)}}{2(1)} = \frac{4 \pm \sqrt{16 - 8}}{2}$$

$$= \frac{4 \pm \sqrt{8}}{2} = \frac{4 \pm 2\sqrt{2}}{2} = \frac{2\,(2 \pm \sqrt{2})}{2} = 2 \pm \sqrt{2}$$

The solution set is $\{2 + \sqrt{2}, 2 - \sqrt{2}\}$. ∎

EXAMPLE 4 Solve $p = -3p^2 - 3$.

Solution: The equation in standard form is $3p^2 + p + 3 = 0$. Thus let $a = 3$, $b = 1$, and $c = 3$ in the quadratic formula.

$$p = \frac{-1 \pm \sqrt{1^2 - 4(3)(3)}}{2 \cdot 3} = \frac{-1 \pm \sqrt{1 - 36}}{6} = \frac{-1 \pm \sqrt{-35}}{6} = \frac{-1 \pm i\sqrt{35}}{6}$$

The solution set is $\left\{\dfrac{-1 + i\sqrt{35}}{6}, \dfrac{-1 - i\sqrt{35}}{6}\right\}$. ∎

2 In the quadratic formula $x = \dfrac{-b \pm \sqrt{b^2 - 4ac}}{2a}$, the radicand $b^2 - 4ac$ is called the **discriminant** because, by knowing its value, we can **discriminate** among the possibilities and number of solutions of a quadratic equation. Possible values of the discriminant and their meanings are summarized next.

Discriminant

The following table corresponds the discriminant $b^2 - 4ac$ of a quadratic equation of the form $ax^2 + bx + c = 0$ with the number and type of solutions of the equation.

$b^2 - 4ac$	Number and Type of Solutions
Positive	Two real solutions
Zero	One real solution
Negative	Two complex but not real solutions

EXAMPLE 5 Use the discriminant to determine the number and type of solutions of each quadratic equation.

a. $x^2 + 2x + 1 = 0$ **b.** $3x^2 + 2 = 0$ **c.** $2x^2 - 7x - 4 = 0$

Solution: **a.** In $x^2 + 2x + 1 = 0$, $a = 1$, $b = 2$, and $c = 1$. Thus

$$b^2 - 4ac = 4 - 4(1)(1) = 0.$$

Since $b^2 - 4ac = 0$, this quadratic equation has one real solution.

b. In this equation, $a = 3$, $b = 0$, $c = 2$. Then $b^2 - 4ac = 0 - 4(3)(2) = -24$. Since $b^2 - 4ac$ is negative, there are two complex solutions.

c. In this equation, $a = 2$, $b = -7$, and $c = -4$. Then

$$b^2 - 4ac = 49 - 4(2)(-4) = 81.$$

Since $b^2 - 4ac$ is positive, there are two real solutions.   ■

EXERCISE SET 8.2

Use the quadratic formula to solve each equation. These equations have real number solutions. See Example 1.

1. $m^2 + 5m - 6 = 0$

2. $p^2 + 11p - 12 = 0$

3. $2y = 5y^2 - 3$

4. $5x^2 - 3 = 14x$

5. $x^2 - 6x + 9 = 0$

6. $y^2 + 10y + 25 = 0$

Use the quadratic formula to solve each equation. These equations have real number solutions. See Example 2.

7. $x^2 + 7x + 4 = 0$

8. $y^2 + 5y + 3 = 0$

9. $8m^2 - 2m = 7$

10. $11n^2 - 9n = 1$

11. $3m^2 - 7m = 3$

12. $x^2 - 13 = 5x$

Use the quadratic formula to solve each equation. These equations have real number solutions. See Example 3.

13. $\dfrac{1}{2}x^2 - x - 1 = 0$

14. $\dfrac{1}{6}x^2 + x + \dfrac{1}{3} = 0$

15. $\dfrac{2}{5}y^2 + \dfrac{1}{5}y = \dfrac{3}{5}$

16. $\dfrac{1}{8}x^2 + x = \dfrac{5}{2}$

17. $\dfrac{1}{3}y^2 - y - \dfrac{1}{6} = 0$

18. $\dfrac{1}{2}y^2 = y + \dfrac{1}{2}$

Use the quadratic formula to solve each equation. See Example 4.

19. $6 = -4x^2 + 3x$

20. $9x^2 + x + 2 = 0$

21. $(x + 5)(x - 1) = 2$

22. $x(x + 6) = 2$

23. $10y^2 + 10y + 3 = 0$

24. $3y^2 + 6y + 5 = 0$

Use the discriminant to determine the number and types of solutions of each equation. See Example 5.

25. $3x = -2x^2 + 7$

26. $3x^2 = 5 - 7x$

27. $6 = 4x - 5x^2$

28. $9x - 2x^2 + 5 = 0$

29. $5 - 4x + 12x^2 = 0$

30. $8x = 3 - 9x^2$

31. $4x^2 + 12x = -9$

32. $9x^2 + 1 = 6x$

Use the quadratic formula to solve each equation. These equations have real number solutions.

33. $x^2 + 5x = -2$

34. $y^2 - 8 = 4y$

35. $2m^2 - 7m + 5 = 0$

36. $7p^2 - 12p + 5 = 0$

37. $\dfrac{x^2}{3} - x = \dfrac{5}{3}$

38. $\dfrac{x^2}{2} - 3 = -\dfrac{9}{2}x$

39. $6x^2 + 2x - 3 = 0$

40. $7x^2 + x = 2$

Use the quadratic formula to solve each equation.

41. $x^2 + 6x + 13 = 0$

42. $x^2 + 2x + 2 = 0$

43. $\dfrac{2}{5}y^2 + \dfrac{1}{5}y + \dfrac{3}{5} = 0$

44. $\dfrac{1}{8}x^2 + x + \dfrac{5}{2} = 0$

45. $\dfrac{1}{2}y^2 = y - \dfrac{1}{2}$

46. $\dfrac{2}{3}x^2 - \dfrac{20}{3}x = -\dfrac{100}{6}$

47. $(n - 2)^2 = 15n$

48. $\left(p - \dfrac{1}{2}\right)^2 = \dfrac{p}{2}$

49. The product of a number and 4 less than the number is 96. Find the number.

50. A whole number increased by its square is two more than twice itself. Find the number.

51. Bill Shaughnessy and his son Billy can clean the house together in 4 hours. When they work alone, it takes the son an hour longer to clean than it takes his dad alone. Find how long to the nearest hundredth hour it takes the son to clean alone.

52. Scratchy and Freckles together eat a 50-pound bag of dog food in 30 days. Scratchy by himself eats a 50-pound bag in 2 weeks less time than Freckles by himself. How many days to the nearest whole day would a 50-pound bag of dog food last Freckles?

A Look Ahead

Use the quadratic formula to solve each quadratic equation. See the following example.

EXAMPLE   Solve $x^2 - 3\sqrt{2}x + 2 = 0$.

Solution: In this equation, $a = 1$, $b = -3\sqrt{2}$, and $c = 2$. By the quadratic formula, we have

$$x = \frac{-b \pm \sqrt{b^2 - 4ac}}{2a}$$

$$= \frac{3\sqrt{2} \pm \sqrt{(-3\sqrt{2})^2 - 4(1)(2)}}{2(1)} = \frac{3\sqrt{2} \pm \sqrt{18 - 8}}{2} = \frac{3\sqrt{2} \pm \sqrt{10}}{2}$$

The solution set is $\left\{ \dfrac{3\sqrt{2} + \sqrt{10}}{2}, \dfrac{3\sqrt{2} - \sqrt{10}}{2} \right\}$.   ∎

53. $3x^2 - \sqrt{12}x + 1 = 0$

54. $5x^2 + \sqrt{20}x + 1 = 0$

55. $x^2 + \sqrt{2}x + 1 = 0$

56. $x^2 - \sqrt{2}x + 1 = 0$

57. $2x^2 - \sqrt{3}x - 1 = 0$

58. $7x^2 + \sqrt{7}x - 2 = 0$

Skill Review

Solve each equation. See Section 7.5.

59. $\sqrt{5x - 2} = 3$

60. $\sqrt{y + 2} + 7 = 12$

Add or subtract as indicated. See Section 7.3.

61. $\sqrt{5} + 4\sqrt{5} - 7\sqrt{5}$

62. $2\sqrt{3x} - 3\sqrt{3x} + 4\sqrt{x}$

63. $\sqrt{12x} + \sqrt{27x}$

64. $\sqrt{45} - \sqrt{80}$

8.3
Solving Equations in Quadratic Form

OBJECTIVES

Tape 24

1   Solve miscellaneous equations that are quadratic in form.

2   Solve applications that lead to equations that are quadratic in form.

1   In this section, we discuss various types of equations that can be solved by the methods of this chapter.

The first example is an equation containing rational expressions that can be written in quadratic form.

EXAMPLE 1   Solve $\dfrac{3x}{x - 2} - \dfrac{x + 1}{x} = \dfrac{6}{x(x - 2)}$.

Solution: In this equation, x cannot be either 2 or 0, because these values cause denominators to equal zero. To solve for x, first multiply both sides of the equation by $x(x - 2)$ to clear fractions. By the distributive property, this means we multiply each term by $x(x - 2)$.

$$x(x - 2) \left(\frac{3x}{x - 2}\right) - x(x - 2) \left(\frac{x + 1}{x}\right) = x(x - 2) \left[\frac{6}{x(x - 2)}\right]$$

$$3x^2 - (x - 2)(x + 1) = 6 \qquad \text{Simplify.}$$

$$3x^2 - (x^2 - x - 2) = 6 \qquad \text{Multiply.}$$

$$3x^2 - x^2 + x + 2 = 6$$

$$2x^2 + x - 4 = 0 \qquad \text{Simplify.}$$

$$x = \frac{-1 \pm \sqrt{1^2 - 4(2)(-4)}}{2 \cdot 2} \qquad \begin{array}{l}\text{Let } a = 2, b = 1, \\ \text{and } c = -4 \text{ in the} \\ \text{quadratic formula.}\end{array}$$

$$= \frac{-1 \pm \sqrt{1 + 32}}{4} \qquad \text{Simplify.}$$

$$= \frac{-1 \pm \sqrt{33}}{4}$$

The solution set is $\left\{\dfrac{-1 + \sqrt{33}}{4}, \dfrac{-1 - \sqrt{33}}{4}\right\}$. ∎

EXAMPLE 2 Solve $x^3 = 8$ for x.

Solution: Begin by subtracting 8 from both sides; then factor the resulting difference of cubes.

$$x^3 = 8$$

$$x^3 - 8 = 0$$

$$(x - 2)(x^2 + 2x + 4) = 0 \qquad \text{Factor.}$$

$$x - 2 = 0 \quad \text{or} \quad x^2 + 2x + 4 = 0 \qquad \text{Set each factor equal to 0.}$$

The solution of $x - 2 = 0$ is 2. Solve the second equation by the quadratic formula.

$$x = \frac{-2 \pm \sqrt{2^2 - 4(1)(4)}}{2 \cdot 1} = \frac{-2 \pm \sqrt{-12}}{2} \qquad \begin{array}{l}\text{Let } a = 1, b = 2, \\ c = 4.\end{array}$$

$$= \frac{-2 \pm 2i\sqrt{3}}{2} = \frac{2(-1 \pm i\sqrt{3})}{2}$$

$$= -1 \pm i\sqrt{3}$$

The solution set is $\{2, -1 + i\sqrt{3}, -1 - i\sqrt{3}\}$. ∎

EXAMPLE 3 Solve $p^4 - 3p^2 - 4 = 0$.

Solution: First, factor the trinomial.

$$p^4 - 3p^2 - 4 = 0$$

$$(p^2 - 4)(p^2 + 1) = 0$$

$$(p - 2)(p + 2)(p^2 + 1) = 0 \qquad \text{Factor.}$$

$$p - 2 = 0 \quad \text{or} \quad p + 2 = 0 \qquad \text{or} \quad p^2 + 1 = 0 \qquad \text{Set each factor equal to 0.}$$
$$p = 2 \quad \text{or} \qquad p = -2 \quad \text{or} \qquad p^2 = -1$$
$$p = \pm\sqrt{-1} = \pm i$$

The solution set is $\{2, -2, i, -i\}$.   ■

EXAMPLE 4   Solve $x^{2/3} - 5x^{1/3} + 6 = 0$.

Solution: The key to solving this equation is recognizing that $x^{2/3} = (x^{1/3})^2$. Replace $x^{1/3}$ with m so that

$$(x^{1/3})^2 - 5x^{1/3} + 6 = 0$$

becomes

$$m^2 - 5m + 6 = 0$$

Now, solve by factoring.

$$m^2 - 5m + 6 = 0$$
$$(m - 3)(m - 2) = 0 \qquad \text{Factor.}$$
$$m - 3 = 0 \quad \text{or} \quad m - 2 = 0 \qquad \text{Set each factor equal to 0.}$$
$$m = 3 \quad \text{or} \qquad m = 2$$

Since $m = x^{1/3}$, $x^{1/3} = 3$ or $x^{1/3} = 2$. To solve for x, cube both sides of each equation: $(x^{1/3})^3 = 3^3 = 27$ or $(x^{1/3})^3 = 2^3 = 8$. The solution set is $\{8, 27\}$.   ■

2   Some applications lead to equations that are quadratic in form.

EXAMPLE 5   The hypotenuse of an isosceles right triangle is 2 centimeters longer than either of its legs. Find the perimeter of the triangle.

Solution: Recall that an isosceles right triangle has legs of equal length. Let $x =$ the length of each leg of the triangle so that $x + 2 =$ the length of the hypotenuse of the triangle. By the Pythagorean theorem, $x^2 + x^2 = (x + 2)^2$.

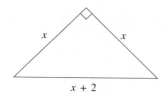

$$x^2 + x^2 = x^2 + 4x + 4 \qquad \text{Square } (x + 2).$$
$$x^2 - 4x - 4 = 0 \qquad \text{Set the equation equal to 0.}$$

Next, substitute $a = 1$, $b = -4$, and $c = -4$ in the quadratic formula.

$$x = \frac{4 \pm \sqrt{16 - 4(1)(-4)}}{2} = \frac{4 \pm \sqrt{32}}{2} = \frac{4 \pm 4\sqrt{2}}{2}$$
$$= \frac{2 \cdot (2 \pm 2\sqrt{2})}{2} = 2 \pm 2\sqrt{2}$$

Since the length of a side cannot be negative, we discard the solution $2 - 2\sqrt{2}$. The length of each leg is $(2 + 2\sqrt{2})$ centimeters. Since the hypotenuse is 2 centimeters longer than the legs, the hypotenuse is $(4 + 2\sqrt{2})$ centimeters. The perimeter of the triangle is

$$(2 + 2\sqrt{2}) \text{ cm} + (2 + 2\sqrt{2}) \text{ cm} + (4 + 2\sqrt{2}) \text{ cm} = (8 + 6\sqrt{2}) \text{ centimeters.} \quad ■$$

EXERCISE SET 8.3

Solve each equation. See Example 1.

1. $\dfrac{2}{x} + \dfrac{3}{x-1} = 1$

2. $\dfrac{6}{x^2} = \dfrac{3}{x+1}$

3. $\dfrac{3}{x} + \dfrac{4}{x+2} = 2$

4. $\dfrac{5}{x-2} + \dfrac{4}{x+2} = 1$

5. $\dfrac{7}{x^2 - 5x + 6} = \dfrac{2x}{x-3} - \dfrac{x}{x-2}$

6. $\dfrac{11}{2x^2 + x - 15} = \dfrac{5}{2x-5} - \dfrac{x}{x+3}$

Solve each equation. See Example 2.

7. $y^3 - 1 = 0$

8. $x^3 + 8 = 0$

9. $x^4 + 27x = 0$

10. $y^5 + y^2 = 0$

11. $z^3 = 64$

12. $z^3 = -125$

Solve each equation. See Example 3.

13. $p^4 - 16 = 0$

14. $x^4 + 2x^2 - 3 = 0$

15. $4x^4 + 11x^2 = 3$

16. $z^4 = 81$

17. $z^4 - 13z^2 + 36 = 0$

18. $9x^4 + 5x^2 - 4 = 0$

Solve each equation. See Example 4.

19. $x^{2/3} - 3x^{1/3} - 10 = 0$

20. $x^{2/3} + 2x^{1/3} + 1 = 0$

21. $2x^{2/3} - 5x^{1/3} = 3$

22. $3x^{2/3} + 11x^{1/3} = 4$

23. $20x^{2/3} - 6x^{1/3} - 2 = 0$

24. $4x^{2/3} + 16x^{1/3} = -15$

Solve. See Example 5.

25. Uri's rectangular dog pen for his Irish setter must have an area of 400 square feet. Also, the length must be $2\dfrac{1}{2}$ times the width. Find the dimensions of the pen.

26. The hypotenuse of an isosceles right triangle is 5 inches longer than either of the legs. Find the length of the legs and the length of the hypotenuse.

Solve each equation.

27. $a^4 - 5a^2 + 6 = 0$

28. $x^4 - 12x^2 + 11 = 0$

29. $\dfrac{2x}{x-2} + \dfrac{x}{x+3} = \dfrac{-5}{x+3}$

30. $\dfrac{5}{x-3} + \dfrac{x}{x+3} = \dfrac{19}{x^2 - 9}$

31. $x^3 + 64 = 0$

32. $y^3 - 27 = 0$

33. $x^{2/3} - 8x^{1/3} + 15 = 0$

34. $x^{2/3} - 2x^{1/3} - 8 = 0$

35. $y^3 + 9y - y^2 - 9 = 0$

36. $x^3 + x - 3x^2 - 3 = 0$

37. $2x^{2/3} + 3x^{1/3} - 2 = 0$

38. $6x^{2/3} - 25x^{1/3} - 25 = 0$

39. $2x^3 - 250 = 0$

40. $8y^3 + 8 = 0$

41. $\dfrac{x}{x-1} + \dfrac{1}{x+1} = \dfrac{2}{x^2-1}$

42. $\dfrac{x}{x-5} + \dfrac{5}{x+5} = \dfrac{-1}{x^2-25}$

43. $p^4 - p^2 - 20 = 0$

44. $x^4 - 10x^2 + 9 = 0$

45. $2x^3 = -54$

46. $y^3 - 216 = 0$

47. $27y^4 + 15y^2 = 2$

48. $8z^4 + 14z^2 = -5$

Solve.

49. The base of a triangle is twice its height. If the area of the triangle is 42 square centimeters, find its base and height.

50. The width of a rectangle is $\dfrac{1}{3}$ its length. If its area is 12 square inches, find its length and width.

51. An entry in the Peach Festival Poster Contest must be rectangular and have an area of 1200 square inches. Furthermore, its length must be $1\dfrac{1}{2}$ times its width. Find the dimensions each entry must have.

52. A holding pen for cattle must be square and have a diagonal length of 100 meters.
  a. Find the length of a side of the pen.
  b. Find the area of the pen.

53. A rectangle is three times longer than it is wide. It has a diagonal of length 50 centimeters.
  a. Find the dimensions of the rectangle.
  b. Find the perimeter of the rectangle.

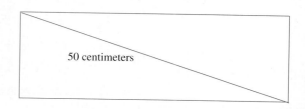

50 centimeters

Skill Review

Solve each inequality. See Section 2.5.

54. $\dfrac{5x}{3} + 2 \le 7$

55. $\dfrac{2x}{3} + \dfrac{1}{6} \ge 2$

56. $\dfrac{y-1}{15} > \dfrac{-2}{5}$

57. $\dfrac{z-2}{12} < \dfrac{1}{4}$

Multiply the following. See Section 7.4.

58. $\sqrt{3} \cdot \sqrt{6}$

59. $\sqrt{8} \cdot \sqrt{2}$

60. $\sqrt[3]{9} \cdot \sqrt[3]{6}$

61. $\sqrt[3]{2} \cdot \sqrt[3]{16}$

8.4
Nonlinear Inequalities in One Variable

1   Solve polynomial inequalities of degree 2 or greater.

2   Solve inequalities containing rational expressions with variables in the denominator.

---

1   Just as we can solve linear inequalities in one variable, so we can solve quadratic inequalities in one variable. A **quadratic inequality** is an inequality that can be written so that one side is a quadratic expression and the other side is 0. Here are examples of quadratic inequalities in one variable. Each is written in **standard form.**

$$x^2 - 10x + 7 \le 0 \qquad 3x^2 + 2x - 6 > 0$$

$$2x^2 + 9x - 2 < 0 \qquad x^2 - 3x + 11 \ge 0$$

A solution of a quadratic inequality in one variable is a value of the variable that makes the inequality a true statement.

The value of an expression such as $x^2 - 3x - 10$ will sometimes be positive, sometimes negative, and sometimes 0, depending on the values substituted for x. To solve the inequality $x^2 - 3x - 10 < 0$, we are looking for all values of x that make the expression $x^2 - 3x - 10$ **less than 0,** or **negative.** It is true that **intervals on the number line where the value of $x^2 - 3x - 10$ is positive are separated from intervals on the number line where the value of $x^2 - 3x - 10$ is negative by values for which the expression is 0.** Find these values for which the expression is 0 by solving the related equation

$$x^2 - 3x - 10 = 0$$

$$(x - 5)(x + 2) = 0 \qquad \text{Factor.}$$

$$x - 5 = 0 \quad \text{or} \quad x + 2 = 0 \qquad \text{Set each factor equal to 0.}$$

$$x = 5 \qquad\qquad x = -2 \qquad \text{Solve.}$$

These two numbers divide the number line into three regions. We will call the regions A, B, and C.

These regions are important because, if the value of $x^2 - 3x - 10$ is negative when one number from a region is substituted for x, then $x^2 - 3x - 10$ is negative for all numbers in the region. The same is true if the value of $x^2 - 3x - 10$ is positive.

To see whether the inequality $x^2 - 3x - 10 < 0$ is true or false in each region, choose a test point from each region and substitute its value for x in the inequality $x^2 - 3x - 10 < 0$. If the resulting inequality is true, the region containing the test point is a solution region.

	Test Point	$(x - 5)(x + 2) < 0$	
Region A	-3	$(-8)(-1) < 0$	False
Region B	0	$(-5)(2) < 0$	True
Region C	6	$(1)(8) < 0$	False

The points in region B satisfy the inequality. The numbers -2 and 5 are not included in the solution set since the inequality symbol is $<$. The solution set is $(-2, 5)$ and its graph is shown.

EXAMPLE 1   Solve $(x + 3)(x - 3) \geq 0$.

Solution: First, solve the related equation $(x + 3)(x - 3) = 0$.

$$(x + 3)(x - 3) = 0$$

$$x + 3 = 0 \quad \text{or} \quad x - 3 = 0$$

$$x = -3 \qquad\qquad x = 3$$

The two numbers -3 and 3 separate the number line into three regions. Substitute the value of a test point from each region. If the value of the test point satisfies the inequality, the region containing the test point is a solution region.

	Test Point	$(x + 3)(x - 3) \geq 0$	
Region A	-4	$(-1)(-7) \geq 0$	True
Region B	0	$(3)(-3) \geq 0$	False
Region C	4	$(7)(1) \geq 0$	True

The points in regions A and C satisfy the inequality. The numbers -3 and 3 are included in the solution since the inequality symbol is $\geq$. The solution set is $(-\infty, -3] \cup [3, \infty)$ and its graph is shown.   ■

The following steps may be used to solve a polynomial inequality.

To Solve a Polynomial Inequality

Step 1   Write the inequality in standard form.

Step 2   Solve the related equation.

Step 3   Separate the number line into intervals using the solutions from *step 2*.

Step 4   For each interval, choose a test point and determine whether its value satisfies the **original inequality.**

Step 5   Write the solution set as the union of intervals whose test point is a solution.

EXAMPLE 2   Solve $x^2 - 4x \geq 0$.

Solution: First, solve the related equation $x^2 - 4x = 0$.

$$x^2 - 4x = 0$$

$$x(x - 4) = 0$$

$$x = 0 \quad \text{or} \quad x = 4$$

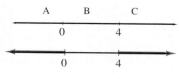

The numbers 0 and 4 separate the number line into three regions. Check test points in each region in the original inequality. The points in regions A and C satisfy the inequality. The numbers 0 and 4 are included in the solution since the inequality symbol is $\geq$. The solution set is $(-\infty, 0] \cup [4, \infty)$ and its graph is shown.   ■

EXAMPLE 3   Solve $(x + 2)(x - 1)(x - 5) \leq 0$.

Solution: First, solve $(x + 2)(x - 1)(x - 5) = 0$. By inspection, the solutions are -2, 1, and 5. They separate the number line into four regions. Next, check test points from each region.

	Test Point	$(x + 2)(x - 1)(x - 5) \leq 0$	
Region A	-3	$(-1)(-4)(-8) \leq 0$	True
Region B	0	$(2)(-1)(-5) \leq 0$	False
Region C	2	$(4)(1)(-3) \leq 0$	True
Region D	6	$(8)(5)(1) \leq 0$	False

The solution set is $(-\infty, -2] \cup [1, 5]$ and its graph is shown. We include the numbers -2, 1, 5 because the inequality symbol is $\leq$. ■

2   Inequalities containing rational expressions with variables in the denominator are solved using a similar procedure.

EXAMPLE 4   Solve $\dfrac{x + 2}{x - 3} \leq 0$.

Solution: First, solve the related equation $\dfrac{x + 2}{x - 3} = 0$.

$$\frac{x + 2}{x - 3} = 0$$

$$x + 2 = 0 \qquad \text{Multiply both sides by the LCD, } x - 3.$$

$$x = -2$$

Also, find all values that make the denominator equal to 0. To do so, solve $x - 3 = 0$ or $x = 3$. Place these numbers on a number line and proceed as before, checking test points in the original inequality.

Choose -3 from region A.   Choose 0 from Region B.

$$\frac{x + 2}{x - 3} \leq 0 \qquad\qquad \frac{x + 2}{x - 3} \leq 0$$

$$\frac{-3 + 2}{-3 - 3} \leq 0 \qquad\qquad \frac{0 + 2}{0 - 3} \leq 0$$

$$\frac{-1}{-6} \leq 0 \qquad\qquad -\frac{2}{3} \leq 0 \qquad \text{True.}$$

$$\frac{1}{6} \leq 0 \qquad \text{False.}$$

Choose 4 from region C.

$$\frac{x + 2}{x - 3} \leq 0$$

$$\frac{4 + 2}{4 - 3} \leq 0$$

$$6 \leq 0 \qquad \text{False.}$$

The solution set is $[-2, 3)$. This interval includes -2 because -2 satisfies the original inequality. This interval does not include 3, because 3 would make the denominator 0. ■

The following steps may be used to solve a rational inequality with variables in the denominator.

To Solve a Rational Inequality

Step 1   Solve the related equation.

Step 2   Solve for values that make the denominator 0.

Step 3   Separate the number line into intervals using the solutions from *steps 1* and *2*.

Step 4   For each interval, choose a test point and determine whether its value satisfies the **original inequality.**

Step 5   Write the solution set as the union of intervals whose test point is a solution.

EXAMPLE 5   Solve $\dfrac{5}{x + 1} < -2$.

Solution: First, solve $\dfrac{5}{x + 1} = -2$.

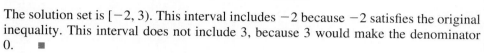

$$5 = -2x - 2 \qquad \text{Simplify.}$$

$$7 = -2x$$

$$-\frac{7}{2} = x$$

Next, find values for x that make the denominator equal to 0.

$$x + 1 = 0$$

$$x = -1$$

Use these two solutions to divide a number line into three intervals and choose test points.

Only a test point from region B satisfies the **original inequality.** The solution set is $\left(-\dfrac{7}{2}, -1\right)$ and its graph is shown. ■

EXERCISE SET 8.4

Solve each quadratic inequality. Write the solution set in interval notation and graph the solution. See Examples 1 and 2.

1. $(x + 1)(x + 5) > 0$

2. $(x + 1)(x + 5) \leq 0$

3. $(x - 3)(x + 4) \leq 0$

4. $(x + 4)(x - 1) > 0$

5. $x^2 - 7x + 10 \leq 0$

6. $x^2 + 8x + 15 \geq 0$

7. $3x^2 + 16x < -5$

8. $2x^2 - 5x < 7$

Solve each inequality. Write the solution set in interval notation and graph the solution. See Example 3.

9. $(x - 6)(x - 4)(x - 2) > 0$

10. $(x - 6)(x - 4)(x - 2) \leq 0$

11. $x(x - 1)(x + 4) \leq 0$

12. $x(x - 6)(x + 2) > 0$

13. $(x^2 - 9)(x^2 - 4) > 0$

14. $(x^2 - 16)(x^2 - 1) \leq 0$

Solve each inequality. Write the solution set in interval notation and graph the solution. See Example 4.

15. $\dfrac{x + 7}{x - 2} < 0$

16. $\dfrac{x - 5}{x - 6} > 0$

17. $\dfrac{5}{x + 1} > 0$

18. $\dfrac{3}{y - 5} < 0$

19. $\dfrac{x + 1}{x - 4} \geq 0$

20. $\dfrac{x + 1}{x - 4} \leq 0$

Solve each inequality. Write the solution set in interval notation and graph the solution. See Example 5.

21. $\dfrac{3}{x - 2} < 4$

22. $\dfrac{-2}{y + 3} > 2$

23. $\dfrac{x^2 + 6}{5x} \geq 1$

24. $\dfrac{y^2 + 15}{8y} \leq 1$

Solve each inequality. Write the solution set in interval notation and graph the solution.

25. $(x - 8)(x + 7) > 0$

26. $(x - 5)(x + 1) < 0$

27. $(2x - 3)(4x + 5) \leq 0$

28. $(6x + 7)(7x - 12) > 0$

29. $x^2 > x$

30. $x^2 < 25$

31. $(2x - 8)(x + 4)(x - 6) \leq 0$

32. $(3x - 12)(x + 5)(2x - 3) \geq 0$

33. $6x^2 - 5x \geq 6$

34. $12x^2 + 11x \leq 15$

35. $x^4 - 26x^2 + 25 \geq 0$

36. $16x^4 - 40x^2 + 9 \leq 0$

37. $(2x - 7)(3x + 5) > 0$

38. $(4x - 9)(2x + 5) < 0$

39. $\dfrac{x}{x - 10} < 0$

40. $\dfrac{x + 10}{x - 10} > 0$

41. $\dfrac{x - 5}{x + 4} \geq 0$

42. $\dfrac{x - 3}{x + 2} \leq 0$

43. $\dfrac{x(x + 6)}{(x - 7)(x + 1)} \geq 0$

44. $\dfrac{(x - 2)(x + 2)}{(x + 1)(x - 4)} \leq 0$

45. $\dfrac{-1}{x - 1} > -1$

46. $\dfrac{4}{y + 2} < -2$

47. $\dfrac{x}{x + 4} \leq 2$

48. $\dfrac{4x}{x - 3} \geq 5$

49. $\dfrac{z}{z - 5} \geq 2z$

50. $\dfrac{p}{p + 4} \leq 3p$

51. $\dfrac{(x + 1)^2}{5x} > 0$

52. $\dfrac{(2x - 3)^2}{x} < 0$

53. A number minus its reciprocal is less than zero. Find the number.

54. Twice a number added to its reciprocal is nonnegative. Find the number.

Writing in Mathematics

55. Explain why $\dfrac{x + 2}{x - 3} > 0$ and $(x + 2)(x - 3) > 0$ have the same solutions.

56. Explain why $\dfrac{x + 2}{x - 3} \geq 0$ and $(x + 2)(x - 3) \geq 0$ do not have the same solutions.

Skill Review

Solve each quadratic equation by completing the square. See Section 8.1.

57. $x^2 + 4x = 12$ **58.** $y^2 + 6y = -5$ **59.** $z^2 + 10z - 1 = 0$ **60.** $x^2 + 14x + 20 = 0$

8.5
Quadratic Functions

OBJECTIVES

Tape 25

1.   Review quadratic functions and their graphs.
2.   Write quadratic functions in the form $y = a(x - h)^2 + k$.
3.   Derive a formula for finding the vertex of a parabola.
4.   Find the minimum or maximum value of a quadratic function.

1   Quadratic functions and their graphs were first introduced in Section 5.6. We discovered that the graph of a quadratic function is a **parabola** opening upward or downward. Remember that, since we are discussing the graphs of quadratic functions, all parabolas will pass the vertical line test. In this section, we introduce the standard form of a parabola,

$$y = a(x - h)^2 + k$$

Recall the definition of a quadratic function.

> **Quadratic Function**
>
> A quadratic function is a function that can be written in the form $f(x) = ax^2 + bx + c$, where a, b, and c are real numbers and $a \neq 0$.

Notice that equations of the form $y = ax^2 + bx + c$, where $a \neq 0$, also define quadratic functions since y is a function of x or $y = f(x)$.

Recall that if $a > 0$ the parabola opens upward, and if $a < 0$, the parabola opens downward. Also, the vertex of a parabola is the lowest point if the parabola opens upward and the highest point if the parabola opens downward. The axis of symmetry is the vertical line that passes through the vertex.

Recall that the x coordinate of the vertex of the graph of $f(x) = ax^2 + bx + c$ can be found by the formula $x = \dfrac{-b}{2a}$ which we derive at the end of this section.

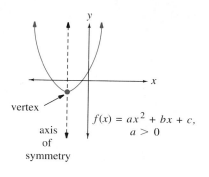

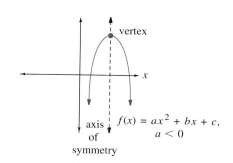

Vertex Formula

The graph of $f(x) = ax^2 + bx + c$, $a \neq 0$, is a parabola with vertex

$$\left(\frac{-b}{2a}, f\left(\frac{-b}{2a}\right)\right)$$

EXAMPLE 1   Graph $f(x) = x^2 - 4x - 12$.

Solution: The graph of this quadratic function is a parabola. To find the x-value of the vertex of this parabola, notice that $a = 1$, $b = -4$, and $c = -12$. Then

$$\frac{-b}{2a} = \frac{-(-4)}{2(1)} = 2$$

The x-value of the vertex is 2. To find the corresponding $f(x)$ or y-value, let $x = 2$. Then

$$f(2) = 2^2 - 4(2) - 12 = 4 - 8 - 12 = -16$$

The vertex is $(2, -16)$, and the axis of symmetry is the line $x = 2$. To find the y-intercept, let $x = 0$ and

$$f(0) = 0^2 - 4(0) - 12 = -12$$

Since $a > 0$, the parabola opens upward. This parabola opening upward with vertex $(2, -16)$ will have two x-intercepts. To find them, let $f(x)$ or $y = 0$.

$$0 = x^2 - 4x - 12$$

$$0 = (x - 6)(x + 2)$$

$$0 = x - 6 \quad \text{or} \quad 0 = x + 2$$

$$6 = x \quad \text{or} \quad -2 = x$$

The two x-intercepts are 6 and -2. The sketch of $f(x) = x^2 - 4x - 12$ is shown.

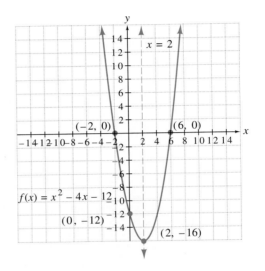

2   Another way to find the vertex of a parabola is to write the quadratic equation in the form $f(x) = a(x - h)^2 + k$, where a, h, and k are constants. The graph of this equation is a parabola with vertex (h, k). To write $f(x) = x^2 - 4x - 12$ in this form, we complete the square on the binomial $x^2 - 4x$.

$$f(x) = (x^2 - 4x) - 12$$

Now, add and subtract the square of half of -4.

$$\frac{1}{2}(-4) = -2 \quad \text{and} \quad (-2)^2 = 4$$

Add and subtract 4.

$$f(x) = (x^2 - 4x \;\; + 4 \;) - 12 \;\; - 4$$

$$f(x) = (x - 2)^2 - 16 \qquad\qquad \text{Factor.}$$

From this equation, we can see that $h = 2$ and $k = -16$, so the vertex of the parabola is $(2, -16)$.
    In general, we have the following.

Graph of a Quadratic Function

The graph of a quadratic function written in the form $f(x) = a(x - h)^2 + k$ is a parabola with vertex (h, k). If $a > 0$, the parabola opens upward, and if $a < 0$, the parabola opens downward. The axis of symmetry is the line whose equation is $x = h$.

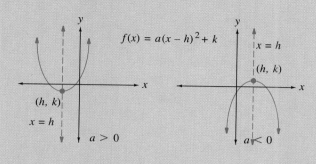

EXAMPLE 2   Graph $f(x) = 4(x - 1)^2 + 3$.

Solution: This equation is written in the form $f(x) = a(x - h)^2 + k$, where $a = 4$, $h = 1$, and $k = 3$. Since $a > 0$, the parabola opens upward. The vertex (h, k) is $(1, 3)$. The axis of symmetry is the line $x = 1$. Because the parabola opens upward and the vertex is in the first quadrant, this parabola has no x-intercepts. To find its y-intercept, let $x = 0$. Then

$$f(0) = 4(0 - 1)^2 + 3 = 7$$

This gives the point with coordinates $(0, 7)$. Use symmetry and these two points to sketch the parabola.

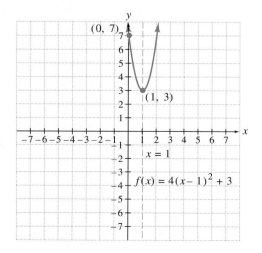

EXAMPLE 3   Graph $f(x) = x^2 + 2x - 3$. Find the vertex and any intercepts.

Solution: Write the equation in the form $f(x) = a(x - h)^2 + k$ to identify the vertex (h, k) by completing the square on x.

$$f(x) = (x^2 + 2x) - 3$$

Add and subtract the square of half of 2.

$$\frac{1}{2}(2) = 1 \quad \text{and} \quad 1^2 = 1$$

Add and subtract 1.

$$f(x) = (x^2 + 2x + 1) - 3 - 1$$
$$= (x + 1)^2 - 4 \qquad \qquad \text{Factor.}$$

Here $a = 1$, $h = -1$, and $k = -4$. This means that the parabola opens upward with vertex $(-1, -4)$ and axis of symmetry $x = -1$.

To find the y-intercept, let $x = 0$ and solve for y. Then

$$f(0) = 0^2 + 2(0) - 3 = -3$$

Thus, -3 is the y-intercept.

To find the x-intercepts, let y or $f(x) = 0$ and solve for x.

$$f(x) = x^2 + 2x - 3$$
$$0 = x^2 + 2x - 3 \qquad \qquad \text{Let } y = 0.$$
$$0 = (x + 3)(x - 1) \qquad \qquad \text{Factor.}$$
$$x + 3 = 0 \quad \text{or} \quad x - 1 = 0 \qquad \text{Set each factor equal to 0.}$$
$$x = -3 \qquad \qquad x = 1 \qquad \text{Solve.}$$

The x-intercepts are -3 and 1. Use these points to sketch the parabola.

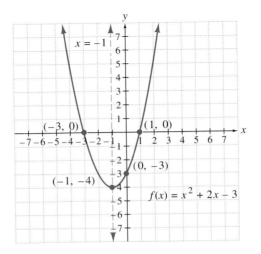

■

EXAMPLE 4   Graph $f(x) = 3x^2 + 3x + 1$. Find the vertex and any intercepts.

Solution: Complete the square on x to write the equation in the form $f(x) = a(x - h)^2 + k$.

$$f(x) = 3x^2 + 3x + 1$$

Factor 3 from the terms $3x^2 + 3x$ so that the coefficient of x^2 is 1.

$$f(x) = 3(x^2 + x) + 1$$

Then $\dfrac{1}{2}(1) = \dfrac{1}{2}$ and $\left(\dfrac{1}{2}\right)^2 = \dfrac{1}{4}$.

$$f(x) = \boxed{3}\left(x^2 + x + \boxed{\dfrac{1}{4}}\right) + 1 - \boxed{3\left(\dfrac{1}{4}\right)}.$$

$$= 3\left(x + \dfrac{1}{2}\right)^2 + \dfrac{4}{4} - \dfrac{3}{4}$$

$$= 3\left(x + \dfrac{1}{2}\right)^2 + \dfrac{1}{4}$$

Then $a = 3$, $h = -\dfrac{1}{2}$, and $k = \dfrac{1}{4}$. This means the parabola opens upward with vertex $\left(-\dfrac{1}{2}, \dfrac{1}{4}\right)$ and axis of symmetry the line $x = -\dfrac{1}{2}$.

To find the y-intercept, let $x = 0$. Then

$$f(0) = 3(0)^2 + 3(0) + 1 = 1$$

This parabola has no x-intercepts since the vertex is in the second quadrant and opens upward. Use the vertex, axis of symmetry, and y-intercept to sketch the parabola.

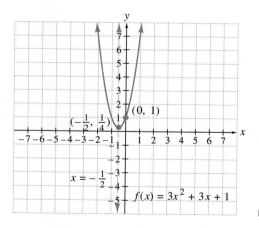

3   Now that we have practiced completing the square, we will now show that the x-coordinate of the vertex of the graph of $f(x) = ax^2 + bx + c$ can be found by the formula $x = \dfrac{-b}{2a}$. To do so, we complete the square on x and write the equation in the form $f(x) = a(x - h)^2 + k$.

First factor a from the terms $ax^2 + bx$.

$$f(x) = ax^2 + bx + c$$

$$f(x) = a\left(x^2 + \frac{b}{a}x\right) + c$$

Next, complete the square by finding the square of half of $\dfrac{b}{a}$.

$$\frac{1}{2}\left(\frac{b}{a}\right) = \frac{b}{2a} \text{ and } \left(\frac{b}{2a}\right)^2 = \frac{b^2}{4a^2}$$

$$f(x) = a\left(x^2 + \frac{b}{a}x + \frac{b^2}{4a^2}\right) + c - a\left(\frac{b^2}{4a^2}\right)$$

$$f(x) = a\left(x + \frac{b}{2a}\right)^2 + c - \frac{b^2}{4a}$$

Compare this form with $f(x) = a(x - h)^2 + k$ and see that h is $\dfrac{-b}{2a}$ which means the x-coordinate of the vertex of the graph of $f(x) = ax^2 + bx + c$ is $\dfrac{-b}{2a}$.

4   The quadratic function whose graph is a parabola that opens upward has a minimum value, and the quadratic function whose graph is a parabola that opens downward has a maximum value. The $f(x)$ or y-value of the vertex is the minimum or maximum value, and the corresponding x-value tells where the minimum or maximum value occurs.

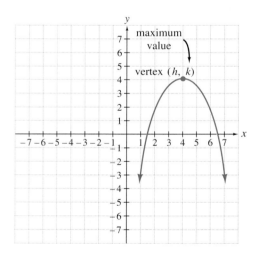

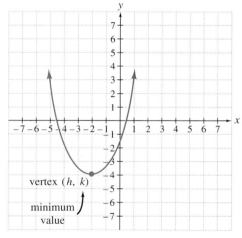

EXAMPLE 5   A rock is thrown upward from the ground. Its height in feet above ground after t seconds is given by the equation $f(t) = -16t^2 + 20t$. Find the maximum height of the rock.

Solution: The maximum height of the rock is the largest value of $f(t)$. Since the equation $f(t) = -16t^2 + 20t$ is a quadratic function, its graph is a parabola. It opens downward since $-16 < 0$. Thus, the maximum value of $f(t)$ is the $f(t)$ or y-value of the vertex of its graph. To find the vertex (h, k), notice that, for $f(t) = -16t^2 + 20t$, $a = -16$, $b = 20$, and $c = 0$. Thus

$$h = \frac{-b}{2a} = \frac{-20}{-32} = \frac{5}{8}$$

$$f\left(\frac{5}{8}\right) = -16\left(\frac{5}{8}\right)^2 + 20\left(\frac{5}{8}\right) = -16\left(\frac{25}{64}\right) + \frac{25}{2} = -\frac{25}{4} + \frac{50}{4} = \frac{25}{4}$$

The graph of $f(t)$ is a parabola opening downward with vertex $\left(\frac{5}{8}, \frac{25}{4}\right)$. This means that the rock's maximum height is $\frac{25}{4}$ feet or $6\frac{1}{4}$ feet, which was reached in $\frac{5}{8}$ second. ■

EXERCISE SET 8.5

Find the vertex of the graph of each quadratic function. See Example 1.

1. $f(x) = x^2 + 8x + 7$ **2.** $f(x) = x^2 + 6x + 5$ **3.** $f(x) = -x^2 + 10x + 5$ **4.** $f(x) = -x^2 - 8x + 2$

5. $f(x) = 5x^2 - 10x + 3$ **6.** $f(x) = -3x^2 + 6x + 4$ **7.** $f(x) = -x^2 + x + 1$ **8.** $f(x) = x^2 - 9x + 8$

Find the vertex of the graph of each quadratic function. Determine whether the graph opens upward or downward, find any intercepts, and sketch the graph. See Example 1.

9. $f(x) = x^2 + 4x - 5$ **10.** $f(x) = x^2 + 2x - 3$ **11.** $f(x) = -x^2 + 2x - 1$ **12.** $f(x) = -x^2 + 4x - 4$

13. $f(x) = x^2 - 4$ **14.** $f(x) = x^2 - 1$ **15.** $f(x) = 4x^2 + 4x - 3$ **16.** $f(x) = 2x^2 - x - 3$

Sketch the graph of each quadratic function. See Example 2.

17. $f(x) = (x - 1)^2 + 3$ **18.** $f(x) = (x - 2)^2 + 1$ **19.** $f(x) = 2(x + 4)^2 + 2$ **20.** $f(x) = 3(x - 3)^2 + 4$

21. $f(x) = -3(x + 2)^2 - 1$ **22.** $f(x) = -(x + 1)^2 - 2$ **23.** $f(x) = \frac{1}{2}(x - 3)^2 - 5$ **24.** $f(x) = \frac{1}{4}\left(x + \frac{1}{2}\right)^2 - 1$

Find the vertex of the graph of each quadratic function by completing the square. Determine whether the graph opens upward or downward, find any intercepts, and sketch the graph. See Example 3.

25. $f(x) = x^2 + 8x + 15$ **26.** $f(x) = x^2 + 10x + 9$ **27.** $f(x) = x^2 - 6x + 5$

28. $f(x) = x^2 - 4x + 3$ **29.** $f(x) = x^2 - 4x + 5$ **30.** $f(x) = x^2 - 6x + 11$

Find the vertex of the graph of each quadratic function by completing the square. Determine whether the graph opens upward or downward, and sketch the graph. See Example 4.

31. $f(x) = 2x^2 + 4x + 5$ **32.** $f(x) = 3x^2 + 12x + 16$ **33.** $f(x) = -2x^2 + 12x$ **34.** $f(x) = -4x^2 + 8x$

Solve. See Example 5.

35. Find two positive numbers whose sum is 60 and whose product is as large as possible. [*Hint:* Let x and $60 - x$ be the two positive numbers. Their product y can be described by the equation $y = x(60 - x)$.]

36. The length and width of a rectangle must have a sum of 40. Find the dimensions of the rectangle that will have maximum area. (Use the hint for Exercise 35.)

Sketch the graph of each quadratic function.

37. $f(x) = x^2 + 2$

38. $f(x) = (x - 3)^2$

39. $f(x) = (x + 1)^2 + 4$

40. $f(x) = (x + 5)^2 + 2$

41. $f(x) = 2(x - 3)^2 + 2$

42. $f(x) = 3(x - 4)^2 + 1$

43. $f(x) = -(x - 4)^2 + \dfrac{3}{2}$

44. $f(x) = -2(x + 7)^2 + \dfrac{1}{2}$

Find the vertex of the graph of each quadratic function. Determine whether the graph opens upward or downward, find any intercepts, and sketch the graph.

45. $f(x) = x^2 + 1$

46. $f(x) = x^2 + 4$

47. $f(x) = x^2 - 2x - 15$

48. $f(x) = x^2 - 4x + 3$

49. $f(x) = -5x^2 + 5x$

50. $f(x) = 3x^2 - 12x$

51. $f(x) = -x^2 + 2x - 12$

52. $f(x) = -x^2 + 8x - 17$

53. $f(x) = 3x^2 - 12x + 15$

54. $f(x) = 2x^2 - 8x + 11$

55. $f(x) = x^2 + x - 6$

56. $f(x) = x^2 + 3x - 18$

57. $f(x) = -2x^2 - 3x + 35$

58. $f(x) = 3x^2 - 13x - 10$

59. If a projectile is fired straight upward from the ground with an initial speed of 96 feet per second, then its height h in feet after t seconds is given by the equation

$$h(t) = -16t^2 + 96t$$

Find the maximum height of the projectile.

60. If Rheam throws a ball upward with an initial speed of 32 feet per second, then its height h in feet after t seconds is given by the equation

$$h(t) = -16t^2 + 32t$$

Find the maximum height of the ball.

61. The cost C in dollars of manufacturing x bicycles at Hol- laday's Production Plant is given by the function $C(x) = 2x^2 - 800x + 92{,}000$.

a. Find the number of bicycles that must be manufactured to minimize the cost.

b. Find the minimum cost.

62. The Utah Ski Club sells calendars to raise money. The profit P, in cents, from selling x calendars is given by the equation

$$P(x) = 360x - x^2$$

a. Find how many calendars must be sold to maximize profit.

b. Find the maximum profit.

A Look Ahead

Find the vertex of the graph of each quadratic function. Determine whether the graph opens upward or downward, find any intercepts, and sketch the graph.

63. $f(x) = x^2 - 4x + 2$

64. $f(x) = x^2 - 2x - 2$

65. $f(x) = 6x^2 - 7$

66. $f(x) = -4x^2 + 5$

67. $f(x) = -4x^2 + 12x - 7$

68. $f(x) = \frac{1}{8}x^2 - \frac{1}{2}x - 1$

Writing in Mathematics

69. Consider a quadratic function whose graph is a parabola opening upward. Explain whether this function has a minimum value, a maximum value, or neither.

Skill Review

Find the distance between each pair of points. See Section 3.1.

70. $(-1, 6), (2, 4)$

71. $(-5, -1), (1, 7)$

72. $(5, 7), (5, 9)$

73. $(-2, -3), (5, -3)$

Determine whether each graph is the graph of a function. See Section 3.5.

74.

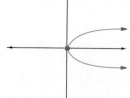

75.

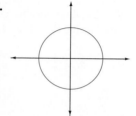

8.6
The Parabola and the Circle

Tape 26

OBJECTIVES		
	1	Graph parabolas of the forms $x = a(y - k)^2 + h$ and $y = a(x - h)^2 + k$.
	2	Graph circles of the form $(x - h)^2 + (y - k)^2 = r^2$.
	3	Write the equation of a circle given its center and radius.
	4	Find the center and the radius of a circle, given its equation.

We have analyzed some of the important connections between a parabola and its equation. Parabolas are interesting in their own right, but more interesting still because they are part of a collection of curves known as **conic sections.** Conic sections derive their name from the fact that each conic section is the intersection of a right circular cone and a plane. The circle, parabola, ellipse, and hyperbola are the conic sections.

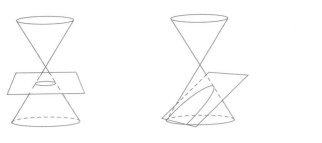

Circle   Parabola   Ellipse
                                                                                Hyperbola

1   Thus far, we have learned that $f(x)$ or $y = a(x - h)^2 + k$ is the equation of a parabola that opens upward if $a > 0$ or downward if $a < 0$. Parabolas can also open left or right, or even on a slant. Equations of these parabolas are not functions of x, of course, since a parabola opening any other way but upward or downward fails the vertical line test. In this section, we introduce parabolas opening to the left and to the right.

Just as $y = a(x - h)^2 + k$ is the equation of a parabola opening upward or downward, $x = a(y - k)^2 + h$ is the equation of a parabola opening to the right or to the left. The parabola opens to the right if $a > 0$ and to the left if $a < 0$. The parabola has vertex (h, k), and its axis of symmetry is the line $y = k$.

Parabolas

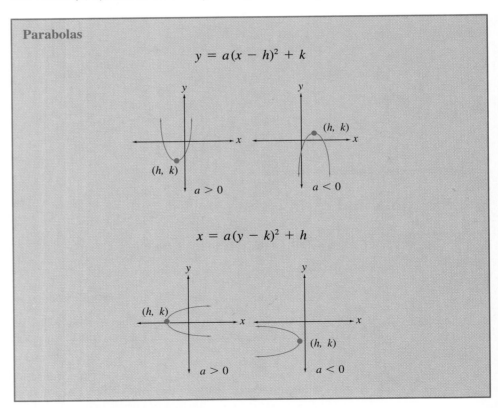

The forms $y = a(x - h)^2 + k$ and $x = a(y - k)^2 + h$ are both called **standard forms.**

EXAMPLE 1   Graph the parabola $x = 2y^2$.

Solution: The equation $x = 2y^2$ written in standard form is $x = 2(y - 0)^2 + 0$ with $a = 2$, $h = 0$, and $k = 0$. Its graph is a parabola with vertex $(0, 0)$ and axis of symmetry the line $y = 0$. Since $a > 0$, this parabola opens to the right. The following table shows a few more ordered pair solutions of $x = 2y^2$. Its graph is also shown.

x	y
8	-2
2	-1
0	0
2	1
8	2

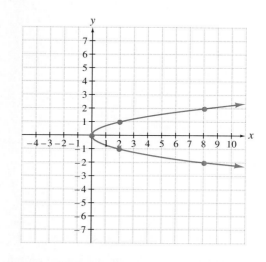

EXAMPLE 2   Graph the parabola $x = -3(y - 1)^2 + 2$.

Solution: The equation $x = -3(y - 1)^2 + 2$ is in the form $x = a(y - k)^2 + h$ with $a = -3$, $k = 1$, and $h = 2$. Since $a < 0$, the parabola opens to the left. The vertex (h, k) is $(2, 1)$, and the axis of symmetry is the line $y = 1$. When $y = 0$, $x = -1$, the x-intercept. The parabola is sketched next.

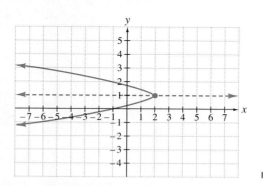

EXAMPLE 3   Graph $y = -x^2 - 2x + 15$.

Solution: Complete the square on x to write the equation in standard form.

$$y = -(x^2 + 2x) + 15$$

$$= -(x^2 + 2x + 1) + 15 + 1 \qquad \text{Subtract and add 1.}$$

$$= -(x + 1)^2 + 16 \qquad \text{Factor.}$$

The equation is now in the form $y = a(x - h)^2 + k$ with $a = -1$, $h = -1$, and $k = 16$. The parabola opens downward since $a < 0$ and has vertex $(-1, 16)$. Its axis of symmetry is the line $x = -1$. The y-intercept is 15.

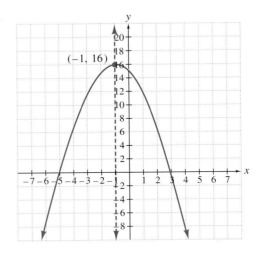

EXAMPLE 4 Graph $x = 2y^2 + 4y + 5$.

Solution: Complete the square on y to write the equation in standard form.

$$x = 2(y^2 + 2y) + 5$$
$$= 2(y^2 + 2y + 1) + 5 - 2(1) \qquad \text{Add and subtract } 2(1) \text{ or } 2.$$
$$= 2(y + 1)^2 + 3 \qquad\qquad \text{Factor.}$$

The equation is now in the form $x = a(y - k)^2 + h$ with $a = 2$, $k = -1$, and $h = 3$. The parabola opens to the right since $a > 0$ and has vertex $(3, -1)$. Its axis of symmetry is the line $y = -1$. The x-intercept is 5.

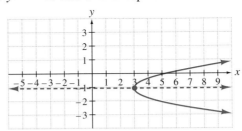

2 Another conic section is the **circle.** A circle is the set of all points in a plane that are the same distance from a fixed point called the **center.** The distance is called the **radius** of the circle. To find a standard equation for a circle, let (h, k) represent the center of the circle, and let (x, y) represent any point on the circle. The distance between (h, k) and (x, y) is defined to be the radius, r units. We can find this distance r by using the distance formula.

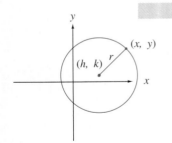

$$r = \sqrt{(x - h)^2 + (y - k)^2}$$
$$r^2 = (x - h)^2 + (y - k)^2 \qquad \text{Square both sides.}$$

Circle

The graph of $(x - h)^2 + (y - k)^2 = r^2$ is a circle with center (h, k) and radius r.

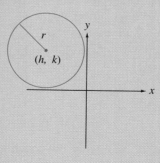

EXAMPLE 5   Graph $x^2 + y^2 = 4$.

Solution: The equation can be written in standard form as

$$(x - 0)^2 + (y - 0)^2 = 2^2$$

The center of the circle is $(0, 0)$, and the radius is 2. Its graph is shown.   ∎

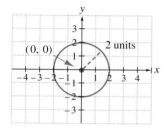

HELPFUL HINT

Notice the difference between the equation of a circle and the equation of a parabola. The equation of a circle contains both x^2 and y^2 terms with equal coefficients. The equation of a parabola has either an x^2 term or a y^2 term, but not both.

EXAMPLE 6   Graph $(x + 1)^2 + y^2 = 8$.

Solution: The equation can be written as $(x + 1)^2 + (y - 0)^2 = 8$ with $h = -1$, $k = 0$, and $r = \sqrt{8}$. The center is $(-1, 0)$, and the radius is $\sqrt{8} = 2\sqrt{2} \approx 2.8$

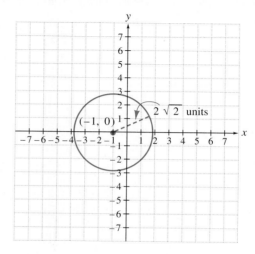

∎

3 Next, we practice writing the equation of a circle given its center and radius.

EXAMPLE 7   Find the equation of the circle with the center $(-7, 3)$ and radius 10.

Solution: We are given that $h = -7$, $k = 3$, and $r = 10$. Using these values, we write the equation

$$(x - h)^2 + (y - k)^2 = r^2$$

or

$$[x - (-7)]^2 + (y - 3)^2 = 10^2$$

or

$$(x + 7)^2 + (y - 3)^2 = 100 \qquad ∎$$

4   To write the equation of a circle in standard form, we complete the square on both x and y.

EXAMPLE 8   Graph $x^2 + y^2 + 4x - 8y = 16$.

Solution: Since this equation contains x^2 and y^2 terms whose coefficients are equal, its graph is a circle. To write the equation in standard form, group the terms involving x and the terms involving y, and then complete the square on each variable.

$$(x^2 + 4x) + (y^2 - 8y) = 16$$

$\frac{1}{2}(4) = 2$ and $2^2 = 4$. Also $\frac{1}{2}(-8) = -4$ and $(-4)^2 = 16$. Add 4 and then 16 to both sides.

$$(x^2 + 4x + 4) + (y^2 - 8y + 16) = 16 + 4 + 16$$

$$(x + 2)^2 + (y - 4)^2 = 36 \qquad \text{Factor.}$$

This circle has the center $(-2, 4)$ and radius 6, as shown.   ∎

MENTAL MATH

The graph of each equation is a parabola. Determine whether the parabola opens upward, downward, to the left, or to the right.

1. $y = x^2 - 7x + 5$ **2.** $y = -x^2 + 16$ **3.** $x = -y^2 - y + 2$

4. $x = 3y^2 + 2y - 5$ **5.** $y = -x^2 + 2x + 1$ **6.** $x = -y^2 + 2y - 6$

EXERCISE SET 8.6

Sketch the graph of each equation. See Examples 1 and 2.

1. $x = 3y^2$ **2.** $x = -2y^2$ **3.** $x = (y - 2)^2 + 3$ **4.** $x = (y - 4)^2 - 1$

5. $y = 3(x - 1)^2 + 5$ **6.** $y = -(x - 2)^2 - 3$ **7.** $x = -3(y - 1)^2 - 2$ **8.** $x = -4(y - 2)^2 + 2$

Find the vertex of the graph of each equation, and sketch its graph. See Examples 3 and 4.

9. $x = y^2 + 6y + 8$ **10.** $x = y^2 - 6y + 6$ **11.** $y = x^2 + 10x + 20$ **12.** $y = x^2 + 4x - 5$

13. $x = -2y^2 + 4y + 6$ **14.** $x = 3y^2 + 6y + 7$ **15.** $x = 4y^2 + 24y + 37$ **16.** $x = -y^2 + 2y + 1$

Sketch the graph of each equation. See Examples 5 and 6.

17. $x^2 + y^2 = 9$ **18.** $x^2 + y^2 = 25$ **19.** $x^2 + (y - 2)^2 = 1$

20. $(x - 3)^2 + y^2 = 9$ **21.** $(x - 5)^2 + (y + 2)^2 = 1$ **22.** $(x + 3)^2 + (y + 3)^2 = 4$

Write an equation of the circle with the given center and radius. See Example 7.

23. $(2, 3)$; 6

24. $(-7, 6)$; 2

25. $(0, 0)$; 2

26. $(0, -6)$; 10

27. $(-5, 4)$; $\sqrt{5}$

28. the origin, $\sqrt{7}$

The graph of each equation is a circle. Find the center and the radius, and then sketch. See Example 8.

29. $x^2 + y^2 + 6y = 0$

30. $x^2 + 10x + y^2 = 0$

31. $x^2 + y^2 + 2x - 4y = 4$

32. $x^2 + 6x - 4y + y^2 = 3$

Sketch the graph of each equation. If the graph is a parabola, find its vertex. If the graph is a circle, find its center and radius.

33. $x = y^2 + 2$

34. $x = y^2 - 3$

35. $y = (x + 3)^2 + 3$

36. $y = (x - 2)^2 - 2$

37. $x^2 + y^2 = 49$

38. $x^2 + y^2 = 1$

39. $x = (y - 1)^2 + 4$

40. $x = (y + 3)^2 - 1$

41. $(x + 3)^2 + (y - 1)^2 = 9$

42. $(x - 2)^2 + (y - 2)^2 = 16$

43. $x = -2(y + 5)^2$

44. $x = -(y - 1)^2$

45. $x^2 + (y + 5)^2 = 5$

46. $(x - 4)^2 + y^2 = 7$

47. $y = 3(x - 4)^2 + 2$

48. $y = 5(x + 5)^2 + 3$

49. $2x^2 + 2y^2 = \dfrac{1}{2}$

50. $\dfrac{x^2}{8} + \dfrac{y^2}{8} = 2$

51. $y = x^2 - 2x - 15$

52. $y = x^2 + 7x + 6$

53. $x^2 + y^2 + 6x + 10y - 2 = 0$

54. $x^2 + y^2 + 2x + 12y - 12 = 0$

55. $x = y^2 + 6y + 2$

56. $x = y^2 + 8y - 4$

57. $x^2 + y^2 - 8y + 5 = 0$ **58.** $x^2 - 10y + y^2 + 4 = 0$ **59.** $x = -2y^2 - 4y$

60. $x = -3y^2 + 30y$ **61.** $\dfrac{x^2}{3} + \dfrac{y^2}{3} = 2$ **62.** $5x^2 + 5y^2 = 25$

63. $y = 4x^2 - 40x + 105$ **64.** $y = 5x^2 - 20x + 16$

65. Doris, an architect, is drawing plans on grid paper for a circular pool with a fountain in the middle. The paper is marked off in centimeters, where each centimeter represents a foot. On the paper, the diameter of the "pool" is 20 centimeters, and "fountain" is the point (0, 0).
   a. Sketch the architect's drawing. Be sure to label the axes.

   b. Write an equation describing the circular pool.
   c. Doris plans to place a circle of lights around the fountain so that each light is 5 feet from the fountain. Write an equation for the circle of lights, and sketch the circle on your drawing.

Writing in Mathematics

66. If you are given a list of equations of circles and parabolas and none are in standard form, explain how you would determine which is an equation of a circle and which is an equation of a parabola. Explain also how you would distinguish the upward or downward parabolas from the left or right parabolas.

Skill Review

Graph each equation. See Section 3.2.

67. $y = 2x + 5$ **68.** $y = -3x + 3$ **69.** $y = 3$ **70.** $x = -2$

Rationalize each denominator and simplify if possible. See Section 7.4.

71. $\dfrac{1}{\sqrt{3}}$ **72.** $\dfrac{\sqrt{5}}{\sqrt{8}}$ **73.** $\dfrac{4\sqrt{7}}{\sqrt{6}}$ **74.** $\dfrac{10}{\sqrt{5}}$

8.7
The Ellipse and the Hyperbola

OBJECTIVES

Tape 26

1   Define the standard form equation for an ellipse.

2   Graph an ellipse.

3   Define the standard form equation for a hyperbola.

4   Graph a hyperbola.

1   An **ellipse** can be thought of as the set of points in a plane the sum of whose distances from two fixed points is constant. The two fixed points are each called a **focus.** The plural of focus is **foci.** The point midway between the foci is called the **center.**

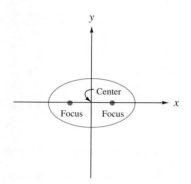

It can be shown that the standard form of an ellipse with center (0, 0) is $\dfrac{x^2}{a^2} + \dfrac{y^2}{b^2} = 1$.

Ellipse with Center (0, 0)

The graph of an equation of the form $\dfrac{x^2}{a^2} + \dfrac{y^2}{b^2} = 1$ is an ellipse with center $(0, 0)$.

The x-intercepts are a and $-a$, and the y-intercepts are b and $-b$.

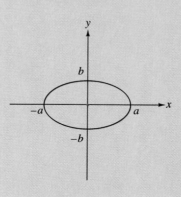

2   Next, we practice graphing ellipses.

EXAMPLE 1   Graph $\dfrac{x^2}{9} + \dfrac{y^2}{16} = 1$.

Solution: The equation is of the form $\dfrac{x^2}{a^2} + \dfrac{y^2}{b^2} = 1$, with $a = 3$ and $b = 4$, so its graph is an ellipse with center $(0, 0)$, x-intercepts 3 and -3, and y-intercepts 4 and -4 as graphed next.

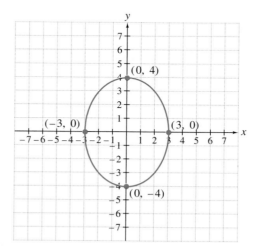

EXAMPLE 2   Graph the equation $4x^2 + 16y^2 = 64$.

Solution: This is not the equation of a circle since the coefficients of x^2 and y^2 are not the same. When this happens, the graph is an ellipse. Since the standard form of the equation of an ellipse has a 1 on one side, we first divide both sides of this equation by 64.

$$4x^2 + 16y^2 = 64$$

$$\frac{4x^2}{64} + \frac{16y^2}{64} = \frac{64}{64} \qquad \text{Divide both sides by 64.}$$

$$\frac{x^2}{16} + \frac{y^2}{4} = 1 \qquad \text{Simplify.}$$

We now recognize the equation of an ellipse with center $(0, 0)$, x-intercepts 4 and -4, and y-intercepts 2 and -2.

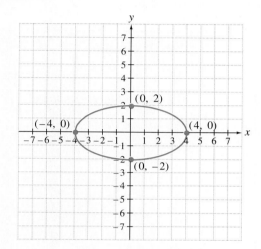

The center of an ellipse is not always $(0, 0)$, as shown in the next example.

EXAMPLE 3   Graph $\dfrac{(x + 3)^2}{25} + \dfrac{(y - 2)^2}{36} = 1$.

Solution: This ellipse has center $(-3, 2)$. Also notice that $a = 5$ and $b = 6$. To find four points on the graph of the ellipse, first graph the center $(-3, 2)$. Since $a = 5$, count 5 units right and 5 units left of the point with coordinates $(-3, 2)$. Next, since $b = 6$, start at $(-3, 2)$ and count 6 units up and then 6 units down to find 2 more points on the ellipse.

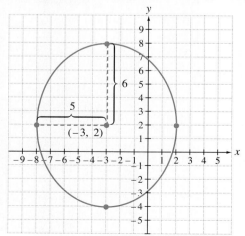

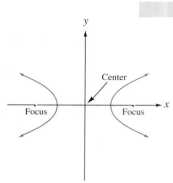

3 A final conic section is the **hyperbola.** A hyperbola is the set of points in a plane such that the absolute value of the difference of the distance from two fixed points is constant. The two fixed points are each called a **focus.** The point midway between the foci is called the **center.**

It can be shown that the graph of $\dfrac{x^2}{a^2} - \dfrac{y^2}{b^2} = 1$ is a hyperbola with x-intercepts a and $-a$. Also, the graph of $\dfrac{y^2}{b^2} - \dfrac{x^2}{a^2} = 1$ is a hyperbola with y-intercepts b and $-b$, as shown next.

Hyperbola with Center (0, 0)

The graph of an equation of the form $\dfrac{x^2}{a^2} - \dfrac{y^2}{b^2} = 1$ is a hyperbola with center $(0, 0)$ and x-intercepts a and $-a$.

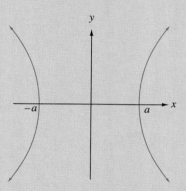

The graph of an equation of the form $\dfrac{y^2}{b^2} - \dfrac{x^2}{a^2} = 1$ is a hyperbola with center $(0, 0)$ and y-intercepts b and $-b$.

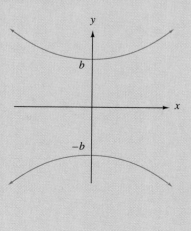

Graphing a hyperbola such as $\dfrac{y^2}{b^2} - \dfrac{x^2}{a^2} = 1$ is made easier by recognizing one of its important characteristics. Examining the following figure, notice how the sides of the branches of the hyperbola extend indefinitely and seem to more and more resemble the dashed lines in the figure. These dashed lines are called the **asymptotes** of the hyperbola.

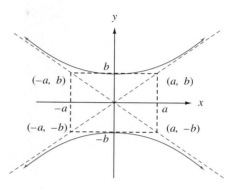

To sketch these lines, or asymptotes, draw a rectangle with vertices (a, b), $(-a, b)$, $(a, -b)$, and $(-a, -b)$. The asymptotes of this graph will be the extended diagonals of this rectangle.

4   Next, we practice graphing hyperbolas.

EXAMPLE 4   Sketch the graph of $\dfrac{x^2}{16} - \dfrac{y^2}{25} = 1$.

Solution: This equation has the form $\dfrac{x^2}{a^2} - \dfrac{y^2}{b^2} = 1$, with $a = 4$ and $b = 5$. Thus its graph is a hyperbola with center $(0, 0)$ and x-intercepts 4 and -4. To aid in graphing the hyperbola, we first sketch its asymptotes. The diagonals of the rectangle with coordinates $(4, 5)$, $(4, -5)$, $(-4, 5)$, and $(-4, -5)$ can be used to sketch the asymptotes of the hyperbola. Then use the asymptotes to aid in sketching the hyperbola.

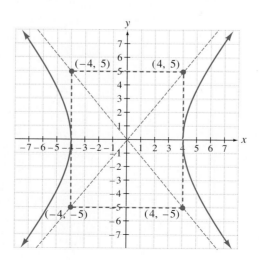

EXAMPLE 5 Sketch the graph of the equation $4y^2 - 9x^2 = 36$.

Solution: Since this is a difference of squared terms in x and y, its graph is a hyperbola, as opposed to an ellipse or a circle. The standard form of the equation of a hyperbola has a 1 on one side, so divide both sides of the equation by 36.

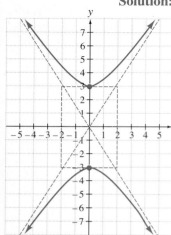

$$4y^2 - 9x^2 = 36$$

$$\frac{4y^2}{36} - \frac{9x^2}{36} = \frac{36}{36} \qquad \text{Divide both sides by 36.}$$

$$\frac{y^2}{9} - \frac{x^2}{4} = 1 \qquad \text{Simplify.}$$

The equation is of the form $\dfrac{y^2}{b^2} - \dfrac{x^2}{a^2} = 1$, with $a = 2$ and $b = 3$, so the hyperbola is centered at $(0, 0)$ with y-intercepts 3 and -3. The sketch of the hyperbola is shown. ∎

The following box provides a summary of conic sections.

Conic Sections

	Standard Form	**Graph**
Parabola	$y = a(x - h)^2 + k$	

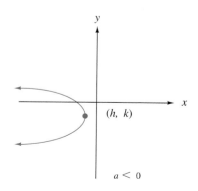

$x = a(y - k)^2 + h$

	Standard Form	**Graph**
Circle	$(x - h)^2 + (y - k)^2 = r^2$	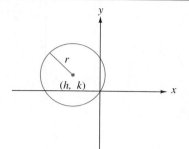
Ellipse	$\dfrac{x^2}{a^2} + \dfrac{y^2}{b^2} = 1$	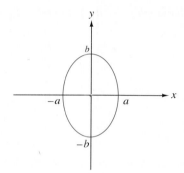
Hyperbola	$\dfrac{x^2}{a^2} - \dfrac{y^2}{b^2} = 1$	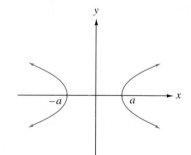
	$\dfrac{y^2}{b^2} - \dfrac{x^2}{a^2} = 1$	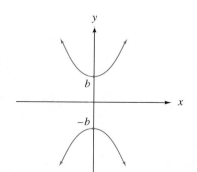

EXERCISE SET 8.7

Sketch the graph of each equation. See Example 1.

1. $\dfrac{x^2}{4} + \dfrac{y^2}{25} = 1$

2. $\dfrac{x^2}{9} + y^2 = 1$

3. $\dfrac{x^2}{16} + \dfrac{y^2}{9} = 1$

4. $x^2 + \dfrac{y^2}{4} = 1$

Sketch the graph of each equation. See Example 2.

5. $9x^2 + 4y^2 = 36$

6. $x^2 + 4y^2 = 16$

7. $4x^2 + 25y^2 = 100$

8. $36x^2 + y^2 = 36$

Sketch the graph of each equation. See Example 3.

9. $\dfrac{(x + 1)^2}{36} + \dfrac{(y - 2)^2}{49} = 1$

10. $\dfrac{(x - 3)^2}{9} + \dfrac{(y + 3)^2}{16} = 1$

11. $\dfrac{(x - 1)^2}{4} + \dfrac{(y - 1)^2}{25} = 1$

12. $\dfrac{(x + 3)^2}{16} + \dfrac{(y + 2)^2}{4} = 1$

Sketch the graph of each equation. See Example 4.

13. $\dfrac{x^2}{4} - \dfrac{y^2}{9} = 1$

14. $\dfrac{x^2}{36} - \dfrac{y^2}{36} = 1$

15. $\dfrac{y^2}{25} - \dfrac{x^2}{16} = 1$

16. $\dfrac{y^2}{25} - \dfrac{x^2}{49} = 1$

Sketch the graph of each equation. See Example 5.

17. $x^2 - 4y^2 = 16$ **18.** $4x^2 - y^2 = 36$ **19.** $16y^2 - x^2 = 16$ **20.** $4y^2 - 25x^2 = 100$

Identify whether each equation when graphed will be a parabola, circle, ellipse, or hyperbola. Sketch the graph of each equation.

21. $(x - 7)^2 + (y - 2)^2 = 4$ **22.** $y = x^2 + 4$ **23.** $y = x^2 + 12x + 36$

24. $\dfrac{x^2}{4} + \dfrac{y^2}{9} = 1$ **25.** $\dfrac{y^2}{9} - \dfrac{x^2}{9} = 1$ **26.** $\dfrac{x^2}{16} - \dfrac{y^2}{4} = 1$

27. $\dfrac{x^2}{16} + \dfrac{y^2}{4} = 1$ **28.** $x^2 + y^2 = 16$ **29.** $x = y^2 + 4y - 1$

30. $x = -y^2 + 6y$ **31.** $9x^2 - 4y^2 = 36$ **32.** $9x^2 + 4y^2 = 36$

33. $\dfrac{(x-1)^2}{49} + \dfrac{(y+2)^2}{25} = 1$

34. $y^2 = x^2 + 16$

35. $\left(x + \dfrac{1}{2}\right)^2 + \left(y - \dfrac{1}{2}\right)^2 = 1$

36. $y = -2x^2 + 4x - 3$

37. A planet's orbit about the sun can be described as an ellipse. Consider the sun as the origin of a rectangular coordinate system. Suppose the x-intercepts of the elliptical path of the planet are $\pm 130,000,000$, and the y-intercepts are $\pm 125,000,000$. Write the equation of the elliptical path of the planet.

38. Comets orbit the sun in elongated ellipses. Again consider the sun as the origin of a rectangular coordinate system. Suppose the equation of the path of a comet is

$$\dfrac{(x - 1,782,000,000)^2}{(3.42)(10^{23})} + \dfrac{(y - 356,400,000)^2}{(1.368)(10^{22})} = 1.$$

Find the center of the path of the comet.

A Look Ahead

Sketch the graph of each equation. See the following example.

EXAMPLE   Sketch the graph of $\dfrac{(x-2)^2}{25} - \dfrac{(y-1)^2}{9} = 1.$

Solution: This hyperbola has center $(2, 1)$. Notice that $a = 5$ and $b = 3$.

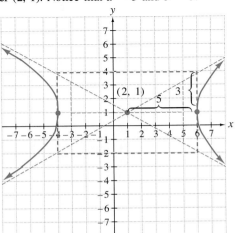

39. $\dfrac{(x-1)^2}{4} - \dfrac{(y+1)^2}{25} = 1$

40. $\dfrac{(x+2)^2}{9} - \dfrac{(y-1)^2}{4} = 1$

41. $\dfrac{y^2}{16} - \dfrac{(x+3)^2}{9} = 1$

42. $\dfrac{(y+4)^2}{4} - \dfrac{x^2}{25} = 1$

43. $\dfrac{(x+5)^2}{16} - \dfrac{(y+2)^2}{25} = 1$

44. $\dfrac{(x-3)^2}{9} - \dfrac{(y-2)^2}{4} = 1$

Skill Review

Solve each inequality. See Section 2.6.

45. $x < 5$ and $x < 1$

46. $x < 5$ or $x < 1$

47. $2x - 1 \geq 7$ or $-3x \leq -6$

48. $2x - 1 \geq 7$ and $-3x \leq -6$

Perform indicated operations. See Sections 4.1 and 4.3.

49. $(2x^3)(-4x^2)$

50. $2x^3 - 4x^3$

51. $-5x^2 + x^2$

52. $(-5x^2)(x^2)$

CRITICAL THINKING   Civil engineers constructing sidewalks use a series of separate concrete blocks, not one continuous concrete slab. The reason is that concrete expands when it gets hot, so by using separate blocks, engineers leave room for expansion.

Ambitiously laying his own patio, Dave estimated he will leave 1/4 foot on each of two sides of a 3x3-foot concrete square to cover an area of 9 1/16 square feet. He reasoned that $3^2 + (1/4)^2 = 9 + 1/16 = 9\ 1/16$. When he followed his plan, leaving 1/4 foot around two sides, he discovered he filled an area roughly 10 1/2 square feet, not the required 9 1/16 square feet. Can you explain the mistake in Dave's thinking, and can you suggest the proper amount of room to leave around each side so that the 9 1/16-square-foot area is accounted for?

CHAPTER 8 GLOSSARY

A **circle** is the set of all points in a plane that are the same distance from a fixed point called the **center.** The distance is called the **radius** of the circle.

Conic sections derive their name from the fact that each conic section is the intersection of a right circular cone and a plane. The circle, parabola, ellipse, and hyperbola are the conic sections.

An **ellipse** can be thought of as the set of points in a plane the sum of whose distances from two fixed points is constant. The two fixed points are each called a **focus.** The plural of focus is **foci.** The point midway between the foci is called the **center.**

A **hyperbola** is the set of points in a plane such that the absolute value of the difference of the distance from two fixed points is constant. The two fixed points are each called a **focus.** The point midway between the foci is called the **center.**

A **quadratic** or **second-degree equation** is an equation that can be written in the form $ax^2 + bx + c = 0$, where a, b, and c are real numbers and $a \neq 0$.

A **quadratic function** is a function that can be defined by the equation $f(x) = ax^2 + bx + c$, where a, b, and c are real numbers and $a \neq 0$.

A **quadratic inequality** is an inequality that can be written so that one side is a quadratic expression and the other side is 0.

The process of writing a quadratic equation so that one side is a perfect square trinomial is called **completing the square.**

In the quadratic formula $x = \dfrac{-b \pm \sqrt{b^2 - 4ac}}{2a}$, the radicand $b^2 - 4ac$ is called the **discriminant** because, by knowing its value, we can **discriminate** among the possibilities and number of solutions of a quadratic equation.

CHAPTER 8 SUMMARY

SQUARE ROOT PROPERTY (8.1)
If b is a real number and if $a^2 = b$, then $a = \pm \sqrt{b}$.

QUADRATIC FORMULA (8.2)
A quadratic equation written in the form $ax^2 + bx + c = 0$ has the solutions
$$x = \frac{-b \pm \sqrt{b^2 - 4ac}}{2a}$$

VERTEX FORMULA (8.5)
The graph of $f(x) = ax^2 + bx + c$, $a \neq 0$, is a parabola with vertex
$$\left(\frac{-b}{2a}, f\left(\frac{-b}{2a} \right) \right)$$

GRAPH OF A QUADRATIC FUNCTION (8.5)
The graph of a quadratic function written in the form $f(x) = a(x - h)^2 + k$ is a parabola with vertex (h, k). If $a > 0$, the parabola opens upward, and if $a < 0$, the parabola opens downward. The axis of symmetry is the line whose equation is $x = h$.

CONIC SECTIONS   (8.7)

	Standard Form	Graph

Parabola $y = a(x - h)^2 + k$

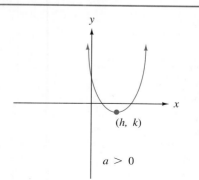

$x = a(y - k)^2 + h$

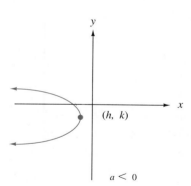

Circle $(x - h)^2 + (y - k)^2 = r^2$

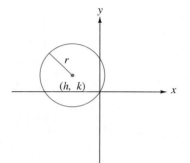

	Standard Form	Graph
Ellipse	$\dfrac{x^2}{a^2} + \dfrac{y^2}{b^2} = 1$	
Hyperbola	$\dfrac{x^2}{a^2} - \dfrac{y^2}{b^2} = 1$	
	$\dfrac{y^2}{b^2} - \dfrac{x^2}{a^2} = 1$	

CHAPTER 8 REVIEW

(8.1) *Solve by factoring.*

1. $x^2 - 15x + 14 = 0$ **2.** $x^2 - x - 30 = 0$ **3.** $10x^2 = 3x + 4$ **4.** $7a^2 = 29a + 30$

Solve using the square root property.

5. $4m^2 = 196$ **6.** $9y^2 = 36$ **7.** $(9n + 1)^2 = 9$ **8.** $(5x - 2)^2 = 2$

Solve by completing the square.

9. $z^2 + 3z + 1 = 0$ **10.** $x^2 + x + 7 = 0$ **11.** $(2x + 1)^2 = x$ **12.** $(3x - 4)^2 = 10x$

(8.2) *If the discriminant of a quadratic equation has the given value, determine the number and type of solutions of the equation.*

13. -8 **14.** 48   **15.** 100   **16.** 0

Solve by using the quadratic formula.

17. $x^2 - 16x + 64 = 0$ **18.** $x^2 + 5x = 0$ **19.** $x^2 + 11 = 0$ **20.** $2x^2 + 3x = 5$

21. $6x^2 + 7 = 5x$ **22.** $9a^2 + 4 = 2a$ **23.** $(5a - 2)^2 - a = 0$ **24.** $(2x - 3)^2 = x$

(8.3)   *Solve each equation for the variable.*

25. $x^3 = 27$

26. $y^3 = -64$

27. $\dfrac{5}{x} + \dfrac{6}{x - 2} = 3$

28. $\dfrac{7}{8} = \dfrac{8}{x^2}$

29. $x^4 - 21x^2 - 100 = 0$

30. $5(x + 3)^2 - 19(x + 3) = 4$

31. $x^{2/3} - 6x^{1/3} + 5 = 0$

32. $x^{2/3} - 6x^{1/3} = -8$

33. $a^6 - a^2 = a^4 - 1$

34. $(m - 4)^5 - 2(m - 4)^3 = (m - 4)^2 - 2$

Solve.

35. Find two consecutive even whole numbers such that the sum of their squares is 100.

36. A daycare center must have a play yard with an area of 1200 square feet. The rectangular shape of the property requires the length to be three times the width. Find the dimensions of the yard.

37. Cadets graduating from military school usually toss their hats high into the air at the end of the ceremony. One cadet threw his hat so that its distance $d(t)$ in feet above the ground t seconds after it was thrown was $d(t) = -16t^2 + 30t + 6$.
  a. Find the distance above the ground of the hat 1 second after it was thrown.
  b. Find the time it takes the hat to hit the ground.

38. The product of a number and 3 less than twice the number is 405. Find the number.

39. The middle number of three consecutive positive even integers is 26 less than $\dfrac{1}{10}$ the product of the other two. Find the integers.

40. The hypotenuse of an isosceles right triangle is 6 centimeters longer than either of the legs. Find the length of the legs.

41. A negative number decreased by its reciprocal is $-\dfrac{24}{5}$. Find the number.

42. One force of 30 pounds pulls an object to the left and another force of 40 pounds pulls the same object downward. Find the resulting force of the pulling. That is, find the length of the diagonal of the rectangle in the sketch.

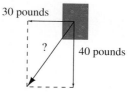

30 pounds

?

40 pounds

(8.4) *Solve each inequality for x. Write each solution set in interval notation and then graph the solution set.*

43. $2x^2 - 50 \leq 0$

44. $\frac{1}{4}x^2 < \frac{1}{16}$

45. $(2x - 3)(4x + 5) \geq 0$

46. $(x^2 - 16)(x^2 - 1) > 0$

47. $\frac{x - 5}{x - 6} < 0$

48. $\frac{x(x + 5)}{4x - 3} \geq 0$

49. $\frac{(4x + 3)(x - 5)}{x(x + 6)} > 0$

50. $(x + 5)(x - 6)(x + 2) \leq 0$

51. $\frac{x + 4}{x - \frac{1}{2}} \leq 0$

52. $\frac{(5x + 6)(x - 3)}{x(6x - 5)} < 0$

53. $\frac{x^2 + 10x + 25}{x^2 - 6x + 9} > 0$

54. $\frac{x^2 - 6x + 9}{x^2 + 10x + 25} < 0$

(8.5) *Sketch the graph of each function.*

55. $f(x) = (x - 4)^2 - 2$

56. $f(x) = -3(x - 1)^2 + 1$

Sketch the graph of each function. Find the vertex and the intercepts.

57. $f(x) = x^2 + 10x + 25$

58. $f(x) = -x^2 + 6x - 9$

59. $f(x) = 4x^2 - 1$

60. $f(x) = -5x^2 + 5$

61. Find two numbers whose product is as large as possible, given that their sum is 420.

62. Write an equation of a quadratic function whose graph is a parabola with vertex $(-3, 7)$ that passes through the origin.

(8.6)   *Write an equation of the circle with the given center and radius.*

63. center $(-4, 4)$, radius 3

64. center $(5, 0)$, diameter 10

65. center $(-7, -9)$, radius $\sqrt{11}$

66. center $(0, 0)$, diameter 7

Sketch the graph of the equation. If the graph is a circle, find its center. If the graph is a parabola, find its vertex.

67. $x^2 + y^2 = 7$

68. $x = 2(y - 5)^2 + 4$

69. $x = -(y + 2)^2 + 3$

70. $(x - 1)^2 + (y - 2)^2 = 4$

71. $y = -x^2 + 4x + 10$

72. $x = -y^2 - 4y + 6$

73. $x = \dfrac{1}{2}y^2 + 2y + 1$

74. $y = -3x^2 + \dfrac{1}{2}x + 4$

75. $x^2 + y^2 + 2x + y = \dfrac{3}{4}$

76. $x^2 + y^2 + 3y = \dfrac{7}{4}$

77. $4x^2 + 4y^2 + 16x + 8y = 1$

78. $3x^2 + 6x + 3y^2 = 9$

79. $y = x^2 + 6x + 9$

80. $x = y^2 + 6y + 9$

81. Write an equation of the circle centered at $(5.6, -2.4)$ with diameter 6.2.

(8.7)   *Sketch the graph of each equation.*

82. $x^2 + \dfrac{y^2}{4} = 1$

83. $x^2 - \dfrac{y^2}{4} = 1$

84. $\dfrac{y^2}{4} - \dfrac{x^2}{16} = 1$

85. $\dfrac{y^2}{4} + \dfrac{x^2}{16} = 1$

86. $\dfrac{x^2}{5} + \dfrac{y^2}{5} = 1$

87. $\dfrac{x^2}{5} - \dfrac{y^2}{5} = 1$

88. $-5x^2 + 25y^2 = 125$

89. $4y^2 + 9x^2 = 36$

90. $\dfrac{(x-2)^2}{4} + (y-1)^2 = 1$

91. $\dfrac{(x+3)^2}{9} + \dfrac{(y-4)^2}{25} = 1$

92. $x^2 - y^2 = 1$

93. $36y^2 - 49x^2 = 1764$

94. $y^2 = x^2 + 9$

95. $x^2 = 4y^2 - 16$

96. $100 - 25x^2 = 4y^2$

Sketch the graph of each equation.

97. $y = x^2 + 4x + 6$

98. $y^2 = x^2 + 6$

99. $y^2 + x^2 = 4x + 6$

100. $y^2 + 2x^2 = 4x + 6$

101. $x^2 + y^2 - 8y = 0$

102. $x - 4y = y^2$

103. $x^2 - 4 = y^2$

104. $x^2 = 4 - y^2$

105. $6(x - 2)^2 + 9(y + 5)^2 = 36$

106. $36y^2 = 576 + 16x^2$

107. $\dfrac{x^2}{16} - \dfrac{y^2}{25} = 1$

108. $3(x - 7)^2 + 3(y + 4)^2 = 1$

CHAPTER 8 TEST

Solve each equation for the variable.

1. $5x^2 - 2x = 7$

2. $x^3 = -8$

3. $m^2 - m + 8 = 0$

4. $u^2 - 6u + 2 = 0$

5. $7x^2 + 8x + 1 = 0$

6. $a^2 - 3a = 5$

7. $\dfrac{4}{x + 2} + \dfrac{2x}{x - 2} = \dfrac{6}{x^2 - 4}$

8. $x^4 - 8x^2 - 9 = 0$

10. $(x + 1)^2 - 15(x + 1) + 56 = 0$

9. $x^6 + 1 = x^4 + x^2$

Solve the equation for the variable by completing the square.

11. $x^2 - 6x = -2$

12. $2a^2 + 5 = 4a$

Solve each inequality for x. Write the solution set in interval notation and then graph the solution set.

13. $(6x + 7)(7x - 12) \leq 0$

14. $(x^2 - 16)(x^2 - 25) > 0$

15. $\dfrac{2x - 11}{3x + 12} \geq 0$

16. $\dfrac{7x - 14}{x^2 - 9} \leq 0$

Sketch the graph of each equation. Label centers, intercepts, vertices, and asymptotes.

17. $x^2 + y^2 = 36$

18. $x^2 - y^2 = 36$

19. $16x^2 + 9y^2 = 144$

20. $y = x^2 - 8x + 16$

21. $x^2 + y^2 + 6x = 16$

22. $x = y^2 + 8y - 3$

23. $\dfrac{(x - 4)^2}{16} + \dfrac{(y - 3)^2}{9} = 1$

24. $y^2 - x^2 = 0$

25. A stone is dropped from a bridge. The height in feet, $s(t)$, of the stone above the water t seconds after it is dropped is a function given by the equation $s(t) = -16t^2 + 256$.
   a. Find the height of the bridge.
   b. Find the time it takes the stone to hit the water.

26. Find three consecutive odd whole numbers such that the product of the first and the third number is thirty-nine more than forty-two times the second number.

CHAPTER 8 CUMULATIVE REVIEW

1. Solve $|p| = 2$.

2. Solve for x: $-(x - 3) + 2 \le 3(2x - 5) + x$.

3. Find the slope of any line perpendicular to the line through $(-2, -2)$ and $(3, -1)$.

4. Evaluate the following:
  a. 7^0 **b.** -7^0 **c.** $(2x + 5)^0$ **d.** $2x^0$

5. Simplify each expression. Write answers using positive exponents.
  a. $\dfrac{x^{-9}}{x^2}$ **b.** $\dfrac{p^4}{p^{-3}}$ **c.** $\dfrac{2^{-3}}{2^{-1}}$

  d. $\dfrac{2x^{-7}y^2}{10xy^{-5}}$ **e.** $\dfrac{(3x^{-3})(x^2)}{x^6}$

6. Subtract $4x^3y^2 - 3x^2y^2 + 2y^2$ from $10x^3y^2 - 7x^2y^2$.

7. Multiply $[3 + (2a + b)]^2$.

8. If $f(x) = x - 1$ and $g(x) = 2x - 3$, find the following:
  a. $(f + g)(x)$ **b.** $(f - g)(x)$

  c. $(f \cdot g)(x)$ **d.** $\left(\dfrac{f}{g}\right)(x)$

9. Factor $17x^3y^2 - 34x^4y^2$.

10. Factor $m^2n^2 + m^2 - 2n^2 - 2$.

11. Factor $5x^4 + 29x^2 - 42$.

12. Factor the following:
  a. $x^2 - 9$ **b.** $16y^2 - 9$ **c.** $50 - 8y^2$

13. Solve $x^3 = 4x$.

14. Graph $f(x) = x^3 - 4x$. Find any intercepts.

15. Write each rational expression in lowest terms.
  a. $\dfrac{x^3 + 8}{2 + x}$ **b.** $\dfrac{2y^2 + 2}{y^3 - 5y^2 + y - 5}$

16. Divide $\dfrac{8x^3 + 125}{x^4 + 5x^2 + 4} \div \dfrac{2x + 5}{2x^2 + 8}$.

17. Perform the indicated operation.
  a. $\dfrac{2}{x^2y} + \dfrac{5}{3x^3y}$

  b. $\dfrac{3z}{z + 2} + \dfrac{2z}{z - 2}$

  c. $\dfrac{5k}{k^2 - 4} - \dfrac{2}{k^2 + k - 2}$

18. Solve $\dfrac{x + 6}{x - 2} = \dfrac{2(x + 2)}{x - 2}$.

19. Simplify. Assume that all variables represent positive numbers.
  a. $\sqrt{25}$ **b.** $\sqrt[3]{-8}$ **c.** $\sqrt{0}$
  d. $\sqrt[4]{-16}$

  e. $\sqrt{\dfrac{1}{36}}$ **f.** $\sqrt[3]{-\dfrac{1}{8}}$

  g. $\sqrt{x^6}$ **h.** $\sqrt[3]{x^{15}}$

20. Rationalize each denominator:
  a. $\dfrac{2}{3\sqrt{2} + 5}$

  b. $\dfrac{\sqrt{6} + 2}{\sqrt{5} - \sqrt{3}}$

  c. $\dfrac{7\sqrt{y}}{\sqrt{12x}}$

  d. $\dfrac{2\sqrt{m}}{3\sqrt{x} + \sqrt{m}}$

21. Solve $3x^2 + 16x + 5 = 0$ for x.

22. Solve $(x + 3)(x - 3) \ge 0$.

23. Graph $x^2 + y^2 = 4$.

24. Sketch the graph of the equation $4y^2 - 9x^2 = 36$.

CHAPTER 9

Systems of Equations

To his Pizza Place menu, Luigi adds a single-person pizza, but he loses money. He makes money on his larger pizzas, and now wonders what fair price he should charge for the single-person size. (See Critical Thinking, page 459.)

INTRODUCTION

In this chapter, two or more equations in two or more variables are solved simultaneously. Such a collection of equations is called a **system of equations.** Systems of
equations are good mathematical models for many real-world problems because these
problems may involve several related patterns.

9.1
Solving Systems of Linear Equations in Two Variables

OBJECTIVES

Tape 27

1   Solve a system by graphing.

2   Solve a system by substitution.

3   Solve a system by elimination.

1   Two linear equations in the same two variables form a **system of linear
equations.**

Systems of Linear Equations in Two Variables

$$\begin{cases} x - 2y = -7 \\ 3x + y = 0 \end{cases} \qquad \begin{cases} x = 5 \\ x + \dfrac{y}{2} = 9 \end{cases} \qquad \begin{cases} x - 3 = 2y + 6 \\ y = 1 \end{cases}$$

Recall that a solution of an equation in two variables is an ordered pair, (x, y),
that makes the equation true. A **solution of a system** of two equations in two variables
is an ordered pair, (x, y), that makes both equations true.

We can estimate the solutions of a system by graphing each equation on the same
coordinate system and estimating the coordinates of any point of intersection. Since
the graph of a linear equation in two variables is a line, graphing two such equations
yields two lines in a plane.

EXAMPLE 1   Solve each system by graphing. If the system has just one solution, estimate the
solution.

a. $\begin{cases} x + y = 2 \\ 3x - y = -2 \end{cases}$

b. $\begin{cases} x - 2y = 4 \\ x = 2y \end{cases}$

c. $\begin{cases} 2x + 4y = 10 \\ x + 2y = 5 \end{cases}$

413

Solution: **a.** $\begin{cases} x + y = 2 \\ 3x - y = -2 \end{cases}$

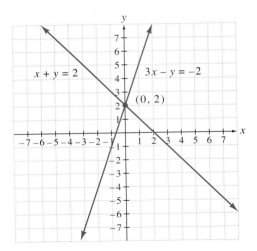

These lines intersect in one point as shown. The coordinates of the point of intersection appear to be (0, 2). Check this estimated solution by replacing x with 0 and y with 2 in **both** equations.

$x + y = 2$ First equation. $3x - y = -2$ Second equation.

$0 + 2 = 2$ Let $x = 0$ and $y = 2$. $3 \cdot 0 - 2 = -2$ Let $x = 0$ and $y = 2$.

$2 = 2$ True. $-2 = -2$ True.

The ordered pair (0, 2) does satisfy both equations. We conclude therefore that (0, 2) is the solution of the system. A system that has at least one solution, such as this one, is said to be **consistent.**

b. $\begin{cases} x - 2y = 4 \\ x = 2y \end{cases}$

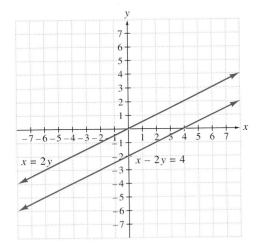

The lines appear to be parallel. To be sure, write each equation in point–slope form, $y = mx + b$.

$x - 2y = 4$ First equation. $x = 2y$ Second equation.

$-2y = -x + 4$ Subtract x from both sides. $\dfrac{1}{2}x = y$ Divide both sides by 2.

$$y = \frac{1}{2}x - 2 \qquad \text{Divide both sides} \qquad y = \frac{1}{2}x$$
by -2.

The graphs of these equations have the same slope, $\frac{1}{2}$, but different y-intercepts, so these lines are parallel. Therefore, the system has no solution. Systems that have no solution are said to be **inconsistent.**

c. $\begin{cases} 2x + 4y = 10 \\ x + 2y = 5 \end{cases}$

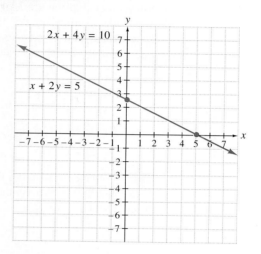

The graph of each equation is the same line. To see this, notice that if both sides of the second equation are multiplied by 2 the result is the first equation. This means that the equations have identical solutions. Any ordered pair solution of one equation will satisfy the other equation also. There are therefore an infinite number of solutions to this system. These equations are said to be **dependent equations.** The solution set of the system is $\{(x, y) \mid x + 2y = 5\}$ or, equivalently, $\{(x, y) \mid 2x + 4y = 10\}$ since the lines describe identical ordered pairs. Written this way, the solution set is read "the set of all ordered pairs (x, y), such that $2x + 4y = 10$." ∎

We can summarize the information discovered in Example 1 as follows.

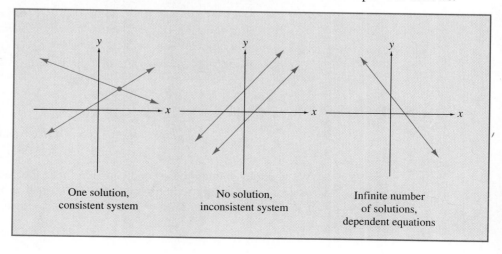

One solution,
consistent system

No solution,
inconsistent system

Infinite number
of solutions,
dependent equations

2   Graphing the equations of a system is not a reliable method for finding solutions of a system. We turn instead to two algebraic methods of solving systems. We use the first method, the **substitution method,** to solve the system

$$\begin{cases} y = x + 5 & \text{First equation.} \\ 3x = 2y - 9 & \text{Second equation.} \end{cases}$$

Remember that we are looking for an ordered pair, if there is one, that satisfies both equations. Satisfying the first equation, $y = x + 5$, means that y must be $x + 5$. **Substituting** the expression $x + 5$ for y in the second equation yields an equation in one variable, which we can solve for x.

$$3x = 2y - 9 \qquad\qquad \text{Second equation.}$$

$$3x = 2(x + 5) - 9 \qquad \text{Replace } y \text{ with } x + 5 \text{ in the second equation.}$$

$$3x = 2x + 10 - 9$$

$$x = 1$$

The x-coordinate of the solution of the system is 1. The y-coordinate is the y-value corresponding to the x-value, 1. Choose either equation of the system and solve for y when x is 1.

$$y = x + 5 \qquad \text{First equation.}$$

$$= 1 + 5 \qquad \text{Let } x = 1.$$

$$= 6$$

The y-coordinate is 6, so the solution of the system is $(1, 6)$. This means that, when both equations are graphed, the one point of intersection occurs at the point with coordinates $(1, 6)$.

To Solve a System of Two Equations by the Substitution Method

Step 1 Solve one of the equations for one of its variables.

Step 2 Substitute the expression for the variable found in *step 1* into the other equation.

Step 3 Find the value of one variable by solving the equation from *step 2*.

Step 4 Find the value of the other variable by substituting the value found in *step 3* in any equation of the system.

Step 5 Check the ordered pair solution in **both** of the original equations.

EXAMPLE 2   Use the substitution method to solve the system

$$\begin{cases} 2x + 4y = -6 & \text{First equation.} \\ x - 2y = -5 & \text{Second equation.} \end{cases}$$

Solution: We begin by solving the second equation for x; the coefficient 1 of the x-term keeps us from introducing tedious fractions. The equation $x - 2y = -5$ solved for x is $x = 2y - 5$. Substitute $2y - 5$ for x in the first equation.

$$2x + 4y = -6 \qquad \text{First equation.}$$

$$2(\overbrace{2y - 5}) + 4y = -6 \qquad \text{Substitute } 2y - 5 \text{ for } x.$$

$$4y - 10 + 4y = -6$$

$$8y = 4$$

$$y = \frac{4}{8} = \frac{1}{2} \qquad \text{Solve for } y.$$

The y-coordinate of the solution is $\frac{1}{2}$. To find the x-coordinate, replace y with $\frac{1}{2}$ in the equation $x = 2y - 5$.

$$x = 2y - 5$$

$$x = 2\left(\frac{1}{2}\right) - 5 = 1 - 5 = -4$$

The solution set is $\left\{\left(-4, \frac{1}{2}\right)\right\}$. To check, see that $\left(-4, \frac{1}{2}\right)$ satisfies both equations of the system. ■

EXAMPLE 3   Use substitution to solve the system

$$\begin{cases} -\dfrac{x}{6} + \dfrac{y}{2} = \dfrac{1}{2} \\[2mm] \dfrac{x}{3} - \dfrac{y}{6} = -\dfrac{3}{4} \end{cases}$$

Solution: First, multiply each equation by its least common denominator in order to write this system as an equivalent system without fractions. We multiply the first equation by 6 and the second equation by 12. Then

$$\begin{cases} 6\left(-\dfrac{x}{6} + \dfrac{y}{2}\right) = 6\left(\dfrac{1}{2}\right) \\[3mm] 12\left(\dfrac{x}{3} - \dfrac{y}{6}\right) = 12\left(-\dfrac{3}{4}\right) \end{cases}$$

simplifies to

$$\begin{cases} -x + 3y = 3 & \text{First equation.} \\ 4x - 2y = -9 & \text{Second equation.} \end{cases}$$

We now solve the first equation for x.

$$-x + 3y = 3 \qquad \text{First equation.}$$

$$3y - 3 = x \qquad \text{Solve for } x.$$

Next, replace x with $3y - 3$ in the second equation.

$$4x - 2y = -9 \qquad \text{Second equation.}$$

$$4(3y - 3) - 2y = -9$$

$$12y - 12 - 2y = -9$$

$$10y = 3$$

$$y = \frac{3}{10} \qquad \text{Solve for } y.$$

The y-coordinate is $\frac{3}{10}$. To find the x-coordinate, replace y with $\frac{3}{10}$ in the equation $x = 3y - 3$. Then

$$x = 3\left(\frac{3}{10}\right) - 3 = \frac{9}{10} - 3 = \frac{9}{10} - \frac{30}{10} = -\frac{21}{10}$$

The solution set is $\left\{\left(-\frac{21}{10}, \frac{3}{10}\right)\right\}$ ∎

3 The **elimination** or **addition method** is a second algebraic technique for solving systems of equations. For this method, we rely on a version of the addition property of equality, which states that "equals added to equals are equal." In symbols,

$$\text{if } A = B \text{ and } C = D, \text{ then } A + C = B + D.$$

EXAMPLE 4 Use the addition method to solve the system

$$\begin{cases} x - 5y = -12 & \text{First equation.} \\ -x + y = 4 & \text{Second equation.} \end{cases}$$

Solution: Since the left side of each equation is equal to the right side, we add equal quantities by adding the left sides of the equations and the right sides of the equations. This sum gives us an equation in one variable, y, which we can solve for y.

$$x - 5y = -12 \qquad \text{First equation.}$$

$$\underline{-x + y = 4} \qquad \text{Second equation.}$$

$$-4y = -8 \qquad \text{Add.}$$

$$y = 2 \qquad \text{Solve for } y.$$

The y-coordinate of the solution is 2. To find the corresponding x-coordinate, replace y with 2 in either original equation of the system.

$$-x + y = 4 \qquad \text{Second equation.}$$

$$-x + 2 = 4 \qquad \text{Let } y = 2.$$

$$-x = 2$$

$$x = -2$$

The solution set is $\{(-2, 2)\}$. Check to see that $(-2, 2)$ satisfies both equations of the system. ∎

> **To Solve a System of Two Linear Equations by the Addition Method**
>
> *Step 1*   Rewrite each equation in standard form, $Ax + By = C$.
> *Step 2*   If necessary, multiply one or both equations by some nonzero number so that the coefficient of one variable in one equation is the opposite of its coefficient in the other equation.
> *Step 3*   Add the equations.
> *Step 4*   Find the value of one variable by solving the equation from *step 3*.
> *Step 5*   Find the value of the second variable by substituting the value found in *step 4* into either of the original equations.
> *Step 6*   Check the proposed solution in both of the original equations.

EXAMPLE 5   Use the addition method to solve the system

$$\begin{cases} 3x + \dfrac{y}{2} = 2 \\ 6x + y = 5 . \end{cases}$$

Solution: If we add the two equations, the sum will still be an equation in two variables. Notice that if we multiply both sides of the first equation by -2 the coefficients of x will be opposites. Then

$$\begin{cases} -2\left(3x + \dfrac{y}{2}\right) = -2\,(2) \\ 6x + y = 5 \end{cases} \quad \text{simplifies to} \quad \begin{cases} -6x - y = -4 \\ 6x + y = 5 \end{cases}$$

Next, add the left sides and add the right sides.

$$\begin{array}{r} -6x - y = -4 \\ \underline{6x + y = 5} \\ 0 = 1 \end{array} \qquad \text{False.}$$

The resulting equation, $0 = 1$, is false for all values of y or x. Thus, the system has no solution. The solution set is $\{\ \}$ or $\emptyset$.

This system is inconsistent and the graphs of the equations are parallel lines.

∎

EXAMPLE 6   Use the addition method to solve the system

$$\begin{cases} 3x - 2y = 10 \\ 4x - 3y = 15 . \end{cases}$$

Solution: To eliminate y when the equations are added, multiply both sides of the first equation by 3 and both sides of the second equation by -2. Then

$$\begin{cases} 3\,(3x - 2y) = 3\,(10) \\ -2\,(4x - 3y) = -2\,(15) \end{cases} \quad \text{simplifies to} \quad \begin{cases} 9x - 6y = 30 \\ \underline{-8x + 6y = -30} \\ x = 0 \quad \text{Add the equations.} \end{cases}$$

To find y, let $x = 0$ in either equation of the system.

$$3x - 2y = 10$$
$$3(0) - 2y = 10 \qquad \text{Let } x = 0.$$
$$-2y = 10$$
$$y = -5$$

The solution set is $\{(0, -5)\}$. Check to see that $(0, -5)$ satisfies both equations. ■

EXAMPLE 7 Use the addition method to solve

$$\begin{cases} -5x - 3y = 9 \\ 10x + 6y = -18. \end{cases}$$

Solution: To eliminate x when the equations are added, multiply both sides of the first equation by 2. Then

$$\begin{cases} 2(-5x - 3y) = 2(9) \\ 10x + 6y = -18 \end{cases} \text{ simplifies to } \begin{cases} -10x - 6y = 18 \\ \underline{10x + 6y = -18} \end{cases}$$

$$0 = 0 \quad \text{Add the equations.}$$

The resulting equation, $0 = 0$, is true for all possible values of y or x. Notice in the original system that, if both sides of the first equation are multiplied by 2, the result is the second equation. This means that the two equations are equivalent and they have the same solution set. Thus the equations of this system are dependent, and the solution set of the system is

$$\{(x, y) \mid -5x - 3y = 9\} \quad \text{or} \quad \{(x, y) \mid 10x + 6y = -18\} \quad ■$$

EXERCISE SET 9.1

Solve each system by graphing. See Example 1.

1. $\begin{cases} x + y = 1 \\ x - 2y = 4 \end{cases}$

2. $\begin{cases} 2x - y = 8 \\ x + 3y = 11 \end{cases}$

3. $\begin{cases} 2y - 4 = 0 \\ x + 2y = 5 \end{cases}$

4. $\begin{cases} 4x - y = 6 \\ x - y = 0 \end{cases}$ $(2, 2)$

5. $\begin{cases} 3x - y = 4 \\ 6x - 2y = 4 \end{cases}$

6. $\begin{cases} -x + 3y = 6 \\ 3x - 9y = 9 \end{cases}$

Solve each system of equations by the substitution method. See Example 2.

7. $\begin{cases} x + y = 10 \\ \quad y = 4x \end{cases}$

8. $\begin{cases} 5x + 2y = -17 \\ \quad x = 3y \end{cases}$

9. $\begin{cases} 4x - y = 9 \\ 2x + 3y = -27 \end{cases}$

10. $\begin{cases} 3x - y = 6 \\ -4x + 2y = -8 \end{cases}$

Solve each system of equations by the substitution method. See Example 3.

11. $\begin{cases} \dfrac{1}{2}x + \dfrac{3}{4}y = -\dfrac{1}{4} \\ \dfrac{3}{4}x - \dfrac{1}{4}y = 1 \end{cases}$

12. $\begin{cases} \dfrac{2}{5}x + \dfrac{1}{5}y = -1 \\ x + \dfrac{2}{5}y = -\dfrac{8}{5} \end{cases}$

13. $\begin{cases} \dfrac{x}{3} + y = \dfrac{4}{3} \\ -x + 2y = 11 \end{cases}$

14. $\begin{cases} \dfrac{x}{8} - \dfrac{y}{2} = 1 \\ \dfrac{x}{3} - y = 2 \end{cases}$

Solve each system of equations by the elimination method. See Example 4.

15. $\begin{cases} 2x - 4y = 0 \\ x + 2y = 5 \end{cases}$

16. $\begin{cases} 2x - 3y = 0 \\ 2x + 6y = 3 \end{cases}$

17. $\begin{cases} 5x + 2y = 1 \\ x - 3y = 7 \end{cases}$

18. $\begin{cases} 6x - y = -5 \\ 4x - 2y = 6 \end{cases}$

Solve each system of equations by the elimination method. See Example 6.

19. $\begin{cases} 5x - 2y = 27 \\ -3x + 5y = 18 \end{cases}$

20. $\begin{cases} 3x + 4y = 2 \\ 2x + 5y = -1 \end{cases}$

21. $\begin{cases} 3x - 5y = 11 \\ 2x - 6y = 2 \end{cases}$

22. $\begin{cases} 6x - 3y = -3 \\ 4x + 5y = -9 \end{cases}$

Solve each system of equations. See Examples 5 and 7.

23. $\begin{cases} x - 2y = 4 \\ 2x - 4y = 4 \end{cases}$

24. $\begin{cases} -x + 3y = 6 \\ 3x - 9y = 9 \end{cases}$

25. $\begin{cases} 3x + y = 1 \\ 2y = 2 - 6x \end{cases}$

26. $\begin{cases} y = 2x - 5 \\ 8x - 4y = 20 \end{cases}$

Solve each system of equations.

27. $\begin{cases} 2x + 5y = 8 \\ 6x + y = 10 \end{cases}$

28. $\begin{cases} x - 4y = -5 \\ -3x - 8y = 0 \end{cases}$

29. $\begin{cases} x + y = 1 \\ x - 2y = 4 \end{cases}$

30. $\begin{cases} 2x - y = 8 \\ x + 3y = 11 \end{cases}$

31. $\begin{cases} \dfrac{1}{3}x + y = \dfrac{4}{3} \\ -\dfrac{1}{4}x - \dfrac{1}{2}y = -\dfrac{1}{4} \end{cases}$

32. $\begin{cases} \dfrac{3}{4}x - \dfrac{1}{2}y = -\dfrac{1}{2} \\ x + y = -\dfrac{3}{2} \end{cases}$

33. $\begin{cases} 2x + 6y = 8 \\ 3x + 9y = 12 \end{cases}$

34. $\begin{cases} x = 3y - 1 \\ 2x - 6y = -2 \end{cases}$

35. $\begin{cases} 4x + 2y = 5 \\ 2x + y = -1 \end{cases}$

36. $\begin{cases} 3x + 6y = 15 \\ 2x + 4y = 3 \end{cases}$

37. $\begin{cases} 10y - 2x = 1 \\ 5y = 4 - 6x \end{cases}$

38. $\begin{cases} 3x + 4y = 0 \\ \quad 7x = 3y \end{cases}$

39. $\begin{cases} \dfrac{3}{4}x + \dfrac{5}{2}y = 11 \\ \dfrac{1}{16}x - \dfrac{3}{4}y = -1 \end{cases}$

40. $\begin{cases} \dfrac{2}{3}x + \dfrac{1}{4}y = -\dfrac{3}{2} \\ \dfrac{1}{2}x - \dfrac{1}{4}y = -2 \end{cases}$

41. $\begin{cases} x = 3y + 2 \\ 5x - 15y = 10 \end{cases}$

42. $\begin{cases} y = \dfrac{1}{7}x + 3 \\ x - 7y = -21 \end{cases}$

43. $\begin{cases} 2x - y = -1 \\ \quad y = -2x \end{cases}$

44. $\begin{cases} x = \dfrac{1}{5}y \\ x - y = -4 \end{cases}$

45. $\begin{cases} 2x = 6 \\ y = 5 - x \end{cases}$

46. $\begin{cases} x = 3y + 4 \\ -y = 5 \end{cases}$

47. $\begin{cases} \dfrac{x + 5}{2} = \dfrac{6 - 4y}{3} \\ \dfrac{3x}{5} = \dfrac{21 - 7y}{10} \end{cases}$

48. $\begin{cases} \dfrac{y}{5} = \dfrac{8 - x}{2} \\ x = \dfrac{2y - 8}{3} \end{cases}$

49. $\begin{cases} 4x - 7y = 7 \\ 12x - 21y = 24 \end{cases}$

50. $\begin{cases} 2x - 5y = 12 \\ -4x + 10y = 20 \end{cases}$

51. $\begin{cases} \dfrac{2}{3}x - \dfrac{3}{4}y = -1 \\ -\dfrac{1}{6}x + \dfrac{3}{8}y = 1 \end{cases}$

52. $\begin{cases} \dfrac{1}{2}x - \dfrac{1}{3}y = -3 \\ \dfrac{1}{8}x + \dfrac{1}{6}y = 0 \end{cases}$

53. $\begin{cases} 0.7x - 0.2y = -1.6 \\ 0.2x - y = -1.4 \end{cases}$

54. $\begin{cases} -0.7x + 0.6y = 1.3 \\ 0.5x - 0.3y = -0.8 \end{cases}$

55. The sum of two numbers is 45 and one number is twice the other. Find the numbers.

56. The difference between two numbers is 5. Twice the smaller number added to five times the larger number is 53. Find the numbers.

A Look Ahead

Solve each system. See the following example.

EXAMPLE Solve the system $\begin{cases} -\dfrac{4}{x} - \dfrac{4}{y} = -11 \\ \dfrac{1}{x} + \dfrac{1}{y} = 1 \end{cases}$

Solution: First, make the following substitution. Let $a = \dfrac{1}{x}$ and $b = \dfrac{1}{y}$ in both equations. Then

$$\begin{cases} -4\left(\dfrac{1}{x}\right) - 4\left(\dfrac{1}{y}\right) = -11 \\ \dfrac{1}{x} + \dfrac{1}{y} = 1 \end{cases} \quad \text{is equivalent to} \quad \begin{cases} -4a - 4b = -11 \\ a + b = 1 \end{cases}$$

We solve by the addition method. Multiplying both sides of the second equation by 4 and adding the left sides and the right sides of the equations,

$$\begin{cases} -4a - 4b = -11 \\ 4\,(a + b) = 4\,(1) \end{cases} \quad \text{simplifies to} \quad \begin{cases} -4a - 4b = -11 \\ 4a + 4b = 4 \\ \hline 0 = -7 \quad \text{False.} \end{cases}$$

The equation $0 = -7$ is false for all values of a and hence for all values of $\dfrac{1}{x}$ and all values of x. This system has no solution. ∎

57. $\begin{cases} \dfrac{1}{x} + y = 12 \\ \dfrac{3}{x} - y = 4 \end{cases}$

58. $\begin{cases} x + \dfrac{2}{y} = 7 \\ 3x + \dfrac{3}{y} = 6 \end{cases}$

59. $\begin{cases} \dfrac{1}{x} + \dfrac{1}{y} = 5 \\ \dfrac{1}{x} - \dfrac{1}{y} = 1 \end{cases}$

60. $\begin{cases} \dfrac{2}{x} + \dfrac{3}{y} = 5 \\ \dfrac{5}{x} - \dfrac{3}{y} = 2 \end{cases}$

61. $\begin{cases} \dfrac{2}{x} + \dfrac{3}{y} = -1 \\ \dfrac{3}{x} - \dfrac{2}{y} = 18 \end{cases}$

62. $\begin{cases} \dfrac{3}{x} - \dfrac{2}{y} = -18 \\ \dfrac{2}{x} + \dfrac{3}{y} = 1 \end{cases}$

63. $\begin{cases} \dfrac{2}{x} - \dfrac{4}{y} = 5 \\ \dfrac{1}{x} - \dfrac{2}{y} = \dfrac{3}{2} \end{cases}$

64. $\begin{cases} \dfrac{5}{x} + \dfrac{7}{y} = 1 \\ -\dfrac{10}{x} - \dfrac{14}{y} = 0 \end{cases}$

Skill Review

Graph. See Sections 8.6 and 8.7.

65. $x^2 + y^2 = 9$

66. $\dfrac{x^2}{9} + \dfrac{y^2}{16} = 1$

67. $\dfrac{x^2}{25} + \dfrac{y^2}{4} = 1$

68. $(x - 2)^2 + (y + 4)^2 = 16$

Simplify. See Section 7.6.

69. $\sqrt{-4}$

70. $\sqrt{-25}$

71. $\sqrt{-20}$

72. $\sqrt{-18}$

9.2
Solving Systems of Linear Equations in Three Variables

Tape 27

OBJECTIVES		
	1	Recognize a linear equation in three variables.
	2	Solve a system of three linear equations in three variables.

1   In this section, the algebraic methods of solving systems of two linear equations in two variables are extended to systems of three linear equations in three variables. We call the equation $3x - y + z = -15$, for example, a **linear equation in three variables** since there are three variables and each variable is raised only to the power 1. A solution of this equation is an **ordered triple, (x, y, z)**, that makes the equation a true statement. For example, the ordered triple $(2, 0, -21)$ is a solution of $3x - y + z = -15$ since replacing x with 2, y with 0, and z with -21 yields the true statement $3(2) - 0 + (-21) = -15$. The graph of this equation is a plane in three-dimensional space, just as the graph of a linear equation in two variables is a line in two-dimensional space.

Although we will not graph equations in three variables, visualizing the possible patterns of intersecting planes gives us insight into the possible patterns of solutions of a system of three three-variable linear equations. There are four possible patterns.

1. Three planes intersect at a single point. This point represents the single solution of the system. The system is **consistent.**

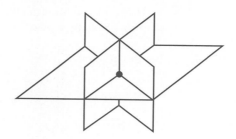

2. Three planes intersect at no points. This system has no solution. A few ways that this can occur are shown. This system is **inconsistent.**

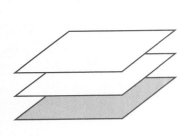 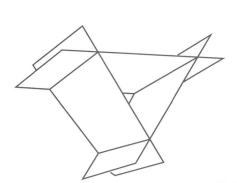 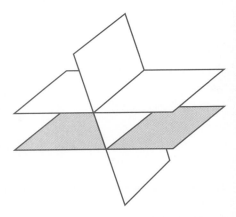

3. Three planes intersect at all the points of a single line. The system has infinitely many solutions. This system is **consistent.**

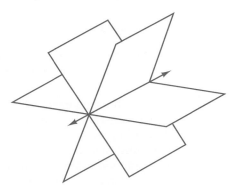

4. Three planes coincide at all points on the plane. The equations are **dependent.**

2   Using the addition method to solve a system in three variables, we reduce the system to a system in two variables.

EXAMPLE 1   Solve the system:

$$\begin{cases} 3x - y + z = -15 & \text{Equation (1)} \\ x + 2y - z = 1 & \text{Equation (2)} \\ 2x + 3y - 2z = 0 & \text{Equation (3)} \end{cases}$$

Solution: Add equations (1) and (2) to eliminate z.

$$\begin{array}{r} 3x - y + z = -15 \\ \underline{x + 2y - z = 1} \\ 4x + y = -14 \end{array} \quad \text{Equation (4)}$$

Next, add two **other** equations and **eliminate z again.** To do so, multiply both sides of equation (1) by 2 and add this resulting equation to equation (3). Then

$$\begin{cases} 2\,(3x - y + z) = 2\,(-15) \\ 2x + 3y - 2z = 0 \end{cases} \quad \text{simplifies to} \quad \begin{cases} 6x - 2y + 2z = -30 \\ \underline{2x + 3y - 2z = 0} \\ 8x + y = -30 \end{cases}$$

$$\text{Equation (5)}$$

Now solve equations (4) and (5) for x and y. To solve by addition, multiply both sides of equation (4) by -1 and add this resulting equation to equation (5). Then

$$\begin{cases} -1\,(4x + y) = -1\,(-14) \\ 8x + y = -30 \end{cases} \quad \text{simplifies to} \quad \begin{cases} -4x - y = 14 \\ \underline{8x + y = -30} \\ 4x = -16 \end{cases}$$

$$\text{Add the equations.}$$

$$x = -4 \quad \text{Solve for } x.$$

Replace x with -4 in equation (4) or (5).

$$4x + y = -14 \quad \text{Equation (4)}$$

$$4(-4) + y = -14 \quad \text{Let } x = -4.$$

$$y = 2 \quad \text{Solve for } y.$$

Finally, replace x with -4 and y with 2 in equation (1), (2), or (3).

$$x + 2y - z = 1 \quad \text{Equation (2).}$$

$$-4 + 2(2) - z = 1 \quad \text{Let } x = -4 \text{ and } y = 2.$$

$$-4 + 4 - z = 1$$

$$-z = 1$$

$$z = -1$$

The solution is $(-4, 2, -1)$. To check, let $x = -4$, $y = 2$, and $z = -1$ in all three original equations of the system.

Equation (1)	*Equation (2)*	*Equation (3)*
$3x - y + z = -15$	$x + 2y - z = 1$	$2x + 3y - 2z = 0$
$3(-4) - 2 + (-1) = -15$	$-4 + 2(2) - (-1) = 1$	$2(-4) + 3(2) - 2(-1) = 0$
$-12 - 2 - 1 = -15$	$-4 + 4 + 1 = 1$	$-8 + 6 + 2 = 0$
$-15 = -15$	$1 = 1$	$0 = 0$
True	True	True

All three statements are true, so the solution set is $\{(-4, 2, -1)\}$. ∎

EXAMPLE 2 Solve the system:

$$\begin{cases} 2x - 4y + 8z = 2 & (1) \\ -x - 3y + z = 11 & (2) \\ x - 2y + 4z = 0 & (3) \end{cases}$$

Solution: If equations (2) and (3) are added, x is eliminated, and the new equation is

$$-5y + 5z = 11 \quad (4)$$

To eliminate x again, multiply both sides of equation (2) by 2, and add the resulting equation to equation (1). Then

$$\begin{cases} 2x - 4y + 8z = 2 \\ 2(-x - 3y + z) = 2(11) \end{cases} \text{ simplifies to } \begin{cases} 2x - 4y + 8z = 2 \\ -2x - 6y + 2z = 22 \end{cases}$$
$$-10y + 10z = 24 \quad (5)$$

Next, solve for y and z using equations (4) and (5). Multiply both sides of equation (4) by -2, and add the resulting equation to equation (5).

$$\begin{cases} -2(-5y + 5z) = -2(11) \\ -10y + 10z = 24 \end{cases} \text{ simplifies to } \begin{cases} 10y - 10z = -22 \\ -10y + 10z = 24 \end{cases}$$
$$0 = 2 \quad \text{False}$$

Since the statement is false, this system is inconsistent and has no solution. The solution set is the empty set $\{\ \}$ or $\emptyset$. ∎

The addition method is summarized next.

To Solve a System of Three Linear Equations by the Addition Method

Step 1 Eliminate any of the three variables from any pair of equations in the system.

Step 2 Choose any other pair of equations and eliminate the **same variable** as in *step 1*.

Step 3 Two equations in two variables should be obtained from *step 1* and *step 2*. Use methods from Section 9.1 to solve this system for both variables.

Step 4 To solve for the third variable, substitute the values of the variables found in *step 3* into any of the original equations.

EXAMPLE 3 Solve the system:

$$\begin{cases} 2x + 4y = 1 & (1) \\ 4x - 4z = -1 & (2) \\ y - 4z = -3 & (3) \end{cases}$$

Solution: Notice that equation (2) has no term containing the variable y. Let us eliminate y using equations (1) and (3). Multiply both sides of equation (3) by -4 and add the resulting equation to equation (1). Then

$$\begin{cases} 2x + 4y = 1 \\ \boxed{-4}\,(y - 4z) = \boxed{-4}\,(-3) \end{cases} \quad \text{simplifies to} \quad \begin{cases} 2x + 4y = 1 \\ -4y + 16z = 12 \\ \hline 2x + 16z = 13 \quad (4) \end{cases}$$

Next, solve for z using equations (4) and (2). Multiply both sides of equation (4) by -2 and add the resulting equation to equation (2).

$$\begin{cases} \boxed{-2}\,(2x + 16z) = \boxed{-2}\,(13) \\ 4x - 4z = -1 \end{cases} \quad \text{simplifies to} \quad \begin{cases} -4x - 32z = -26 \\ 4x - 4z = -1 \\ \hline -36z = -27 \end{cases}$$

$$z = \frac{3}{4}$$

Replace z with $\dfrac{3}{4}$ in equation (3) and solve for y.

$$y - 4\left(\frac{3}{4}\right) = -3 \qquad \text{Let } z = \frac{3}{4} \text{ in equation (3).}$$

$$y - 3 = -3$$

$$y = 0$$

Replace y with 0 in equation (1) and solve for x.

$$2x + 4(0) = 1$$

$$2x = 1$$

$$x = \frac{1}{2}$$

The solution set is $\left\{\left(\dfrac{1}{2},\, 0,\, \dfrac{3}{4}\right)\right\}$. Check to see that this solution satisfies all three equations of the system.   ■

EXAMPLE 4   Solve the system:

$$\begin{cases} x - 5y - 2z = 6 \quad (1) \\ -2x + 10y + 4z = -12 \quad (2) \\ \dfrac{1}{2}x - \dfrac{5}{2}y - z = 3 \quad (3) \end{cases}$$

Solution: Multiply both sides of equation (3) by 2 to eliminate fractions, and multiply both sides of equation (2) by $-\dfrac{1}{2}$ so that the coefficient of x is 1. The resulting system is then:

$$\begin{cases} x - 5y - 2z = 6 \quad (1) \\ x - 5y - 2z = 6 \quad \text{Multiply (2) by } -\dfrac{1}{2}. \\ x - 5y - 2z = 6 \quad \text{Multiply (3) by 2.} \end{cases}$$

All three equations are identical, and therefore equations (1), (2), and (3) are all equivalent. There are infinitely many solutions of this system. The equations are dependent. The solution set can be written as $\{(x, y, z) \mid x - 5y - 2z = 6\}$. ∎

EXERCISE SET 9.2

Solve each system. See Examples 1 and 3.

1. $\begin{cases} x + y = 3 \\ 2y = 10 \\ 3x + 2y - 3z = 1 \end{cases}$
2. $\begin{cases} 5x = 5 \\ 2x + y = 4 \\ 3x + y - 4z = -15 \end{cases}$
3. $\begin{cases} 2x + 2y + z = 1 \\ -x + y + 2z = 3 \\ x + 2y + 4z = 0 \end{cases}$
4. $\begin{cases} 2x - 3y + z = 5 \\ x + y + z = 0 \\ 4x + 2y + 4z = 4 \end{cases}$

Solve each system. See Examples 2 and 4.

5. $\begin{cases} x - 2y + z = -5 \\ -3x + 6y - 3z = 15 \\ 2x - 4y + 2z = -10 \end{cases}$
6. $\begin{cases} 3x + y - 2z = 2 \\ -6x - 2y + 4z = -2 \\ 9x + 3y - 6z = 6 \end{cases}$

7. $\begin{cases} 4x - y + 2z = 5 \\ 2y + z = 4 \\ 4x + y + 3z = 10 \end{cases}$
8. $\begin{cases} 5y - 7z = 14 \\ 2x + y + 4z = 10 \\ 2x + 6y - 3z = 30 \end{cases}$

Solve each system.

9. $\begin{cases} x + 5z = 0 \\ 5x + y = 0 \\ y - 3z = 0 \end{cases}$
10. $\begin{cases} x - 5y = 0 \\ x - z = 0 \\ -x + 5z = 0 \end{cases}$

11. $\begin{cases} 6x - 5z = 17 \\ 5x - y + 3z = -1 \\ 2x + y = -41 \end{cases}$
12. $\begin{cases} x + 2y = 6 \\ 7x + 3y + z = -33 \\ x - z = 16 \end{cases}$

13. $\begin{cases} x + y + z = 8 \\ 2x - y - z = 10 \\ x - 2y - 3z = 22 \end{cases}$
14. $\begin{cases} 5x + y + 3z = 1 \\ x - y + 3z = -7 \\ -x + y = 1 \end{cases}$

15. $\begin{cases} x + 2y - z = 5 \\ 6x + y + z = 7 \\ 2x + 4y - 2z = 5 \end{cases}$
16. $\begin{cases} 4x - y + 3z = 10 \\ x + y - z = 5 \\ 8x - 2y + 6z = 10 \end{cases}$

17. $\begin{cases} 2x - 3y + z = 2 \\ x - 5y + 5z = 3 \\ 3x - y - 3z = 1 \end{cases}$
18. $\begin{cases} 4x + y - z = 8 \\ x - y + 2z = 3 \\ 3x - y + z = 6 \end{cases}$

19. $\begin{cases} -2x - 4y + 6z = -8 \\ x + 2y - 3z = 4 \\ 4x + 8y - 12z = 16 \end{cases}$
20. $\begin{cases} -6x + 12y + 3z = -6 \\ 2x - 4y - z = 2 \\ -x + 2y + \dfrac{z}{2} = -1 \end{cases}$

21. $\begin{cases} 2x + 2y - 3z = 1 \\ y + 2z = -14 \\ 3x - 2y = -1 \end{cases}$
22. $\begin{cases} 7x + 4y = 10 \\ x - 4y + 2z = 6 \\ y - 2z = -1 \end{cases}$

23. $\begin{cases} \dfrac{3}{4}x - \dfrac{1}{3}y + \dfrac{1}{2}z = 9 \\ \dfrac{1}{6}x + \dfrac{1}{3}y - \dfrac{1}{2}z = 2 \\ \dfrac{1}{2}x - \ y + \dfrac{1}{2}z = 2 \end{cases}$

24. $\begin{cases} \dfrac{1}{3}x - \dfrac{1}{4}y + \ z = -9 \\ \dfrac{1}{2}x - \dfrac{1}{3}y - \dfrac{1}{4}z = -6 \\ \ x - \dfrac{1}{2}y - \ z = -8 \end{cases}$

A Look Ahead

Solve each system.

25. $\begin{cases} x + y \qquad - w = \ 0 \\ \quad y + 2z + w = \ 3 \\ x \qquad + z \qquad = \ 1 \\ 2x - y \qquad - w = -1 \end{cases}$

26. $\begin{cases} 5x + 4y \qquad\qquad = 29 \\ \quad y + z - w = -2 \\ 5x \qquad + z \qquad = 23 \\ \quad y - z + w = 4 \end{cases}$

27. $\begin{cases} x + y + z + w = 5 \\ 2x + y + z + w = 6 \\ x + y + z \qquad = 2 \\ x + y \qquad\quad = 0 \end{cases}$

28. $\begin{cases} 2x \qquad - z \qquad = -1 \\ \quad y + z \ + w = 9 \\ \quad y \qquad - 2w = -6 \\ x + y \qquad\qquad = 3 \end{cases}$

Skill Review

Use the quadratic formula to solve each quadratic equation. See Section 8.2.

29. $x^2 + 4x + 1 = 0$ 30. $2x^2 + x - 1 = 0$ 31. $3x^2 - x + 2 = 0$ 32. $x^2 - 2x + 3 = 0$

Simplify each complex fraction. See Section 6.4.

33. $\dfrac{\dfrac{1}{x}}{\dfrac{5}{x^2} - \dfrac{3}{x}}$

34. $\dfrac{\dfrac{2}{a} + \dfrac{1}{b}}{\dfrac{3}{b} - \dfrac{5}{a}}$

9.3
Applications

OBJECTIVE   **1**   Solve problems that can be modeled by a system of linear equations.

Tape 27

1   We have solved problems by writing one-variable equations and solving for the variable. Some of these problems can be solved, perhaps more easily, by writing a system of equations, which we illustrate in this section.

EXAMPLE 1   A number is 4 less than a second number. Four times the first is 6 more than twice the second. Find the numbers.

Solution: Let $x =$ first number
$y =$ second number

Use the first two sentences in the problem to write a system of two equations.

$\begin{cases} \text{A number is 4 less than a second number.} \\ \text{Four times the first number is 6 more than twice the second.} \end{cases}$ or $\begin{cases} x = y - 4 \\ 4x = 2y + 6 \end{cases}$

Now we solve this system using any of the methods presented. The first equation expresses x in terms of y, so we will use substitution.

Substitute $y - 4$ for x in the second equation and solve for y.

$$4x = 2y + 6$$
$$4(y - 4) = 2y + 6 \qquad \text{Let } x = y - 4.$$
$$4y - 16 = 2y + 6$$
$$2y = 22$$
$$y = 11$$

Now replace y with 11 in the equation $x = y - 4$ and solve for x. Then $x = y - 4$ becomes $x = 11 - 4 = 7$. The solution of the system is $(7, 11)$, corresponding to the numbers 7 and 11. Checking, notice that 7 **is** 4 less than 11, and 4 times 7 **is** 6 more than twice 11. The proposed numbers, 7 and 11, are correct. ∎

EXAMPLE 2 Two cars leave Indianapolis, one traveling east and the other west. After 3 hours they are 297 miles apart. If one car is traveling 5 mph faster than the other, what is the speed of each?

Solution: Let x = speed of one car
y = speed of the other car

The cars have each traveled 3 hours. Since distance = rate · time, we have that their distances are $3x$ and $3y$ miles, respectively. This information is organized as follows:

	rate ·	time	= distance
One car	x	3	$3x$
Other car	y	3	$3y$

$\begin{cases} \text{The sum of their distances is 297.} \\ \text{One car is 5 mph faster than the other.} \end{cases}$ or $\begin{cases} 3x + 3y = 297 \\ x = y + 5 \end{cases}$

Again, the substitution method is appropriate. Replace x with $y + 5$ in the first equation and solve for y.

$$3x + 3y = 297$$
$$3(y + 5) + 3y = 297 \qquad \text{Let } x = y + 5.$$
$$3y + 15 + 3y = 297$$
$$6y = 282$$
$$y = 47$$

To find x, replace y with 47 in the equation $x = y + 5$. Then $x = 47 + 5 = 52$. The cars are traveling at 52 mph and 47 mph. To check, notice that if one car travels 52 mph for 3 hours the distance is $3(52) = 156$ miles. Also, the other car traveling for 3 hours at 47 mph travels a distance of $3(47) = 141$ miles. The sum of the distances $156 + 141$ is 297 miles, the required distance. ∎

EXAMPLE 3   Lynn Pike, a pharmacist, needs 70 liters of 50% alcohol solution. He has available a 30% alcohol solution and an 80% alcohol solution. How many liters of each solution should he mix to obtain 70 liters of a 50% alcohol solution?

Solution: Let x = number of liters of 30% solution
y = number of liters of 80% solution
Use a table to organize the given data.

% Alcohol in Solution	Liters of Solution	Liters of Alcohol
30	x	$0.30x$
80	y	$0.80y$
50	70	$(0.50)(70)$

$\begin{cases} \text{The total number of liters should be 70.} \\ \text{The liters of alcohol in the 50\% solution is the} \\ \text{sum of the liters in the 30\% and 80\% solutions.} \end{cases}$ or $\begin{cases} x + y = 70 \\ 0.30x + 0.80y = (0.50)(70) \end{cases}$

To solve this system, use the addition method. Multiply both sides of the first equation by -3 and both sides of the second equation by 10. Then

$\begin{cases} -3\ (x + y) = -3\ (70) \\ 10\ (0.30x + 0.80y) = 10\ (0.50)(70) \end{cases}$ simplifies to $\begin{cases} -3x - 3y = -210 \\ \underline{3x + 8y = 350} \\ \qquad\ 5y = 140 \\ \qquad\ \ y = 28 \end{cases}$

Replace y with 28 in the equation $x + y = 70$ and find that $x + 28 = 70$ or $x = 42$. The pharmacist needs to mix 42 liters of 30% solution and 28 liters of 80% solution to obtain 70 liters of 50% solution. Check this solution in the originally stated problem. ∎

EXAMPLE 4   A rectangular garden is to be completely fenced to keep animals out. The length of the garden is four times the width, and the garden requires 210 meters of fencing. Find the dimensions of the garden.

Solution: Let W = width of garden
L = length of garden

We use the formula $P = 2L + 2W$, where P is perimeter. The system is

$\begin{cases} P = 2L + 2W \\ \text{Length is four times the width.} \end{cases}$ or $\begin{cases} 210 = 2W + 2L \\ \quad L = 4W \end{cases}$

The substitution method is appropriate here. Replace L with $4W$ in the first equation.

$$2W + 2L = 210 \qquad \text{First equation.}$$
$$2W + 2(4W) = 210 \qquad \text{Let } L = 4W.$$
$$10W = 210$$
$$W = 21$$

Replace W with 21 in $L = 4W$ and $L = 4(21) = 84$. The garden is 21 meters by 84 meters. Notice that the length is 4 times the width since $4(21) = 84$. Also, the perimeter of the resulting rectangle is $2(21) + 2(84) = 42 + 168 = 210$ meters, the required perimeter. ■

The next problem is solved by using three variables.

EXAMPLE 5   The measure of the largest angle of a triangle is 80° more than the measure of the smallest angle, and the measure of the remaining angle is 10° more than the measure of the smallest angle. Find the measure of each angle.

Solution: Recall that the sum of the measures of the angles of a triangle is 180°.

Let $x =$ degree measure of the smallest angle.
Let $y =$ degree measure of the largest angle.
Let $z =$ degree measure of the remaining angle.

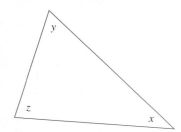

Then the system is

$$\begin{cases} x + y + z = 180 \\ y = x + 80 \\ z = x + 10 \end{cases}$$

Since y and z are both expressed in terms of x, we will solve using the substitution method.

Substitute $y = x + 80$ and $z = x + 10$ in the first equation. Then

$$x + y + z = 180$$

becomes

$$x + (x + 80) + (x + 10) = 180$$
$$3x + 90 = 180$$
$$3x = 90$$
$$x = 30$$

Then $y = x + 80 = 30 + 80 = 110$ and $z = x + 10 = 30 + 10 = 40$. The angles measure 30°, 40°, and 110°. To check, notice that $30° + 40° + 110° = 180°$. Also, the measure of the largest angle, 110°, is 80° more than the measure of the smallest angle, 30°. The measure of the remaining angle, 40°, is 10° more than the measure of the smallest angle, 30°. ■

EXERCISE SET 9.3

Solve. See Example 1.

1. One number is two more than a second number. Twice the first is 4 less than 3 times the second. Find the numbers.

2. Three times one number minus a second is 8, and the sum of the numbers is 12. Find the numbers.

Solve. See Example 2.

3. A Delta 727 traveled 560 mph with the wind and 480 mph against the wind. Find the speed of the plane in still air and the speed of the wind.

4. Terry Watkins can row about 10.6 kilometers in one hour downstream and 6.8 kilometers upstream in one hour. Find how fast he can row in still water, and find how fast the current is.

Solve. See Example 3.

5. Find how many quarts of 4% butterfat milk and 1% butterfat milk should be mixed to yield 60 quarts of 2% butterfat milk.

6. A pharmacist needs 500 milliliters of a 20% phenobarbi-

tal solution, but has only 5% and 25% phenobarbital solutions available. Find how many milliliters of each she should mix to get the desired solution.

Solve. See Example 4.

7. Megan has 156 feet of fencing to make a rectangular pen for her German shepherds. If she wants the pen to be twice as long as it is wide, find the length and width of the pen.

8. A rectangular swimming pool is 22 feet longer than it is wide. The perimeter of the pool is 80 feet. Find the length and width of the pool.

Solve. See Example 5.

9. Chris Peckaitis bought some large frames for $15 each and some small frames for $8 each at a closeout sale. If she bought 22 frames for $239, find how many of each type she bought.

10. Hilton University Drama Club sold 311 tickets for a play. Student tickets cost 50 cents each; nonstudent tickets cost $1.50. If total receipts were $385.50, find how many tickets of each type were sold.

Solve.

11. One number is two less than a second number. Twice the first is 4 more than 3 times the second. Find the numbers.

12. Twice one number plus a second number is 42, and the one number minus the second number is −6. Find the numbers.

13. Seven tablets and 4 pens cost $6.40. Two tablets and 19 pens cost $5.40. Find the price of each.

14. A candy shop manager mixes M&M's worth $2.00 per pound with trail mix worth $1.50 per pound. Find how many pounds of each she should use to get 50 pounds of a party mix worth $1.80 per pound.

15. An airplane takes 3 hours to travel a distance of 2160 miles with the wind. The return trip takes 4 hours against the wind. Find the speed of the plane in still air and the speed of the wind.

16. Two cyclists start at the same point and travel in opposite

directions. One travels 4 mph faster than the other. In 4 hours they are 112 miles apart. Find how fast each is traveling.

17. The perimeter of a quadrilateral (four-sided polygon) is 29 inches. The longest side is twice as long as the shortest side, and the other two sides are equally long and each are 2 inches longer than the shortest side. Find the length of all four sides.

18. The perimeter of a triangle is 93 centimeters. If two sides are equally long and the third side is 9 centimeters longer than the others, find the lengths of the three sides.

19. May Jones's change purse contains $2.70 in nickels and dimes. If she has 39 coins in all, find how many of each type she has.

20. Gerhart Moore has a coin collection with a face value of $5.52. If he has 48 coins in quarters and pennies, find how many of each type he has.

21. The sum of three numbers is 40. One number is five more than a second and twice the third. Find the numbers.

22. The sum of the digits of a three-digit number is 15. The tens-place digit is twice the hundreds place digit, and the ones-place digit is 1 less than the hundreds-place digit. Find the three-digit number.

23. Jack Reinholt, a car salesman, has a choice of two pay arrangements: a weekly salary of $200 plus 5% commission on sales, or a straight 15% commission. Find the amount of sales for which Jack's earnings are the same regardless of the pay arrangement.

24. Hertz car rental agency charges $25 daily plus 10 cents per mile. Budget charges $20 daily plus 25 cents per mile. Find the daily mileage for which the Budget charge for the day is twice that of the Hertz charge for the day.

25. Allan Little has 85 bills, some 10-dollar bills, and some 20-dollar bills, totaling $1480. Find how many of each type there are.

26. A bank teller has 155 bills of $1 and $5 denominations with a total value of $471. Find how many of each type of bill he has.

27. Mary Dooley has a collection of dimes, nickels, and pennies worth $3.42. She has twice as many dimes as nickels and four more pennies than nickels. Find how many of each she has.

28. Carroll Blakemore, a drafting student, bought 3 templates and a pencil one day for $6.45. Another day he bought 2 pads of paper and 4 pencils for $7.50. If the price of a pad of paper is three times the price of a pencil, find the price of each type of item.

29. Rabbits in a lab are to be kept on a strict daily diet to include 30 grams of protein, 16 grams of fat, and 24 grams of carbohydrates. The scientist has only three food mixes available with the following grams of nutrients per unit.

	Protein	Fat	Carbohydrate
Mix A	4	6	3
Mix B	6	1	2
Mix C	4	1	12

Find how many units of each mix are needed daily to meet each rabbit's dietary needs.

30. Gerry Gundersen mixes different solutions with concentrations of 25%, 40%, and 50% to get 200 liters of a 32% solution. If he uses twice as much of the 25% solution as of the 40% solution, find how many liters of each kind he uses.

Skill Review

Factor completely. See Section 5.1.

31. $x^2 + xy + 3x + 3y$

32. $ab + 4a - 2b - 8$

Multiply. See Section 4.4.

33. $(x + 5)^2$

34. $(y - 3)^2$

35. $(2x - y)^2$

36. $(3x + 4)^2$

9.4
Solving Systems of Equations by Determinants

OBJECTIVES

Tape 28

1. Define and evaluate a 2 × 2 determinant.
2. Use Cramer's rule to solve a system of two linear equations in two variables.
3. Define and evaluate a 3 × 3 determinant.
4. Use Cramer's rule to solve a linear system of three equations in three variables.

1   Three methods for solving systems of two linear equations in two variables have been shown: graphically, by substitution, and by elimination. Now we will analyze another method called **Cramer's rule.** First, we introduce determinants. A **determinant** is a real number associated with a square array of numbers written between two vertical bars.

$$\begin{vmatrix} 1 & 6 \\ 5 & 2 \end{vmatrix} \qquad \begin{vmatrix} 2 & 4 & 1 \\ 0 & 5 & 2 \\ 3 & 6 & 9 \end{vmatrix}$$

Each number in the array is called an **element.** The numbers in a horizontal line form a **row;** those in a vertical line form a **column.** A second-order determinant or a 2 × 2 (read as 2 by 2) determinant has 2 rows and 2 columns. The value of a 2 × 2 determinant is defined next.

Value of a 2 × 2 Determinant

$$\begin{vmatrix} a & b \\ c & d \end{vmatrix} = ad - bc$$

EXAMPLE 1   Find the value of each determinant.

a. $\begin{vmatrix} -1 & 2 \\ 3 & -4 \end{vmatrix}$ **b.** $\begin{vmatrix} 2 & 0 \\ 7 & -5 \end{vmatrix}$

Solution: First identify the values of a, b, c, and d.

a. Here $a = -1$, $b = 2$, $c = 3$, and $d = -4$.

$$\begin{vmatrix} -1 & 2 \\ 3 & -4 \end{vmatrix} = ad - bc = (-1)(-4) - (2)(3) = -2$$

b. In this example, $a = 2$, $b = 0$, $c = 7$, and $d = -5$.

$$\begin{vmatrix} 2 & 0 \\ 7 & -5 \end{vmatrix} = ad - bc = 2(-5) - (0)(7) = -10 \qquad ■$$

2   To develop Cramer's rule, we solve by elimination the system $\begin{cases} ax + by = h \\ cx + dy = k \end{cases}$.

First, eliminate y by multiplying both sides of the first equation by d and both sides of the second equation by $-b$ so that the coefficients of y are opposite. The result is the following system. We then add the two equations and solve for x.

$$\begin{cases} d\,(ax + by) = d \cdot h \\ -b\,(cx + dy) = -b \cdot k \end{cases} \quad \text{simplifies to} \quad \begin{cases} adx + bdy = hd \\ -bcx - bdy = -kb \end{cases}$$

$$adx - bcx = hd - kb$$

Add the equations.

$$(ad - bc)\,x = hd - kb$$

$$x = \frac{hd - kb}{ad - bc}$$

Solve for x.

When we replace x with $\dfrac{hd - kb}{ad - bc}$ in the equation $ax + by = h$ and solve for y, we find that $y = \dfrac{ak - ch}{ad - bc}$.

Notice that the numerator of the value of x can be written as a determinant.

$$hd - kb = \begin{vmatrix} h & b \\ k & d \end{vmatrix}$$

The numerator of the value of y can be written as the determinant

$$ak - ch = \begin{vmatrix} a & h \\ c & k \end{vmatrix}$$

The denominator of the values of x and y is the same and can be written as

$$ad - bc = \begin{vmatrix} a & b \\ c & d \end{vmatrix}$$

This means that the values of x and y can be written as

$$x = \dfrac{\begin{vmatrix} h & b \\ k & d \end{vmatrix}}{\begin{vmatrix} a & b \\ c & d \end{vmatrix}} \quad \text{and} \quad y = \dfrac{\begin{vmatrix} a & h \\ c & k \end{vmatrix}}{\begin{vmatrix} a & b \\ c & d \end{vmatrix}}$$

For convenience, we will call the determinants

x-coefficients
↓
$$\begin{vmatrix} a & b \\ c & d \end{vmatrix} = D$$
↑
y-coefficients

$$\begin{vmatrix} h & b \\ k & d \end{vmatrix} = D_x$$
↑
x-column replaced
by constants

$$\begin{vmatrix} a & h \\ c & k \end{vmatrix} = D_y$$
↑
y-column replaced by
constants

These determinant formulas for the coordinates of the solution of a system are known as **Cramer's rule.**

Cramer's Rule for Two Linear Equations in Two Variables

The solution of the system $\begin{cases} ax + by = h \\ cx + dy = k \end{cases}$ is given by

$$x = \dfrac{\begin{vmatrix} h & b \\ k & d \end{vmatrix}}{\begin{vmatrix} a & b \\ c & d \end{vmatrix}} = \dfrac{D_x}{D} ; \quad y = \dfrac{\begin{vmatrix} a & h \\ c & k \end{vmatrix}}{\begin{vmatrix} a & b \\ c & d \end{vmatrix}} = \dfrac{D_y}{D}$$

as long as $D = ad - bc$ is not 0.

When $D = 0$, the system is either inconsistent or the equations are dependent. When this happens, use another method to see which.

EXAMPLE 2   Use Cramer's rule to solve each system.

a. $\begin{cases} 3x + 4y = -7 \\ x - 2y = -9 \end{cases}$

b. $\begin{cases} 5x + y = 5 \\ -7x - 2y = -7 \end{cases}$

Solution: **a.** Find D, D_x, and D_y.

$$\begin{array}{ccc} a & b & h \\ \downarrow & \downarrow & \downarrow \end{array}$$
$$\begin{cases} 3x + 4y = -7 \\ x - 2y = -9 \end{cases}$$
$$\begin{array}{ccc} \uparrow & \uparrow & \uparrow \\ c & d & k \end{array}$$

$$D = \begin{vmatrix} 3 & 4 \\ 1 & -2 \end{vmatrix} = 3(-2) - 4(1) = -10$$

$$D_x = \begin{vmatrix} -7 & 4 \\ -9 & -2 \end{vmatrix} = (-7)(-2) - 4(-9) = 50$$

$$D_y = \begin{vmatrix} 3 & -7 \\ 1 & -9 \end{vmatrix} = 3(-9) - (-7)(1) = -20$$

Then $x = \dfrac{D_x}{D} = \dfrac{50}{-10} = -5$ and $y = \dfrac{D_y}{D} = \dfrac{-20}{-10} = 2$. The solution set is $\{(-5, 2)\}$. As always, check the solution in both original equations.

b. $\begin{cases} 5x + y = 5 \\ -7x - 2y = -7 \end{cases}$ Find D, D_x, and D_y.

$$D = \begin{vmatrix} 5 & 1 \\ -7 & -2 \end{vmatrix} = 5(-2) - (-7)(1) = -3$$

$$D_x = \begin{vmatrix} 5 & 1 \\ -7 & -2 \end{vmatrix} = 5(-2) - (-7)(1) = -3$$

$$D_y = \begin{vmatrix} 5 & 5 \\ -7 & -7 \end{vmatrix} = 5(-7) - 5(-7) = 0$$

$$x = \frac{D_x}{D} = \frac{-3}{-3} = 1, \qquad y = \frac{D_y}{D} = \frac{0}{-3} = 0$$

The solution set is $\{(1, 0)\}$.   ∎

3   Three-by-three determinants can be used to solve systems of three equations in three variables. Finding the value of a 3×3 determinant, however, is considerably more complex than the 2×2 case.

> **The Value of a 3 × 3 Determinant**
>
> $$\begin{vmatrix} a_1 & b_1 & c_1 \\ a_2 & b_2 & c_2 \\ a_3 & b_3 & c_3 \end{vmatrix} = a_1 \cdot \begin{vmatrix} b_2 & c_2 \\ b_3 & c_3 \end{vmatrix} - a_2 \cdot \begin{vmatrix} b_1 & c_1 \\ b_3 & c_3 \end{vmatrix} + a_3 \cdot \begin{vmatrix} b_1 & c_1 \\ b_2 & c_2 \end{vmatrix}$$

The value of a 3 × 3 determinant, then, is based on three 2 × 2 determinants. Each of these 2 × 2 determinants is called a **minor,** and every element of a 3 × 3 determinant has a minor associated with it. For example, the minor of c_2 has no row or column containing c_2.

$$\begin{array}{ccc} a_1 & b_1 & c_1 \\ a_2 & b_2 & c_2 \\ a_3 & b_3 & c_3 \end{array}$$

The minor of c_2 is

$$\begin{vmatrix} a_1 & b_1 \\ a_3 & b_3 \end{vmatrix}$$

Also, the minor of element a_1 is the 2 × 2 determinant that has no row or column containing a_1.

$$\begin{array}{ccc} a_1 & b_1 & c_1 \\ a_2 & b_2 & c_2 \\ a_3 & b_3 & c_3 \end{array}$$

The minor of a_1 is

$$\begin{vmatrix} b_2 & c_2 \\ b_3 & c_3 \end{vmatrix}$$

So the value of a 3 × 3 determinant can be written as

$$a_1 \cdot (\text{minor of } a_1) - a_2(\text{minor of } a_2) + a_3(\text{minor of } a_3)$$

We call this finding the value of the determinant by **expanding** by the minors of the first row. The value of a determinant can be found by expanding by the minors of any row or column. The following array of signs is helpful in determining whether to add or subtract the product of an element and its minor.

$$\begin{vmatrix} + & - & + \\ - & + & - \\ + & - & + \end{vmatrix}$$

If an element is in a position marked $+$, we add. If marked $-$, we subtract.

EXAMPLE 3   Find the value of the determinant by expanding by the minors of the given row or column.

$$\begin{vmatrix} 0 & 5 & 1 \\ 1 & 3 & -1 \\ -2 & 2 & 4 \end{vmatrix}$$

a. First column   **b.** Second row

Solution: **a.** The elements of the first column are 0, 1, and -2. The first column of the array of signs is $+, -, +$.

$$0 \cdot \begin{vmatrix} 3 & -1 \\ 2 & 4 \end{vmatrix} - 1 \cdot \begin{vmatrix} 5 & 1 \\ 2 & 4 \end{vmatrix} + (-2) \cdot \begin{vmatrix} 5 & 1 \\ 3 & -1 \end{vmatrix}$$

$$= 0(12 + 2) - 1(20 - 2) + (-2)(-5 - 3)$$

$$= 0 - 18 + 16 = -2$$

b. The elements of the second row are 1, 3, and -1. This time, the signs begin with $-$ and again alternate.

$$-1 \cdot \begin{vmatrix} 5 & 1 \\ 2 & 4 \end{vmatrix} + 3 \cdot \begin{vmatrix} 0 & 1 \\ -2 & 4 \end{vmatrix} - (-1) \cdot \begin{vmatrix} 0 & 5 \\ -2 & 2 \end{vmatrix}$$

$$= -1(20 - 2) + 3(0 - (-2)) - (-1)(0 - (-10))$$

$$= -18 + 6 + 10 = -2$$

Notice that the determinant has the same value regardless of the row or column you select to expand by. ∎

4   A system of three equations in three variables may be solved with Cramer's rule also. Using the elimination process to solve a system with unknown constants as coefficients leads to the following.

Cramer's Rule for Three Equations in Three Variables

The solution of the system $\begin{cases} a_1x + b_1y + c_1z = k_1 \\ a_2x + b_2y + c_2z = k_2 \\ a_3x + b_3y + c_3z = k_3 \end{cases}$ is given by $x = \dfrac{D_x}{D}, y = \dfrac{D_y}{D}$,

and $z = \dfrac{D_z}{D}$, where

$$D = \begin{vmatrix} a_1 & b_1 & c_1 \\ a_2 & b_2 & c_2 \\ a_3 & b_3 & c_3 \end{vmatrix} \quad D_x = \begin{vmatrix} k_1 & b_1 & c_1 \\ k_2 & b_2 & c_2 \\ k_3 & b_3 & c_3 \end{vmatrix}$$

$$D_y = \begin{vmatrix} a_1 & k_1 & c_1 \\ a_2 & k_2 & c_2 \\ a_3 & k_3 & c_3 \end{vmatrix} \quad D_z = \begin{vmatrix} a_1 & b_1 & k_1 \\ a_2 & b_2 & k_2 \\ a_3 & b_3 & k_3 \end{vmatrix}$$

as long as D is not 0.

EXAMPLE 4   Use Cramer's rule to solve the system.

$$\begin{cases} x - 2y + z = 4 \\ 3x + y - 2z = 3 \\ 5x + 5y + 3z = -8 \end{cases}$$

Solution: First, evaluate D, D_x, D_y, and D_z. We will expand D by the minors of the first column.

$$D = \begin{vmatrix} 1 & -2 & 1 \\ 3 & 1 & -2 \\ 5 & 5 & 3 \end{vmatrix}$$

$$= 1 \cdot \begin{vmatrix} 1 & -2 \\ 5 & 3 \end{vmatrix} - 3 \cdot \begin{vmatrix} -2 & 1 \\ 5 & 3 \end{vmatrix} + 5 \cdot \begin{vmatrix} -2 & 1 \\ 1 & -2 \end{vmatrix}$$

$$= 1(3 - (-10)) - 3(-6 - 5) + 5(4 - 1)$$

$$= 13 + 33 + 15 = 61$$

$$D_x = \begin{vmatrix} 4 & -2 & 1 \\ 3 & 1 & -2 \\ -8 & 5 & 3 \end{vmatrix}$$

$$= 4 \cdot \begin{vmatrix} 1 & -2 \\ 5 & 3 \end{vmatrix} - 3 \cdot \begin{vmatrix} -2 & 1 \\ 5 & 3 \end{vmatrix} + (-8) \cdot \begin{vmatrix} -2 & 1 \\ 1 & -2 \end{vmatrix}$$

$$= 4(3 - (-10)) - 3(-6 - 5) + (-8)(4 - 1)$$

$$= 52 + 33 - 24 = 61$$

$$D_y = \begin{vmatrix} 1 & 4 & 1 \\ 3 & 3 & -2 \\ 5 & -8 & 3 \end{vmatrix}$$

$$= 1 \cdot \begin{vmatrix} 3 & -2 \\ -8 & 3 \end{vmatrix} - 3 \cdot \begin{vmatrix} 4 & 1 \\ -8 & 3 \end{vmatrix} + 5 \cdot \begin{vmatrix} 4 & 1 \\ 3 & -2 \end{vmatrix}$$

$$= 1(9 - 16) - 3(12 + 8) + 5(-8 - 3)$$

$$= -7 - 60 - 55 = -122$$

$$D_z = \begin{vmatrix} 1 & -2 & 4 \\ 3 & 1 & 3 \\ 5 & 5 & -8 \end{vmatrix}$$

$$= 1 \cdot \begin{vmatrix} 1 & 3 \\ 5 & -8 \end{vmatrix} - 3 \cdot \begin{vmatrix} -2 & 4 \\ 5 & -8 \end{vmatrix} + 5 \cdot \begin{vmatrix} -2 & 4 \\ 1 & 3 \end{vmatrix}$$

$$= 1(-8 - 15) - 3(16 - 20) + 5(-6 - 4)$$

$$= -23 + 12 - 50 = -61$$

From these determinants, we calculate the solution:

$$x = \frac{D_x}{D} = \frac{61}{61} = 1, \qquad y = \frac{D_y}{D} = \frac{-122}{61} = -2, \qquad z = \frac{D_z}{D} = \frac{-61}{61} = -1$$

The solution set of the system is $\{(1, -2, -1)\}$. Check this solution by verifying that it satisfies each equation of the system. ∎

EXERCISE SET 9.4

Find the value of each determinant. See Example 1.

1. $\begin{vmatrix} 3 & 5 \\ -1 & 7 \end{vmatrix}$

2. $\begin{vmatrix} -5 & 1 \\ 0 & -4 \end{vmatrix}$

3. $\begin{vmatrix} 9 & -2 \\ 4 & -3 \end{vmatrix}$

4. $\begin{vmatrix} 4 & 0 \\ 9 & 8 \end{vmatrix}$

5. $\begin{vmatrix} -2 & 9 \\ 4 & -18 \end{vmatrix}$

6. $\begin{vmatrix} -40 & 8 \\ 70 & -14 \end{vmatrix}$

Use Cramer's rule, if possible, to solve each system of linear equations. See Example 2.

7. $\begin{cases} 2y - 4 = 0 \\ x + 2y = 5 \end{cases}$

8. $\begin{cases} 4x - y = 5 \\ 3x - 3 = 0 \end{cases}$

9. $\begin{cases} 3x + y = 1 \\ 2y = 2 - 6x \end{cases}$

10. $\begin{cases} y = 2x - 5 \\ 8x - 4y = 20 \end{cases}$

11. $\begin{cases} 5x - 2y = 27 \\ -3x + 5y = 18 \end{cases}$

12. $\begin{cases} 4x - y = 9 \\ 2x + 3y = -27 \end{cases}$

Find the value of each determinant. See Example 3.

13. $\begin{vmatrix} 2 & 1 & 0 \\ 0 & 5 & -3 \\ 4 & 0 & 2 \end{vmatrix}$

14. $\begin{vmatrix} -6 & 4 & 2 \\ 1 & 0 & 5 \\ 0 & 3 & 1 \end{vmatrix}$

15. $\begin{vmatrix} 4 & -6 & 0 \\ -2 & 3 & 0 \\ 4 & -6 & 1 \end{vmatrix}$

16. $\begin{vmatrix} 5 & 2 & 1 \\ 3 & -6 & 0 \\ -2 & 8 & 0 \end{vmatrix}$

17. $\begin{vmatrix} 3 & 6 & -3 \\ -1 & -2 & 3 \\ 4 & -1 & 6 \end{vmatrix}$

18. $\begin{vmatrix} 2 & -2 & 1 \\ 4 & 1 & 3 \\ 3 & 1 & 2 \end{vmatrix}$

Use Cramer's rule, if possible, to solve each system of linear equations. See Example 4.

19. $\begin{cases} 3x \quad\quad + z = -1 \\ -x - 3y + z = 7 \\ \quad\quad 3y + z = 5 \end{cases}$

20. $\begin{cases} 4y - 3z = -2 \\ 8x - 4y \quad\quad = 4 \\ -8x + 4y + z = -2 \end{cases}$

21. $\begin{cases} x + y + z = 8 \\ 2x - y - z = 10 \\ x - 2y + 3z = 22 \end{cases}$

22. $\begin{cases} 5x + y + 3z = 1 \\ x - y - 3z = -7 \\ -x + y \quad\quad = 1 \end{cases}$

Find the value of each determinant.

23. $\begin{vmatrix} 10 & -1 \\ -4 & 2 \end{vmatrix}$

24. $\begin{vmatrix} -6 & 2 \\ 5 & -1 \end{vmatrix}$

25. $\begin{vmatrix} 1 & 0 & 4 \\ 1 & -1 & 2 \\ 3 & 2 & 1 \end{vmatrix}$

26. $\begin{vmatrix} 0 & 1 & 2 \\ 3 & -1 & 2 \\ 3 & 2 & -2 \end{vmatrix}$

27. $\begin{vmatrix} \dfrac{3}{4} & \dfrac{5}{2} \\ -\dfrac{1}{6} & \dfrac{7}{3} \end{vmatrix}$

28. $\begin{vmatrix} \dfrac{5}{7} & \dfrac{1}{3} \\ \dfrac{6}{7} & \dfrac{2}{3} \end{vmatrix}$

29. $\begin{vmatrix} 4 & -2 & 2 \\ 6 & -1 & 3 \\ 2 & 1 & 1 \end{vmatrix}$

30. $\begin{vmatrix} 1 & 5 & 0 \\ 7 & 9 & -4 \\ 3 & 2 & -2 \end{vmatrix}$

31. $\begin{vmatrix} -2 & 5 & 4 \\ 5 & -1 & 3 \\ 4 & 1 & 2 \end{vmatrix}$

32. $\begin{vmatrix} 5 & -2 & 4 \\ -1 & 5 & 3 \\ 1 & 4 & 2 \end{vmatrix}$

Use Cramer's rule, if possible, to solve each system of linear equations.

33. $\begin{cases} 2x - 5y = 4 \\ x + 2y = -7 \end{cases}$

34. $\begin{cases} 3x - y = 2 \\ -5x + 2y = 0 \end{cases}$

35. $\begin{cases} 4x + 2y = 5 \\ 2x + y = -1 \end{cases}$

36. $\begin{cases} 3x + 6y = 15 \\ 2x + 4y = 3 \end{cases}$

37. $\begin{cases} 2x + 2y + z = 1 \\ -x + y + 2z = 3 \\ x + 2y + 4z = 0 \end{cases}$

38. $\begin{cases} 2x - 3y + z = 5 \\ x + y + z = 0 \\ 4x + 2y + 4z = 4 \end{cases}$

39. $\begin{cases} \dfrac{2}{3}x - \dfrac{3}{4}y = -1 \\ -\dfrac{1}{6}x + \dfrac{3}{4}y = \dfrac{5}{2} \end{cases}$

40. $\begin{cases} \dfrac{1}{2}x - \dfrac{1}{3}y = -3 \\ \dfrac{1}{8}x + \dfrac{1}{6}y = 0 \end{cases}$

41. $\begin{cases} 0.7x - 0.2y = -1.6 \\ 0.2x - y = -1.4 \end{cases}$

42. $\begin{cases} -0.7x + 0.6y = 1.3 \\ 0.5x - 0.3y = -0.8 \end{cases}$

43. $\begin{cases} -2x + 4y - 2z = 6 \\ x - 2y + z = -3 \\ 3x - 6y + 3z = -9 \end{cases}$

44. $\begin{cases} -x - y + 3z = 2 \\ 4x + 4y - 12z = -8 \\ -3x - 3y + 9z = 6 \end{cases}$

45. $\begin{cases} x - 2y + z = -5 \\ 3y + 2z = 4 \\ 3x - y = -2 \end{cases}$

46. $\begin{cases} 4x + 5y = 10 \\ 3y + 2z = -6 \\ x + y + z = 3 \end{cases}$

A Look Ahead

Find the value of each determinant. See the following example.

EXAMPLE Evaluate the determinant.

$$\begin{vmatrix} 2 & 0 & -1 & 3 \\ 0 & 5 & -2 & -1 \\ 3 & 1 & 0 & 1 \\ 4 & 2 & -2 & 0 \end{vmatrix}$$

Solution: To evaluate a 4×4 determinant, select any row or column and expand by the minors. The array of signs for a 4×4 determinant is the same as for a 3×3 determinant except expanded. We expand using the fourth row.

$$\rightarrow \begin{vmatrix} 2 & 0 & -1 & 3 \\ 0 & 5 & -2 & -1 \\ 3 & 1 & 0 & 1 \\ 4 & 2 & -2 & 0 \end{vmatrix} = -4 \cdot \begin{vmatrix} 0 & -1 & 3 \\ 5 & -2 & -1 \\ 1 & 0 & 1 \end{vmatrix} + 2 \cdot \begin{vmatrix} 2 & -1 & 3 \\ 0 & -2 & -1 \\ 3 & 0 & 1 \end{vmatrix} - (-2) \cdot \begin{vmatrix} 2 & 0 & 3 \\ 0 & 5 & -1 \\ 3 & 1 & 1 \end{vmatrix} + 0 \cdot \begin{vmatrix} 2 & 0 & -1 \\ 0 & 5 & -2 \\ 3 & 1 & 0 \end{vmatrix}$$

Now find the value of each 3×3 determinant. The value of the 4×4 determinant is

$$-4(12) + 2(17) + 2(-33) + 0 = -80 \quad \blacksquare$$

47. $\begin{vmatrix} 5 & 0 & 0 & 0 \\ 0 & 4 & 2 & -1 \\ 1 & 3 & -2 & 0 \\ 0 & -3 & 1 & 2 \end{vmatrix}$

48. $\begin{vmatrix} 1 & 7 & 0 & -1 \\ 1 & 3 & -2 & 0 \\ 1 & 0 & -1 & 2 \\ 0 & -6 & 2 & 4 \end{vmatrix}$

49. $\begin{vmatrix} 4 & 0 & 2 & 5 \\ 0 & 3 & -1 & 1 \\ 0 & 0 & 2 & 0 \\ 0 & 0 & 0 & 1 \end{vmatrix}$

50. $\begin{vmatrix} 2 & 0 & -1 & 4 \\ 6 & 0 & 4 & 1 \\ 2 & 4 & 3 & -1 \\ 4 & 0 & 5 & -4 \end{vmatrix}$

Writing in Mathematics

51. If the elements in a single row of a determinant are all zero, what is the value of the determinant? Explain your answer. Is the answer the same if "row" is replaced by "column"?

Skill Review

Multiply both sides of equation (1) by 2, and add the resulting equation to equation (2). See Section 9.2.

52. $\begin{array}{l} 3x - y + z = 2 \quad (1) \\ -x + 2y + 3z = 6 \quad (2) \end{array}$

53. $\begin{array}{l} 2x + y + 3z = 7 \quad (1) \\ -4x + y + 2z = 4 \quad (2) \end{array}$

Multiply. See Section 7.6.

54. $(5 + i)(2 - 3i)$ **55.** $(7 - 2i)(1 + i)$ **56.** $(3 - 4i)(3 + 4i)$ **57.** $(2 + 9i)(2 + 9i)$

9.5
Solving Systems of Equations by Matrices

OBJECTIVES

Tape 28

1 Use matrices to solve a system of two equations.

2 Use matrices to solve a system of three equations.

By now, you have seen that the solution of a system of equations depends on the coefficients of the equations in the system. Cramer's rule gives formulas for the coordinates of the solution in terms of these coefficients. In this section, we introduce solving a system of equations by a **matrix** (plural **matrices**).

1 A matrix is a rectangular array of numbers. The following are examples of matrices.

$$\begin{pmatrix} 1 & 0 \\ 0 & 1 \end{pmatrix}$$

$$\begin{pmatrix} 2 & 1 & 3 & -1 \\ 0 & -1 & 4 & 5 \\ -6 & 2 & 1 & 0 \end{pmatrix}$$

$$\begin{pmatrix} a & b & c \\ d & e & f \end{pmatrix}$$

2 × 2 matrix
2 rows, 2 columns

3 × 4 matrix
3 rows, 4 columns

2 × 3 matrix
2 rows, 3 columns

To see the relationship between systems of equations and matrices, consider the system of equations, written in standard form.

$$\begin{cases} 2x - 3y = 6 \\ x + y = 0 \end{cases}$$

A corresponding matrix associated with this system is

$$\begin{pmatrix} 2 & -3 & \vdots & 6 \\ 1 & 1 & \vdots & 0 \end{pmatrix}$$

The coefficients of each variable are placed to the left of a vertical dashed line. The constants are placed to the right. This 2 × 3 matrix is called the **augmented matrix of the system.** Observe that the rows of this augmented matrix correspond to the equations in the system. The first equation corresponds to the first row; the second equation corresponds to the second row.

The method of solving systems by matrices is to write the augmented matrix as an equivalent matrix from which we can easily find the solution. Two matrices are equivalent if they represent systems that have the same solution set. The following **row operations** can be performed on matrices, and the result is an equivalent matrix.

Elementary Row Operations

1. Any two rows in a matrix may be interchanged.

2. The elements of any row may be multiplied (or divided) by the same nonzero number.

3. The elements of any row may be multiplied (or divided) by a nonzero number and added to its corresponding elements in any other row.

EXAMPLE 1   Solve the system using matrices.

$$\begin{cases} x + 3y = 5 \\ 2x - y = -4 \end{cases}$$

Solution: The augmented matrix is $\begin{pmatrix} 1 & 3 & \vdots & 5 \\ 2 & -1 & \vdots & -4 \end{pmatrix}$

Use elementary row operations to write an equivalent matrix that has 1's along the main diagonal and 0's below each 1 in the main diagonal. The main diagonal of a matrix is the left-to-right diagonal starting with row 1, column 1. For the matrix given, the element in the first row, first column is already 1, as desired. Next, we write an equivalent matrix with a 0 below the 1. To do this, multiply row 1 by -2 and add to row 2. **We will only change row 2.**

$$\begin{pmatrix} 1 & 3 & \vdots & 5 \\ -2(1) + 2 & -2(3) + (-1) & \vdots & -2(5) + (-4) \end{pmatrix} \text{ simplifies to } \begin{pmatrix} 1 & 3 & \vdots & 5 \\ 0 & -7 & \vdots & -14 \end{pmatrix}$$

$$\begin{array}{cccccc} \uparrow & \uparrow & & \uparrow & \uparrow & & \uparrow & \uparrow \\ \text{row 1} & \text{row 2} & & \text{row 1} & \text{row 2} & & \text{row 1} & \text{row 2} \\ \text{element} & \text{element} & & \text{element} & \text{element} & & \text{element} & \text{element} \end{array}$$

Next, continue down the main diagonal and change the -7 to a 1 by use of an elementary row operation. Divide row 2 by -7. Then

$$\begin{pmatrix} 1 & 3 & \vdots & 5 \\ \dfrac{0}{-7} & \dfrac{-7}{-7} & \vdots & \dfrac{-14}{-7} \end{pmatrix} \text{ simplifies to } \begin{pmatrix} 1 & 3 & \vdots & 5 \\ 0 & 1 & \vdots & 2 \end{pmatrix}$$

This last matrix corresponds to the system

$$\begin{cases} 1x + 3y = 5 \\ 0x + 1y = 2 \end{cases} \text{ or } \begin{cases} x + 3y = 5 \\ y = 2 \end{cases}$$

To find x, let $y = 2$ in the first equation, $x + 3y = 5$.

$$x + 3y = 5 \qquad \text{First equation.}$$

$$x + 3(2) = 5 \qquad \text{Let } y = 2.$$

$$x = -1$$

The solution set is $\{(-1, 2)\}$. Check to see that this ordered pair satisfies both equations. ∎

2 Solving a system of three equations in three variables using matrices means writing the corresponding matrix and finding an equivalent matrix that has 1's along the main diagonal and 0's below the 1's.

EXAMPLE 2 Solve the system using matrices.

$$\begin{cases} x + 2y + z = 2 \\ -2x - y + 2z = 5 \\ x + 3y - 2z = -8 \end{cases}$$

Solution: The corresponding matrix is $\begin{pmatrix} 1 & 2 & 1 & \vdots & 2 \\ -2 & -1 & 2 & \vdots & 5 \\ 1 & 3 & -2 & \vdots & -8 \end{pmatrix}$. Our goal is to write an equivalent matrix with 1's on the main diagonal and 0's below the 1's. The element in row 1, column 1 is already 1. Next, get 0's for each element in the rest of column 1. To do this, first multiply the elements of row 1 by 2, and add the new elements to row 2. Also, we multiply the elements of row 1 by -1, and add the new elements to the elements of row 3. We **do not change row 1.** Then

$$\begin{pmatrix} 1 & 2 & 1 & \vdots & 2 \\ 2(1)-2 & 2(2)-1 & 2(1)+2 & \vdots & 2(2)+5 \\ -1(1)+1 & -1(2)+3 & -1(1)-2 & \vdots & -1(2)-8 \end{pmatrix} \text{ simplifies to } \begin{pmatrix} 1 & 2 & 1 & \vdots & 2 \\ 0 & 3 & 4 & \vdots & 9 \\ 0 & 1 & -3 & \vdots & -10 \end{pmatrix}$$

Next, continue down the diagonal and use elementary row operations to get 1 where the element 3 is now. To do this, interchange rows 2 and 3.

$$\begin{pmatrix} 1 & 2 & 1 & \vdots & 2 \\ 0 & 3 & 4 & \vdots & 9 \\ 0 & 1 & -3 & \vdots & -10 \end{pmatrix} \text{ is equivalent to } \begin{pmatrix} 1 & 2 & 1 & \vdots & 2 \\ 0 & 1 & -3 & \vdots & -10 \\ 0 & 3 & 4 & \vdots & 9 \end{pmatrix}$$

Next, we want the row 3, column 2 element to be 0. Multiply the elements of row 2 by -3, and add the new elements to the elements of row 3.

$$\begin{pmatrix} 1 & 2 & 1 & \vdots & 2 \\ 0 & 1 & -3 & \vdots & -10 \\ -3(0)+0 & -3(1)+3 & -3(-3)+4 & \vdots & -3(-10)+9 \end{pmatrix} \text{ simplifies to } \begin{pmatrix} 1 & 2 & 1 & \vdots & 2 \\ 0 & 1 & -3 & \vdots & -10 \\ 0 & 0 & 13 & \vdots & 39 \end{pmatrix}$$

Finally, we divide the elements of row 3 by 13, so that the final main diagonal element is 1.

$$\begin{pmatrix} 1 & 2 & 1 & \vdots & 2 \\ 0 & 1 & -3 & \vdots & -10 \\ \dfrac{0}{13} & \dfrac{0}{13} & \dfrac{13}{13} & \vdots & \dfrac{39}{13} \end{pmatrix} \text{ simplifies to } \begin{pmatrix} 1 & 2 & 1 & \vdots & 2 \\ 0 & 1 & -3 & \vdots & -10 \\ 0 & 0 & 1 & \vdots & 3 \end{pmatrix}$$

This matrix corresponds to the system

$$\begin{cases} x + 2y + z = 2 \\ y - 3z = -10 \\ z = 3 \end{cases}$$

We identify the z-coordinate of the solution as 3. Replace z with 3 in the second equation and solve for y.

$$y - 3z = -10 \qquad \text{Second equation.}$$

$$y - 3(3) = -10 \qquad \text{Let } z = 3.$$

$$y = -1$$

To find x, let $z = 3$ and $y = -1$ in the first equation.

$$x + 2y + z = 2 \qquad \text{First equation.}$$

$$x + 2(-1) + 3 = 2 \qquad \text{Let } z = 3 \text{ and } y = -1.$$

$$x = 1$$

The solution set is $\{(1, -1, 3)\}$. Check to see that it satisfies the original system.   ■

EXAMPLE 3   Solve the system using matrices.

$$\begin{cases} 2x - y = 3 \\ 4x - 2y = 5 \end{cases}$$

Solution: The corresponding augmented matrix is $\begin{pmatrix} 2 & -1 & \vdots & 3 \\ 4 & -2 & \vdots & 5 \end{pmatrix}$. To get 1 in the row 1, column 1 position, divide the elements of row 1 by 2.

$$\begin{pmatrix} \dfrac{2}{2} & -\dfrac{1}{2} & \vdots & \dfrac{3}{2} \\ 4 & -2 & \vdots & 5 \end{pmatrix} \quad \text{simplifies to} \quad \begin{pmatrix} 1 & -\dfrac{1}{2} & \vdots & \dfrac{3}{2} \\ 4 & -2 & \vdots & 5 \end{pmatrix}$$

To get 0 under the 1, multiply the elements of row 1 by -4 and add the new elements to the elements of row 2.

$$\begin{pmatrix} 1 & -\dfrac{1}{2} & \vdots & \dfrac{3}{2} \\ -4\,(1) + 4 & -4\left(-\dfrac{1}{2}\right) - 2 & \vdots & -4\left(\dfrac{3}{2}\right) + 5 \end{pmatrix} \quad \text{simplifies to} \quad \begin{pmatrix} 1 & -\dfrac{1}{2} & \vdots & \dfrac{3}{2} \\ 0 & 0 & \vdots & -1 \end{pmatrix}$$

The corresponding system is $\begin{cases} x - \dfrac{1}{2}y = \dfrac{3}{2} \\ 0 = -1 \end{cases}$. The equation $0 = -1$ is false for all values of y or x and hence the system is inconsistent and has no solution.   ■

EXERCISE SET 9.5

Solve each system of linear equations using matrices. See Example 1.

1. $\begin{cases} x + y = 1 \\ x - 2y = 4 \end{cases}$

2. $\begin{cases} 2x - y = 8 \\ x + 3y = 11 \end{cases}$

3. $\begin{cases} 2y - 4 = 0 \\ x + 2y = 0 \end{cases}$

4. $\begin{cases} 4x - y = 5 \\ 3x - 3 = 0 \end{cases}$

Solve each system of linear equations using matrices. See Example 2.

5. $\begin{cases} x + y = 3 \\ 2y = 10 \\ 3x + 2y - 4z = 12 \end{cases}$

6. $\begin{cases} 5x = 5 \\ 2x + y = 4 \\ 3x + y - 5z = -15 \end{cases}$

7. $\begin{cases} 2y - z = -7 \\ x + 4y + z = -4 \\ 5x - y + 2z = 13 \end{cases}$

8. $\begin{cases} 4y + 3z = -2 \\ 5x - 4y = 1 \\ -5x + 4y + z = -3 \end{cases}$

Solve each system of linear equations using matrices. See Example 3.

9. $\begin{cases} x - 2y = 4 \\ 2x - 4y = 4 \end{cases}$

10. $\begin{cases} -x + 3y = 6 \\ 3x - 9y = 9 \end{cases}$

11. $\begin{cases} 3x - 3y = 9 \\ 2x - 2y = 6 \end{cases}$

12. $\begin{cases} 9x - 3y = 6 \\ -18x + 6y = -12 \end{cases}$

Solve each system of linear equations using matrices.

13. $\begin{cases} x - 4 = 0 \\ x + y = 1 \end{cases}$

14. $\begin{cases} 3y = 6 \\ x + y = 7 \end{cases}$

15. $\begin{cases} x + y + z = 2 \\ 2x - z = 5 \\ 3y + z = 2 \end{cases}$

16. $\begin{cases} x + 2y + z = 5 \\ x - y - z = 3 \\ 3y + 2z = 2 \end{cases}$

17. $\begin{cases} 5x - 2y = 27 \\ -3x + 5y = 18 \end{cases}$

18. $\begin{cases} 4x - y = 9 \\ 2x + 3y = -27 \end{cases}$

19. $\begin{cases} 4x - 7y = 7 \\ 12x - 21y = 24 \end{cases}$

20. $\begin{cases} 2x - 5y = 12 \\ -4x + 10y = 20 \end{cases}$

21. $\begin{cases} 4x - y + 2z = 5 \\ 2y + z = 4 \\ 4x + y + 3z = 10 \end{cases}$

22. $\begin{cases} 5y - 7z = 14 \\ 2x + y + 4z = 10 \\ 2x + 6y - 3z = 30 \end{cases}$

23. $\begin{cases} 4x + y + z = 3 \\ -x + y - 2z = -11 \\ x + 2y + 2z = -1 \end{cases}$

24. $\begin{cases} x + y + z = 9 \\ 3x - y + z = -1 \\ -2x + 2y - 3z = -2 \end{cases}$

Skill Review

Solve. See Section 2.3.

25. Twice a number subtracted from 25 is 13. Find the number.

26. Three-fourths of a number is 21. Find the number.

27. Five less than four times a number is three more than twice the same number. Find the number.

28. Eight more than a number is 2 less than 3 times the number. Find the number.

29. A rectangle is 4 kilometers longer than it is wide. The perimeter is 28 kilometers. Find its dimensions.

30. The perimeter of a painting shaped like an equilateral triangle (three sides equal) is 81 inches. Find the length of a side.

9.6
Solving Nonlinear Systems of Equations

Tape 29

OBJECTIVES

1   Solve a nonlinear system by substitution.

2   Solve a nonlinear system by elimination.

In Section 9.1, we used graphing, substitution, and elimination methods to find solutions of systems of linear equations. We now apply these same methods to nonlinear equations. A nonlinear system of equations is a system of equations at least one of which is not linear. Since we will be graphing the equations in each system, we are interested in real number solutions only.

1

EXAMPLE 1   Solve the system

$$\begin{cases} x^2 - 3y = 1 \\ x \ - \ y = 1 \end{cases}$$

Solution: We can solve this system by substitution if we solve one equation for one of the variables. Solving the first equation for x would be a poor choice since we would then need to take a square root. We solve the second equation for y.

$$x - y = 1 \qquad \text{Second equation.}$$
$$x - 1 = y \qquad \text{Solve for } y.$$

Replace y with $x - 1$ in the first equation, and solve for x.

$$x^2 - 3y = 1 \qquad \text{First equation.}$$
$$x^2 - 3(x - 1) = 1 \qquad \text{Replace } y \text{ with } x - 1.$$
$$x^2 - 3x + 3 = 1$$
$$x^2 - 3x + 2 = 0$$
$$(x - 2)(x - 1) = 0$$
$$x = 2 \quad \text{or} \quad x = 1$$

Let $x = 2$ and then $x = 1$ in the equation $y = x - 1$ to find corresponding y-values.

Let $x = 2$.	Let $x = 1$.
$y = x - 1$	$y = x - 1$
$y = 2 - 1 = 1$	$y = 1 - 1 = 0$

The solution set is $\{(2, 1), (1, 0)\}$. Check both solutions in both equations. Both solutions will satisfy both equations, so both are solutions of the system. The graph of each equation in the system is next.

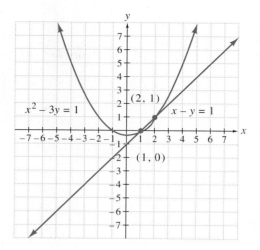

EXAMPLE 2 Solve the system

$$\begin{cases} y = \sqrt{x} \\ x^2 + y^2 = 6 \end{cases}.$$

Solution: This system is ideal for substitution since y is expressed in terms of x in the first equation. Substitute $\sqrt{x}$ for y in the second equation, and solve for x.

$$x^2 + y^2 = 6$$
$$x^2 + (\sqrt{x})^2 = 6 \qquad \text{Let } y = \sqrt{x}.$$
$$x^2 + x = 6$$
$$x^2 + x - 6 = 0$$
$$(x + 3)(x - 2) = 0$$
$$x = -3 \quad \text{or} \quad x = 2$$

Let $x = -3$ and then $x = 2$ in the first equation to find corresponding y-values.

Let $x = -3$.   Let $x = 2$.

$$y = \sqrt{x} \qquad\qquad y = \sqrt{x}$$
$$y = \sqrt{-3} \qquad\qquad y = \sqrt{2}$$

Not a real number.

Since we are interested in real number solutions, the only solution is $(2, \sqrt{2})$. The solution set is $\{(2, \sqrt{2})\}$. Check to see that this solution satisfies both equations. The graph of each equation in this system is shown next.

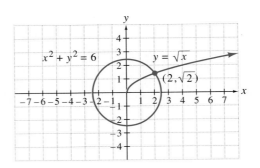

2

EXAMPLE 3   Solve the system

$$\begin{cases} x^2 + 2y^2 = 10 \\ x^2 - y^2 = 1 \end{cases}.$$

Solution: Use addition or elimination to solve this system. To eliminate x^2 when we add the two equations, multiply both sides of the second equation by -1. Then

$$\begin{cases} x^2 + 2y^2 = 10 \\ \boxed{-1}\,(x^2 - y^2) = \boxed{-1} \cdot 1 \end{cases} \quad \text{simplifies to} \quad \begin{cases} x^2 + 2y^2 = 10 \\ \underline{-x^2 + y^2 = -1} \end{cases}$$

$$3y^2 = 9$$
$$y^2 = 3$$
$$y = \pm\sqrt{3}$$

To find corresponding x-values, let $y = \sqrt{3}$ and $y = -\sqrt{3}$ in either original equation. We choose the second equation.

Let $y = \sqrt{3}$.   Let $y = -\sqrt{3}$.

$$x^2 - y^2 = 1 \qquad\qquad\qquad x^2 - y^2 = 1$$
$$x^2 - (\sqrt{3})^2 = 1 \qquad\qquad x^2 - (-\sqrt{3})^2 = 1$$
$$x^2 - 3 = 1 \qquad\qquad\qquad x^2 - 3 = 1$$
$$x^2 = 4 \qquad\qquad\qquad\qquad x^2 = 4$$
$$x = \pm\sqrt{4} = \pm 2 \qquad\qquad x = \pm\sqrt{4} = \pm 2$$

The solution set is $\{(2, \sqrt{3}), (-2, \sqrt{3}), (2, -\sqrt{3}), (-2, -\sqrt{3})\}$. Check all four ordered pairs in both equations of the system. The graph of each equation in this system is shown at the top of p. 451. ■

EXAMPLE 4   Solve the system

$$\begin{cases} x^2 + y^2 = 4 \\ x + y = 3 \end{cases}.$$

Solution: The addition method is not a good choice here, since x and y are squared in the first equation but not in the second equation. Use the substitution method and solve the second equation for x.

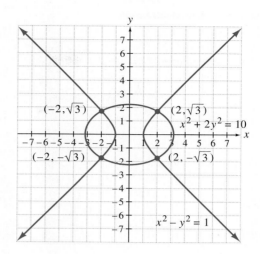

$$x + y = 3 \qquad \text{Second equation.}$$

$$x = 3 - y$$

Let $x = 3 - y$ in the first equation.

$$x^2 + y^2 = 4 \qquad \text{First equation.}$$

$$(3 - y)^2 + y^2 = 4$$

$$9 - 6y + y^2 + y^2 = 4$$

$$2y^2 - 6y + 5 = 0$$

By the quadratic formula, where $a = 2$, $b = -6$, and $c = 5$, we have

$$y = \frac{6 \pm \sqrt{6^2 - 4 \cdot 2 \cdot 5}}{2 \cdot 2} = \frac{6 \pm \sqrt{-4}}{4}$$

Since $\sqrt{-4}$ is not a real number, there is no solution. Graphically, the circle and the line do not intersect, as shown.

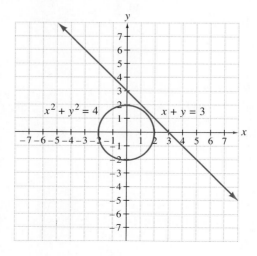

EXERCISE SET 9.6

Solve each nonlinear system of equations. See Examples 1 and 2.

1. $\begin{cases} x^2 + y^2 = 25 \\ 4x + 3y = 0 \end{cases}$

2. $\begin{cases} x^2 + y^2 = 25 \\ 3x + 4y = 0 \end{cases}$

3. $\begin{cases} x^2 + 4y^2 = 10 \\ y = x \end{cases}$

4. $\begin{cases} 4x^2 + y^2 = 10 \\ y = x \end{cases}$

5. $\begin{cases} y^2 = 4 - x \\ x - 2y = 4 \end{cases}$

6. $\begin{cases} x^2 + y^2 = 4 \\ x + y = -2 \end{cases}$

Solve each system of equations. See Example 3.

7. $\begin{cases} x^2 + y^2 = 9 \\ 16x^2 - 4y^2 = 64 \end{cases}$

8. $\begin{cases} 4x^2 + 3y^2 = 35 \\ 5x^2 + 2y^2 = 42 \end{cases}$

Solve each system of equations. See Example 4.

9. $\begin{cases} x^2 + 2y^2 = 2 \\ x - y = 2 \end{cases}$

10. $\begin{cases} x^2 + 2y^2 = 2 \\ x^2 - 2y^2 = 6 \end{cases}$

Solve each system of equations.

11. $\begin{cases} y = x^2 - 3 \\ 4x - y = 6 \end{cases}$

12. $\begin{cases} y = x + 1 \\ x^2 - y^2 = 1 \end{cases}$

13. $\begin{cases} y = x^2 \\ 3x + y = 10 \end{cases}$

14. $\begin{cases} 6x - y = 5 \\ xy = 1 \end{cases}$

15. $\begin{cases} y = 2x^2 + 1 \\ x + y = -1 \end{cases}$

16. $\begin{cases} x^2 + y^2 = 9 \\ x + y = 5 \end{cases}$

17. $\begin{cases} y = x^2 - 4 \\ y = x^2 - 4x \end{cases}$

18. $\begin{cases} x = y^2 - 3 \\ x = y^2 - 3y \end{cases}$

19. $\begin{cases} 2x^2 + 3y^2 = 14 \\ -x^2 + y^2 = 3 \end{cases}$

20. $\begin{cases} 4x^2 - 2y^2 = 2 \\ -x^2 + y^2 = 2 \end{cases}$

21. $\begin{cases} x^2 + y^2 = 1 \\ x^2 + (y + 3)^2 = 4 \end{cases}$

22. $\begin{cases} x^2 + 2y^2 = 4 \\ x^2 - y^2 = 4 \end{cases}$

23. $\begin{cases} y = x^2 + 2 \\ y = -x^2 + 4 \end{cases}$

24. $\begin{cases} x = -y^2 - 3 \\ x = y^2 - 5 \end{cases}$

25. $\begin{cases} 3x^2 + y^2 = 9 \\ 3x^2 - y^2 = 9 \end{cases}$

26. $\begin{cases} x^2 + y^2 = 25 \\ x = y^2 - 5 \end{cases}$

27. $\begin{cases} x^2 + 3y^2 = 6 \\ x^2 - 3y^2 = 10 \end{cases}$

28. $\begin{cases} x^2 + y^2 = 1 \\ y = x^2 - 9 \end{cases}$

29. $\begin{cases} x^2 + y^2 = 36 \\ y = \dfrac{1}{6}x^2 - 6 \end{cases}$

30. $\begin{cases} x^2 + y^2 = 16 \\ y = -\dfrac{1}{4}x^2 + 4 \end{cases}$

Skill Review

Graph each inequality in two variables. See Section 3.6.

31. $x > -3$

32. $y \le 1$

33. $y < 2x - 1$

34. $3x - y \le 4$

Find the perimeter of each geometric figure. See Section 4.3.

35.

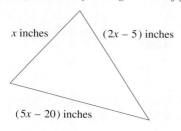

x inches $(2x - 5)$ inches

$(5x - 20)$ inches

36.

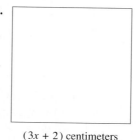

$(3x + 2)$ centimeters

37. $(x^2 + 3x + 1)$ meters

x^2 meters

38.

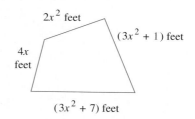

$2x^2$ feet

$(3x^2 + 1)$ feet

$4x$ feet

$(3x^2 + 7)$ feet

9.7
Nonlinear Inequalities and Systems of Inequalities

OBJECTIVES

Tape 29

1   Sketch the graph of a nonlinear inequality.

2   Sketch the solution set of a system of nonlinear inequalities.

1   We can graph a nonlinear inequality in two variables such as $\frac{x^2}{9} + \frac{y^2}{16} \leq 1$ in a way similar to the way we graphed a linear inequality in two variables in Section 3.6. First, we graph the related equation $\frac{x^2}{9} + \frac{y^2}{16} = 1$. The graph of the equation is our boundary. Then, using test points, we shade the region whose points satisfy the inequality.

EXAMPLE 1   Graph $\frac{x^2}{9} + \frac{y^2}{16} \leq 1$.

Solution: First, graph the equation $\frac{x^2}{9} + \frac{y^2}{16} = 1$. We graph using a solid line since the graph of $\frac{x^2}{9} + \frac{y^2}{16} \leq 1$ includes the graph of $\frac{x^2}{9} + \frac{y^2}{16} = 1$. The graph is an ellipse and it divides the plane into two regions, the "inside" and the "outside" of the ellipse. To determine which region contains the solutions, select a test point in one of the regions and determine whether the coordinates of the point satisfy the inequality. We choose $(0, 0)$ as the test point.

$$\frac{x^2}{9} + \frac{y^2}{16} \leq 1$$

$$\frac{0^2}{9} + \frac{0^2}{16} \leq 1 \qquad \text{Let } x = 0 \text{ and } y = 0.$$

$$0 \leq 1 \qquad \text{True}$$

Since this statement is true, we shade the region containing $(0, 0)$. The graph of the solution set is the set of points on and within the ellipse, as shaded in the figure.

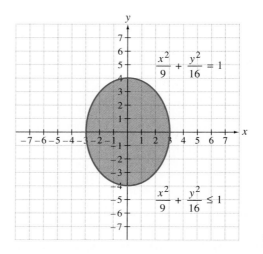

EXAMPLE 2   Graph $4y^2 > x^2 + 16$.

Solution: The related equation is $4y^2 = x^2 + 16$ or $\dfrac{y^2}{4} - \dfrac{x^2}{16} = 1$, which is a hyperbola. Graph the hyperbola using a dashed line since the graph of $4y^2 > x^2 + 16$ does **not** include the graph of $4y^2 = x^2 + 16$. The hyperbola divides the plane into three regions. Select a test point in each region to determine whether that region contains solutions of the inequality.

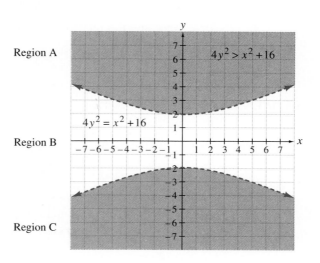

Test region A with $(0, 4)$	Test region B with $(0, 0)$	Test region C with $(0, -4)$
$4y^2 > x^2 + 16$	$4y^2 > x^2 + 16$	$4y^2 > x^2 + 16$
$4(4)^2 > 0^2 + 16$	$4(0)^2 > 0^2 + 16$	$4(-4)^2 > 0^2 + 16$
$64 > 16$ True	$0 > 16$ False	$64 > 16$ True

The graph of the solution set includes the shaded regions only, not the boundary.   ∎

2   In Section 3.6, we looked at the intersection of graphs of inequalities in two variables. Although we did not identify them as such, we now can recognize these sets of inequalities as systems of inequalities. The graph of the solution set of a system of inequalities is the intersection of the graphs of the inequalities.

EXAMPLE 3   Graph the system

$$\begin{cases} x \le 1 - 2y \\ y \le x^2 \end{cases}.$$

Solution: Graph each inequality on the same set of axes. The intersection is the darkest shaded region along with its boundary lines. The coordinates of the points of intersection can be found by solving the related system

$$\begin{cases} x = 1 - 2y \\ y = x^2 \end{cases}$$

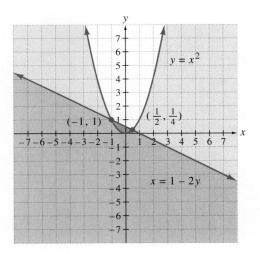

∎

EXAMPLE 4   Graph the system.

$$\begin{cases} x^2 + y^2 < 25 \\ \dfrac{x^2}{9} - \dfrac{y^2}{25} < 1 \\ y < x + 3 \end{cases}$$

Solution: Graph each inequality. The graph of $x^2 + y^2 < 25$ contains points "inside" the circle that has center $(0, 0)$ and radius 5. The graph of $\dfrac{x^2}{9} - \dfrac{y^2}{25} < 1$ is the region between the two branches of the hyperbola with x-intercepts -3 and 3 and center $(0, 0)$. The graph of $y < x + 3$ is the region "below" the line with slope 1 and y-intercept 3. The graph of the solution set of the system is the intersection of all the graphs, the darkest shaded region shown. The boundary of this region is not part of the solution.

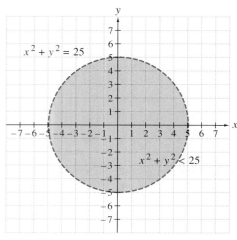

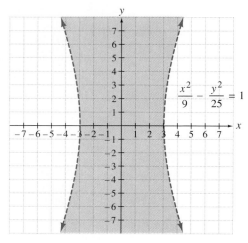

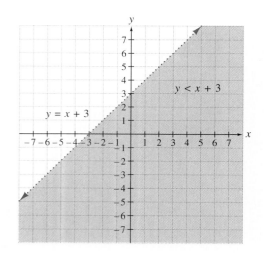

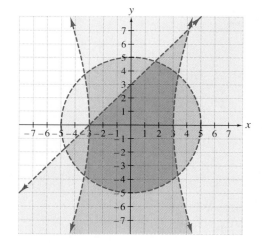

EXERCISE SET 9.7

Graph each inequality. See Examples 1 and 2.

1. $y < x^2$

2. $y < -x^2$

3. $x^2 + y^2 \geq 16$

4. $x^2 + y^2 < 36$

5. $\dfrac{x^2}{4} - y^2 < 1$

6. $x^2 - \dfrac{y^2}{9} \geq 1$

Graph the solution of each system. See Examples 3 and 4.

7. $\begin{cases} 2x - y < 2 \\ y \leq -x \end{cases}$

8. $\begin{cases} x - 2y > 4 \\ y > -x^2 \end{cases}$

9. $\begin{cases} 4x + 3y \geq 12 \\ x^2 + y^2 < 16 \end{cases}$

10. $\begin{cases} 3x - 4y \leq 12 \\ x^2 + y^2 < 16 \end{cases}$

11. $\begin{cases} x^2 + y^2 \leq 9 \\ x^2 + y^2 \geq 1 \end{cases}$

12. $\begin{cases} x^2 + y^2 \geq 9 \\ x^2 + y^2 \geq 16 \end{cases}$

Graph each inequality.

13. $y > (x - 1)^2 - 3$

14. $y > (x + 3)^2 + 2$

15. $x^2 + y^2 \leq 9$

16. $x^2 + y^2 > 4$

17. $y > -x^2 + 5$

18. $y < -x^2 + 5$

19. $\dfrac{x^2}{4} + \dfrac{y^2}{9} \leq 1$

20. $\dfrac{x^2}{25} + \dfrac{y^2}{4} \geq 1$

21. $\dfrac{y^2}{4} - x^2 \le 1$

22. $\dfrac{y^2}{16} - \dfrac{x^2}{9} > 1$

23. $y < (x - 2)^2 + 1$

24. $y > (x - 2)^2 + 1$

25. $y \le x^2 + x - 2$

26. $y > x^2 + x - 2$

Graph the solution of each system.

27. $\begin{cases} y > x^2 \\ y \ge 2x + 1 \end{cases}$

28. $\begin{cases} y \le -x^2 + 3 \\ y \le 2x - 1 \end{cases}$

29. $\begin{cases} x > y^2 \\ y > 0 \end{cases}$

30. $\begin{cases} x < (y + 1)^2 + 2 \\ x + y \ge 3 \end{cases}$

31. $\begin{cases} x^2 + y^2 > 9 \\ y > x^2 \end{cases}$

32. $\begin{cases} x^2 + y^2 \le 9 \\ y < x^2 \end{cases}$

33. $\begin{cases} \dfrac{x^2}{4} + \dfrac{y^2}{9} \ge 1 \\ x^2 + y^2 \ge 4 \end{cases}$

34. $\begin{cases} x^2 + (y - 2)^2 \ge 9 \\ \dfrac{x^2}{4} + \dfrac{y^2}{25} < 1 \end{cases}$

35. $\begin{cases} x^2 - y^2 \ge 1 \\ y \ge 0 \end{cases}$

36. $\begin{cases} x^2 - y^2 \ge 1 \\ x \ge 0 \end{cases}$

37. $\begin{cases} x + y \ge 1 \\ 2x + 3y < 1 \\ x > -3 \end{cases}$

38. $\begin{cases} x - y < -1 \\ 4x - 3y > 0 \\ y > 0 \end{cases}$

39. $\begin{cases} x^2 - y^2 < 1 \\ \dfrac{x^2}{16} + y^2 \leq 1 \\ x \geq -2 \end{cases}$

40. $\begin{cases} x^2 - y^2 \geq 1 \\ \dfrac{x^2}{16} + \dfrac{y^2}{4} \leq 1 \\ y \geq 1 \end{cases}$

Skill Review

Solve for y. See Section 2.2.

41. $x + 3y = 5x$

42. $2y + 7x = 3x$

43. $a - 2y = c$

44. $ax + by = c$

Solve each inequality. See Section 8.4.

45. $(x - 1)(x + 2) > 0$

46. $(x + 5)(x - 3) \leq 0$

47. $\dfrac{x}{x - 3} < 0$

48. $\dfrac{5}{x + 2} > 0$

CRITICAL THINKING

Luigi's Pizza Place has always been a favorite with the local college crowd. Some of Luigi's customers, though, have requested he include on his menu a smaller pizza suitable for one. Luigi is more than happy to accommodate by offering a $3.00 8-inch pizza in addition to his $6.50 10-inch pizza and his $10.00 12-inch pizza. Luigi figured the price of $3.00 was fair because the 8-inch diameter is 2 inches less than the 10-inch diameter, which in turn is 2 inches less than the 12-inch diameter. Since the 10-inch pizza costs $3.50 less than the 12-inch pizza, Luigi figured the 8-inch pizza also should cost $3.50 less than the 10-inch pizza.

As reasonable as this seems, Luigi lost money on the 8-inch pizzas.

What is wrong with Luigi's reasoning? What factors did Luigi fail to account for? Assuming that Luigi's prices for the 10-inch and 12-inch pizzas are appropriate, can you suggest the appropriate cost for his 8-inch pizza?

CHAPTER 9 GLOSSARY

A **determinant** is a real number associated with a square array of numbers written between two vertical bars.

Each number in the array of a determinant is called an **element.**

A **matrix** is a rectangular array of numbers.

Systems in which at least one equation is not linear are **nonlinear systems.**

A **solution of a system** of two equations in two variables is an ordered pair, (x, y), that makes both equations true.

Two or more equations in two or more variables considered simultaneously form a **system of equations.**

When a system has at least one solution, the system is said to be **consistent.**

Systems that have no solution are said to be **inconsistent.**

Equations whose graphs are the same are **dependent equations.**

CHAPTER 9 SUMMARY

(9.1)

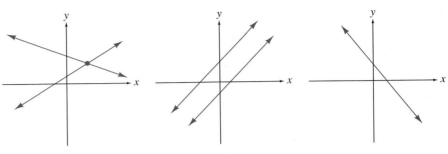

One solution,
consistent system

No solution,
inconsistent system

Infinite number
of solutions,
dependent equations

THE VALUE OF A 2 × 2 DETERMINANT (9.4)

$$\begin{vmatrix} a & b \\ c & d \end{vmatrix} = ad - bc$$

CRAMER'S RULE FOR TWO LINEAR EQUATIONS IN TWO VARIABLES (9.4)

The solution of the system $\begin{cases} ax + by = h \\ cx + dy = k \end{cases}$ is given by

$$x = \frac{\begin{vmatrix} h & b \\ k & d \end{vmatrix}}{\begin{vmatrix} a & b \\ c & d \end{vmatrix}} = \frac{D_x}{D}; \qquad y = \frac{\begin{vmatrix} a & h \\ c & k \end{vmatrix}}{\begin{vmatrix} a & b \\ c & d \end{vmatrix}} = \frac{D_y}{D}$$

as long as $D = ad - bc$ is not 0.

VALUE OF A 3 × 3 DETERMINANT (9.4)

$$\begin{vmatrix} a_1 & b_1 & c_1 \\ a_2 & b_2 & c_2 \\ a_3 & b_3 & c_3 \end{vmatrix} = a_1 \cdot \begin{vmatrix} b_2 & c_2 \\ b_3 & c_3 \end{vmatrix} - a_2 \cdot \begin{vmatrix} b_1 & c_1 \\ b_3 & c_3 \end{vmatrix} + a_3 \cdot \begin{vmatrix} b_1 & c_1 \\ b_2 & c_2 \end{vmatrix}$$

CRAMER'S RULE FOR THREE EQUATIONS IN THREE VARIABLES (9.4)

The solution of the system $\begin{cases} a_1x + b_1y + c_1z = k_1 \\ a_2x + b_2y + c_2z = k_2 \\ a_3x + b_3y + c_3z = k_3 \end{cases}$ is given by $x = \dfrac{D_x}{D}$, $y = \dfrac{D_y}{D}$, and

$z = \dfrac{D_z}{D}$, where

$$D = \begin{vmatrix} a_1 & b_1 & c_1 \\ a_2 & b_2 & c_2 \\ a_3 & b_3 & c_3 \end{vmatrix} \qquad D_x = \begin{vmatrix} k_1 & b_1 & c_1 \\ k_2 & b_2 & c_2 \\ k_3 & b_3 & c_3 \end{vmatrix}$$

$$D_y = \begin{vmatrix} a_1 & k_1 & c_1 \\ a_2 & k_2 & c_2 \\ a_3 & k_3 & c_3 \end{vmatrix} \qquad D_z = \begin{vmatrix} a_1 & b_1 & k_1 \\ a_2 & b_2 & k_2 \\ a_3 & b_3 & k_3 \end{vmatrix}$$

as long as D is not 0.

ELEMENTARY ROW OPERATIONS   (9.5)

1. Any two rows in a matrix may be interchanged.
2. The elements of any row may be multiplied (or divided) by the same nonzero number.
3. The elements of any row may be multiplied (or divided) by a nonzero number and added to its corresponding elements in any other row.

CHAPTER 9 REVIEW

(9.1)   *Solve each system of equations in two variables by each of three methods: (1) graphing, (2) substitution, and (3) elimination.*

1. $\begin{cases} 3x + 10y = 1 \\ x + 2y = -1 \end{cases}$

2. $\begin{cases} y = \dfrac{1}{2}x + \dfrac{2}{3} \\ 4x + 6y = 4 \end{cases}$

3. $\begin{cases} 2x - 4y = 22 \\ 5x - 10y = 16 \end{cases}$

4. $\begin{cases} 3x - 6y = 12 \\ 2y = x - 4 \end{cases}$

5. $\begin{cases} \dfrac{1}{2}x - \dfrac{3}{4}y = -\dfrac{1}{2} \\ \dfrac{1}{8}x + \dfrac{3}{4}y = \dfrac{19}{8} \end{cases}$

(9.2)   *Solve each system of equations in three variables.*

6. $\begin{cases} x + z = 4 \\ 2x - y = 4 \\ x + y - z = 0 \end{cases}$

7. $\begin{cases} 2x + 5y = 4 \\ x - 5y + z = -1 \\ 4x - z = 11 \end{cases}$

8. $\begin{cases} 4y + 2z = 5 \\ 2x + 8y = 5 \\ 6x + 4z = 1 \end{cases}$

9. $\begin{cases} 5x + 7y = 9 \\ 14y - z = 28 \\ 4x + 2z = -4 \end{cases}$

10. $\begin{cases} 3x - 2y + 2z = 5 \\ -x + 6y + z = 4 \\ 3x + 14y + 7z = 20 \end{cases}$

11. $\begin{cases} x + 2y + 3z = 11 \\ y + 2z = 3 \\ 2x - 2z = 10 \end{cases}$

12. $\begin{cases} 7x - 3y + 2z = 0 \\ 4x - 4y - z = 2 \\ 5x + 2y + 3z = 1 \end{cases}$

13. $\begin{cases} x - 3y - 5z = -5 \\ 4x - 2y + 3z = 13 \\ 5x + 3y + 4z = 22 \end{cases}$

(9.3) *Solve the following applications using systems of equations.*

14. The sum of three numbers is 98. The sum of the first and second is two more than the third number, and the second is four times the first. Find the numbers.

15. Alice's coin purse has 95 coins in it, all dimes, nickels, and pennies, worth $4.03 total. There are twice as many pennies as dimes and one fewer nickels than dimes. Find how many of each type of coin the purse contains.

16. One number is 3 times a second number, and twice the sum of the numbers is 168. Find the numbers.

17. Sue is 16 years older than Pat. In 15 years Sue will be twice as old as Pat is then. How old is each now?

18. Two cars leave Chicago, one traveling east and the other west. After 4 hours they are 492 miles apart. If one car is traveling 7 mph faster than the other, find the speed of each.

19. The foundation for a rectangular Hardware Warehouse has a length three times the width and is 296 feet around. Find the dimensions of the building.

20. James has available a 10% alcohol solution and a 60% alcohol solution. Find how many liters of each solution he should mix to make 50 liters of 40% alcohol solution.

21. An employee at See's Candy Store needs a special mixture of candy. She has creme-filled chocolates that sell for $3.00 per pound, chocolate-covered nuts that sell for $2.70 per pound, and chocolate-covered raisins that sell for $2.25 per pound. She wants to have twice as many raisins as nuts in the mixture. Find how many pounds of each she should use to make 45 pounds worth $2.80 per pound.

22. Chris has $2.77 in his coin jar, all in pennies, nickels, and dimes. If he has 53 coins in all and four more nickels than dimes, find how many of each type of coin he has.

23. If $10,000 and $4000 are invested so that $1250 is earned in one year, and if the rate of interest on the larger investment is 2% more than that of the smaller investment, find the rates of interest.

24. The perimeter of an isosceles (two sides equal) triangle is 73 centimeters. If two sides are of equal length and the third side is 7 centimeters longer than the others, find the lengths of the three sides.

25. The sum of three numbers is 295. One number is five more than a second and twice the third. Find the numbers.

(9.4) *Evaluate.*

26. $\begin{vmatrix} -1 & 3 \\ 5 & 2 \end{vmatrix}$

27. $\begin{vmatrix} 3 & -1 \\ 2 & 5 \end{vmatrix}$

28. $\begin{vmatrix} 2 & -1 & -3 \\ 1 & 2 & 0 \\ 3 & -2 & 2 \end{vmatrix}$

29. $\begin{vmatrix} -2 & 3 & 1 \\ 4 & 4 & 0 \\ 1 & -2 & 3 \end{vmatrix}$

Use Cramer's rule to solve each system of equations.

30. $\begin{cases} 3x - 2y = -8 \\ 6x + 5y = 11 \end{cases}$

31. $\begin{cases} 6x - 6y = -5 \\ 10x - 2y = 1 \end{cases}$

32. $\begin{cases} 3x + 10y = 1 \\ x + 2y = -1 \end{cases}$

33. $\begin{cases} y = \dfrac{1}{2}x + \dfrac{2}{3} \\ 4x + 6y = 4 \end{cases}$

34. $\begin{cases} 2x - 4y = 22 \\ 5x - 10y = 16 \end{cases}$

35. $\begin{cases} 3x - 6y = 12 \\ 2y = x - 4 \end{cases}$

36. $\begin{cases} x + z = 4 \\ 2x - y = 0 \\ x + y - z = 0 \end{cases}$

37. $\begin{cases} 2x + 5y = 4 \\ x - 5y + z = -1 \\ 4x - z = 11 \end{cases}$

38. $\begin{cases} x + 3y - z = 5 \\ 2x - y - 2z = 3 \\ x + 2y + 3z = 4 \end{cases}$

39. $\begin{cases} 2x - z = 1 \\ 3x - y + 2z = 3 \\ x + y + 3z = -2 \end{cases}$

40. $\begin{cases} x + 2y + 3z = 14 \\ y + 2z = 3 \\ 2x - 2z = 10 \end{cases}$

41. $\begin{cases} 5x + 7y = 9 \\ 14y - z = 28 \\ 4x + 2z = -4 \end{cases}$

(9.5) *Solve each system using matrices.*

42. $\begin{cases} 3x + 10y = 1 \\ x + 2y = -1 \end{cases}$

43. $\begin{cases} 3x - 6y = 12 \\ 2y = x - 4 \end{cases}$

44. $\begin{cases} 3x - 2y = -8 \\ 6x + 5y = 11 \end{cases}$

45. $\begin{cases} 6x - 6y = -5 \\ 10x - 2y = 1 \end{cases}$

46. $\begin{cases} 3x - 6y = 0 \\ 2x + 4y = 5 \end{cases}$

47. $\begin{cases} 5x - 3y = 10 \\ -2x + y = -1 \end{cases}$

48. $\begin{cases} 0.2x - 0.3y = -0.7 \\ 0.5x + 0.3y = 1.4 \end{cases}$

49. $\begin{cases} 3x + 2y = 8 \\ 3x - y = 5 \end{cases}$

50. $\begin{cases} x + z = 4 \\ 2x - y = 0 \\ x + y - z = 0 \end{cases}$

51. $\begin{cases} 2x + 5y = 4 \\ x - 5y + z = -1 \\ 4x - z = 11 \end{cases}$

52. $\begin{cases} 3x - y = 11 \\ x + 2z = 13 \\ y - z = -7 \end{cases}$

53. $\begin{cases} 5x + 7y + 3z = 9 \\ 14y - z = 28 \\ 4x + 2z = -4 \end{cases}$

54. $\begin{cases} 7x - 3y + 2z = 0 \\ 4x - 4y - z = 2 \\ 5x + 2y + 3z = 1 \end{cases}$

55. $\begin{cases} x + 2y + 3z = 14 \\ y + 2z = 3 \\ 2x - 2z = 10 \end{cases}$

(9.6) *Solve each system of equations.*

56. $\begin{cases} y = 2x - 4 \\ y^2 = 4x \end{cases}$

57. $\begin{cases} x^2 + y^2 = 4 \\ x - y = 4 \end{cases}$

58. $\begin{cases} y = x + 2 \\ y = x^2 \end{cases}$

59. $\begin{cases} y = x^2 - 5x + 1 \\ y = -x + 6 \end{cases}$

60. $\begin{cases} 4x - y^2 = 0 \\ 2x^2 + y^2 = 16 \end{cases}$

61. $\begin{cases} x^2 + 4y^2 = 16 \\ x^2 + y^2 = 4 \end{cases}$

62. $\begin{cases} x^2 + y^2 = 10 \\ 9x^2 + y^2 = 18 \end{cases}$

63. $\begin{cases} x^2 + 2y = 9 \\ 5x - 2y = 5 \end{cases}$

64. $\begin{cases} y = 3x^2 + 5x - 4 \\ y = 3x^2 - x + 2 \end{cases}$

65. $\begin{cases} x^2 - 3y^2 = 1 \\ 4x^2 + 5y^2 = 21 \end{cases}$

(9.7) *Graph the inequality or system of inequalities.*

66. $y \le -x^2 + 3$

67. $x^2 + y^2 < 9$

68. $x^2 - y^2 < 1$

69. $\dfrac{x^2}{4} + \dfrac{y^2}{9} \ge 1$

70. $\begin{cases} 2x \le 4 \\ x + y \ge 1 \end{cases}$

71. $\begin{cases} 3x + 4y \le 12 \\ x - 2y > 6 \end{cases}$

72. $\begin{cases} y > x^2 \\ x + y \ge 3 \end{cases}$

73. $\begin{cases} x^2 + y^2 \le 16 \\ x^2 + y^2 \ge 4 \end{cases}$

74. $\begin{cases} x^2 + y^2 < 4 \\ x^2 - y^2 \le 1 \end{cases}$

75. $\begin{cases} x^2 + y^2 < 4 \\ y \ge x^2 - 1 \\ x \ge 0 \end{cases}$

CHAPTER 9 TEST

Evaluate each determinant.

1. $\begin{vmatrix} 4 & -7 \\ 2 & 5 \end{vmatrix}$

2. $\begin{vmatrix} 4 & 0 & 2 \\ 1 & -3 & 5 \\ 0 & -1 & 2 \end{vmatrix}$

Solve each system of equations graphically and then solve by the addition method or the substitution method.

3. $\begin{cases} 2x - y = -1 \\ 5x + 4y = 17 \end{cases}$

4. $\begin{cases} 7x - 14y = 5 \\ x = 2y \end{cases}$

Solve each system.

5. $\begin{cases} 4x - 7y = 29 \\ 2x + 5y = -11 \end{cases}$

6. $\begin{cases} 15x + 6y = 15 \\ 10x + 4y = 10 \end{cases}$

7. $\begin{cases} 2x - 3y = 4 \\ 3y + 2z = 2 \\ x - z = -5 \end{cases}$

8. $\begin{cases} 3x - 2y - z = -1 \\ 2x - 2y = 4 \\ 2x - 2z = -12 \end{cases}$

Solve each system using Cramer's rule.

9. $\begin{cases} 3x - y = 7 \\ 2x + 5y = -1 \end{cases}$

10. $\begin{cases} 4x - 3y = -6 \\ -2x + y = 0 \end{cases}$

11. $\begin{cases} x + y + z = 4 \\ 2x + 5y = 1 \\ x - y - 2z = 0 \end{cases}$

12. $\begin{cases} 3x + 2y + 3z = 3 \\ x - z = 9 \\ 4y + z = -4 \end{cases}$

Solve each system using matrices.

13. $\begin{cases} x - y = -2 \\ 3x - 3y = -6 \end{cases}$

14. $\begin{cases} x + 2y = -1 \\ 2x + 5y = -5 \end{cases}$

15. $\begin{cases} x - y - z = 0 \\ 3x - y - 5z = -2 \\ 2x + 3y = -5 \end{cases}$

16. $\begin{cases} 2x - y + 3z = 4 \\ 3x - 3z = -2 \\ -5x + y = 0 \end{cases}$

Solve each system.

17. $\begin{cases} x^2 + y^2 = 169 \\ 5x + 12y = 0 \end{cases}$

18. $\begin{cases} x^2 + y^2 = 26 \\ x^2 - y^2 = 24 \end{cases}$

19. $\begin{cases} y = x^2 - 5x + 6 \\ y = 2x \end{cases}$

20. $\begin{cases} x^2 + 4y^2 = 5 \\ y = x \end{cases}$

21. $\begin{cases} 2x + 5y \geq 10 \\ y \geq x^2 + 1 \end{cases}$

22. $\begin{cases} \dfrac{x^2}{4} + y^2 \leq 1 \\ x + y > 1 \end{cases}$

23. $\begin{cases} x^2 + y^2 > 1 \\ \dfrac{x^2}{4} - y^2 \geq 1 \end{cases}$

24. $\begin{cases} x^2 + y^2 \geq 4 \\ x^2 + y^2 < 16 \\ y \geq 0 \end{cases}$

Solve.

25. Dean jogs clockwise around a 4-mile track at the same time Tom bicycles counterclockwise, going 10 mph faster than Dean. They meet after 10 minutes. Find how fast each person travels and find how far Dean jogs.

26. A student whose average score is between 70 and 79 inclusive in a mathematics class receives a C. If Jamie has grades of 70, 64, 85, and 73 on the first four tests, find the lowest score possible on the fifth test so that Jamie gets a C grade. (Assume no rounding off in averaging.)

CHAPTER 9 CUMULATIVE REVIEW

1. Solve $2 < 4 - x < 7$.

2. Find the slope of the line containing the points $(5, -7)$ and $(-3, 6)$.

3. Find the equation of the line through $(0, -5)$ and perpendicular to the line $x = 2y$.

4. Write each number in standard notation.
   a. 7.7×10^8 **b.** 1.025×10^{-3}

5. Simplify each expression. Write answers using positive exponents.
   a. $\left(\dfrac{3x^2y}{y^{-9}z}\right)^{-2}$ **b.** $\left(\dfrac{3a^2}{2x^{-1}}\right)^3 \left(\dfrac{x^{-3}}{4a^{-2}}\right)^{-1}$

6. Multiply:
   a. $(2x^3)(5x^6)$ **b.** $(7y^4z^4)(-xy^{11}z^5)$

7. Find the quotient: $\dfrac{7a^2b - 2ab^2}{2ab^2}$

8. Factor $12x^3y - 22x^2y + 8xy$.

9. Solve $x^3 + 5x^2 = x + 5$.

10. The sum of the squares of two consecutive odd negative integers is 130. Find the numbers.

11. Perform the indicated operation.
    a. $\dfrac{x}{x-3} - 5$ **b.** $\dfrac{7}{x-y} + \dfrac{3}{y-x}$

12. Solve $\dfrac{z}{2z^2 + 3z - 2} - \dfrac{1}{2z} = \dfrac{3}{z^2 + 2z}$.

13. Multiply. Assume that all variables represent positive numbers.
    a. $z^{2/3}(z^{1/3} - z^5)$
    b. $(x^{1/3} - 5)(x^{1/3} + 2)$

14. Simplify the following. Assume that all variables represent positive integers.
    a. $\dfrac{\sqrt{27}}{\sqrt{5}}$ **b.** $\dfrac{2\sqrt{16}}{\sqrt{9x}}$ **c.** $\sqrt[3]{\dfrac{1}{2}}$

15. Solve $m^2 - 7m - 1 = 0$ for m by completing the square.

16. Solve $\dfrac{3x}{x-2} - \dfrac{x+1}{x} = \dfrac{6}{x(x-2)}$.

17. Solve $(x + 2)(x - 1)(x - 5) \le 0$.

18. Graph $x^2 + y^2 + 4x - 8y = 16$.

19. Solve the system:
$$\begin{cases} 3x - y + z = -15 \\ x + 2y - z = 1 \\ 2x + 3y - 2z = 0 \end{cases}$$

20. Find the value of the determinant by expanding the minors of the given row or column.
$$\begin{vmatrix} 0 & 5 & 1 \\ 1 & 3 & -1 \\ -2 & 2 & 4 \end{vmatrix}$$
    a. First column   **b.** Second row

21. Solve the system using matrices: $\begin{cases} 2x - y = 3 \\ 4x - 2y = 5 \end{cases}$.

22. Solve the system: $\begin{cases} x^2 + 2y^2 = 10 \\ x^2 - y^2 = 1 \end{cases}$.

CHAPTER # 10

Exponential and Logarithmic Functions

Like many "third-world" countries, Kenya faces surging population growth. Populations typically grow "exponentially," so that the growth can become unmanageable. (See Critical Thinking, page 505.)

INTRODUCTION

In this chapter, we discuss two closely related functions: exponential and logarithmic functions. These functions are vital in applications in economics, finance, engineering, the sciences, education, and other fields. Models of tumor growth and learning curves are two examples of the uses of exponential and logarithmic functions.

10.1
Inverse Functions

Tape 30

OBJECTIVES

1. Determine whether a function is a one-to-one function.
2. Use the horizontal line test to test whether a function is a one-to-one function.
3. Define the inverse of a function.
4. Find the equation of the inverse of a function.

1 The set $f = \{(0, 1), (2, 2), (-3, 5), (7, 6)\}$ is a function since each x-value corresponds to a unique y-value. For this particular function f, each y-value also corresponds to a unique x-value. When this happens, we call the function a **one-to-one function.**

> **One-to-One Function**
>
> If f is a function, then f is a **one-to-one-function** if each y-value corresponds to a unique x-value.

EXAMPLE 1 Determine whether each function is one-to-one.
 a. $f = \{(6, 2), (5, 4), (-1, 0), (7, 3)\}$
 b. $g = \{(3, 9), (-4, 2), (-3, 9), (0, 0)\}$
 c. $h = \{(1, 1), (2, 2), (10, 10), (-5, -5)\}$

Solution: **a.** f is one-to-one since each y-value corresponds to only one x-value.

 b. g is not one-to-one because in $(3, 9)$ and $(-3, 9)$ the y-value 9 corresponds to two different x-values.

 c. h is a one-to-one function since each y-value corresponds to only one x-value. ∎

2 Recall that we recognize the graph of a function when it passes the vertical line test. Since every x-value of the function corresponds to exactly one y-value, each vertical line, for which all the x-values are the same, intersects the function's graph at

467

most once. Is the function shown below a one-to-one function? The answer is no. To see why, notice that the y-value of the ordered pair $(-3, 3)$, for example, is the same as y-value of the ordered pair $(3, 3)$. This function is therefore not one-to-one.

To test whether a graph is the graph of a one-to-one function, apply the vertical line test to see if it is a function, and then apply a similar **horizontal line test** to see if it is a one-to-one function.

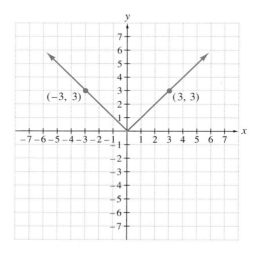

Horizontal Line Test

If every horizontal line intersects the graph of a function at most once, then the function is a one-to-one function.

EXAMPLE 2   Determine whether each graph is the graph of a one-to-one function.

a.

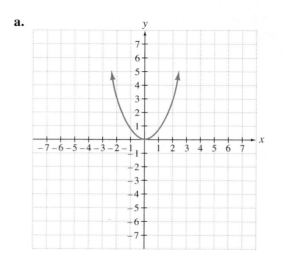

b.

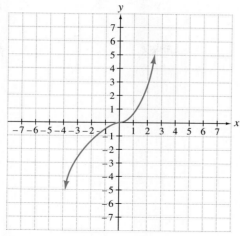

c.

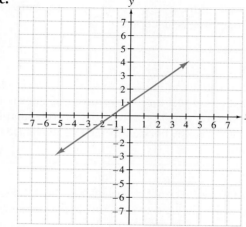

d.

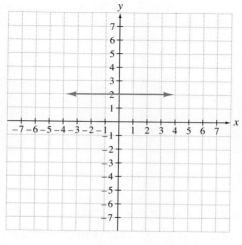

e.

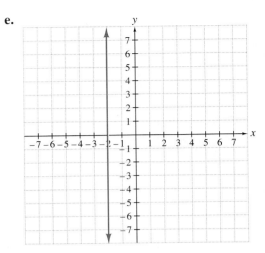

Solution: Graphs **a, b, c, d** all pass the vertical line test so only these graphs are graphs of functions. But, of these, only **b** and **c** pass the horizontal line test, so only **b** and **c** are graphs of one-to-one-functions. ■

HELPFUL HINT

All linear equations are one-to-one functions except those whose graphs are horizontal or vertical lines. A vertical line does not pass the vertical line test and hence is not the graph of a function. A horizontal line is the graph of a function, but does not pass the horizontal line test and hence is not the graph of a one-to-one function.

---

3 One-to-one functions are special in that their graphs pass the vertical and horizontal line tests. They are special, too, in that for each one-to-one function we can find its **inverse function** by switching the order of the coordinates of the ordered pairs of the function. If the function f is $\{(2, -3), (5, 10), (9, 1)\}$, for example, then its inverse function is $\{(-3, 2), (10, 5), (1, 9)\}$. For a function f, we use the notation f^{-1}, read "f inverse," to denote its inverse function. Notice that since the coordinates of each ordered pair have been switched the domain of f is the range of f^{-1}, and the range of f is the domain of f^{-1}.

Inverse Function

The inverse of a one-to-one function f is the one-to-one function f^{-1} that is the set of all ordered pairs (y, x) where (x, y) belongs to f.

HELPFUL HINT

The symbol f^{-1} is the single symbol used to denote the inverse of the function f. It is read as "f inverse". This symbol **does not mean** $\dfrac{1}{f}$.

4   If a one-to-one function f is defined as a set of ordered pairs, we can find f^{-1} by interchanging the x and y coordinates of the ordered pairs. If a one-to-one function f is given in the form of an equation, we can find f^{-1} using a similar procedure.

> **To Find the Inverse of a One-to-One Function $f(x)$**
>
> *Step 1*   Replace $f(x)$ by y.
> *Step 2*   Interchange x and y.
> *Step 3*   Solve the equation for y.
> *Step 4*   Replace y with the notation $f^{-1}(x)$.

EXAMPLE 3   Find the equation of the inverse of $f(x) = 3x - 5$. Graph f and f^{-1} on the same set of axes.

Solution: First, let $y = f(x)$.

$$f(x) = 3x - 5$$
$$y = 3x - 5$$

Next, interchange x and y and solve for y.

$$x = 3y - 5 \qquad \text{Interchange } x \text{ and } y.$$
$$3y = x + 5$$
$$y = \frac{x + 5}{3} \qquad \text{Solve for } y.$$

Let $y = f^{-1}(x)$.

$$f^{-1}(x) = \frac{x + 5}{3}$$

Now graph $f(x)$ and $f^{-1}(x)$ on the same set of axes. Both $f(x) = 3x - 5$ and $f^{-1}(x) = \dfrac{x + 5}{3}$ are linear functions, so each graph is a line.

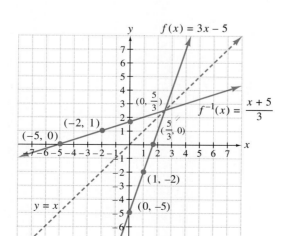

$$f(x) = 3x - 5$$

x	$y = f(x)$
1	-2
0	-5
$\dfrac{5}{3}$	0

$$f^{-1}(x) = \frac{x + 5}{3}$$

x	$y = f^{-1}(x)$
0	$\dfrac{5}{3}$
-5	0
-2	1

Notice that the graphs of f and f^{-1} are mirror images of each other, and the "mirror" is the line $y = x$. This is true for every function and its inverse. For this reason, we say the graphs of f and f^{-1} are **symmetric about the line** $y = x$.

EXAMPLE 4 Find the equation of the inverse of $g(x) = \dfrac{-2x + 1}{3}$. Graph g and g^{-1} on the same set of axes.

Solution: The function g is a one-to-one function since its graph is a line that is neither vertical nor horizontal. Let $y = g(x)$ and obtain

$$y = \frac{-2x + 1}{3}$$

Next, switch variables and solve for y.

$$x = \frac{-2y + 1}{3} \qquad \text{Interchange } x \text{ and } y.$$

$$3x = -2y + 1 \qquad \text{Multiply both sides by 3.}$$

$$2y = -3x + 1$$

$$y = \frac{-3x + 1}{2} \qquad \text{Solve for } y.$$

Let $y = g^{-1}(x)$.

$$g^{-1}(x) = \frac{-3x + 1}{2}$$

The graphs of g and g^{-1} are two lines symmetric about the line $y = x$.

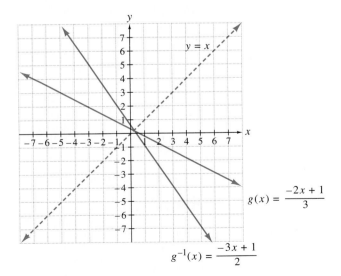

$$g(x) = \frac{-2x + 1}{3}$$

$$g^{-1}(x) = \frac{-3x + 1}{2}$$

EXERCISE SET 10.1

Determine whether each function is a one-to-one function. See Example 1.

1. $f = \{(-1, -1), (1, 1), (0, 2), (2, 0)\}$

2. $g = \{(8, 6), (9, 6), (3, 4), (-4, 4)\}$

3. $h = \{(10, 10)\}$

4. $r = \{(1, 2), (3, 4), (5, 6), (6, 7)\}$

5. $f = \{(11, 12), (4, 3), (3, 4), (6, 6)\}$

6. $g = \{(0, 3), (3, 7), (6, 7), (-2, -2)\}$

Determine whether each graph is the graph of a one-to-one function. See Example 2.

7.

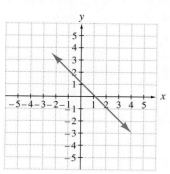

8.

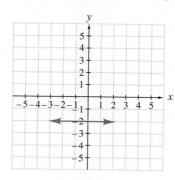

9.

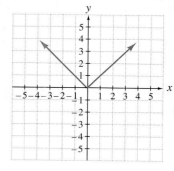

10.

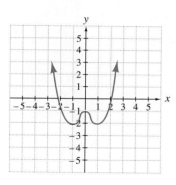

11.

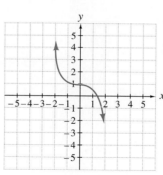

12.

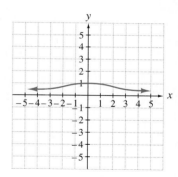

13.

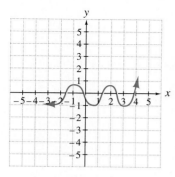

14.

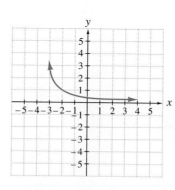

Each function is one-to-one. Find the inverse of each function and graph the function and its inverse on the same set of axes. See Examples 3 and 4.

15. $f(x) = x + 4$

16. $f(x) = x - 5$

17. $f(x) = 2x - 3$

18. $f(x) = 4x + 9$

19. $f(x) = \dfrac{12x - 4}{3}$

20. $f(x) = \dfrac{7x + 5}{11}$

Determine whether each function is one-to-one. If a function is one-to-one, list the elements of its inverse function.

21. $h = \{(-3, 9), (-2, 4), (-1, 1), (0, 0), (1, 1)\}$

22. $g = \{(-5, 2), (0, 7), (2, 9), (1, 8)\}$

23. $f = \{(1, 1), (2, 1), (3, 2), (4, 2)\}$

24. $g = \{(7, 3), (4, 0), (3, -1), (-3, -7)\}$

25. $f = \{(-4, -8), (-6, -12), (-8, -16), (-9, -18)\}$

26. $h = \{(-4, -3), (-3, -2), (-2, -1), (-1, -1)\}$

Sketch the graph of each equation and decide if it is a one-to-one function.

27. $x = y^2$

28. $y = x^2$

29. $y = 5x$

30. $y = -4x + 1$

31. $x = 7$

32. $y = -3$

Each function is one-to-one. Find the inverse of each function and graph the function and its inverse on the same set of axes.

33. $f(x) = 3x + 1$

34. $f(x) = -2x + 5$

35. $f(x) = \dfrac{x - 2}{5}$

36. $f(x) = \dfrac{4x - 3}{2}$

37. $g(x) = \dfrac{1}{2}x - 4$

38. $g(x) = \dfrac{3}{2}x - \dfrac{5}{4}$

Writing in Mathematics

39. Discuss the purpose of the vertical line test and the horizontal line test.

40. Explain whether this statement is true: Every straight line is the graph of a one-to-one function.

Skill Review

Simplify each expression. See Section 4.2.

41. 5^0

42. -2^0

43. $3x^0$

44. $(6x)^0 + 6x^0$

Find the unknown side of each right triangle. See Section 7.5.

45.

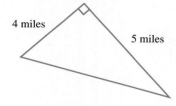

4 miles

5 miles

46.

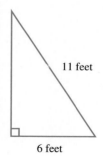

11 feet

6 feet

10.2
Exponential Functions

OBJECTIVES

Tape 30

1 Identify exponential functions.

2 Graph exponential functions.

3 Solve equations of the form $b^x = b^y$.

1 In earlier chapters, we gave meaning to exponential expressions such as 2^x, where x is a rational number. For example,

$$2^3 = 2 \cdot 2 \cdot 2 = 8$$

It is beyond the scope of this book to give meaning to 2^x if x is irrational. As long as the base b is positive, we will assume that b^x is a real number for all real numbers x. We also assume that rules of exponents apply whether x is rational or irrational as long as b is positive. The equation $y = b^x$ is called an **exponential function.**

> **Exponential Function**
>
> A function of the form
> $$f(x) = b^x$$
> is an exponential function where $b > 0$, b is not 1, and x is a real number.

2

EXAMPLE 1   Graph the exponential functions defined by $f(x) = 2^x$ and $g(x) = 3^x$ on the same set of axes.

Solution: Graph each function by plotting points. Set up a table of values for each of the two functions:

$$f(x) = 2^x \qquad g(x) = 3^x$$

x	$f(x)$
0	1
1	2
2	4
3	8
-1	$\dfrac{1}{2}$
-2	$\dfrac{1}{4}$

x	$g(x)$
0	1
1	3
2	9
3	27
-1	$\dfrac{1}{3}$
-2	$\dfrac{1}{9}$

If each set of points is plotted and connected with a smooth curve, the following graphs result:

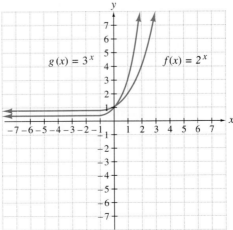

A number of things should be noted about these two graphs of exponential functions. First, the graphs confirm that $f(x) = 2^x$ and $g(x) = 3^x$ define one-to-one functions since each graph passes the vertical and horizontal line tests. The y-intercept of each graph is 1, but neither graph has an x-intercept. From the graph, we can also see that the domain of each function is all real numbers and the range is $\{y \mid y > 0\}$. We can also see that as x increases, y increases also.

EXAMPLE 2   Graph the exponential functions $y = \left(\dfrac{1}{2}\right)^x$ and $y = \left(\dfrac{1}{3}\right)^x$ on the same set of axes.

Solution: As before, plot points and connect them with a smooth curve.

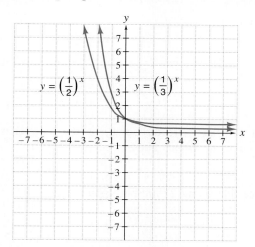

$y = \left(\dfrac{1}{2}\right)^x$		$y = \left(\dfrac{1}{3}\right)^x$	
x	y	x	y
0	1	0	1
1	$\dfrac{1}{2}$	1	$\dfrac{1}{3}$
2	$\dfrac{1}{4}$	2	$\dfrac{1}{9}$
3	$\dfrac{1}{8}$	3	$\dfrac{1}{27}$
-1	2	-1	3
-2	4	-2	9

Each function again is a one-to-one function. The y-intercept of both is 1. The domain is the set of all real numbers and the range is $\{y \mid y > 0\}$.   ∎

Notice the difference between the graphs of Example 1 and the graphs of Example 2. An exponential function is always increasing if the base is larger than 1. When the base is between 0 and 1, the graph is decreasing. The following figures summarize these characteristics of exponential functions.

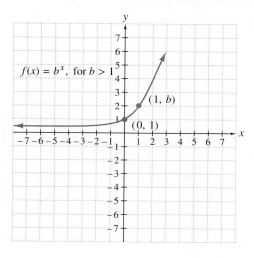

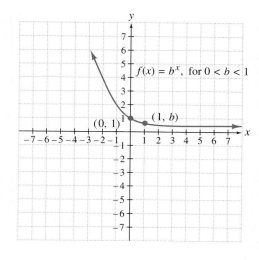

3 Because an exponential function $y = b^x$ is a one-to-one function, we have the following property.

Uniqueness of b^x

Let $b > 0$ and $b \neq 1$. If

$$b^x = b^y, \quad \text{then} \quad x = y$$

We can use this property to solve exponential equations.

EXAMPLE 3 Solve each equation for x.

    a. $2^x = 16$ **b.** $9^x = 27$ **c.** $4^{x+3} = 8^x$

Solution: **a.** We write 16 as a power of 2 and then use the uniqueness of b^x to solve.

$$2^x = 16$$
$$2^x = 2^4$$

Since the bases are the same, by the uniqueness of b^x, we have then that the exponents are equal. Thus,

$$x = 4$$

b. Notice that 9 and 27 are both powers of 3.

$$9^x = 27$$
$$(3^2)^x = 3^3 \qquad \text{Write 9 and 27 as powers of 3.}$$
$$3^{2x} = 3^3$$
$$2x = 3 \qquad \text{Apply the uniqueness of } b^x.$$
$$x = \frac{3}{2} \qquad \text{Divide by 2.}$$

To check, replace x with $\dfrac{3}{2}$ in the original expression $9^x = 27$.

c. Write both 4 and 8 as powers of 2.

$$4^{x+3} = 8^x$$
$$(2^2)^{x+3} = (2^3)^x$$
$$2^{2x+6} = 2^{3x}$$
$$2x + 6 = 3x \qquad \text{Apply the uniqueness of } b^x.$$
$$6 = x \qquad \text{Subtract } 2x \text{ from both sides.} \quad \blacksquare$$

There is one major problem with the preceding technique. Often the two expressions in the equation cannot easily be written as powers of a common base. We explore how to solve an equation like $4 = 3^x$ with the help of **logarithms** later.

The natural world abounds with patterns that can be modeled by exponential functions. To make these applications realistic, numbers are used that warrant a calculator. The first application has to do with exponential decay.

EXAMPLE 4   As a result of the Chernobyl nuclear accident, radioactive debris was carried through the atmosphere. One immediate concern was the impact it had on the milk supply. The percent y of radioactive material in raw milk after t days is estimated by $y = 100\,(2.7)^{-0.1t}$. Estimate the expected percent of radioactive material in the milk after 30 days.

Solution: Replace t with 30 in the given equation.

$$y = 100(2.7)^{-0.1t}$$
$$= 100(2.7)^{-3}$$

To **approximate** the percentage y, press the following keys on your calculator.

| 2.7 | y^x | 3 | +/− | = | × | 100 | = |

The display should read

| 5.0805263 |

Thus nearly 5% of the radioactive material still contaminates the milk supply after 30 days. ■

The exponential function defined by $A = P\left(1 + \dfrac{r}{n}\right)^{nt}$ models the pattern relating the dollars A accrued (or owed) after P dollars are invested (or loaned) at a rate of interest r compounded n times each year for t years.

EXAMPLE 5   Find the amount owed at the end of 5 years if $1600 is loaned at a rate of 9% compounded monthly.

Solution: Use the formula $A = P\left(1 + \dfrac{r}{n}\right)^{nt}$, with the following values:

$P = \$1600$ (the amount of the loan)

$r = 9\% = 0.09$ (the rate of interest)

$n = 12$ (the number of times interest is compounded each year)

$t = 5$ (the duration of the loan)

$$A = P\left(1 + \frac{r}{n}\right)^{nt} \qquad \text{Compound interest formula.}$$
$$= 1600\left(1 + \frac{0.09}{12}\right)^{12(5)} \qquad \text{Substitute known values.}$$
$$= 1600(1.0075)^{60}$$

To **approximate** A, press the following keys on your calculator.

| 1.0075 | y^x | 60 | = | × | 1600 | = |

The display should read

$$\boxed{2505.0896}$$

Thus the amount A owed is approximately \$2505.09. ∎

EXERCISE SET 10.2

Graph each exponential function. See Example 1.

1. $y = 4^x$

2. $y = 5^x$

3. $y = 1 + 2^x$

4. $y = 3^x - 1$

Graph each exponential function. See Example 2.

5. $y = \left(\dfrac{1}{4}\right)^x$

6. $y = \left(\dfrac{1}{5}\right)^x$

7. $y = \left(\dfrac{1}{2}\right)^x - 2$

8. $y = \left(\dfrac{1}{3}\right)^x + 2$

Solve each equation for x. See Example 3.

9. $3^x = 27$

10. $6^x = 36$

11. $16^x = 8$

12. $64^x = 16$

13. $32^{2x-3} = 2$

14. $9^{2x+1} = 81$

15. $\dfrac{1}{4} = 2^{3x}$

16. $\dfrac{1}{27} = 3^{2x}$

Solve. See Example 4.

17. One type of uranium has a daily radioactive decay rate of 0.4%. If 30 pounds of this uranium is available today, find how much will still remain after 50 days. Use $y = 30(2.7)^{-0.004t}$.

18. The nuclear waste from an atomic energy plant decays at a rate of 3% each century. If 150 pounds of nuclear waste is disposed of, find how much of it will still remain after 10 centuries. Use $y = 150(2.7)^{-0.03t}$.

Solve. Use $A = P\left(1 + \dfrac{r}{n}\right)^{nt}$. See Example 5.

19. Find the amount Erica owes at the end of 3 years if \$6000 is loaned to her at a rate of 8% compounded monthly.

20. Find the amount owed at the end of 5 years if \$3000 is loaned at a rate of 10% compounded quarterly.

Graph each exponential function.

21. $y = -2^x$ **22.** $y = -3^x$ **23.** $y = 3^x - 2$ **24.** $y = 2^x - 3$

25. $y = -\left(\dfrac{1}{4}\right)^x$ **26.** $y = -\left(\dfrac{1}{5}\right)^x$ **27.** $y = \left(\dfrac{1}{3}\right)^x + 1$ **28.** $y = \left(\dfrac{1}{2}\right)^x - 2$

Solve each equation for x.

29. $5^x = 625$ **30.** $2^x = 64$ **31.** $4^x = 8$ **32.** $32^x = 4$

33. $27^{x+1} = 9$ **34.** $125^{x-2} = 25$ **35.** $81^{x-1} = 27^{2x}$ **36.** $4^{3x-7} = 32^{2x}$

Write an exponential equation defining the function whose graph is given.

37.

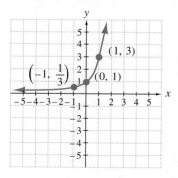

38.

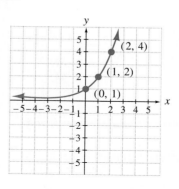

39.

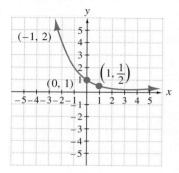

40.

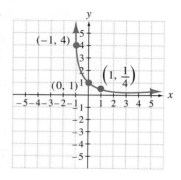

Solve.

41. The size of the rat population of a wharf area grows at a rate of 8% monthly. If there are 200 rats in January, find how many rats should be expected by next January. Use $y = 200(2.7)^{0.08t}$

42. National Park Service personnel are trying to increase the size of the bison population of Theodore Roosevelt National Park. If 260 bison currently live in the park, and if the population's rate of growth is 2.5% annually, find how many bison there should be in 10 years. Use $y = 260(2.7)^{0.025t}$.

43. A rare isotope of a nuclear material is very unstable, decaying at a rate of 15% each second. Find the percent of this isotope remaining 10 seconds after 5 grams of the isotope is created. Use $y = 5(2.7)^{-0.15t}$.

44. An accidental spill of 75 grams of radioactive material in a local stream has led to the presence of radioactive debris decaying at a rate of 4% each day. Find the percent of this debris still remaining after 14 days. Use $y = 75(2.7)^{-0.04t}$

45. Due to economic conditions, Mexico City is growing at a rate of 5.2% annually. If there were 7,000,000 residents of Mexico City in 1990, find how many will be living in the city in 1995 (to the nearest ten-thousand). Use $y = 7,000,000(2.7)^{0.052t}$.

46. An unusually wet spring has caused the size of the Cape Cod mosquito population to increase by 8% each day. If an estimated 200,000 mosquitoes are on Cape Cod on May 12, find how many thousands of mosquitoes will inhabit the Cape on May 25. Use $y = 200,000(2.7)^{0.08t}$.

47. Find the total amount Janina has in a college savings account if $2000 was invested and earned 6% compounded semiannually for 12 years.

48. Find the amount accrued if $500 is invested and earns 7% compounded monthly for 4 years.

Writing in Mathematics

49. Explain why the graph of an exponential function $y = b^x$ contains the point $(1, b)$.

50. Explain why an exponential function $y = b^x$ has a y-intercept of 1.

Skill Review

Solve each equation. See Sections 2.1 and 5.5.

51. $5x - 2 = 18$

54. $2 - 6x = 6(1 - x)$

52. $3x - 7 = 11$

55. $x^2 + 6 = 5x$

53. $3x - 4 = 3(x + 1)$

56. $18 = 11x - x^2$

10.3
Logarithmic Functions

OBJECTIVES

Tape 31

1	Write exponential equations using logarithmic notation, and write logarithmic equations using exponential notation.
2	Solve logarithmic equations by using exponential notation.
3	Identify and graph logarithmic functions.

1 Recall from the last section that $y = 2^x$ is a one-to-one function. Let's begin this section by finding the inverse of this one-to-one function. To find its inverse, interchange variables and then solve for y.

$$y = 2^x$$

$$x = 2^y$$

At this point, we have no method for solving for y until we develop in this section a new notation. We use the symbol $\log_b x$ to mean the **exponent that raises b to x.**

Logarithmic Definition

If $b > 0$ and $b \neq 1$, then

$$x = b^y \quad \text{is equivalent to} \quad y = \log_b x$$

Before returning to the function $x = 2^y$ and solving it for y in terms of x, let's practice using the new notation $\log_b x$. The expression $\log_b x = y$ is read "the logarithm of x to the base b is y" or "the logarithm base b of x is y."

It is important to be able to write exponential equations using logarithmic notation, and vice versa. The following table shows examples of both forms.

Logarithmic Form		*Exponential Form*
$\log_3 9 = 2$	is equivalent to	$3^2 = 9$
$\log_6 1 = 0$	is equivalent to	$6^0 = 1$
$\log_2 8 = 3$	is equivalent to	$2^3 = 8$
$\log_4 \dfrac{1}{16} = -2$	is equivalent to	$4^{-2} = \dfrac{1}{16}$
$\log_8 2 = \dfrac{1}{3}$	is equivalent to	$8^{1/3} = 2$

HELPFUL HINT

Notice that the base of the logarithmic expression is the base of the exponential expression.

EXAMPLE 1   Find the value of each logarithmic expression.

a. $\log_4 16$ **b.** $\log_{10} \dfrac{1}{10}$ **c.** $\log_9 3$

Solution: **a.** $\log_4 16 = 2$ because $4^2 = 16$.   **b.** $\log_{10} \dfrac{1}{10} = -1$ because $10^{-1} = \dfrac{1}{10}$.

c. $\log_9 3 = \dfrac{1}{2}$ because $9^{1/2} = \sqrt{9} = 3$.   ∎

2   The ability to interchange the logarithmic and exponential forms of an expression can be very helpful when solving logarithmic equations.

EXAMPLE 2   Solve each equation for x.

a. $\log_4 \dfrac{1}{4} = x$ **b.** $\log_5 x = 3$ **c.** $\log_x 25 = 2$

d. $\log_3 1 = x$ **e.** $\log_b 1 = x$

Solution: **a.** $\log_4 \dfrac{1}{4} = x$ is equivalent to $4^x = \dfrac{1}{4}$. Solve $4^x = \dfrac{1}{4}$ for x.

$$4^x = \frac{1}{4}$$

$$4^x = 4^{-1}$$

Since the bases are the same, by the uniqueness of b^x, we have that

$$x = -1$$

The solution set is $\{-1\}$. To check, see that $\log_4 \frac{1}{4} = -1$, since $4^{-1} = \frac{1}{4}$.

b. $\log_5 x = 3$ is equivalent to $5^3 = x$ or

$$x = 125$$

The solution set is $\{125\}$.

c. $\log_x 25 = 2$ is equivalent to $x^2 = 25$.

$$x^2 = 25$$

$$x = \pm \sqrt{25} \quad \text{or} \quad \pm 5$$

Since the base b of a logarithm must be positive, the solution set contains 5 only. The solution set is $\{5\}$.

d. $\log_3 1 = x$ is equivalent to $3^x = 1$. Either solve this equation by inspection or solve by writing 1 as 3^0 as shown.

$$3^x = 3^0 \qquad \text{Since } 3^0 \text{ equals } 1.$$

$$x = 0 \qquad \text{Apply the uniqueness of } b^x.$$

The solution set is $\{0\}$.

e. $\log_b 1 = x$ is equivalent to $b^x = 1$.

$$b^x = b^0 \qquad \text{Write } 1 \text{ as } b^0.$$

$$x = 0 \qquad \text{Apply the uniqueness of } b^x.$$

The solution set is $\{0\}$. ∎

In Example 3e we proved an important property of logarithms. That is, $\log_b 1$ is always 0. This property as well as two important others are summarized next.

Properties of Logarithms

If b is a real number, $b > 0$ and $b \neq 1$, then:

1. $\log_b 1 = 0$
2. $\log_b b^x = x$
3. $b^{\log_b x} = x$

To see that $\log_b b^x = x$, change the logarithmic form to exponential form. Then, $\log_b b^x = x$ only when $b^x = b^x$. In exponential form, the expressions are identical, so in logarithmic form, the expressions are identical.

EXAMPLE 3   Simplify.

a. $\log_3 3^2$ **b.** $\log_7 7^{-1}$ **c.** $5^{\log_5 3}$ **d.** $2^{\log_2 6}$

Solution: **a.** From property 2, $\log_3 3^2 = 2$.   **b.** From property 2, $\log_7 7^{-1} = -1$.

c. From property 3, $5^{\log_5 3} = 3$.   **d.** From property 3, $2^{\log_2 6} = 6$.   ∎

3   Having gained proficiency with the notation $\log_b a$, we return to the function $x = 2^y$ and others like it. We know the function $x = 2^y$ is the inverse of the function $y = 2^x$, and we would like to solve $x = 2^y$ for y in terms of x. We use logarithmic notation to do so.

$$2^y = x$$

is equivalent to

$$\log_2 x = y$$

because $\log_2 x = y$ only when $2^y = x$. We now have an expression of y in terms of x. This equation, $y = \log_2 x$, defines a function that is the inverse function of the function $y = 2^x$. The function $y = \log_2 x$ is called a **logarithmic function.**

Logarithmic Function

If x is a positive real number, b is a constant positive real number, and b is not 1, then a **logarithmic function** is a function that can be defined by

$$f(x) = \log_b x$$

The domain of f is the positive real numbers, and the range of f is the real numbers.

We can explore logarithmic functions by graphing them.

EXAMPLE 4   Graph the logarithmic function $y = \log_2 x$.

Solution: Write the equation using exponential notation as $2^y = x$. Find some points that satisfy this equation. Plot the points and connect them with a smooth curve. The domain of this function is $\{x \mid x > 0\}$ and the range of the function is all real numbers, **R**.

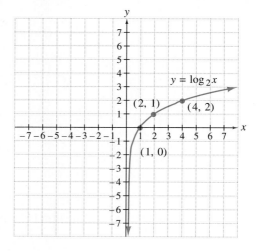

Recall that the graph of a function and its inverse are symmetric about the line $y = x$. Next, we graph the exponential function $y = 2^x$ and its inverse function $y = \log_2 x$ or $x = 2^y$ on the same set of axes to illustrate this.

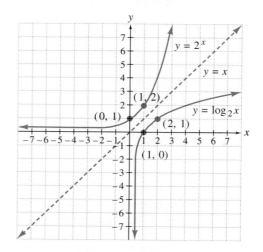

EXAMPLE 5   Graph the logarithmic function $f(x) = \log_{1/3} x$.

Solution: Replace $f(x)$ with y, and write the result using exponential notation.

$$f(x) = \log_{1/3} x$$

$$y = \log_{1/3} x \qquad \text{Replace } f(x) \text{ with } y.$$

$$\left(\frac{1}{3}\right)^y = x \qquad \text{Write in exponential form.}$$

Find points that satisfy $\left(\dfrac{1}{3}\right)^y = x$, plot these points, and connect them with a smooth curve as shown in the figure.

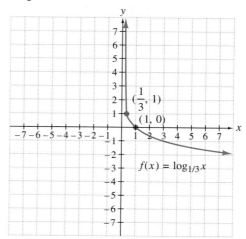

The domain of this function is $\{x \mid x > 0\}$ and the range is the set of all real numbers.

■

The figure shown summarizes characteristics of logarithmic functions.

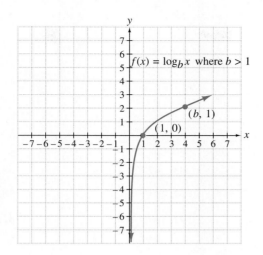

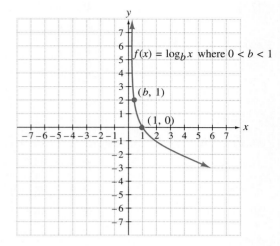

EXERCISE SET 10.3

Find the value of each logarithmic expression. See Example 1.

1. $\log_2 8$ **2.** $\log_4 64$ **3.** $\log_3 \dfrac{1}{9}$ **4.** $\log_2 \dfrac{1}{32}$

5. $\log_{25} 5$ **6.** $\log_8 \dfrac{1}{2}$ **7.** $\log_{1/2} 2$ **8.** $\log_{2/3} \dfrac{4}{9}$

Solve each equation for x. See Example 2.

9. $\log_3 9 = x$ **10.** $\log_2 8 = x$ **11.** $\log_3 x = 4$ **12.** $\log_2 x = 3$

13. $\log_x 49 = 2$ **14.** $\log_x 8 = 3$ **15.** $\log_2 \dfrac{1}{8} = x$ **16.** $\log_3 \dfrac{1}{81} = x$

Simplify. See Example 3.

17. $\log_5 5^3$ **18.** $\log_6 6^2$ **19.** $2^{\log_2 3}$

20. $7^{\log_7 4}$ **21.** $\log_9 9$ **22.** $\log_8 (8)^{-1}$

Graph each logarithmic function. Label any intercepts. See Example 4.

23. $y = \log_3 x$ **24.** $y = \log_2 x$ **25.** $f(x) = \log_{1/4} x$ **26.** $f(x) = \log_{1/2} x$

Graph each function and its inverse function on the same set of axes. Label any intercepts. See Example 5.

27. $y = 4^x$; $y = \log_4 x$

28. $y = 3^x$; $y = \log_3 x$

29. $y = \left(\frac{1}{3}\right)^x$; $y = \log_{1/3} x$

30. $y = \left(\frac{1}{2}\right)^x$; $y = \log_{1/2} x$

Find the value of each expression.

31. $\log_7 1$

32. $\log_9 9$

33. $\log_2 2^4$

34. $\log_6 6^{-2}$

35. $\log_{10} 100$

36. $\log_{10} \frac{1}{10}$

37. $3^{\log_3 5}$

38. $5^{\log_5 7}$

39. $\log_3 81$

40. $\log_2 16$

41. $\log_4 \frac{1}{64}$

42. $\log_3 \frac{1}{9}$

Solve each equation for x.

43. $\log_3 \frac{1}{27} = x$

44. $\log_5 \frac{1}{125} = x$

45. $\log_8 x = \frac{1}{3}$

46. $\log_9 x = \frac{1}{2}$

47. $\log_4 16 = x$

48. $\log_2 16 = x$

49. $\log_{3/4} x = 3$

50. $\log_{2/3} x = 2$

51. $\log_x 100 = 2$

52. $\log_x 27 = 3$

Graph each logarithmic function. Label any intercepts.

53. $f(x) = \log_5 x$

54. $f(x) = \log_6 x$

55. $f(x) = \log_{1/6} x$

56. $f(x) = \log_{1/5} x$

57. The formula $\log_{10} (1 - k) = \frac{-0.3}{H}$ models the relationship between the half-life H of a radioactive material and its rate of decay k. Find the rate of decay of the iodine isotope I-131 if its half-life is 8 days.

Writing in Mathematics

58. Explain why the graph of the function $y = \log_b x$ contains the point $(1, 0)$ no matter what b is.

59. Explain why $\log_b a$ is not defined if a is negative.

Skill Review

Simplify each rational expression. See Section 6.1.

60. $\frac{x + 3}{3 + x}$

61. $\frac{x - 5}{5 - x}$

62. $\frac{x^2 - 8x + 16}{2x - 8}$

63. $\frac{x^2 - 3x - 10}{2 + x}$

Add or subtract as indicated. See Section 6.3.

64. $\dfrac{2}{x} + \dfrac{3}{x^2}$

65. $\dfrac{3x}{x+3} + \dfrac{9}{x+3}$

66. $\dfrac{m^2}{m+1} - \dfrac{1}{m+1}$

67. $\dfrac{5}{y+1} - \dfrac{4}{y-1}$

10.4
Properties of Logarithms

OBJECTIVES

1 Apply the product property of logoarithms.

2 Apply the quotient property of logarithms.

3 Apply the power property of logarithms.

Tape 31

In the previous section we explored some basic properties of logarithms. We now introduce and explore additional properties.

1 The first of these properties is called the **product property of logarithms,** because it deals with the logarithm of a product.

> **Product Property of Logarithms**
>
> If x, y, and b are positive real numbers, $b \neq 1$, then
>
> $$\log_b xy = \log_b x + \log_b y.$$

To prove this, let $\log_b x = M$ and $\log_b y = N$. Now write each logarithm using exponential notation.

$$\log_b x = M \quad \text{is equivalent to} \quad b^M = x$$

$$\log_b y = N \quad \text{is equivalent to} \quad b^N = y$$

Multiply the left sides and the right sides of the exponential equations and we have that

$$xy = (b^M)(b^N) = b^{M+N}$$

Now, write the equation $xy = b^{M+N}$ in equivalent logarithmic form.

$$xy = b^{M+N} \quad \text{is equivalent to} \quad \log_b xy = M + N$$

But since $M = \log_b x$ and $N = \log_b y$, we can write

$$\log_b xy = M + N$$

as

$$\log_b xy = \log_b x + \log_b y \qquad \text{Let } M = \log_b x \text{ and } N = \log_b y.$$

The logarithm of a product is the sum of the logarithms of the factors. This property is sometimes used to simplify logarithmic expressions.

EXAMPLE 1 Use the product rule to simplify. Assume that variables represent positive numbers.

a. $\log_{11} 10 + \log_{11} 3$ **b.** $\log_3 \dfrac{1}{2} + \log_3 12$ **c.** $\log_2 (x + 2) + \log_2 x$

Solution: All terms have a common logarithmic base.

a. $\log_{11} 10 + \log_{11} 3 = \log_{11} (10 \cdot 3)$ Apply the product property.

$$= \log_{11} 30$$

b. $\log_3 \dfrac{1}{2} + \log_3 12 = \log_3 \left(\dfrac{1}{2} \cdot 12 \right) = \log_3 6$

c. $\log_2 (x + 2) + \log_2 x = \log_2 [(x + 2) \cdot x] = \log_2 (x^2 + 2x)$ ■

2 The second property is the **quotient property of logarithms.**

Quotient Property of Logarithms

If x, y, and b are positive real numbers, $b \neq 1$, then

$$\log_b \frac{x}{y} = \log_b x - \log_b y$$

The proof of the quotient property of logarithms is similar to the proof of the product rule. Notice that the quotient property says that the logarithm of a quotient is the difference of the logarithms of the dividend and divisor.

EXAMPLE 2 Use the quotient property to simplify. Assume that x represents positive numbers only.

a. $\log_{10} 27 - \log_{10} 3$ **b.** $\log_5 8 - \log_5 x$ **c.** $\log_4 25 + \log_4 3 - \log_4 5$
d. $\log_3 (x^2 + 5) - \log_3 (x^2 + 1)$

Solution: All terms have a common logarithmic base.

a. $\log_{10} 27 - \log_{10} 3 = \log_{10} \dfrac{27}{3} = \log_{10} 9$

b. $\log_5 8 - \log_5 x = \log_5 \dfrac{8}{x}$

c. Use both the product and quotient properties.

$\log_4 25 + \log_4 3 - \log_4 5 = \log_4 (25 \cdot 3) - \log_4 5$ Apply the product property.

$$= \log_4 75 - \log_4 5$$ Simplify.

$$= \log_4 \dfrac{75}{5}$$ Apply the quotient property.

$$= \log_4 15$$ Simplify.

d. $\log_3 (x^2 + 5) - \log_3 (x^2 + 1) = \log_3 \dfrac{x^2 + 5}{x^2 + 1}$ Apply the quotient rule.

■

3 The third and final property we introduce is called the **power property of logarithms.**

> **Power Property of Logarithms**
>
> If x and b are positive real numbers, $b \neq 1$, and r is a real number, then
>
> $$\log_b x^r = r \log_b x$$

For example,

$$\log_3 2^4 = 4 \log_3 2, \qquad \log_5 x^3 = 3 \log_5 x, \qquad \log_4 \sqrt{x} = \log_4 (x)^{1/2} = \frac{1}{2} \log_4 x$$

EXAMPLE 3   Write each as the logarithm of a single expression. Assume that x is a positive number.
a. $2 \log_5 3 + 3 \log_5 2$ **b.** $2 \log_9 x + \log_9 (x + 1)$

Solution: Notice that the terms have a common logarithmic base.

a. $2 \log_5 3 + 3 \log_5 2 = \log_5 3^2 + \log_5 2^3$ Apply the power property.

$$= \log_5 9 + \log_5 8$$

$$= \log_5 (9 \cdot 8) \qquad \text{Apply the product property.}$$

$$= \log_5 72$$

b. $2 \log_9 x + \log_9 (x + 1) = \log_9 x^2 + \log_9 (x + 1)$ Apply the power property.

$$= \log_9 [x^2 (x + 1)] \qquad \text{Apply the product property.}$$

■

EXAMPLE 4   Use properties of logarithms to write each expression as a sum or difference of multiples of logarithms.
a. $\log_3 \dfrac{5 \cdot 7}{4}$ **b.** $\log_2 \dfrac{x^5}{y^2}$

Solution: **a.** $\log_3 \dfrac{5 \cdot 7}{4} = \log_3 (5 \cdot 7) - \log_3 4$

$$= \log_3 5 + \log_3 7 - \log_3 4$$

b. $\log_2 \dfrac{x^5}{y^2} = \log_2 (x^5) - \log_2 (y^2)$ Apply the quotient property.

$$= 5 \log_2 x - 2 \log_2 y \qquad \text{Apply the power property.} \quad ■$$

HELPFUL HINT

Notice that we are not able to simplify further a logarithmic expression such as $\log_5 (2x - 1)$. None of the basic properties gives a way to write the logarithm of a difference in some equivalent form.

EXAMPLE 5   If $\log_b 2 = 0.43$ and $\log_b 3 = 0.68$, use the properties of logarithms to evaluate.
a. $\log_b 6$ **b.** $\log_b 9$ **c.** $\log_b \sqrt{2}$

Solution: **a.** $\log_b 6 = \log_b (2 \cdot 3)$ Write 6 as $2 \cdot 3$.

$= \log_b 2 + \log_b 3$ Apply the product property.

$= 0.43 + 0.68$ Substitute given values.

$= 1.11$

b. $\log_b 9 = \log_b 3^2$ Write 9 as 3^2.

$= 2 \log_b 3$

$= 2(0.68)$ Substitute 0.68 for $\log_b 3$.

$= 1.36$

c. First, recall that $\sqrt{2} = 2^{1/2}$. Then

$$\log_b \sqrt{2} = \log_b 2^{1/2} = \frac{1}{2} \log_b 2$$

$$= \frac{1}{2}(0.43)$$

$$= 0.215 \quad \blacksquare$$

Next we summarize the basic properties of logarithms that we developed so far.

Properties of Logarithms

If x, y, and b are positive real numbers, $b \neq 1$, and r is a real number, then:

1. $\log_b 1 = 0$
2. $\log_b b^x = x$
3. $b^{\log_b x} = x$
4. $\log_b xy = \log_b x + \log_b y$ Product property.
5. $\log_b \dfrac{x}{y} = \log_b x - \log_b y$ Quotient property.
6. $\log_b x^r = r \log_b x$ Power property.

EXERCISE SET 10.4

Write each as the logarithm of a single expression. Assume that variables represent positive numbers. See Example 1.

1. $\log_5 2 + \log_5 7$

2. $\log_3 8 + \log_3 4$

3. $\log_4 9 + \log_4 x$

4. $\log_2 x + \log_2 y$

5. $\log_{10} 5 + \log_{10} 2 + \log_{10} (x^2 + 2)$

6. $\log_6 3 + \log_6 (x + 4) + \log_6 5$

Write each as the logarithm of a single expression. Assume that variables represent positive numbers. See Example 2.

7. $\log_5 12 - \log_5 4$

8. $\log_7 20 - \log_7 4$

9. $\log_2 x - \log_2 y$

10. $\log_3 12 - \log_3 z$

11. $\log_4 2 + \log_4 10 - \log_4 5$

12. $\log_6 18 + \log_6 2 - \log_6 9$

Write each as the logarithm of a single expression. Assume that variables represent positive numbers. See Example 3.

13. $2 \log_2 5$

14. $3 \log_5 2$

15. $3 \log_5 x + 6 \log_5 z$

16. $2 \log_7 y + 6 \log_7 z$

17. $\log_{10} x - \log_{10} (x + 1) + \log_{10} (x^2 - 2)$

18. $\log_9 (4x) - \log_9 (x - 3) + \log_9 (x^3 + 1)$

Write each expression as a sum or difference of multiples of logarithms. Assume that variables represent positive numbers. See Example 4.

19. $\log_3 \dfrac{4y}{5}$

20. $\log_4 \dfrac{2}{9z}$

21. $\log_2 \left(\dfrac{x^3}{y} \right)$

22. $\log_5 \left(\dfrac{x}{y^4} \right)$

23. $\log_b \sqrt{7x}$

24. $\log_b \sqrt{\dfrac{3}{y}}$

If $\log_b 3 \approx 0.5$ and $\log_b 5 \approx 0.7$, approximate the following. See Example 5.

25. $\log_b \dfrac{5}{3}$

26. $\log_b 25$

27. $\log_b 15$

28. $\log_b \dfrac{3}{5}$

29. $\log_b \sqrt[3]{5}$

30. $\log_b \sqrt[4]{3}$

Write each as the logarithm of a single expression. Assume that variables represent positive numbers.

31. $\log_4 5 + \log_4 7$

32. $\log_3 2 + \log_3 5$

33. $\log_3 8 - \log_3 2$

34. $\log_5 12 - \log_5 3$

35. $\log_7 6 + \log_7 3 - \log_7 4$

36. $\log_8 5 + \log_8 15 - \log_8 20$

37. $3 \log_4 2 + \log_4 6$

38. $2 \log_3 5 + \log_3 2$

39. $3 \log_2 x + \dfrac{1}{2} \log_2 x - 2 \log_2 (x + 1)$

40. $2 \log_5 x + \dfrac{1}{3} \log_5 x - 3 \log_5 (x + 5)$

41. $2 \log_8 x - \dfrac{2}{3} \log_8 x + 4 \log_8 x$

42. $5 \log_6 x - \dfrac{3}{4} \log_6 x + 3 \log_6 x$

Write each expression as a sum or difference of multiples of logarithms. Assume that variables represent positive numbers.

43. $\log_7 \dfrac{5x}{4}$

44. $\log_9 \dfrac{7}{y}$

45. $\log_5 x^3 (x + 1)$

46. $\log_2 y^3 z$

47. $\log_6 \dfrac{x^2}{x + 3}$

48. $\log_3 \dfrac{(x + 5)^2}{x}$

If $\log_b 2 = 0.43$ and $\log_b 3 = 0.68$, evaluate the following.

49. $\log_b 8$

50. $\log_b 81$

51. $\log_b \dfrac{3}{9}$

52. $\log_b \dfrac{4}{32}$

53. $\log_b \sqrt{\dfrac{2}{3}}$

54. $\log_b \sqrt{\dfrac{3}{2}}$

Writing in Mathematics

55. Explain whether the quotient property of logarithms can be applied to $\dfrac{\log_4 18}{\log_4 6}$.

56. Explain whether the product property of logarithms can be applied to $(\log_3 6) \cdot (\log_3 4)$.

Skill Review

57. Graph the functions $y = 10^x$ and $y = \log_{10} x$ on the same set of axes. See Section 10.3.

Solve each equation. See Section 7.5.

58. $\sqrt{3x + 1} = 2$ **59.** $\sqrt{2x - 3} = 5$ **60.** $\sqrt{x^2 + 9} = x + 3$ **61.** $\sqrt{x^2 - 2} = x - 2$

10.5
Common Logarithms, Natural Logarithms, and Change of Base

OBJECTIVES

Tape 32

1   Identify common logarithms and approximate them by calculator.

2   Evaluate common logarithms of powers of 10.

3   Identify natural logarithms and approximate them by calculator.

4   Evaluate natural logarithms of powers of e.

5   Apply the change of base formula.

In this section we look closely at two particular logarithmic bases. These two logarithmic bases are used so frequently that logarithms to their bases are given special names. **Common logarithms** are logarithms to base 10. **Natural logarithms** are logarithms to base e, which we introduce in this section. Because of the wide availability and low cost of calculators today, the work in this section is based on using calculators, which typically have both the common "log" and the natural "log" keys.

1   Logarithms to base 10, common logarithms, are used frequently because our number system is a base 10 decimal system. The notation $\log x$ means the same as $\log_{10} x$.

> **Common Logarithms**
>
> $$\log x \text{ means } \log_{10} x$$

EXAMPLE 1   Use a calculator to approximate log 7 to 4 decimal places.

Solution: Press the following sequence of keys:

$$\boxed{7}\ \boxed{\log}$$

(Some calculators require $\boxed{7}\ \boxed{\log}\ \boxed{=}$.) The number $\boxed{0.845098}$ should appear in the display. To four decimal places,

$$\log 7 \approx 0.8451$$

Some scientific calculators do not have a $\boxed{\log}$ key, but do have a $\boxed{10^x}$ key. If this is the case, then log 7 is approximated by pressing

$$\boxed{7}\ \boxed{\text{INV}}\ \boxed{10^x}$$

This sequence is based on the fact that the functions $y = \log_{10} x$ or $y = \log x$ and $y = 10^x$ are inverses of each other. ■

2   To evaluate the common log of a power of 10, a calculator is not needed. According to the property of logarithms

$$\log_b b^x = x$$

it follows that

$$\log 10^x = x$$

because the base of this logarithm is understood to be 10.

EXAMPLE 2   Find the exact value of each logarithm.

a. $\log 10$ **b.** $\log 1000$ **c.** $\log \dfrac{1}{10}$ **d.** $\log \sqrt{10}$

Solution: **a.** $\log 10 = \log 10^1 = 1$ **b.** $\log 1000 = \log 10^3 = 3$

c. $\log \dfrac{1}{10} = \log 10^{-1} = -1$ **d.** $\log \sqrt{10} = \log 10^{1/2} = \dfrac{1}{2}$ ■

As we will soon see, equations containing common logs are useful models of many natural phenomena.

EXAMPLE 3   Solve $\log x = 1.2$ for x. Give an exact solution, and then approximate the solution to four decimal places.

Solution: Write the logarithmic equation using exponential notation. Keep in mind that the base of a common log is understood to be 10.

$$\log x = 1.2$$

$$10^{1.2} = x \qquad \text{Write using exponential notation.}$$

$$15.848932 \approx x \qquad \text{Press } \boxed{10}\ \boxed{y^x}\ \boxed{1.2}\ \boxed{=}.$$

The exact solution is $10^{1.2}$. To four decimal places, $x \approx 15.8489$. ■

3   **Natural logarithms** are also frequently used, especially to describe natural events; hence the label "natural logarithm." Natural logarithms are logarithms to the base e, which is a constant approximately equal to 2.7183. The number e is an irrational number like π. The notation $\log_e x$ is usually abbreviated to $\ln x$.

Natural Logarithms
$\ln x$ means $\log_e x$

EXAMPLE 4   Use a calculator to approximate ln 8 to four decimal places.

Solution: Press the following sequence of keys:

$$\boxed{8}\ \boxed{\ln}$$

The display should show $\boxed{2.0794415}$. To four decimal places,

$$\ln 8 \approx 2.0794$$

Some scientific calculators do not have a $\boxed{\ln}$ key, but they have an $\boxed{e^x}$ key instead. If this is the case, then ln 8 is approximated by pressing

$$\boxed{8}\ \boxed{\text{INV}}\ \boxed{e^x}$$

This sequence is based on the fact that the functions $y = \log_e x$ or $y = \ln x$ and $y = e^x$ are inverses of each other. ∎

> **4**   As a result of the property $\log_b b^x = x$, we know that $\log_e e^x = x$ or $\ln e^x = x$.

EXAMPLE 5   Find the exact value of each natural logarithm.
a. $\ln e^3$ **b.** $\ln \sqrt[5]{e}$

Solution: **a.** $\ln e^3 = 3$ **b.** $\ln \sqrt[5]{e} = \ln e^{1/5} = \dfrac{1}{5}$ ∎

EXAMPLE 6   Solve the equation $\ln 3x = 5$ for x. Give an exact solution and then approximate the solution to four decimal places.

Solution: Write the equation using exponential notation. Keep in mind that the base of a natural logarithm is understood to be e.

$$\ln 3x = 5$$

$$e^5 = 3x \qquad \text{Write using exponential notation.}$$

$$\frac{e^5}{3} = x$$

The exact solution is $\dfrac{e^5}{3}$. To four decimal places, $x \approx 49.4711$.   ∎

The Richter scale measures the intensity or magnitude of an earthquake. The formula for the magnitude R of an earthquake is $R = \log\left(\dfrac{a}{T}\right) + B$, where a is the amplitude in micrometers of the vertical motion of the ground at the recording station, T is the number of seconds between successive seismic waves, and B is an adjustment factor that takes into account the weakening of the seismic wave as the distance increases from the epicenter of the earthquake.

EXAMPLE 7   Find the magnitude of an earthquake on the Richter scale if a recording station measures an amplitude of 300 micrometers and 2.5 seconds between waves. Assume that B is 4.2. Approximate the solution to the nearest tenth.

Solution: Substitute the known values into the formula for earthquake intensity.

$$R = \log\left(\frac{a}{T}\right) + B \qquad \text{Richter scale formula.}$$

$$= \log\left(\frac{300}{2.5}\right) + 4.2 \qquad \text{Let } a = 300, \ T = 2.5, \text{ and } B = 4.2.$$

$$= \log(120) + 4.2$$

$$\approx 2.1 + 4.2 \qquad \text{Approximate log 120 by 2.1.}$$

$$= 6.3$$

This earthquake had a magnitude of 6.3 on the Richter scale.   ■

Recall from Section 10.2 the formula $A = P\left(1 + \dfrac{r}{n}\right)^{nt}$ for compound interest, where n represents the number of compoundings per year. When interest is compounded continuously, the formula $A = Pe^{rt}$ is used.

EXAMPLE 8   Find the amount owed at the end of 5 years if $1600 is loaned at a rate of 9% compounded continuously.

Solution: Use the formula $A = Pe^{rt}$, where

$$P = \$1600 \ (\text{the size of the loan})$$

$$r = 9\% = 0.09 \ (\text{the rate of interest})$$

$$t = 5 \ (\text{the 5-year duration of the loan})$$

$$A = Pe^{rt}$$

$$= 1600e^{0.09(5)} \qquad \text{Substitute in known values.}$$

$$= 1600e^{0.45}$$

Now use a calculator to approximate the solution. Press the keys

$$\boxed{0.45}\ \boxed{e^x}\ \boxed{\times}\ \boxed{1600}\ \boxed{=}$$

Thus

$$A \approx 2509.30$$

The total amount of money owed is approximately $2509.30.   ■

5   Calculators are handy tools for approximating natural and common logarithms. Unfortunately, some calculators cannot be used to approximate logarithms to bases other than e or 10, at least not directly. In such cases, we use the change of base formula.

Change of Base

If a, b, and c are positive real numbers and neither b nor c is 1, then

$$\log_b a = \frac{\log_c a}{\log_c b}$$

EXAMPLE 9 Approximate $\log_5 3$ to four decimal places.

Solution: Use the change of base property to write $\log_5 3$ as a quotient of logarithms to base 10.

$$\log_5 3 = \frac{\log 3}{\log 5} \qquad \text{Use the change of base property.}$$

$$\approx \frac{0.4771213}{0.69897} \qquad \text{Approximate logarithms by calculator.}$$

$$\approx 0.6826063 \qquad \text{Simplify by calculator.}$$

To four decimal places, $\log_5 3 \approx 0.6826$. ∎

EXERCISE SET 10.5

Use a calculator to approximate each logarithm to four decimal places. See Example 1.

1. log 8   **2.** log 6   **3.** log 2.31   **4.** log 4.86

Find the exact value. See Example 2.

5. log 100   **6.** log 10,000   **7.** $\log\left(\frac{1}{1000}\right)$ **8.** $\log\left(\frac{1}{100}\right)$

Solve each equation for x. Give an exact solution and a four-decimal-place approximation. See Example 3.

9. $\log x = 1.3$ **10.** $\log x = 2.1$ **11.** $\log 2x = 1.1$ **12.** $\log 3x = 1.3$

Use a calculator to approximate each logarithm to four decimal places. See Example 4.

13. ln 2   **14.** ln 3   **15.** ln 0.0716   **16.** ln 0.0032

Find the exact value. See Example 5.

17. $\ln e^2$ **18.** $\ln e^4$ **19.** $\ln \sqrt[4]{e}$ **20.** $\ln \sqrt[5]{e}$

Solve each equation for x. Give an exact solution and a four-decimal-place approximation. See Example 6.

21. $\ln x = 1.4$ **22.** $\ln x = 2.1$ **23.** $\ln (3x - 4) = 2.3$ **24.** $\ln (2x + 5) = 3.4$

Use the formula $R = \log\left(\frac{a}{T}\right) + B$ to find the intensity R on the Richter scale of the earthquakes fitting the descriptions given. See Example 7.

25. Amplitude a is 200 micrometers, time T between waves is 1.6 seconds, and B is 2.1.

26. Amplitude a is 150 micrometers, time T between waves is 3.6 seconds, and B is 1.9.

Solve. See Example 8.

27. Find the amount of money Paul Banks owes at the end of 4 years if 6% interest is compounded continuously on his $2000 debt.

28. Find the amount of money a $2500 certificate of deposit is redeemable for if it has been paying 10% interest compounded continuously for 3 years.

Approximate each logarithm to four decimal places. See Example 9.

29. $\log_2 3$ **30.** $\log_3 2$ **31.** $\log_{1/2} 5$ **32.** $\log_{1/3} 2$

Use a calculator to approximate each logarithm to four decimal places.

33. $\log 12.6$ **34.** $\log 25.9$ **35.** $\ln 5$ **36.** $\ln 7$
37. $\log 41.5$ **38.** $\ln 41.5$

Find the exact value.

39. $\log 10^3$ **40.** $\ln e^5$ **41.** $\ln e^2$ **42.** $\log 10^7$
43. $\log 0.0001$ **44.** $\log 0.001$ **45.** $\ln \sqrt{e}$ **46.** $\log \sqrt{10}$

Solve each equation for x. Give an exact solution and a four-decimal-place approximation.

47. $\log x = 2.3$ **48.** $\log x = 3.1$
49. $\ln x = -2.3$ **50.** $\ln x = -3.7$
51. $\log (2x + 1) = -0.5$ **52.** $\log (3x - 2) = -0.8$

53. $\ln 4x = 0.18$ **54.** $\ln 3x = 0.76$

Approximate each logarithm to four decimal places.

55. $\log_4 9$ **56.** $\log_9 4$
57. $\log_3 \frac{1}{6}$ **58.** $\log_6 \frac{2}{3}$
59. $\log_8 6$ **60.** $\log_6 8$

Use the formula $R = \log\left(\frac{a}{T}\right) + B$ to find the intensity R on the Richter scale of the earthquakes fitting the descriptions given.

61. Amplitude a is 400 micrometers, time T between waves is 2.6 seconds, and B is 3.1.

62. Amplitude a is 450 micrometers, time T between waves is 4.2 seconds, and B is 2.7.

63. Find how much money Dana Jones has after 12 years if $1400 is invested at 8% interest compounded continuously.

64. Determine the size of an account, where $3500 earns 6% interest compounded continuously for 1 year.

Psychologists call the graph of the formula

$$t = \frac{1}{c}\ln\left(\frac{A}{A - N}\right)$$

the learning curve, since the formula relates time t passed, in weeks, to a measure N of learning achieved, to a measure A of maximum learning possible, to a measure c of an individual's learning style.

65. Norman is learning to type. If he wants to type at a rate of 50 words per minute (N is 50), and his expected maximum rate is 75 words per minute (A is 75), find how many weeks it takes him to achieve his goal. Assume that c is 0.09.

66. Janine is working on her dictation skills. She wants to take dictation at a rate of 150 words per minute and believes that the maximum rate she could ever hope for is 210 words per minute. Find how many weeks it takes her to achieve the 150-word level if c is 0.07.

67. An experiment with teaching chimpanzees sign language shows that a typical chimp can master a maximum of 65 signs. Find how many weeks it should take a chimpanzee to achieve mastery of 30 signs if c is 0.03.

68. A psychologist is measuring human capability to memorize nonsense syllables. Find how many minutes it should take a subject to learn 15 nonsense syllables if the maximum possible to learn is 24 syllables and c is 0.17.

Writing in Mathematics

69. On a calculator, press $\boxed{-3}\ \boxed{\log}$. Describe what happens and explain why it does.

70. Without using a calculator, explain which of log 50 and ln 50 must be larger.

Skill Review

Solve each equation for x. See Sections 2.1 and 2.2.

71. $6x - 3(2 - 5x) = 6$

72. $2x + 3 = 5 - 2(3x - 1)$

73. $2x + 3y = 6x$

74. $4x - 8y = 10x \quad x = -\dfrac{4y}{3}$

Solve each system of equations. See Section 9.1.

75. $\begin{cases} x + 2y = -4 \\ 3x - y = 9 \end{cases}$

76. $\begin{cases} 5x + y = 5 \\ -3x - 2y = -10 \end{cases}$

10.6
Exponential and Logarithmic Equations and Applications

OBJECTIVES

Tape 32

1 Solve exponential equations.

2 Solve logarithmic equations.

3 Solve problems that can be modeled by exponential and logarithmic equations.

---

1 In Section 10.2, we solved exponential equations like $2^x = 16$ by writing 16 as a power of 2 and applying the uniqueness of b^x.

$$2^x = 16$$

$$2^x = 2^4 \qquad \text{Write 16 as } 2^4.$$

$$x = 4 \qquad \text{Apply the uniqueness of } b^x.$$

To solve an equation like $3^x = 7$, we use the following.

> **Uniqueness of $\log_b a$ as a Logarithm to Base b**
>
> If a, b, and c are real numbers such that $\log_b a$ and $\log_b c$ are real numbers and b is not 1, then
>
> $$\log_b a = \log_b c \text{ only when } a = c$$

EXAMPLE 1   Solve $3^x = 7$.

Solution: To solve, use the uniqueness of logarithms and take the logarithm of both sides. For this example, we use the common logarithm.

$$3^x = 7$$

$$\log 3^x = \log 7 \qquad \text{Take the common log of both sides.}$$

$$x \log 3 = \log 7 \qquad \text{Apply the power property of logarithms.}$$

$$x = \frac{\log 7}{\log 3} \qquad \text{Divide both sides by } \log 3.$$

The exact solution is $\dfrac{\log 7}{\log 3}$. If a decimal approximation is preferred, $\dfrac{\log 7}{\log 3} \approx \dfrac{0.845098}{0.4771213} \approx 1.7712$ to four decimal places. The solution set is $\left\{ \dfrac{\log 7}{\log 3} \right\}$ or **approximately** $\{1.7712\}$.   ■

  2   By applying the appropriate properties of logarithms, a broad variety of logarithmic equations can be solved.

EXAMPLE 2   Solve $\log_4 (x - 2) = 2$ for x.

Solution: First, write the equation using exponential notation.

$$\log_4 (x - 2) = 2$$

$$4^2 = x - 2$$

$$16 = x - 2$$

$$18 = x \qquad \text{Add 2 to both sides.}$$

To check, replace x with 18 in the **original equation.**

$$\log_4 (x - 2) = 2$$

$$\log_4 (18 - 2) = 2 \qquad \text{Let } x = 18.$$

$$\log_4 16 = 2$$

$$4^2 = 16 \qquad \text{True.}$$

The solution set is $\{18\}$.   ■

EXAMPLE 3 Solve $\log_2 x + \log_2 (x - 1) = 1$ for x.

Solution: Apply the product rule to the left side of the equation.

$$\log_2 x + \log_2 (x - 1) = 1$$

$$\log_2 x(x - 1) = 1 \qquad \text{Use the product rule.}$$

$$\log_2 (x^2 - x) = 1$$

Next, write the equation using exponential notation and solve for x.

$$2^1 = x^2 - x$$

$$0 = x^2 - x - 2 \qquad \text{Subtract 2 from both sides.}$$

$$0 = (x - 2)(x + 1) \qquad \text{Factor.}$$

$$0 = x - 2 \quad \text{or} \quad 0 = x + 1 \qquad \text{Set each factor equal to 0.}$$

$$2 = x \qquad\qquad -1 = x$$

Verify that 2 satisfies the original equation. Now check -1 by replacing x with -1 in the original equation.

$$\log_2 x + \log_2 (x - 1) = 1$$

$$\log_2 (-1) + \log_2 (-1 - 1) = 1 \qquad \text{Let } x = -1.$$

Because the logarithm of a negative number is undefined, -1 is an extraneous solution. The solution set is $\{2\}$. ∎

EXAMPLE 4 Solve $\log(x + 2) - \log x = 2$ for x.

Solution: Apply the quotient property of logarithms to the left side of the equation.

$$\log(x + 2) - \log x = 2$$

$$\log \frac{x + 2}{x} = 2 \qquad \text{Apply the quotient property.}$$

$$10^2 = \frac{x + 2}{x} \qquad \text{Write using exponential notation.}$$

$$100 = \frac{x + 2}{x}$$

$$100x = x + 2 \qquad \text{Multiply both sides by } x.$$

$$99x = 2 \qquad \text{Subtract } x \text{ from both sides.}$$

$$x = \frac{2}{99} \qquad \text{Divide both sides by 99.}$$

Verify that the solution set is $\left\{ \dfrac{2}{99} \right\}$. ∎

3 Throughout this chapter we have emphasized that logarithmic and exponential equations are used in a variety of scientific, technical, and business settings. A few examples are shown.

EXAMPLE 5 The population of lemmings varies according to the relationship $y = y_0 e^{0.15t}$. In this formula, t is time in months, and y_0 is some initial population at time 0. Estimate the population in 6 months if there are originally 5000 lemmings.

Solution: Substitute 5000 for y_0 and 6 for t.

$$y = y_0 e^{0.15t}$$

$$= 5000 e^{0.15(6)} \qquad \text{Let } t = 6 \text{ and } y_0 = 5000.$$

$$= 5000 e^{0.9} \qquad \text{Multiply.}$$

To find an approximation for $5000 e^{0.9}$, press these keys on your calculator:

$$\boxed{0.9}\ \boxed{e^x}\ \boxed{\times}\ \boxed{5000}\ \boxed{=}$$

Then $y \approx 12{,}298.016$. In 6 months the population should be approximately 12,300 lemmings. ∎

EXAMPLE 6 How long does it take an investment of $2000 to double if it is invested at 5% interest compounded quarterly? The necessary formula to use is $A = P\left(1 + \dfrac{r}{n}\right)^{nt}$, where A is the accrued (or owed) amount, P is the principal invested, r is the rate of interest, n is the number of compounding periods per year, and t is the number of years.

Solution: We are given that $P = \$2000$ and $r = 5\% = 0.05$. Compounding quarterly means 4 times a year so that $n = 4$. The investment is to double, so A must be $4000. Substitute these values and solve for t.

$$A = P\left(1 + \frac{r}{n}\right)^{nt}$$

$$4000 = 2000\left(1 + \frac{0.05}{4}\right)^{4t} \qquad \text{Substitute in known values.}$$

$$4000 = 2000(1.0125)^{4t} \qquad \text{Simplify } 1 + \frac{0.05}{4}.$$

$$2 = (1.0125)^{4t} \qquad \text{Divide both sides by 2000.}$$

$$\log 2 = \log 1.0125^{4t} \qquad \text{Apply the uniqueness of logarithms.}$$

$$\log 2 = 4t(\log 1.0125) \qquad \text{Apply the power property.}$$

$$\frac{\log 2}{(4 \log 1.0125)} = t \qquad \text{Divide both sides by 4 log 1.0125.}$$

$$13.949408 = t \qquad \text{Approximate by calculator.}$$

Thus it takes nearly 14 years for the money to double in value. ∎

EXERCISE SET 10.6

Solve each equation. Give an exact solution, and also approximate the solution to four decimal places. See Example 1.

1. $3^x = 6$

2. $4^x = 7$

3. $3^{2x} = 3.8$

4. $5^{3x} = 5.6$

5. $2^{x-3} = 5$

6. $8^{x-2} = 12$

Solve each equation. See Example 2.

7. $\log_2 (x + 5) = 4$

8. $\log_6 (x^2 - x) = 1$

9. $\log_3 x^2 = 4$

10. $\log_2 x^2 = 6$

Solve each equation. See Examples 3 and 4.

11. $\log_4 2 + \log_4 x = 0$

12. $\log_3 5 + \log_3 x = 1$

13. $\log_2 6 - \log_2 x = 3$

14. $\log_4 10 - \log_4 x = 2$

15. $\log_4 x + \log_4 (x + 6) = 2$

16. $\log_3 x + \log_3 (x + 6) = 3$

17. $\log_5 (x + 3) - \log_5 x = 2$

18. $\log_6 (x + 2) - \log_6 x = 2$

Solve. See Example 5.

19. The size of the wolf population at Isle Royale National Park increases at a rate of 4.3% per year. If the size of the current population is 83 wolves, find how many there should be in 5 years. Use $y = y_0 \, e^{0.043t}$.

20. The number of victims of a flu epidemic is increasing at a rate of 7.5% per week. If 20,000 persons are currently infected, find in how many days we can expect 45,000 to have the flu. Use $y = y_0 \, e^{0.075t}$.

Solve. Use the formula $A = P\left(1 + \dfrac{r}{n}\right)^{nt}$ to solve the compound interest problem. See Example 6.

21. Find how long it takes $1000 to double if it is invested at 8% interest compounded semiannually.

22. Find how long it takes $1000 to double if it is invested at 8% interest compounded monthly.

Solve each equation. Give an exact solution, and also approximate the solution to four decimal places.

23. $9^x = 5$

24. $3^x = 11$

25. $4^{x+7} = 3$

26. $6^{x+3} = 2$

27. $7^{3x-4} = 11$

28. $5^{2x-6} = 12$

29. $e^{6x} = 5$

30. $e^{2x} = 8$

Solve each equation.

31. $\log_3 (x - 2) = 2$

32. $\log_2 (x - 5) = 3$

33. $\log_4 (x^2 - 3x) = 1$

34. $\log_8 (x^2 - 2x) = 1$

35. $\log_3 5 + \log_3 x = 2$

36. $\log_5 2 + \log_5 x = 3$

37. $3 \log_8 x - \log_8 x^2 = 2$

38. $2 \log_6 x - \log_6 x = 3$

39. $\log_2 x + \log_2 (x + 5) = 1$

40. $\log_4 x + \log_4 (x + 7) = 1$

41. $\log_4 x - \log_4 (2x - 3) = 3$

42. $\log_2 x - \log_2 (3x + 5) = 4$

43. $\log_2 x + \log_2 (3x + 1) = 1$

44. $\log_3 x + \log_3 (x - 8) = 2$

Solve.

45. The size of the population of Senegal is increasing at a rate of 2.6% per year. If 7,000,000 people lived in Senegal in 1986, find how many inhabitants there will be by 1995, rounded to the nearest ten thousand. Use $y = y_0 e^{0.026t}$.

46. In 1986, 784 million people were citizens of India. Find how long it will take India's population to reach a size of 1000 million (that is, 1 billion) if the population size is growing at a rate of 2.1% per year. Round to the nearest tenth. Use $y = y_0 e^{0.021t}$.

Use the formula $A = P\left(1 + \dfrac{r}{n}\right)^{nt}$ to solve these compound interest problems. See Example 6.

47. Find how long it takes $600 to double if it is invested at 7% interest compounded monthly.

48. Find how long it takes $600 to double if it is invested at 12% interest compounded monthly.

49. Find how long it takes a $1200 investment to earn $200 interest if it is invested at 9% interest compounded quarterly.

50. Find how long it takes a $1500 investment to earn $200 interest if it is invested at 10% compounded semiannually.

The formula $w = 0.00185h^{2.67}$ is used to estimate the normal weight w of a boy h inches tall. Use this formula to solve the height–weight problem.

51. Find the expected height of a boy who normally should weigh 85 pounds.

52. Find the expected height of a boy who normally should weigh 140 pounds.

The formula $P = 14.7e^{-0.21x}$ gives the average atmospheric pressure P, in pounds per square inch, at an altitude x, in miles above sea level. Use this formula to solve these pressure problems.

53. Find the average atmospheric pressure of Denver, which is 1 mile above sea level.

54. Find the average atmospheric pressure of Pikes Peak, which is 2.7 miles above sea level.

55. Find the elevation of a Delta jet if the atmospheric pressure outside the jet is 7.5 lb/in.^{2}.

56. Find the elevation of a remote Himalayan peak if the atmospheric pressure atop the peak is 6.5 lb/in.

Skill Review

If $x = -2$, $y = 0$, and $z = 3$, find the value of each expression. See Section 1.4.

57. $\dfrac{x^2 - y + 2z}{3x}$

58. $\dfrac{x^3 - 2y + z}{2z}$

59. $\dfrac{3z - 4x + y}{x + 2z}$

60. $\dfrac{4y - 3x + z}{2x + y}$

Find the inverse function of each one-to-one function. See Section 10.1.

61. $f(x) = 5x + 2$

62. $f(x) = \dfrac{x - 3}{4}$

CRITICAL THINKING Kenya is currently the world's fastest-growing country. Its population is increasing at a rate of 4% annually. If this pattern of growth continues, Kenya will more than double the size of its population in less than 20 years. Clearly, that kind of growth will strain the country's resources. If the government of Kenya determines that it can be prepared only for 10% more people after 20 years, what steady rate of growth does this represent?

Even at this reduced rate of growth, theoretically Kenya's population will be infinitely large (why?). Can you suggest factors that naturally limit the size of a population? How might these factors be accounted for in a mathematical model of population growth?

CHAPTER 10 GLOSSARY

Common logarithms are logarithms to base 10.

A function of the form $f(x) = b^x$ is an **exponential function,** where $b > 0$, b is not 1, and x is a real number.

If x is a positive real number, b is a constant positive real number, and b is not 1, then a **logarithmic function** is a function that can be defined by $f(x) = \log_b x$.

The **inverse of a one-to-one function** f is the one-to-one function f^{-1}, which is the set of all ordered pairs (y, x), where (x, y) belongs to f.

Natural logarithms are logarithms to base e.

If f is a function, then f is a **one-to-one-function** if each y-value corresponds to a unique x-value.

CHAPTER 10 SUMMARY

HORIZONTAL LINE TEST (10.1)

If every horizontal line intersects the graph of a function at most once, then the function is a one-to-one function.

TO FIND THE INVERSE OF A ONE-TO-ONE FUNCTION $f(x)$ (10.1)

1. Replace $f(x)$ by y.
2. Interchange x and y.
3. Solve the equation for y.
4. Replace y with the notation $f^{-1}(x)$.

LOGARITHMIC DEFINITION (10.3)

If $b > 0$ and $b \neq 1$, then

$$x = b^y \quad \text{is equivalent to} \quad y = \log_b x.$$

PROPERTIES OF LOGARITHMS (10.4)

If x, y, and b are positive real numbers, $b \neq 1$, and r is a real number, then:

1. $\log_b 1 = 0$
2. $\log_b b^x = x$
3. $b^{\log_b x} = x$
4. $\log_b xy = \log_b x + \log_b y$ Product property.
5. $\log_b \dfrac{x}{y} = \log_b x - \log_b y$ Quotient property.
6. $\log_b x^r = r \log_b x$ Power property.

CHANGE OF BASE (10.5)

If a, b, and c are positive real numbers and neither b nor c is 1, then

$$\log_b a = \frac{\log_c a}{\log_c b}$$

CHAPTER 10 REVIEW

(10.1)   *Determine whether each function is a one-to-one function. If so, list the elements of its inverse.*

1. $h = \{(-9, 14), (6, 8), (-11, 12), (15, 15)\}$

2. $f = \{(-5, 5), (0, 4), (13, 5), (11, -6)\}$

Determine whether each function is a one-to-one function.

3.

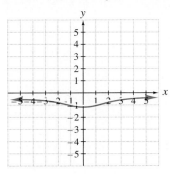

4.

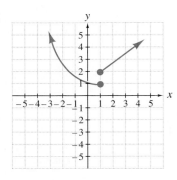

5.

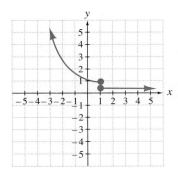

6.

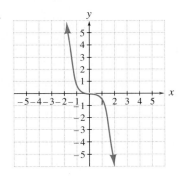

Find an equation defining the inverse function of the given function.

7. $f(x) = 6x + 11$

8. $f(x) = 12x$

9. $q(x) = mx + b$

10. $g(x) = \dfrac{12x - 7}{6}$

11. $r(x) = \dfrac{13}{2}x - 4$

On the same set of axes, graph the given one-to-one function and its inverse.

12. $g(x) = \sqrt{x}$

13. $h(x) = 5x - 5$

(10.2)   *Solve each equation for x.*

14. $4^x = 64$

15. $3^x = \dfrac{1}{9}$

16. $2^{3x} = \dfrac{1}{16}$

17. $5^{2x} = 125$

18. $9^{x+1} = 243$

19. $8^{3x-2} = 4$

Graph each exponential function.

20. $y = 3^x$

21. $y = \left(\dfrac{1}{3}\right)^x$

22. $y = 4 \cdot 2^x$

23. $y = 2^x + 4$

Use the formula $A = P\left(1 + \dfrac{r}{n}\right)^{nt}$ to solve the interest problems. In this formula:

$$A = \text{amount accrued (or owed)}$$
$$P = \text{principal invested (or loaned)}$$
$$r = \text{rate of interest}$$
$$n = \text{number of compounding periods per year}$$
$$t = \text{time in years}$$

24. Find the amount if $1600 is invested at 9% interest compounded semiannually for 7 years.

25. $800 is invested in a 7% certificate of deposit for which interest is compounded quarterly. Find the value of this certificate at the end of 5 years.

(10.3) *Write each equation using logarithmic notation.*

26. $49 = 7^2$

27. $2^{-4} = \dfrac{1}{16}$ $\log_2$

Write each logarithmic equation using exponential notation.

28. $\log_{\frac{1}{2}} 16 = -4$

29. $\log_{0.4} 0.064 = 3$

Solve for x.

30. $\log_4 x = -3$

31. $\log_3 x = 2$

32. $\log_3 1 = x$

33. $\log_4 64 = x$

34. $\log_x 64 = 2$

35. $\log_x 81 = 4$

36. $\log_4 4^5 = x$

37. $\log_7 7^{-2} = x$

38. $5^{\log_5 4} = x$

39. $2^{\log_2 9} = x$

40. $\log_2 (3x - 1) = 4$

41. $\log_3 (2x + 5) = 2$

42. $\log_4 (x^2 - 3x) = 1$

43. $\log_8 (x^2 + 7x) = 1$

Graph each pair of equations on the same coordinate system.

44. $y = 2^x$ and $y = \log_2 x$

45. $y = \left(\dfrac{1}{2}\right)^x$ and $y = \log_{\frac{1}{2}} x$

(10.4) *Write each of the following as single logarithms.*

46. $\log_3 8 + \log_3 4$

47. $\log_2 6 + \log_2 3$

48. $\log_7 15 - \log_7 20$

49. $\log 18 - \log 12$

50. $\log_{11} 8 + \log_{11} 3 - \log_{11} 6$

51. $\log_5 14 + \log_5 3 - \log_5 21$

52. $2\log_5 x - 2\log_5 (x + 1) + \log_5 x$

53. $4\log_3 x - \log_3 x + \log_3 (x + 2)$

Use properties of logarithms to write each expression as a sum or difference of multiples of logarithms.

54. $\log_3 \dfrac{x^3}{x + 2}$

55. $\log_4 \dfrac{x + 5}{x^2}$

56. $\log_2 \dfrac{3x^2 y}{z}$

57. $\log_7 \dfrac{yz^3}{x}$

If $\log_b 2 = 0.36$ and $\log_b 5 = 0.83$, find the following.

58. $\log_b 50$

59. $\log_b \dfrac{4}{5}$

(10.5) *Use a calculator to approximate the logarithm to four decimal places.*

60. $\log 3.6$

61. $\log 0.15$

62. $\ln 1.25$

63. $\ln 4.63$

Find the exact value.

64. $\log 1000$

65. $\log \dfrac{1}{10}$

66. $\ln \left(\dfrac{1}{e}\right)$

67. $\ln (e^4)$

Solve each equation for x.

68. $\ln (2x) = 2$

69. $\ln (3x) = 1.6$

70. $\ln (2x - 3) = -1$

71. $\ln (3x + 1) = 2$

Use the formula $\ln \dfrac{I}{I_0} = -kx$ to solve radiation problems. In this formula:

$$x = \text{depth in millimeters}$$
$$I = \text{intensity of radiation}$$
$$I_0 = \text{initial intensity}$$
$$k = \text{a constant measure dependent on the material}$$

72. Find the depth at which the intensity of the radiation passing through a lead shield is reduced to 3% of the original intensity if the value of k is 2.1.

73. If k is 3.2, find the depth at which 2% of the original radioactivity will penetrate.

Approximate the logarithm to four decimal places.

74. $\log_5 1.6$

75. $\log_3 4$

Use the formula $A = Pe^{rt}$ to solve the interest problems in which interest is compounded continuously. In this formula:

$$A = \text{amount accrued (or owed)}$$

$$P = \text{principal invested (or loaned)}$$

$$r = \text{rate of interest}$$

$$t = \text{time in years}$$

76. Chase Manhattan Bank offers a 5-year 6% continuously compounded investment option. Find the amount accrued if $1450 is invested.

77. Find the amount that a $940 investment grows to if it is invested at 11% compounded continuously for 3 years.

(10.6)   *Solve each exponential equation for x. Give an exact solution and also approximate the solution to four decimal places.*

78. $3^{2x} = 7$

79. $6^{3x} = 5$

80. $3^{2x+1} = 6$

81. $4^{3x+2} = 9$

82. $5^{3x-5} = 4$

83. $8^{4x-2} = 3$

84. $2 \cdot 5^{x-1} = 1$

85. $3 \cdot 4^{x+5} = 2$

Solve the equation for x.

86. $\log_5 2 + \log_5 x = 2$

87. $\log_3 x + \log_3 10 = 2$

88. $\log (5x) - \log (x + 1) = 4$

89. $\ln (3x) - \ln (x - 3) = 2$

90. $\log_2 x + \log_2 2x - 3 = 1$

91. $-\log_6 (4x + 7) + \log_6 x = 1$

Use the formula $y = y_0 e^{kt}$ to solve the population growth problems. In this formula:

$$y = \text{size of population}$$

$$y_0 = \text{initial count of population}$$

$$k = \text{rate of growth}$$

$$t = \text{time}$$

92. The population of mallard ducks in Nova Scotia is expected to grow at a rate of 6% per week during the spring migration. If 155,000 ducks are already in Nova Scotia, find how many are expected by the end of 4 weeks.

93. The population of Indonesia is growing at a rate of 1.7% per year. If the population in 1986 was 176,800,000, find the expected population by the year 2000.

94. Anaheim, California, is experiencing an annual growth rate of 3.16%. If 230,000 people now live in Anaheim, find how long it will take for the size of the population to be 500,000.

95. Memphis, Tennessee, is growing at a rate of 0.36% per year. Find how long it will take the population of Memphis to increase from 650,000 to 700,000.

96. Egypt's population is increasing at a rate of 2.1% per year. Find how long it will take for its 50,500,000-person population to double in size.

97. The greater Mexico City area had a population of 16.9 million in 1985. How long will it take the city to triple in population if its growth rate is 3.4% annually?

Use the compound interest equation $A = P \left(1 + \dfrac{r}{n}\right)^{nt}$ to solve the following. (See directions for Exercises 24 and 25 for an explanation of this formula.)

98. Find how long it will take a $5000 investment to grow to $10,000 if it is invested at 8% interest compounded quarterly.

99. An investment of $6000 has grown to $10,000 while the money was invested at 6% interest compounded monthly. Find how long it was invested.

CHAPTER 10 TEST

On the same set of axes, graph the given one-to-one function and its inverse.

1. $7x - 14 = f(x)$

Determine whether the given graph is the graph of a one-to-one function.

2.

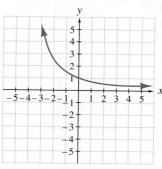

3.

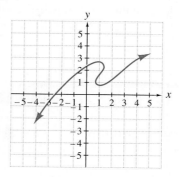

Find a set of ordered pairs or an equation that defines the inverse function of the given function.

4. $y = 6 - 2x$

5. $f = \{(0, 0), (2, 3), (-1, 5)\}$

Use the properties of logarithms to write each expression as a single logarithm.

6. $\log_3 6 + \log_3 4$

7. $\log_5 x + 3 \log_5 x - \log_5 (x + 1)$

8. Write the expression $\log_6 \dfrac{2x}{y^3}$ as the sum or difference of multiples of logarithms.

9. If $\log_b 3 = 0.79$ and $\log_b 5 = 1.16$, find the value of $\log_b \left(\dfrac{3}{25}\right)$.

10. Approximate $\log_7 8$ to four decimal places.

11. Solve $8^{x-1} = \dfrac{1}{64}$ for x. Give an exact solution.

12. Solve $3^{2x+5} = 4$ for x. Give an exact solution, and also approximate the solution to four decimal places.

Solve each logarithmic equation for x. Give an exact solution.

13. $\log_3 x = -2$

14. $\ln \sqrt{e} = x$

15. $\log_8 (3x - 2) = 2$

16. $\log_5 x + \log_5 3 = 2$

17. $\log_4 (x + 1) - \log_4 (x - 2) = 3$

18. Solve $\ln (3x + 7) = 1.31$ accurate to four decimal places.

19. Graph $y = \left(\dfrac{1}{2}\right)^x + 1$.

20. Graph the functions $y = 3^x$ and $y = \log_3 x$ on the same coordinate system.

Use the formula $A = P\left(1 + \dfrac{r}{n}\right)^{nt}$ to solve Exercises 21 and 22.

21. Find the amount in the account if $4000 is invested for 3 years at 9% interest compounded monthly.

22. Find how long it will take $2000 to grow to $3000 if the money is invested at 7% interest compounded semiannually.

Use the population growth formula $y = y_0 e^{kt}$ to solve Exercises 23 and 24.

23. The prairie dog population of the Grand Rapids area now stands at 57,000 animals. If the population is growing at a rate of 2.6% annually, find how many prairie dogs there will be 5 years from now.

24. In an attempt to save an endangered species of wood duck, naturalists would like to increase their population from 400 to 1000 ducks. If the annual population growth rate is 6.2%, find how long it will take the naturalists to reach their goal.

25. The formula $\log(1 + k) = \dfrac{0.3}{D}$ relates the doubling time D, in days, and growth rate k for a population of mice. Find the rate at which the population is increasing if the doubling time is 56 days.

CHAPTER 10 CUMULATIVE REVIEW

1. Simplify the following expressions.
  a. $11 + 2 - 7$
  b. $-5 - 4 + 2$
  c. $-5 - (4 + 2)$
  d. $|2 - 6| + |5 - 4|$
  e. $(2 - 6) + (5 - 4)$

2. Solve $5x - 3 \leq 10$ or $x + 1 \geq 5$.

3. Find the x- and y-intercepts of $2x + 3y = 5$ and graph the equation.

4. Find the slope of the line $y = 2$.

5. Simplify the following expressions. Write all answers using positive exponents.
  a. $(x^5)^7$
  b. $(2^2)^3$
  c. $(5^{-1})^2$
  d. $(y^{-3})^{-4}$

6. Add $11x^3 - 12x^2 + x - 3$ and $x^3 - 10x + 5$.

7. Find the following products.
  a. $(x + 5)^2$
  b. $(x - 9)^2$
  c. $(3x + 2z)^2$
  d. $(4m^2 - 3n)^2$

8. Factor $x^2 - 12x + 35$.

9. Factor $125q^2 - n^3 q^2$.

10. Multiply.
  a. $\dfrac{2x^2 + 3x - 2}{-4x - 8} \cdot \dfrac{16x^2}{4x^2 - 1}$
  b. $(ac - ad + bc - bd) \cdot \dfrac{a + b}{d - c}$

11. Simplify each complex fraction.
  a. $\dfrac{\dfrac{5x}{x + 2}}{\dfrac{10}{x - 2}}$
  b. $\dfrac{x + \dfrac{1}{y}}{y + \dfrac{1}{x}}$

12. Evaluate the following.
  a. $16^{-3/4}$ **b.** $(-27)^{-2/3}$

13. Use the quotient rule to simplify. Assume that all variables represent positive numbers.
  a. $\sqrt{\dfrac{5}{49}}$
  b. $\sqrt[3]{-\dfrac{8}{27}}$
  c. $\sqrt{\dfrac{36x^4}{25y^2}}$
  d. $\sqrt[3]{\dfrac{16x^3}{z^6}}$

14. Multiply the complex numbers. Write the product in the form $a + bi$.
  a. $(2 - 5i)(4 + i)$
  b. $(2 - i)^2$
  c. $(7 + 3i)(7 - 3i)$

15. Solve $p = -3p^2 - 3$.

16. Graph $y = 4(x - 1)^2 + 3$.

17. Graph $(x + 1)^2 + y^2 = 8$.

18. Solve the system:
$$\begin{cases} 2x - 4y + 8z = 2 \\ -x - 3y + z = 11 \\ x - 2y + 4z = 0 \end{cases}$$

19. Graph $4y^2 > x^2 + 16$.

20. Find the equation of the inverse of $f(x) = 3x - 5$. Graph f and f^{-1} on the same set of axes.

21. Write each as the logarithm of a single expression. Assume that x is a positive number.
 a. $2 \log_5 3 + 3 \log_5 2$
 b. $2 \log_9 x + \log_9 (x + 1)$

22. Solve $3^x = 7$.

CHAPTER **11**

Sequences, Series, and the Binomial Theorem

Drone bees originate from single-parent families, in the literal sense of the word, creating family trees quite distinct from peoples'. Both kinds of family trees, though, are described by sequences. (See Critical Thinking, page 540.)

INTRODUCTION

Having explored in some depth the concept of function, we turn now in this final chapter to sequences. In one sense, a sequence is simply an ordered list of numbers. In another sense, a sequence is itself a function. Phenomena modeled by sequences are everywhere around us in the mathematical world. The starting place for all mathematics is, after all, the sequence of natural numbers: 1, 2, 3, 4, and so on.

Sequences lead us to **series,** which are a sum of ordered numbers. Through series we gain new insight, for example, about the expansion of a binomial $(a + b)^n$, the concluding topic for this book.

11.1
Sequences

OBJECTIVES

Tape 33

1 Write the terms of a sequence given its general term.

2 Find the general term of a sequence.

3 Solve applications involving sequences.

A town has a present population of 100,000 and is growing by 5% each year. After the first year, the town's population is

$$100,000 + 0.05(100,000) = 105,000$$

After the second year, the town's population is

$$105,000 + 0.05(105,000) = 110,250$$

After the third year, the town's population is

$$110,250 + 0.05(110,250) = 115,762.5$$

The town's population can be written as the **sequence** of numbers

$$105,000, 110,250, 115,762.5, \ldots$$

Another sequence of numbers is 2, 4, 8, 16, . . .

> **Infinite Sequence**
>
> An infinite sequence is a function whose domain is the set of natural numbers 1, 2, 3, 4, and so on.

1 The **general term** of the sequence 2, 4, 8, 16, . . . is 2^n, where n is a natural number. Since 2^n is a function, we can write it as $f(n) = 2^n$, where $n = 1, 2, 3, \ldots$. Instead, we use the notation

$$a_n = 2^n$$

515

The domain of this function or sequence is the set of natural numbers. The range of this sequence is the set of function values $a_1, a_2, a_3, \ldots$. These values are called the **terms** of the sequence.

EXAMPLE 1   Write the first five terms of the sequence whose general term is

$$a_n = n^2 - 1$$

Solution: Evaluate a_n, when n is 1, 2, 3, 4, and 5.

$$a_n = n^2 - 1$$

$$a_1 = 1^2 - 1 = 0 \qquad \text{Replace } n \text{ with 1.}$$

$$a_2 = 2^2 - 1 = 3 \qquad \text{Replace } n \text{ with 2.}$$

$$a_3 = 3^2 - 1 = 8 \qquad \text{Replace } n \text{ with 3.}$$

$$a_4 = 4^2 - 1 = 15 \qquad \text{Replace } n \text{ with 4.}$$

$$a_5 = 5^2 - 1 = 24 \qquad \text{Replace } n \text{ with 5.}$$

Thus the first five terms of the sequence $a_n = n^2 - 1$ are 0, 3, 8, 15, and 24.   ∎

EXAMPLE 2   If the general term of a sequence is $a_n = \dfrac{(-1)^n}{3n}$, find:

a. the first term of the sequence   **b.** a_8
c. the one-hundredth term of the sequence   **d.** a_{15}

Solution: **a.** $a_1 = \dfrac{(-1)^1}{3(1)} = -\dfrac{1}{3}$ $\qquad$ Replace n with 1.

b. $a_8 = \dfrac{(-1)^8}{3(8)} = \dfrac{1}{24}$ $\qquad$ Replace n with 8.

c. $a_{100} = \dfrac{(-1)^{100}}{3(100)} = \dfrac{1}{300}$ $\qquad$ Replace n with 100.

d. $a_{15} = \dfrac{(-1)^{15}}{3(15)} = -\dfrac{1}{45}$ $\qquad$ Replace n with 15.   ∎

2   Suppose we know the first few terms of a sequence and want to find the general term.

EXAMPLE 3   Find the general term a_n for the sequence whose first four terms are $\dfrac{1}{2}, \dfrac{1}{4}, \dfrac{1}{8}, \dfrac{1}{16}$.

Solution: Notice that the denominators double each time.

$$\frac{1}{2}, \quad \frac{1}{2 \cdot 2}, \quad \frac{1}{2(2 \cdot 2)}, \quad \frac{1}{2(2 \cdot 2 \cdot 2)}$$

$$= \frac{1}{2}, \quad \frac{1}{2^2}, \quad \frac{1}{2^3}, \quad \frac{1}{2^4}$$

We might suppose then that the general term is $a_n = \dfrac{1}{2^n}$.   ∎

3 Sequences model many phenomena of the physical world, as illustrated by the following example.

EXAMPLE 4 The amount of weight, in pounds, a puppy gains in each month of its first year is modeled by a sequence whose general term is $a_n = n + 4$, where n is the number of the month. Write the first five terms of the sequence, and find how much weight the puppy should gain in its fifth month.

Solution: Evaluate $a_n = n + 4$ when n is 1, 2, 3, 4, and 5.

$$a_1 = 1 + 4 = 5$$
$$a_2 = 2 + 4 = 6$$
$$a_3 = 3 + 4 = 7$$
$$a_4 = 4 + 4 = 8$$
$$a_5 = 5 + 4 = 9$$

The puppy should gain 9 pounds in its fifth month. ∎

EXERCISE SET 11.1

Write the first five terms of each sequence whose general term is given. See Example 1.

1. $a_n = n + 4$

2. $a_n = 5 - n$

3. $a_n = (-1)^n$

4. $a_n = (-2)^n$

5. $a_n = \dfrac{1}{n + 3}$

6. $a_n = \dfrac{1}{7 - n}$

Find the indicated term for each sequence whose general term is given. See Example 2.

7. $a_n = 3n^2$; a_5

8. $a_n = -n^2$; a_{15}

9. $a_n = 6n - 2$; a_{20}

10. $a_n = 100 - 7n$; a_{50}

11. $a_n = \dfrac{n + 3}{n}$; a_{15}

12. $a_n = \dfrac{n}{n + 4}$; a_{24}

Find a general term a_n for each sequence whose five terms are given. See Example 3.

13. 3, 7, 11, 15

14. 2, 7, 12, 17

15. $-2, -4, -8, -16$

16. $-4, 16, -64, 256$

17. $\dfrac{1}{3}, \dfrac{1}{9}, \dfrac{1}{27}, \dfrac{1}{81}$

18. $\dfrac{2}{5}, \dfrac{2}{25}, \dfrac{2}{125}, \dfrac{2}{625}$

Solve. See Example 4.

19. The distance, in feet, that a thermos dropped from a cliff falls in each consecutive second is modeled by a sequence whose general term is $a_n = 32n - 16$, when n is the number of seconds. Find the distance the thermos falls in the second, third, and fourth seconds.

20. A culture of bacteria triples every hour so that its size is modeled by the sequence $a_n = 50(3)^{n-1}$, where n is the number of the hour just beginning. Find the size of the culture at the beginning of the fourth hour. Find the size of the culture originally.

21. Mrs. Laser agrees to give her son Mark an allowance of $0.10 on the first day of his 14-day vacation, $0.20 on the second day, $0.40 on the third day, and so on. Write an equation of a sequence whose terms correspond to Mark's allowance. Find the allowance Mark receives on the last day of his vacation.

22. A small theater has 10 rows with 12 seats in the first row, 15 seats in the second row, 18 seats in the third row, and so on. Write an equation of a sequence whose terms correspond to the seats in each row. Find the number of seats in the eighth row.

Write the first five terms of each sequence whose general term is given.

23. $a_n = 2n$

24. $a_n = -6n$

25. $a_n = -n^2$

26. $a_n = n^2 + 2$

27. $a_n = 2^n$

28. $a_n = 3^{n-2}$

29. $a_n = 2n + 5$

30. $a_n = 1 - 3n$

31. $a^n = (-1)^n n^2$

32. $a_n = (-1)^{n+1}(n$

Find the indicated term for each sequence whose general term is given.

33. $a_n = (-3)^n$; a_6

34. $a_n = 5^{n+1}$; a_3

35. $a_n = \dfrac{n-2}{n+1}$; a_6

36. $a_n = \dfrac{n+3}{n+4}$; a_8

37. $a_n = \dfrac{(-1)^n}{n}$; a_8

38. $a_n = \dfrac{(-1)^n}{2n}$; a_{100}

39. $a_n = -n^2 + 5$; a_{10}

40. $a_n = 8 - n^2$; a_{20}

41. $a_n = \dfrac{(-1)^n}{n+6}$; a_{19}

42. $a_n = \dfrac{n-4}{(-2)^n}$; a_6

43. The number of cases of a new infectious disease is doubling every year so that the number of cases is modeled by a sequence whose general term is $a_n = 75(2)^{n-1}$, where n is the number of the year just beginning. Find how many cases there are at the beginning of the sixth year. Find how many cases there were when the disease was first discovered.

44. A new college had an initial enrollment of 2700 students in 1988, and each year the enrollment increases by 150 students. Find the enrollment for each of 5 years, beginning with 1988.

45. An endangered species of sparrow had an estimated population numbering 800 in 1986, and scientists predict that its population will decrease by half each year. Estimate the population in 1990. Estimate the year the sparrow will be extinct.

46. A **Fibonacci sequence** is a special type of sequence in which the first two terms are 1 and each term thereafter is the sum of the two previous terms: 1, 1, 2, 3, 5, 8. Many plants and animals seem to grow according to a Fibonacci sequence, including pine cones, pineapple scales, nautilus shells, and certain flowers. Write the first 15 terms of the Fibonacci sequence.

11.2
Arithmetic and Geometric Sequences

OBJECTIVES

Tape 33

1 Identify arithmetic sequences and their common differences.

2 Identify geometric sequences and their common ratios.

1 Find the first four terms of the sequence whose general term is

$$a_n = 5 + (n - 1)3.$$

$$a_1 = 5 + (1 - 1)3 = 5 \qquad \text{Replace } n \text{ with 1.}$$

$$a_2 = 5 + (2 - 1)3 = 8 \qquad \text{Replace } n \text{ with 2.}$$

$$a_3 = 5 + (3 - 1)3 = 11 \qquad \text{Replace } n \text{ with 3.}$$

$$a_4 = 5 + (4 - 1)3 = 14 \qquad \text{Replace } n \text{ with 4.}$$

The first four terms are 5, 8, 11, and 14. Notice that each term after the first is the sum of 3 and the previous term; that is, $a_n = 3 + \underline{a_{n-1}}$. When this happens, we

$$\text{previous term}$$

call the sequence an **arithmetic sequence,** or an **arithmetic progression.** The constant difference d in successive terms is called the **common difference.** In this example, d is 3.

Arithmetic Sequence and Common Difference

An **arithmetic sequence** is a sequence in which each term after the first differs from the preceding term by a constant amount d. The constant d is called the **common difference.**

The sequence 2, 6, 10, 14, 18, . . . is an arithmetic sequence. Its common difference is 4. Given the first term, a_1, and the common difference d of an arithmetic sequence, we can find any term of the sequence.

EXAMPLE 1 Write the first five terms of the arithmetic sequence whose first term is 7 and whose common difference is 2.

Solution:

$$a_1 = 7$$

$$a_2 = 7 + 2 = 9 \qquad a_2 = a_1 + d$$

$$a_3 = 9 + 2 = 11 \qquad a_3 = a_1 + 2d$$

$$a_4 = 11 + 2 = 13 \qquad a_4 = a_1 + 3d$$

$$a_5 = 13 + 2 = 15 \qquad a_5 = a_1 + 4d$$

The first five terms are 7, 9, 11, 13, 15. ∎

The pattern on the right suggests that the general term a_n of an arithmetic sequence is given by

$$a_n = a_1 + (n - 1)d$$

General Term of an Arithmetic Sequence

The general term a_n of an arithmetic sequence is given by

$$a_n = a_1 + (n - 1)d$$

where a_1 is the first term and d is the common difference.

EXAMPLE 2 Consider the arithmetic sequence whose first term is 3 and common difference is -5.
a. Write an expression for the general term a_n.
b. Find the twentieth term of this sequence.

Solution: **a.** Since this is an arithmetic sequence, the general term a_n is given by $a_n = a_1 + (n - 1)d$. Here, $a_1 = 3$ and $d = -5$ so that

$$a_n = 3 + (n - 1)(-5)$$

b. $a_{20} = 3 + (20 - 1)(-5) = 3 + 19(-5) = -92$

The twentieth term is -92.   ∎

EXAMPLE 3   Find the eleventh term of the arithmetic sequence whose first three terms are 2, 9, 16.

Solution: Since the sequence is arithmetic, the eleventh term is

$$a_{11} = a_1 + (11 - 1)d = a_1 + 10d$$

We know a_1 is the first term of the sequence, so $a_1 = 2$. Also, d is the constant difference of terms, so $d = a_2 - a_1 = 9 - 2 = 7$. Thus

$$a_{11} = a_1 + 10d$$
$$= 2 + 10 \cdot 7$$
$$= 72   ∎$$

EXAMPLE 4   If the third term of an arithmetic progression is 12 and the eighth term is 27, find the fifth term.

Solution: We need to find a_1 and d to write the general term, which then enables us to find a_5, the fifth term. The two given terms a_3 and a_8 lead to a system of linear equations.

$$\begin{cases} a_3 = a_1 + (3 - 1)d \\ a_8 = a_1 + (8 - 1)d \end{cases} \quad \text{or} \quad \begin{cases} 12 = a_1 + 2d \\ 27 = a_1 + 7d \end{cases}$$

Next, we solve the system $\begin{cases} 12 = a_1 + 2d \\ 27 = a_1 + 7d \end{cases}$ by addition. Multiply both sides of the second equation by -1 so that

$$\begin{cases} 12 = a_1 + 2d \\ -1 \ (27) = -1 \ (a_1 + 7d) \end{cases} \quad \text{simplifies to} \quad \begin{cases} 12 = a_1 + 2d \\ -27 = -a_1 - 7d \\ \overline{-15 = -5d} \end{cases}$$

$$3 = d \qquad \text{Divide both sides by } -5.$$

To find a_1, let $d = 3$ in $12 = a_1 + 2d$. Then

$$12 = a_1 + 2(3)$$
$$12 = a_1 + 6$$
$$6 = a_1$$

Thus $a_1 = 6$ and $d = 3$, so

$$a_n = 6 + (n - 1)(3)$$
$$a_5 = 6 + (5 - 1)(3) = 18   ∎$$

EXAMPLE 5   Donna has an offer for a job starting at $20,000 per year and guaranteeing her a raise of $800 per year for the next 5 years. Write the general term for the arithmetic sequence modeling Donna's potential annual salaries and find her salary for the fourth year.

Solution: The first term $a_1 = 20{,}000$ and $d = 800$. So

$$a_n = 20{,}000 + (n - 1)(800) \quad \text{and} \quad a_4 = 20{,}000 + (3)(800) = 22{,}400$$

Her salary for the fourth year will be $22,400.   ∎

2   We now investigate a **geometric sequence**, also called a **geometric progression.** In the sequence 5, 15, 45, 135, . . . , each term after the first is the **product** of 3 and the preceding term. This pattern of multiplying by a constant to get the next term defines a geometric sequence.

> **Geometric Sequence and Common Ratio**
>
> A **geometric sequence** is a sequence in which each term after the first is obtained by multiplying the preceding term by a constant r. The constant r is called the **common ratio.**

The sequence $12, 6, 3, \frac{3}{2}, \ldots$ is geometric since each term after the first is the product of the previous term and $\frac{1}{2}$.

EXAMPLE 6 Write the first five terms of a geometric sequence whose first term is 7 and whose common ratio is 2.

Solution:

$$a_1 = 7$$

$$a_2 = 7(2) = 14 \qquad a_2 = a_1(r)$$

$$a_3 = 14(2) = 28 \qquad a_3 = a_1(r^2)$$

$$a_4 = 28(2) = 56 \qquad a_4 = a_1(r^3)$$

$$a_5 = 56(2) = 112 \qquad a_5 = a_1(r^4)$$

The first five terms are 7, 14, 28, 56, and 112. ∎

Notice that the pattern on the right suggests that the general term of a geometric sequence is given by $a_n = a_1 r^{n-1}$.

> **General Term of a Geometric Sequence**
>
> The general term a_n of a geometric sequence is given by
>
> $$a_n = a_1 r^{n-1}$$
>
> where a_1 is the first term and r is the common ratio.

EXAMPLE 7 Find the eighth term of the geometric sequence whose first term is 12 and whose common ratio is $\frac{1}{2}$.

Solution: Since this is a geometric sequence, the general term a_n is given by

$$a_n = a_1 r^{n-1}$$

Here $a_1 = 12$ and $r = \frac{1}{2}$, so $a_n = 12\left(\frac{1}{2}\right)^{n-1}$. Evaluate a_n when $n = 8$.

$$a_8 = 12\left(\frac{1}{2}\right)^{8-1} = 12\left(\frac{1}{2}\right)^7 = 12\left(\frac{1}{128}\right) = \frac{3}{32} \qquad ∎$$

EXAMPLE 8 Find the fifth term of the geometric sequence whose first three terms are 2, −6, and 18.

Solution: Since the sequence is geometric, the fifth term must be $a_1 r^{5-1}$, or $2r^4$. We know that r is the common ratio of terms, so r must be $\dfrac{-6}{2}$, or −3. Thus

$$a_5 = 2r^4$$
$$a_5 = 2(-3)^4 = 162 \quad \blacksquare$$

EXAMPLE 9 If the second term of a geometric sequence is $\dfrac{5}{4}$ and the third term is $\dfrac{5}{16}$, find the first term and the common ratio.

Solution: Notice that $\dfrac{5}{16} \div \dfrac{5}{4} = \dfrac{1}{4}$, so $r = \dfrac{1}{4}$. Then

$$a_2 = a_1 \left(\frac{1}{4}\right)^1$$
$$\frac{5}{4} = a_1 \left(\frac{1}{4}\right) \quad \text{or} \quad a_1 = 5 \qquad \text{Replace } a_2 \text{ with } \frac{5}{4}.$$

The first term is 5. $\quad \blacksquare$

EXAMPLE 10 A bacteria culture growing under controlled conditions doubles each day. Find how large the culture is at the beginning of day 7 if it measured 10 units at the beginning of day 1.

Solution: Since the culture doubles its size each day, the sizes are modeled by a geometric sequence. Here $a_1 = 10$ and $r = 2$. Thus

$$a_n = a_1 r^{n-1} = 10(2)^{n-1} \quad \text{and} \quad a_7 = 10(2)^{7-1} = 640$$

The bacteria measures 640 units at the beginning of day 7. $\quad \blacksquare$

EXERCISE SET 11.2

Write the first five terms of the arithmetic or geometric sequence whose first term a_1 and common difference d or common ratio r are given. See Examples 1 and 6.

1. $a_1 = 4; d = 2$ **2.** $a_1 = 3; d = 10$ **3.** $a_1 = 6; d = -2$ **4.** $a_1 = -20; d = 3$

5. $a_1 = 1; r = 3$ **6.** $a_1 = -2; r = 2$ **7.** $a_1 = 48; r = \dfrac{1}{2}$ **8.** $a_1 = 1; r = \dfrac{1}{3}$

Find the indicated term of each sequence. See Examples 2 and 7.

9. The eighth term of the arithmetic sequence whose first term is 12 and whose common difference is 3.

10. The twelfth term of the arithmetic sequence whose first term is 32 and whose common difference is −4.

11. The fourth term of the geometric sequence whose first term is 7 and whose common ratio is −5.

12. The fifth term of the geometric sequence whose first term is 3 and whose common ratio is 3.

13. The fifteenth term of the arithmetic sequence whose first term is -4 and whose common difference is -4.

14. The sixth term of the geometric sequence whose first term is 5 and whose common ratio is -4.

Find the indicated term of each sequence. See Examples 3 and 8.

15. The ninth term of the arithmetic sequence $0, 12, 24, \ldots$.

16. The thirteenth term of the arithmetic sequence $-3, 0, 3, \ldots$.

17. The twenty-fifth term of the arithmetic sequence $20, 18, 16, \ldots$.

18. The ninth term of the geometric sequence $5, 10, 20, \ldots$.

19. The fifth term of the geometric sequence $2, -10, 50, \ldots$.

20. The sixth term of the geometric sequence $\dfrac{1}{2}, \dfrac{3}{2}, \dfrac{9}{2} \ldots$.

Find the indicated term of each sequence. See Examples 4 and 9.

21. The eighth term of the arithmetic sequence whose fourth term is 19 and whose fifteenth term is 52.

22. If the second term of an arithmetic sequence is 6 and the tenth term is 30, find the twenty-fifth term.

23. If the second term of an arithmetic progression is -1 and the fourth term is 5, find the ninth term.

24. If the second term of a geometric progression is 15 and the third term is 3, find a_1 and r.

25. If the second term of a geometric progression is $-\dfrac{4}{3}$ and the third term is $\dfrac{8}{3}$, find a_1 and r.

26. If the third term of a geometric sequence is 4 and the fourth term is -12, find a_1 and r.

Solve. See Examples 5 and 10.

27. An auditorium has 54 seats in the first row, 58 seats in the second row, 62 seats in the third row, and so on. Find the general term of this arithmetic sequence and the number of seats in the twentieth row.

28. A triangular display of cans in a grocery store has 20 cans in the first row, 17 cans in the next row, and so on, in an arithmetic sequence. Find the general term and the number of cans in the fifth row. Find how many rows there are in the display and how many cans are in the top row.

29. The initial size of a virus culture is 6 units and it triples its size every day. Find the general term of the geometric sequence modeling the culture's size.

30. A real estate investment broker predicts that a certain property will increase in value 15% each year. Thus the yearly property values can be modeled by a geometric sequence whose common ratio r is 1.15. If the initial property value was $500,000, write the first four terms of the sequence and predict the value at the end of the third year.

Given are the first three terms of a sequence that is either arithmetic or geometric. Based on these terms, if a sequence is arithmetic, find a_1 and d. If a sequence is geometric, find a_1 and r.

31. $2, 4, 6$

32. $8, 16, 24$

33. $5, 10, 20$

34. $2, 6, 18$

35. $\dfrac{1}{2}, \dfrac{1}{10}, \dfrac{1}{50}$

36. $\dfrac{2}{3}, \dfrac{4}{3}, 2$

37. $x, 5x, 25x$

38. $y, -3y, 9y$

39. $p, p + 4, p + 8$

40. $t, t - 1, t - 2$

Find the indicated term of each sequence.

41. The twenty-first term of the arithmetic sequence whose first term is 14 and whose common difference is $\dfrac{1}{4}$.

42. The fifth term of the geometric sequence whose first term is 8 and whose common ratio is -3.

43. The fourth term of the geometric sequence whose first term is 3 and whose common ratio is $-\dfrac{2}{3}$.

44. The fourth term of the arithmetic sequence whose first term is 9 and whose common difference is 5.

45. The fifteenth term of the arithmetic sequence $\dfrac{3}{2}, 2, \dfrac{5}{2}, \ldots$.

46. The eleventh term of the arithmetic sequence $2, \dfrac{5}{3}, \dfrac{4}{3}, \ldots$.

47. The sixth term of the geometric sequence $24, 8, \dfrac{8}{3}, \ldots$.

48. The eighteenth term of the arithmetic sequence $5, 2, -1, \ldots$.

49. If the third term of an arithmetic progression is 2 and the seventeenth term is −40, find the tenth term.

50. If the third term of a geometric sequence is −28 and the fourth term is −56, find a_1 and r.

51. A rubber ball is dropped from a height of 486 feet, and each time it bounces back one-third the height from which it last fell. Write out the first five terms of this geometric sequence and find the general term. Find how many bounces it takes for the ball to rebound less than 1 foot.

52. On the first swing, the length of the arc through which a pendulum swings is 50 inches. The length of each successive swing is 80% of the preceding swing. Determine whether this sequence is arithmetic or geometric. Find the length of the fourth swing.

53. Jose takes a job with a monthly starting salary of $1000 and guaranteeing him a monthly raise of $125 during his first year of training. Find the general term of this arithmetic sequence and his salary at the end of his training.

54. At the beginning of Claudia's exercise program, she rides 15 minutes on the Lifecycle. Each week she increases her riding time by 5 minutes. Write the general term of this arithmetic sequence, and find her riding time after 7 weeks. Find how many weeks it takes her to reach a riding time of 1 hour.

55. If an element has a half-life of 3 hours, then x grams of the element dwindles to $\dfrac{x}{2}$ grams after 3 hours. If a nuclear reactor has 400 grams of a radioactive material with a half-life of 3 hours, find the amount of radioactive material after 12 hours.

Writing in Mathematics

56. Explain why 14, 10, 6 may be the first three terms of an arithmetic sequence when it appears we are subtracting instead of adding to get the next term.

57. Explain why 80, 20, 5 may be the first three terms of a geometric sequence when it appears we are dividing instead of multiplying to get the next term.

58. Describe a situation in your life that can be modeled by a geometric sequence. Write an equation for the sequence.

11.3
Series

Tape 34

OBJECTIVES		
	1	Identify series.
	2	Use summation notation.
	3	Find partial sums.

1   A person who conscientiously saves money by saving first $100, and then saving $10 more each month than he saved the preceding month, is saving money according to the arithmetic sequence

$$a_n = 100 + 10(n - 1)$$

Following this sequence, he can predict how much money he should save for any particular month. But if he also wants to know how much money **in total** he has saved, say on the fifth month, he must find the **sum** of the first five terms of the sequence

$$\underbrace{100}_{a_1} + \underbrace{100 + 10}_{a_2} + \underbrace{100 + 20}_{a_3} + \underbrace{100 + 30}_{a_4} + \underbrace{100 + 40}_{a_5}$$

A sum of the terms of a sequence is called a **series** (plural is also series). As our example here suggests, series are frequently used to model financial and natural phenomena.

A series is a **finite series** if it is the sum of only the first k terms, for some natural number k. A series is an **infinite series** if it is the sum of all the terms. For example,

Sequence	Series	
5, 9, 13	$5 + 9 + 13$	Finite, k is 3
5, 9, 13, . . .	$5 + 9 + 13 + \cdots$	Infinite
$4, -2, 1, -\dfrac{1}{2}, \dfrac{1}{4}$	$4 + (-2) + 1 + \left(-\dfrac{1}{2}\right) + \left(\dfrac{1}{4}\right)$	Finite, k is 5
$4, -2, 1, \ldots$	$4 + (-2) + 1 + \cdots$	Infinite
$3, 6, \ldots, 99$	$3 + 6 + \cdots + 99$	Finite, k is 33

2   A shorthand notation for denoting a series when the general term of the sequence is known is called **summation notation.** The Greek uppercase letter **sigma** Σ is used to mean "sum." The expression $\displaystyle\sum_{n=1}^{5} (3n + 1)$ is read "the sum of $3n + 1$ as n goes from 1 to 5" and tells us to find the sum of the first five terms of the sequence whose general term is $a_n = 3n + 1$. Often, the variable i is used instead of n when we use summation notation: $\displaystyle\sum_{i=1}^{5} (3i + 1)$. Whether we use n, i, k, or some other variable, the variable is called the **index of summation.** The equation $i = 1$ below Σ indicates the beginning value for i, and the number 5 above the Σ indicates the ending value for i. Thus the terms of the sequence are found by successively replacing i with the natural numbers 1, 2, 3, 4, 5. To find the sum, we write out the terms and then add:

$$\sum_{i=1}^{5} (3i + 1) = (3 \cdot \boxed{1} + 1) + (3 \cdot \boxed{2} + 1) + (3 \cdot \boxed{3} + 1)$$
$$+ (3 \cdot \boxed{4} + 1) + (3 \cdot \boxed{5} + 1)$$
$$= 4 + 7 + 10 + 13 + 16 = 50$$

EXAMPLE 1   Evaluate.

a. $\displaystyle\sum_{i=0}^{6} \frac{i - 2}{2}$ **b.** $\displaystyle\sum_{i=3}^{5} 2^i$

Solution: **a.** $\displaystyle\sum_{i=0}^{6} \frac{i - 2}{2} = \frac{0 - 2}{2} + \frac{1 - 2}{2} + \frac{2 - 2}{2} + \frac{3 - 2}{2} + \frac{4 - 2}{2} + \frac{5 - 2}{2} + \frac{6 - 2}{2}$

$$= (-1) + \left(-\frac{1}{2}\right) + 0 + \frac{1}{2} + 1 + \frac{3}{2} + 2$$

$$= \frac{7}{2} \quad \text{or} \quad 3\frac{1}{2}$$

b. $\displaystyle\sum_{i=3}^{5} 2^i = 2^3 + 2^4 + 2^5$

$$= 8 + 16 + 32$$
$$= 56 \quad \blacksquare$$

EXAMPLE 2   Write each series using summation notation.

a. $3 + 6 + 9 + 12 + 15$ **b.** $\dfrac{1}{2} + \dfrac{1}{4} + \dfrac{1}{8} + \dfrac{1}{16}$

Solution: **a.** Since each term is the **sum** of the preceding term and 3, the terms correspond to the first five terms of an arithmetic sequence with $a_1 = 3$, $d = 3$, and $a_n = 3 + (n - 1)3$. Thus

$$3 + 6 + 9 + 12 + 15 = \sum_{i=1}^{5} 3 + (i - 1)3$$

b. Since each term is the **product** of the preceding term and $\frac{1}{2}$, these terms correspond to the first four terms of a geometric sequence. Here $a_1 = \frac{1}{2}$, $r = \frac{1}{2}$, and $a_n = \left(\frac{1}{2}\right)\left(\frac{1}{2}\right)^{n-1} = \left(\frac{1}{2}\right)^n$. So

$$\frac{1}{2} + \frac{1}{4} + \frac{1}{8} + \frac{1}{16} = \sum_{i=1}^{4} \left(\frac{1}{2}\right)^i \quad \blacksquare$$

3   If we want the sum of the first n terms of a sequence, we find a **partial sum, S_n.** Thus

$$S_1 = a_1$$
$$S_2 = a_1 + a_2$$
$$S_3 = a_1 + a_2 + a_3$$

and so on. In general, S_n is the sum of the first n terms of a sequence.

EXAMPLE 3   Find the sum of the first three terms of the sequence whose general term is $a_n = \dfrac{n + 3}{2n}$.

Solution: $S_3 = \displaystyle\sum_{i=1}^{3} \frac{i + 3}{2i} = 2 + \frac{5}{4} + 1 = 4\frac{1}{4}.$ $\blacksquare$

The next example illustrates how these sums model real-life phenomena.

EXAMPLE 4   The number of babies born per year in the San Diego Zoo's gorilla house is a sequence defined by $a_n = n(n - 1)$, where n is the number of the year. Find the **total** number of baby gorillas born in the **first 4 years.**

Solution: To solve, find the sum,

$$\sum_{i=1}^{4} i(i - 1)$$

$$\sum_{i=1}^{4} i(i - 1) = 0 + 2 + 6 + 12 = 20$$

There are 20 gorillas born in the first 4 years. $\blacksquare$

EXERCISE SET 11.3

Evaluate. See Example 1.

1. $\displaystyle\sum_{i=1}^{4} (i - 3)$

2. $\displaystyle\sum_{i=1}^{5} (i + 6)$

3. $\displaystyle\sum_{i=4}^{7} (2i + 4)$

4. $\displaystyle\sum_{i=2}^{3} (5i - 1)$

5. $\displaystyle\sum_{i=2}^{4} (i^2 - 3)$

6. $\displaystyle\sum_{i=3}^{5} i^3$

7. $\displaystyle\sum_{i=1}^{3} \left(\frac{1}{i + 5}\right)$

8. $\displaystyle\sum_{i=2}^{4} \left(\frac{2}{i + 3}\right)$

Write each series using summation notation. See Example 2.

9. $1 + 3 + 5 + 7 + 9$

10. $4 + 7 + 10 + 13$

11. $4 + 12 + 36 + 108$

12. $5 + 10 + 20 + 40 + 80 + 160$

13. $12 + 9 + 6 + 3 + 0 + (-3)$

14. $5 + 1 + (-3) + (-7)$

Find each partial sum. See Example 3.

15. Find the sum of the first four terms of the sequence whose general term is $a_n = -2n$.

16. Find the sum of the first three terms of the sequence whose general term is $a_n = -\dfrac{n}{3}$.

17. Find the sum of the first five terms of the sequence whose general term is $a_n = (n - 1)^2$.

18. Find the sum of the first three terms of the sequence whose general term is $a_n = (n + 4)^2$.

Solve. See Example 4.

19. A gardener is making a triangular planting with 1 tree in the first row, 2 trees in the second row, 3 trees in the third row, and so on for 10 rows. Write the sequence describing the trees in each row. Find the total number of trees planted.

20. Some surfers at the beach form a human pyramid with 2 surfers in the top row, 3 surfers in the second row, 4 surfers in third row, and so on. If there are 6 rows in the pyramid, write the sequence describing the number of surfers in each row in the pyramid. Find the total number of surfers.

21. A fungus culture starts with 6 units and doubles every day. Write the sequence describing the growth of this fungus. Find the number of fungus units at the beginning of the fifth day.

22. A bacteria colony begins with 100 bacteria and doubles every 6 hours. Write the sequence describing the growth of the bacteria. Find the number of bacteria after 24 hours.

Evaluate.

23. $\displaystyle\sum_{i=1}^{3} \frac{1}{6i}$

24. $\displaystyle\sum_{i=1}^{3} \frac{1}{3i}$

25. $\displaystyle\sum_{i=2}^{6} 3i$

26. $\displaystyle\sum_{i=3}^{6} -4i$

27. $\displaystyle\sum_{i=3}^{5} i(i + 2)$

28. $\displaystyle\sum_{i=2}^{4} i(i - 3)$

29. $\displaystyle\sum_{i=1}^{5} 2^i$

30. $\displaystyle\sum_{i=1}^{4} 3^{i-1}$

31. $\displaystyle\sum_{i=1}^{4} \frac{4i}{i + 3}$

32. $\displaystyle\sum_{i=2}^{5} \frac{6 - i}{6 + i}$

Write each series using summation notation.

33. $12 + 4 + \dfrac{4}{3} + \dfrac{4}{9}$

34. $80 + 20 + 5 + \dfrac{5}{4} + \dfrac{5}{16}$

35. $1 + 4 + 9 + 16 + 25 + 36 + 49$

36. $1 + (-4) + 9 + (-16)$

Find each partial sum.

37. Find the sum of the first two terms of the sequence whose general term is $a_n = (n + 2)(n - 5)$.

38. Find the sum of the first two terms of the sequence whose general term is $a_n = n(n - 6)$.

39. Find the sum of the first six terms of the sequence whose general term is $a_n = (-1)^n$.

40. Find the sum of the first seven terms of the sequence whose general term is $a_n = (-1)^{n-1}$.

41. Find the sum of the first four terms of the sequence whose general term is $a_n = (n + 3)(n + 1)$.

42. Find the sum of the first five terms of the sequence whose general term is $a_n = \dfrac{(-1)^n}{2n}$.

43. A bacteria colony begins with 50 bacteria and doubles every 12 hours. Write the sequence describing the growth of the bacteria. Find the number of bacteria after 48 hours.

44. The number of otters born each year in a new aquarium forms a sequence whose general term is $a_n = (n - 1)(n + 3)$. Find the number of otters born in the third year, and find the total number of otters born in the first three years.

45. The number of opossums killed each month on a new highway describes the sequence whose general term is $a_n = (n + 1)(n + 2)$, where n is the number of the months. Find the number of opossums killed in the fourth month, and find the total number killed in the first four months.

46. In 1988 the population of an endangered fish was estimated by environmentalists to be decreasing each year. The size of the population in a given year is $24 - 4n$ thousand fish fewer than the previous year. Find the decrease in population in 1990, if year 1 is 1988. Find how many total fish died from 1988 through 1990.

47. The amount of decay in pounds of a radioactive isotope each year is given by the sequence whose general term is $a_n = 100(0.5)^n$, where n is the number of the year. Find the amount of decay in the fourth year, and find the total amount of decay in the first four years.

48. Susan has a choice between two job offers. Job A has an annual starting salary of $20,000 with guaranteed annual raises of $1200 for the next four years, while job B has an annual starting salary of $18,000 with guaranteed annual raises of $2500 for the next four years. Compare the fifth partial sums for each sequence to determine which job would pay Susan more money over the next 5 years.

49. A pendulum swings a length of 40 inches on its first swing. Each successive swing is $\dfrac{4}{5}$ of the preceding swing. Find the length of the fifth swing, and find the total length swung during the first five swings. (Round to the nearest tenth of an inch.)

Writing in Mathematics

50. Explain the difference between a sequence and a series.

11.4
Partial Sums of Arithmetic and Geometric Sequences

OBJECTIVES

Tape 34

1 Find the partial sum of an arithmetic sequence.

2 Find the partial sum of a geometric sequence.

3 Find the infinite series of a geometric sequence.

1 Partial sums S_n are relatively easy to find when n is small, that is, when the number of terms to add is small. But when n is large, finding S_n can be tedious. For large n, S_n is still relatively easy to find if the addends are terms of an arithmetic sequence or a geometric sequence.

For an arithmetic sequence, $a_n = a_1 + (n - 1)d$ for some first term a_1 and some common difference d. So S_n, the sum of the first n terms, is

$$S_n = a_1 + (a_1 + d) + (a_1 + 2d) + \cdots + a_n$$

We might also find S_n by "working backward" from the nth term a_n, finding the preceding term a_{n-1}, by subtracting d each time.

$$S_n = a_n + (a_n - d) + (a_n - 2d) + \cdots + a_1$$

Now add left sides of these two equations, and add right sides:

$$2S_n = (a_1 + a_n) + (a_1 + a_n) + (a_1 + a_n) + \cdots + (a_1 + a_n)$$

The d terms subtract out, leaving n sums of the first term a_1 and last term a_n. Thus we write

$$2S_n = n(a_1 + a_n)$$

or

$$S_n = \frac{n}{2}(a_1 + a_n)$$

Partial Sum S_n of an Arithmetic Sequence

The partial sum S_n of the first n terms of an arithmetic sequence is given by

$$S_n = \frac{n}{2}(a_1 + a_n)$$

where a_1 is the first term of the sequence and a_n is the nth term.

EXAMPLE 1   Use the partial sum formula to find the sum of the first six terms of the arithmetic sequence 2, 5, 8, 11, 14, 17, ...

Solution: Use the formula for S_n of an arithmetic sequence, replacing n with 6, a_1 with 2, and a_n with 17.

$$S_n = \frac{n}{2}(a_1 + a_n) = \frac{6}{2}(2 + 17) = 3(19) = 57 \quad \blacksquare$$

EXAMPLE 2   Find the sum of the first 30 positive integers.

Solution: Because 1, 2, 3, . . . , 30 is an arithmetic sequence, use the formula for S_n with $n = 30$, $a_1 = 1$, and $a_n = 30$. Thus

$$S_n = \frac{n}{2}(a_1 + a_n) = \frac{30}{2}(1 + 30) = 15(31) = 465 \quad \blacksquare$$

EXAMPLE 3   Rolls of carpet are stacked in 20 rows with 3 rolls in the top row, 4 rolls in the next row, and so on, forming an arithmetic sequence. Find the total number of carpet rolls if there are 22 rolls in the last row.

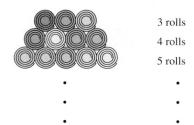

3 rolls

4 rolls

5 rolls

Solution: The list 3, 4, 5, . . . , 22 is the first 20 terms of an arithmetic sequence. Use the formula for S_n with $a_1 = 3$, $a_n = 22$, and $n = 20$ terms. Thus,

$$S_{20} = \frac{20}{2}(3 + 22) = 10(25) = 250$$

There are a total of 250 rolls of carpet in the display.   ■

2   It is also useful to have a formula for the partial sum of the first n terms of a geometric series. To derive the formula, we write

$$S_n = a_1 + a_1 r + a_1 r^2 + \cdots + a_1 r^{n-1}$$

$$\uparrow \quad \uparrow \quad \uparrow \qquad\qquad \uparrow$$

1st   2nd   3rd nth
term   term   term   term

Multiply each side of the equation by $-r$:

$$-rS_n = -a_1 r - a_1 r^2 - a_1 r^3 - \cdots - a_1 r^n$$

Add the two equations.

$$S_n - rS_n = a_1 + (a_1 r - a_1 r) + (a_1 r^2 - a_1 r^2) + (a_1 r^3 - a_1 r^3) + \cdots - a_1 r^n$$

$$S_n - rS_n = a_1 - a_1 r^n$$

Now factor each side.

$$(1 - r)S_n = a_1 (1 - r^n)$$

Solve for S_n by dividing both sides by $1 - r$.

$$S_n = \frac{a_1(1 - r^n)}{1 - r}$$

as long as r is not 1.

Partial Sum S_n of a Geometric Sequence

The partial sum S_n of the first n terms of a geometric sequence, whose first term is a_1, is given by

$$S_n = \frac{a_1(1 - r^n)}{1 - r}$$

as long as the common ratio of the sequence r is not 1.

EXAMPLE 4   Find the sum of the first six terms of the geometric sequence 5, 10, 20, 40, 80, 160.

Solution: Use the S_n formula for the sum of the terms of a geometric sequence. Here, $n = 6$, the first term $a_1 = 5$, and the common ratio $r = 2$.

$$S_n = \frac{a_1(1 - r^n)}{1 - r}$$

$$S_6 = \frac{5(1 - 2^6)}{1 - 2} = \frac{5(-63)}{-1} = 315 \quad \blacksquare$$

EXAMPLE 5   A grant from an alumnus to a university specified that the university will receive $800,000 during the first year and 75% of the preceding year's donation during each of the next five years. Find the total amount donated during the 6 years.

Solution: The donations model the first six terms of a geometric sequence. Evaluate S_n when $n = 6$, $a_1 = 800{,}000$, and $r = 0.75$.

$$S_6 = \frac{800{,}000[1 - (0.75)^6]}{1 - 0.75}$$

$$= \$2{,}630{,}468.75$$

The total amount donated during the 6 years is $2,630,468.75. $\blacksquare$

3   Is it possible to find the sum of all the terms of an infinite sequence? Examine the partial sums of the geometric sequence $\dfrac{1}{2}, \dfrac{1}{4}, \dfrac{1}{8}, \ldots$.

$$S_1 = \frac{1}{2}$$

$$S_2 = \frac{1}{2} + \frac{1}{4} = \frac{3}{4}$$

$$S_3 = \frac{1}{2} + \frac{1}{4} + \frac{1}{8} = \frac{7}{8}$$

$$S_4 = \frac{1}{2} + \frac{1}{4} + \frac{1}{8} + \frac{1}{16} = \frac{15}{16}$$

$$S_5 = \frac{1}{2} + \frac{1}{4} + \frac{1}{8} + \frac{1}{16} + \frac{1}{32} = \frac{31}{32}$$

$$\vdots$$

$$S_{10} = \frac{1}{2} + \frac{1}{4} + \frac{1}{8} + \cdots + \frac{1}{2^{10}} = \frac{1023}{1024}$$

Even though each partial sum is larger than the preceding partial sum, we see that each partial sum is closer to 1 than the preceding partial sum. If n gets larger and larger, then S_n gets closer and closer to 1. In general, if $|r| < 1$, the following formula will give the sum of an infinite geometric series.

> **Sum of the Terms of an Infinite Geometric Sequence**
>
> The sum of the terms of an infinite geometric sequence with first term a_1 and common ratio r with $|r| < 1$ is given by S_∞, where
>
> $$S_\infty = \frac{a_1}{1 - r}$$
>
> If $|r| \geq 1$, S_∞ does not exist.

What happens for other values of r? For example, in the following geometric sequence $r = 3$.

$$6, 18, 54, 162, \ldots$$

Here, as n increases, the sum S_n increases also. This time, though, S_n does not approach a number but will increase without bound.

EXAMPLE 6   Find the sum of the terms of the geometric sequence $2, \dfrac{2}{3}, \dfrac{2}{9}, \dfrac{2}{27}, \ldots$.

Solution: For this geometric sequence, $r = \dfrac{1}{3}$. Since $|r| < 1$, we may use the formula S_∞ of a geometric sequence with $a_1 = 2$ and $r = \dfrac{1}{3}$.

$$S_\infty = \frac{a_1}{1 - r} = \frac{2}{1 - \dfrac{1}{3}} = \frac{2}{\dfrac{2}{3}} = 3 \qquad \blacksquare$$

The formula for the sum of the terms of an infinite geometric sequence can be used to write a repeating decimal as a fraction. For example,

$$0.33\overline{3} = \frac{3}{10} + \frac{3}{100} + \frac{3}{1000} + \cdots$$

This sum is the sum of the terms of an infinite geometric sequence whose first term $a_1 = \dfrac{3}{10}$ and whose common ratio $r = \dfrac{1}{10}$. Using the formula for S_∞,

$$S_\infty = \frac{a_1}{1 - r} = \frac{\dfrac{3}{10}}{1 - \dfrac{1}{10}} = \frac{1}{3}$$

So $0.33\overline{3} = \dfrac{1}{3}$.

EXAMPLE 7   A pendulum swings through an arc of 24 inches on its first pass. On each pass thereafter, the length of the arc is 75% of the length of the arc on the preceding pass. Find the total distance the pendulum travels before it comes to rest.

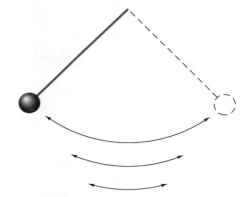

Solution: We must find the sum of the terms of an infinite geometric sequence whose first term a_1 is 24 and whose common ratio r is 0.75. Since $|r| < 1$, we may use the formula for S_∞.

$$S_\infty = \frac{a_1}{1 - r} = \frac{24}{1 - 0.75} = \frac{24}{0.25} = 96$$

The pendulum travels a total distance of 96 inches before it comes to rest. ■

EXERCISE SET 11.4

Use the partial sum formula to find the partial sum of the given arithmetic sequence. See Example 1.

1. Find the sum of the first six terms of the arithmetic sequence 1, 3, 5, 7, ...

2. Find the sum of the first seven terms of the arithmetic sequence $-7, -11, -15, ...$

3. Find the sum of the first six terms of the arithmetic sequence 3, 6, 9, ...

4. Find the sum of the first four terms of the arithmetic sequence $-4, -8, -12, ...$

Solve. See Example 2.

5. Find the sum of the first ten positive integers.

6. Find the sum of the first eight negative integers.

7. Find the sum of the first four positive odd integers.

8. Find the sum of the first five negative odd integers.

Solve. See Example 3.

9. Modern Car Company has come out with a new car model. Market analysts predict 4000 cars will be sold in the first month and that sales will drop by 50 cars per month after that during the first year. Write out the first five terms of the sequence, and find the number of sold cars predicted for the twelfth month. Find the total number of sold cars predicted for the first year.

10. A fax company charges $3 for the first page sent and $0.10 less than the preceding for each additional page sent. The cost per page forms an arithmetic sequence. Write the first five terms of this sequence, and use a partial sum to find the cost of sending a nine-page document.

11. Sal has two job offers: Firm *A* starts at $22,000 per year and guarantees raises of $1000 per year, while Firm *B* starts at $20,000 and guarantees raises of $1200 per year. Over a 10-year period, determine the more profitable offer.

12. The game of pool uses 15 balls numbered 1 to 15. In the game of rotation, when a player sinks a ball, the player receives as many points as the number on the ball. Use an arithmetic series to find the score of a player who sinks all 15 balls.

Use the partial sum formula to find the partial sum of the given geometric sequence. See Example 4.

13. Find the sum of the first five terms of the geometric sequence 4, 12, 36, ...

14. Find the sum of the first eight terms of the geometric sequence -1, 2, -4, ...

15. Find the sum of the first four terms of the geometric sequence $2, \dfrac{2}{5}, \dfrac{2}{25}, \ldots$.

16. Find the sum of the first five terms of the geometric sequence $\dfrac{1}{3}, -\dfrac{2}{3}, \dfrac{4}{3}, \ldots$.

Solve. See Example 5.

17. A woman made $30,000 during the first year she owned her business and made an additional 10% over the previous year in each subsequent year. Find how much she made during her fourth year of business. Find her total earnings during the first four years.

18. A parachutist in free fall falls 16 feet during the first second, 48 feet during the second second, 80 feet during the third second, and so on. Find how far she falls during the eighth second. Find the total distance she falls during the first 8 seconds.

19. A trainee in a computer company takes 0.9 times as long to assemble each computer as he took to assemble the preceding computer. If it took him 30 minutes to assemble the first computer, find how long it takes him to assemble the fifth computer. Find the total time he takes to assemble the first five computers (round to the nearest minute).

20. On a gambling trip to Reno, Carol doubled her bet each time she lost. If her first losing bet was $5 and she lost six consecutive bets, find how much she lost on the sixth bet. Find the total amount lost on these six bets.

Find the sum of the terms of each infinite geometric sequence. See Example 6.

21. 12, 6, 3, . . .

22. 45, 15, 5, . . .

23. $\dfrac{1}{10}, \dfrac{1}{100}, \dfrac{1}{1000}, \ldots$

24. $\dfrac{3}{5}, \dfrac{3}{20}, \dfrac{3}{80}, \ldots$

Solve. See Example 7.

25. A ball is dropped from a height of 20 feet and repeatedly rebounds to a height that is $\dfrac{4}{5}$ of its previous height. Find the total distance the ball covers before it comes to rest.

26. A rotating flywheel coming to rest makes 300 revolutions in the first minute, and in each minute thereafter makes $\dfrac{2}{5}$ as many revolutions as in the preceding minute. Find how many revolutions the wheel makes before coming to rest.

Solve.

27. Find the sum of the first ten terms of the sequence -4, 1, 6, ...

28. Find the sum of the first twelve terms of the sequence -3, -13, -23, ...

29. Find the sum of the first seven terms of the sequence $3, \dfrac{3}{2}, \dfrac{3}{4}, \ldots$.

30. Find the sum of the first five terms of the sequence -2, -6, -18, ...

31. Find the sum of the first five terms of the sequence -12, 6, -3, ...

32. Find the sum of the first four terms of the sequence $-\dfrac{1}{4}, -\dfrac{3}{4}, -\dfrac{9}{4}, \ldots$.

33. Find the sum of the first twenty terms of the sequence $\dfrac{1}{2}, \dfrac{1}{4}, 0, \ldots$.

34. Find the sum of the first fifteen terms of the sequence -5, -9, -13, ...

35. If a_1 is 8 and r is $-\dfrac{2}{3}$, find S_3.

36. If a_1 is 10 and d is $-\dfrac{1}{2}$, find S_{18}.

Find the sum of the terms of each infinite geometric sequence.

37. $-10, -5, -\dfrac{5}{2}, \ldots$

38. $-16, -4, -1, \ldots$

39. $2, -\dfrac{1}{4}, \dfrac{1}{32}, \ldots$

40. $-3, \dfrac{3}{5}, -\dfrac{3}{25}, \ldots$

41. $\dfrac{2}{3}, -\dfrac{1}{3}, \dfrac{1}{6}, \ldots$

42. $6, -4, \dfrac{8}{3}, \ldots$

Solve.

43. In the pool game of rotation, player A sinks balls numbered 1 to 9, and player B sinks the rest of the balls. Use arithmetic series to find each player's score (see Exercise 12).

44. A godfather deposited \$250 in a savings account on the day his godchild was born. On each subsequent birthday he deposited \$50 more than he deposited the previous year. Find how much money he deposited on his godchild's twenty-first birthday. Find the total amount deposited over the 21 years.

45. During the holiday rush a business can rent a computer system for \$200 the first day, with the rent decreasing \$5 for each additional day. Find how much rent is paid for 20 days during the holiday rush.

46. Spraying a field with insecticide killed 6400 weevils the first day, 1600 the second day, 400 the third day, and so

on. Find the total number of weevils killed during the first 5 days.

47. A college student humorously asks his parents to charge him room and board according to this geometric sequence: \$0.01 for the first day of the month, \$0.02 for the second day, \$0.04 for the third day, and so on. Find the total room and board he would pay for 30 days.

48. A bank attracted 80 new customers the first day following its television advertising campaign, 120 the second day, 160 the third day, and so on, in an arithmetic sequence. Find how many new customers were attracted during the first 5 days following its television campaign.

49. Write $0.88\overline{8}$ as an infinite geometric series and use the formula for S_∞ to write it as a rational number.

50. Write 0.5454 as an infinite geometric series and use the formula S_∞ to write it as a rational number.

Writing in Mathematics

51. Explain whether the sequence $5, 5, 5, \ldots$ is arithmetic, geometric, neither, or both.

52. Describe a situation in everyday life that can be modeled by an infinite geometric series.

11.5
The Binomial Theorem

Tape 35

OBJECTIVES

1 Use Pascal's triangle to expand binomials.

2 Evaluate factorials.

3 Use the binomial theorem to expand binomials.

4 Find the nth term in the expansion of a binomial raised to a positive power.

In this section, we learn how to easily expand binomials of the form $(a + b)^n$. First, we review the patterns in the expansions of $(a + b)^n$.

$$(a + b)^0 = 1 \qquad\qquad\qquad \text{1 term}$$

$$(a + b)^1 = a + b \qquad\qquad\quad \text{2 terms}$$

$$(a + b)^2 = a^2 + 2ab + b^2 \qquad \text{3 terms}$$

$$(a + b)^3 = a^3 + 3a^2b + 3ab^2 + b^3 \qquad\qquad\qquad\qquad \text{4 terms}$$

$$(a + b)^4 = a^4 + 4a^3b + 6a^2b^2 + 4ab^3 + b^4 \qquad\qquad \text{5 terms}$$

$$(a + b)^5 = a^5 + 5a^4b + 10a^3b^2 + 10a^2b^3 + 5ab^4 + b^5 \qquad \text{6 terms}$$

Notice the following patterns:

1. The expansion of $(a + b)^n$ contains $n + 1$ terms. For example, for $(a + b)^3$, $n = 3$, and the expansion contains $3 + 1$ or 4 terms.
2. The first term of the expansion of $(a + b)^n$ is a^n and the last term is b^n.
3. The powers of a decrease by 1 for each term, while the powers of b increase by 1 for each term.
4. The sum of the variable exponents for each term is n.

1   There are more patterns in the coefficients of the terms as well. Written in a triangular array, the coefficients are called **Pascal's triangle.**

$$
\begin{array}{lccccccc}
(a + b)^0: & & & & 1 & & & \\
(a + b)^1: & & & 1 & & 1 & & \\
(a + b)^2: & & 1 & & 2 & & 1 & \\
(a + b)^3: & 1 & & 3 & & 3 & & 1 \\
(a + b)^4: & 1 & 4 & & 6 & & 4 & 1 \\
(a + b)^5: & 1 & 5 & 10 & & 10 & 5 & 1 \\
\end{array}
$$

Each row in Pascal's triangle begins and ends with 1. Any other number is the sum of the two closest numbers above it. For example, we can write the next row, the seventh row, by beginning with 1, adding the two numbers closest above in row 6, and ending with 1.

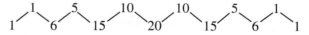

We can use Pascal's triangle and the patterns noted to expand $(a + b)^n$ without actually multiplying.

EXAMPLE 1   Expand $(a + b)^6$.

Solution: Using the seventh row of Pascal's triangle as the coefficients and following the patterns noted, $(a + b)^6$ can be expanded as

$$a^6 + 6a^5b + 15a^4b^2 + 20a^3b^3 + 15a^2b^4 + 6ab^5 + b^6 \qquad \blacksquare$$

2   For large n, using Pascal's triangle to find coefficients for $(a + b)^n$ can be tedious. An alternative method for determining these coefficients is based on the concept of a **factorial.** A **factorial of n,** written $n!$ (read "n factorial") is the product of the first n consecutive natural numbers.

> **Factorial of n: $n!$**
>
> If n is a natural number, then $n! = n(n-1)(n-2)(n-3) \cdot \cdots \cdot 3 \cdot 2 \cdot 1$. The factorial of 0, written 0!, is defined to be 1.

For example, $3! = 3 \cdot 2 \cdot 1 = 6$, $5! = 5 \cdot 4 \cdot 3 \cdot 2 \cdot 1 = 120$, and $0! = 1$.

EXAMPLE 2   Evaluate each expression.

    a. $\dfrac{5!}{6!}$ **b.** $\dfrac{10!}{7!3!}$

    c. $\dfrac{3!}{2!1!}$ **d.** $\dfrac{7!}{7!0!}$

Solution: **a.** $\dfrac{5!}{6!} = \dfrac{5 \cdot 4 \cdot 3 \cdot 2 \cdot 1}{6 \cdot 5 \cdot 4 \cdot 3 \cdot 2 \cdot 1} = \dfrac{1}{6}$

    b. $\dfrac{10!}{7!3!} = \dfrac{10 \cdot 9 \cdot 8 \cdot \boxed{7!}}{\boxed{7!} \cdot 3 \cdot 2 \cdot 1} = \dfrac{10 \cdot 9 \cdot 8}{3 \cdot 2 \cdot 1}$

         $= 10 \cdot 3 \cdot 4 = 120$

    c. $\dfrac{3!}{2!1!} = \dfrac{3 \cdot 2 \cdot 1}{2 \cdot 1 \cdot 1} = 3$

    d. $\dfrac{7!}{7!0!} = \dfrac{7!}{7! \cdot 1} = 1$ ∎

> **HELPFUL HINT**
>
> We can use a calculator with a factorial key to evaluate the factorial. A calculator uses scientific notation for large results.

3   It can be proved, though we won't do so here, that the coefficients of terms in the expansion of $(a + b)^n$ can be expressed in terms of factorials. Following patterns 1 through 4 given earlier and using the factorial expressions of the coefficients, we have the **binomial theorem.**

> **Binomial Theorem**
>
> If n is a positive integer, then
>
> $$(a + b)^n = a^n + \frac{n}{1!}a^{n-1}b^1 + \frac{n(n-1)}{2!}a^{n-2}b^2 + \frac{n(n-1)(n-2)}{3!}a^{n-3}b^3$$
> $$+ \cdots + b^n$$

We call the formula for $(a + b)^n$ given by the binomial theorem the **binomial formula.**

EXAMPLE 3   Use the binomial theorem to expand $(x + y)^{10}$.

Solution: Let $a = x$, $b = y$, and $n = 10$ in the binomial formula.

$$(x + y)^{10} = x^{10} + \frac{10}{1!}x^9 y + \frac{10 \cdot 9}{2!}x^8 y^2 + \frac{10 \cdot 9 \cdot 8}{3!}x^7 y^3 + \frac{10 \cdot 9 \cdot 8 \cdot 7}{4!}x^6 y^4$$

$$+ \frac{10 \cdot 9 \cdot 8 \cdot 7 \cdot 6}{5!}x^5 y^5 + \frac{10 \cdot 9 \cdot 8 \cdot 7 \cdot 6 \cdot 5}{6!}x^4 y^6$$

$$+ \frac{10 \cdot 9 \cdot 8 \cdot 7 \cdot 6 \cdot 5 \cdot 4}{7!}x^3 y^7$$

$$+ \frac{10 \cdot 9 \cdot 8 \cdot 7 \cdot 6 \cdot 5 \cdot 4 \cdot 3}{8!}x^2 y^8$$

$$+ \frac{10 \cdot 9 \cdot 8 \cdot 7 \cdot 6 \cdot 5 \cdot 4 \cdot 3 \cdot 2}{9!}xy^9 + y^{10}$$

$$= x^{10} + 10x^9 y + 45x^8 y^2 + 120x^7 y^3 + 210x^6 y^4 + 252x^5 y^5 + 210x^4 y^6$$

$$+ 120x^3 y^7 + 45x^2 y^8 + 10xy^9 + y^{10} \quad \blacksquare$$

EXAMPLE 4   Use the binomial theorem to expand $(x + 2y)^5$.

Solution: Let $a = x$ and $b = 2y$ in the binomial formula.

$$(x + 2y)^5 = x^5 + \frac{5}{1!}x^4(2y) + \frac{5 \cdot 4}{2!}x^3(2y)^2 + \frac{5 \cdot 4 \cdot 3}{3!}x^2(2y)^3$$

$$+ \frac{5 \cdot 4 \cdot 3 \cdot 2}{4!}x(2y)^4 + (2y)^5$$

$$= x^5 + 10x^4 y + 40x^3 y^2 + 80x^2 y^3 + 80xy^4 + 32y^5 \quad \blacksquare$$

EXAMPLE 5   Use the binomial theorem to expand $(3m - n)^4$.

Solution: Let $a = 3m$ and $b = -n$ in the binomial formula.

$$(3m - n)^4 = (3m)^4 + \frac{4}{1!}(3m)^3(-n) + \frac{4 \cdot 3}{2!}(3m)^2(-n)^2$$

$$+ \frac{4 \cdot 3 \cdot 2}{3!}(3m)(-n)^3 + (-n)^4$$

$$= 81m^4 - 108m^3 n + 54m^2 n^2 - 12mn^3 + n^4 \quad \blacksquare$$

4   Sometimes it is convenient to find a specific term of a binomial expansion without writing out the entire expansion. By studying the expansion of binomials, a pattern forms for each term. This pattern is most easily stated for the $(r + 1)$st term.

> **$(r + 1)$st Term in a Binomial Expansion**
>
> The $(r + 1)$st term of the expansion of $(a + b)^n$ is $\dfrac{n!}{r!(n - r)!}a^{n-r}b^r$.

EXAMPLE 6   Find the eighth term in the expansion of $(2x - y)^{10}$.

Solution: Use the formula, with $n = 10$, $a = 2x$, $b = -y$, and $r + 1 = 8$. Notice that, since $r + 1 = 8$, $r = 7$.

$$\frac{n!}{r!(n - r)!}a^{n-r}b^r = \frac{10!}{7!3!}(2x)^3(-y)^7$$
$$= 120(8x^3)(-y)^7$$
$$= -960x^3y^7 \quad \blacksquare$$

EXERCISE SET 11.5

Use Pascal's triangle to expand the binomial. See Example 1.

1. $(m + n)^3$ **2.** $(x + y)^4$ **3.** $(c + d)^5$

4. $(a + b)^6$ **5.** $(y - x)^5$ **6.** $(q - r)^7$

Evaluate each expression. See Example 2.

7. $\dfrac{8!}{7!}$ **8.** $\dfrac{6!}{0!}$ **9.** $\dfrac{7!}{5!}$ **10.** $\dfrac{8!}{5!}$

11. $\dfrac{10!}{7!2!}$ **12.** $\dfrac{9!}{5!3!}$ **13.** $\dfrac{8!}{6!0!}$ **14.** $\dfrac{10!}{4!6!}$

Use the binomial formula to expand each binomial. See Examples 3 and 4.

15. $(a + b)^7$ **16.** $(x + y)^8$ **17.** $(a + 2b)^5$ **18.** $(x + 3y)^6$

Use the binomial formula to expand each binomial. See Example 5.

19. $(2a - b)^5$ **20.** $(5x - y)^4$ **21.** $(c - 2d)^6$ **22.** $(m - 3n)^4$

Find the indicated term. See Example 6.

23. The third term of the expansion of $(x + y)^4$

24. The fourth term of the expansion of $(a + b)^8$

25. The second term of the expansion of $(a + 3b)^{10}$

26. The third term of the expansion of $(m + 5n)^7$

Expand each binomial.

27. $(q + r)^9$

28. $(b + c)^6$

29. $(4a + b)^5$

30. $(3m + n)^4$

31. $(5a - 2b)^4$

32. $(m - 4)^6$

33. $(2a + 3b)^3$

34. $(4 - 3x)^5$

35. $(x + 2)^5$

36. $(3 + 2a)^4$

Find the indicated term.

37. The fifth term of the expansion of $(c - d)^5$

38. The fourth term of the expansion of $(x - y)^6$

39. The eighth term of the expansion of $(2c + d)^7$

40. The tenth term of the expansion of $(5x - y)^9$

41. The fourth term of the expansion of $(2r - s)^5$

42. The first term of the expansion of $(3q - 7r)^6$

Writing in Mathematics

43. Explain how to generate a row of Pascal's triangle.

CRITICAL THINKING   People who investigate their family trees often underestimate the enormity of the task. For example, the number of your ancestors in any generation preceding you describes a sequence 2, 4, 8, 16, 32, . . . so that in the *n*th generation preceding you there are 2^n ancestors to investigate.

This sequence is merely a result of the fact that each person has two parents. The ancestors of a honeybee describe quite a different sequence because a drone bee hatches from an unfertilized queen bee egg, and therefore has only a single parent. The queen bee, however, hatches from a fertilized egg, and therefore has two parents, one a drone and one a queen.

Can you find an equation that describes the number a_n of ancestors of a single drone in the *n*th generation preceding the drone?

CHAPTER 11 GLOSSARY

An **arithmetic sequence** is a sequence in which each term after the first differs from the preceding term by a constant amount d. The constant d is called the **common difference**.

The **domain** of a sequence is the set of natural numbers.

A series is a **finite series** if it is the sum of only the first k terms, for some natural number k.

A **geometric sequence** is a sequence for which each term after the first is obtained by multiplying the preceding term by a constant r. The constant r is called the **common ratio**.

A series is an **infinite series** if it is the sum of all the terms of a sequence.

An **infinite sequence** or sequence is a function whose domain is the set of natural numbers.

Written in a triangular array, the coefficients of the expansions of the binomial $(a + b)^n$ for $n = 1, 2, 3$, and so on, are called **Pascal's triangle.**

A shorthand notation for denoting a series when the general term of the sequence is known is called **summation notation.** The Greek uppercase letter **sigma** Σ is used to mean "sum."

The **range** of a sequence is the set of function values a_1, a_2, a_3, ... These values are called the **terms** of the sequence.

A sum of the terms of a sequence is called a **series** (plural is also series).

CHAPTER 11 SUMMARY

GENERAL TERM OF AN ARITHMETIC SEQUENCE   (8.2)

The general term a_n of an arithmetic sequence is given by

$$a_n = a_1 + (n - 1)d$$

where a_1 is the first term and d is the common difference.

GENERAL TERM OF A GEOMETRIC SEQUENCE   (8.2)

The general term of a geometric sequence is given by

$$a_n = a_1 r^{n-1}$$

where a_1 is the first term and r is the common ratio.

PARTIAL SUM S_n OF AN ARITHMETIC SEQUENCE   (8.4)

The partial sum S_n of the first n terms of an arithmetic sequence is given by

$$S_n = \frac{n}{2}(a_1 + a_n)$$

PARTIAL SUM S_n OF A GEOMETRIC SEQUENCE   (8.4)

The partial sum S_n of the first n terms of a geometric series whose first term is a_1 is given by

$$S_n = \frac{a_1(1 - r^n)}{1 - r}$$

as long as the common ratio of the sequence r is not 1.

SUM OF THE TERMS OF AN INFINITE GEOMETRIC SEQUENCE   (8.4)

The sum of the terms of an infinite geometric sequence with first term a_1 and common ratio r with $|r| < 1$ is given by S_∞, where

$$S_\infty = \frac{a_1}{1 - r}$$

If $|r| \geq 1$, S_∞ is not a real number.

FACTORIAL OF n: $n!$ (8.5)

If n is a natural number, then $n! = n(n-1)(n-2)(n-3) \cdot \cdots \cdot 3 \cdot 2 \cdot 1$. The factorial of 0, written 0!, is defined to be 1.

BINOMIAL THEOREM   (8.5)

If n is a positive integer, then

$$(a + b)^n = a^n + \frac{n}{1!}a^{n-1}b^1 + \frac{n(n-1)}{2!}a^{n-2}b^2 + \frac{n(n-1)(n-2)}{3!}a^{n-3}b^3 + \cdots + b^n$$

CHAPTER 11 REVIEW

(11.1)   *Find the indicated term(s) of the given sequence.*

1. The first five terms of the sequence $a_n = -3n^2$.

2. The first five terms of the sequence $a_n = n^2 + 2n$.

3. The one-hundredth term of the sequence $a_n = \dfrac{(-1)^n}{100}$.

4. The fiftieth term of the sequence $a_n = \dfrac{2n}{(-1)^2}$.

5. The general term a_n of the sequence $\dfrac{1}{6}, \dfrac{1}{12}, \dfrac{1}{18}, \ldots$.

6. The general term a_n of the sequence $-1, 4, -9, 16, \ldots$.

Solve the following applications.

7. The distance in feet that an olive falling from rest in a vacuum will travel during each second is given by an arithmetic sequence whose general term is $a_n = 32n - 16$, where n is the number of the second. Find the distance the olive will fall during the fifth, sixth, and seventh seconds.

8. A culture of yeast doubles every day in a geometric progression whose general term is $a_n = 100(2)^{n-1}$, where n is the number of the day just beginning. Find how many days it takes the yeast culture to measure at least 10,000. Find the original measure of the yeast culture.

9. A Center for Disease Control (CDC) reported that a new type of immune system virus infected approximately 450 people during 1988, the year it was first discovered. The CDC predicts that during the next decade the virus will infect three times as many people each year as the year before. Write out the first five terms of this geometric sequence, and predict the number of infected people in 1992.

10. The first row of an amphitheater contains 50 seats, and each row thereafter contains 8 additional seats. Write the first ten terms of this arithmetic progression, and find the number of seats in the tenth row.

(11.2)

11. Find the first five terms for the geometric sequence whose first term is -2 and whose common ratio is $\dfrac{2}{3}$.

12. Find the first five terms for the arithmetic sequence whose first term is 12 and whose common difference is -1.5.

13. Find the thirtieth term of the arithmetic sequence whose first term is -5 and whose common difference is 4.

14. Find the eleventh term of the arithmetic sequence whose first term is 2 and whose common difference is $\dfrac{3}{4}$.

15. Find the twentieth term of the arithmetic sequence whose first three terms are 12, 7, and 2.

16. Find the sixth term of the geometric sequence whose first three terms are 4, 6, and 9.

17. If the fourth term of an arithmetic sequence is 18 and the twentieth term is 98, find the first term and the common difference.

18. If the third term of a geometric sequence is -48 and the fourth term is 192, find the first term and the common ratio.

19. Find the general term of the sequence $\dfrac{3}{10}$, $\dfrac{3}{100}$, $\dfrac{3}{1000}$, ...

20. Find a general term that satisfies the terms shown for the sequence 50, 58, 66, ...

Determine which of the following sequences are arithmetic or geometric. If a sequence is arithmetic, find a_1 and d. If a sequence is geometric, find a_1 and r. If neither, write neither.

21. $\dfrac{8}{3}$, 4, 6, . . .

22. -10.5, -6.1, -1.7

23. $7x$, $-14x$, $28x$

24. $3x^2$, $9x^4$, $81x^8$, . . .

Solve the following applications.

25. To test the bounce of a racquetball, the ball is dropped from a height of 8 feet. The ball is judged "good" if it rebounds at least 75% of its previous height with each bounce. Write out the first six terms of this geometric sequence (round to the nearest tenth). Determine if a ball is "good" that rebounds to a height of 2.5 feet after the fifth bounce.

26. A display of oil cans in an auto parts store has 25 cans in the bottom row, 21 cans in the next row, and so on, in an arithmetic progression. Find the general term, and the number of cans in the top row.

27. Suppose that you save $1 the first day of a month, $2 the second day, $4 the third day, continuing to double your savings each day. Write the general term of this geometric sequence and find the amount you will save on the tenth day. Estimate the amount you will save on the thirtieth day of the month, and check your estimate with a calculator.

28. On the first swing, the length of an arc through which a pendulum swings is 30 inches. The length of the arc for each successive swing is 70% of the preceding swing. Find the length of the arc for the fifth swing.

29. Rosa takes a job that has a monthly starting salary of $900 and guarantees her a monthly raise of $150 during her 6-month training period. Find the general term of this sequence and her salary at the end of her training.

30. A sheet of paper is $\dfrac{1}{512}$-inch thick. By folding the sheet in half, the total thickness will be $\dfrac{1}{256}$ inch: a second fold produces a total thickness of $\dfrac{1}{128}$ inch. Estimate the thickness of the stack after 15 folds, and then check your estimate with a calculator.

(11.3) *Write out the terms and find the sum for each of the following.*

31. $\displaystyle\sum_{i=1}^{5} 2i - 1$

32. $\displaystyle\sum_{i=1}^{5} i(i + 2)$

33. $\displaystyle\sum_{i=2}^{4} \dfrac{(-1)^i}{2i}$

34. $\displaystyle\sum_{i=3}^{5} 5(-1)^{i-1}$

Find the partial sum of the given sequence.

35. S_4 of the sequence $a_n = (n - 3)(n + 2)$

36. S_6 of the sequence $a_n = n^2$

37. S_5 of the sequence $a_n = -8 + (n - 1)3$

38. S_3 of the sequence $a_n = 5(4)^{n-1}$

Write the sum using Σ notation.

39. $1 + 3 + 9 + 27 + 81 + 243$

40. $6 + 2 + (-2) + (-6) + (-10) + (-14) + (-18)$

41. $\frac{1}{4} + \frac{1}{16} + \frac{1}{64} + \frac{1}{256}$

42. $1 + \left(-\frac{3}{2}\right) + \frac{9}{4}$

Solve.

43. A yeast colony begins with 20 yeast and doubles every 8 hours. Write the sequence that describes the growth of the yeast, and find the total yeast after 48 hours.

44. The number of cranes born each year in a new aviary forms a sequence whose general term is $a_n = n^2 + 2n - 1$. Find the number of cranes born in the fourth year, and the total number of cranes born in the first four years.

45. Harold has a choice between two job offers. Job *A* has an annual starting salary of $19,500 with guaranteed annual raises of $1100 for the next four years, while job *B* has an annual starting salary of $21,000 with guaranteed annual raises of $700 for the next four years. Compare the salaries for the fifth year under each job offer.

46. A sample of radioactive waste is decaying so that the amount decaying in kilograms during year n is $a_n = 200(0.5)^n$. Find the amount of decay in the third year, and the total amount of decay in the first three years.

(11.4) *Find the partial sum of the given sequence.*

47. The sixth partial sum of the sequence 15, 19, 23, ...

48. The ninth partial sum of the sequence 5, −10, 20, ...

49. The sum of the first 30 odd positive integers.

50. The sum of the first 20 positive multiples of 7.

51. The sum of the first 20 terms of the sequence 8, 5, 2, ...

52. The sum of the first eight terms of the sequence $\frac{3}{4}, \frac{9}{4}, \frac{27}{4}, \ldots$

53. S_4 if $a_1 = 6$ and $r = 5$.

54. S_{100} if $a_1 = -3$ and $d = -6$.

Find the sum of each infinite geometric sequence.

55. $5, \frac{5}{2}, \frac{5}{4}, \ldots$

56. $18, -2, \frac{2}{9}, \ldots$

57. $-20, -4, -\frac{4}{5}, \ldots$

58. 0.2, 0.02, 0.002, . . .

Solve.

59. A frozen yogurt store owner cleared $20,000 the first year he owned his business and made an additional 15% over the previous year in each subsequent year. Find how much he made during his fourth year of business. Find his total earnings during the first 4 years (round to the nearest dollar).

60. On his first morning in a television assembly factory, a trainee takes 0.8 times as long to assemble each television as he took to assemble the one before. If it took him 40 minutes to assemble the first television, find how long it takes him to assemble the fourth television. Find the total time he takes to assemble the first four televisions (round to the nearest minute).

61. During the harvest season a farmer can rent a combine machine for $100 the first day, with the rent decreasing $7 for each additional day. Find how much rent the farmer pays for the seventh day. Find how much total rent the farmer pays for 7 days.

62. A rubber ball is dropped from a height of 15 feet and rebounds 80% of its previous height after each bounce. Find the total distance the ball travels before it comes to rest.

63. Spraying a pond once with insecticide killed 1800 mosquitoes the first day, 600 the second day, 200 the third day, and so on. Find the total number of mosquitoes killed during the first 6 days after the spraying (round to the nearest unit).

64. See Exercise 63. Find the day on which the insecticide is no longer effective, and find the total number of mosquitoes killed (round to the nearest mosquito).

65. Use the formula S_∞ to write $0.55\overline{5}$ as a fraction.

66. A movie theater has 27 seats in the first row, 30 seats in the second row, 33 seats in the third row, and so on. Find the total number of seats in the theater if there are 20 rows.

(11.5)   *Use Pascal's triangle to expand the binomial.*

67. $(x + z)^5$ **68.** $(y - r)^6$ **69.** $(2x + y)^4$ **70.** $(3y - z)^4$

Use the binomial formula to expand the following.

71. $(b + c)^8$ **72.** $(x - w)^7$ **73.** $(4m - n)^4$ **74.** $(p - 2r)^5$

Find the indicated term.

75. The fourth term of the expansion of $(a + b)^7$.   **76.** The eleventh term of the expansion of $(y + 2z)^{10}$.

CHAPTER 11 TEST

Find the indicated term(s) of the given sequence.

1. The first five terms of the sequence $a_n = \dfrac{(-1)^n}{n + 4}$.

2. The first five terms of the sequence $a_n = \dfrac{3}{(-1)^n}$.

3. The eightieth term of the sequence $a_n = 10 + 3(n - 1)$.

4. The two-hundredth term of the sequence $a_n = (n + 1)(n - 1)(-1)^n$.

5. The general term of the sequence $\dfrac{2}{5}, \dfrac{2}{25}, \dfrac{2}{125}, \ldots$.

6. The general term of the sequence $-9, 18, -27, 36, \ldots$.

Find the partial sum of the given sequence.

7. S_5 of the sequence $a_n = 5(2)^{n-1}$

8. S_{30} of the sequence $a_n = 18 + (n - 1)(-2)$

9. S_∞ of the sequence $a_1 = 24$ and $r = \dfrac{1}{6}$

10. S_∞ of the sequence $\dfrac{3}{2}, -\dfrac{3}{4}, \dfrac{3}{8}, \ldots$.

11. $\displaystyle\sum_{i=1}^{4} i(i - 2)$

12. $\displaystyle\sum_{i=2}^{4} 5(2)^i (-1)^{i-1}$

Expand the binomial using Pascal's triangle.

13. $(a - b)^6$ **14.** $(2x + y)^5$

Expand the binomial using the binomial formula.

15. $(y + z)^8$

16. $(2p + r)^7$

Solve the following applications.

17. The population of a small town is growing yearly according to the sequence defined by $a_n = 250 + 75(n - 1)$, where n is the number of the year just beginning. Predict the population at the beginning of the tenth year. Find the town's initial population.

18. A gardener is making a triangular planting with one shrub in the first row, three shrubs in the second row, five shrubs in the third row, and so on, for eight rows. Write the finite series of this sequence and find the total number of shrubs planted.

19. A pendulum swings through an arc of length 80 centimeters on its first swing. On each successive swing, the length of the arc is $\frac{3}{4}$ the length of the arc on the preceding swing. Find the length of the arc on the fourth swing, and find the total arc lengths for the first four swings.

20. See Exercise 19. Find the total arc lengths before the pendulum comes to rest.

21. A parachutist in free fall falls 16 feet during the first second, 48 feet during the second second, 80 feet during the third second, and so on. Find how far he falls during the tenth second. Find the total distance he falls during the first 10 seconds.

22. Use the formula S_∞ to write $0.42\overline{42}$ as a fraction.

CHAPTER 11 CUMULATIVE REVIEW

1. Solve $A = \frac{1}{2}(B + b)h$ for b.

2. Solve $|2x + 9| + 5 > 3$.

3. Find the equation of the line with slope $\frac{1}{4}$ and y-intercept $-\frac{2}{3}$. Write the equation in standard form.

4. Simplify each expression. Write answers using positive exponents.

   a. $(2x^0 y^{-3})^{-2}$

   b. $\left(\dfrac{x^{-5}}{x^{-2}}\right)^{-3}$

   c. $\left(\dfrac{2}{7}\right)^{-2}$

   d. $\dfrac{5^{-2} x^{-3} y^{11}}{x^2 y^{-5}}$

5. Find the following products.
   a. $(x - 3)(x + 3)$
   b. $(4y + 1)(4y - 1)$
   c. $(x^2 + 2y)(x^2 - 2y)$

6. Factor $7x(x^2 + 5y) - (x^2 + 5y)$.

7. Factor the following:
   a. $p^4 - 16$

   b. $(x + 3)^2 - 36$

8. Add $\dfrac{2x - 1}{2x^2 - 9x - 5} + \dfrac{x + 3}{6x^2 - x - 2}$.

9. Evaluate the following:
   a. $4^{1/2}$ **b.** $64^{1/3}$ **c.** $81^{1/4}$ **d.** $0^{1/6}$
   e. $-9^{1/2}$

10. Multiply.
   a. $\sqrt{3}(5 + \sqrt{30})$
   b. $(\sqrt{5} - \sqrt{6})(\sqrt{7} + 1)$
   c. $(7\sqrt{x} + 5)(3\sqrt{x} - \sqrt{5})$
   d. $(4\sqrt{3} - 1)^2$
   e. $(\sqrt{2x} - 5)(\sqrt{2x} + 5)$

11. Solve $\sqrt{2x + 5} + \sqrt{2x} = 3$.

12. Solve $x^2 - 4x \geq 0$.

13. Graph $f(x) = 3x^2 + 3x + 1$. Find the vertex and any intercepts.

14. Graph $\dfrac{(x + 3)^2}{25} + \dfrac{(y - 2)^2}{36} = 1$.

15. Solve the system:
$$\begin{cases} 2x + 4y = 1 \\ 4x - 4z = -1 \\ y - 4z = -3 \end{cases}$$

16. Solve the system:
$$\begin{cases} y = \sqrt{x} \\ x^2 + y^2 = 6 \end{cases}.$$

17. Simplify.

    a. $\log_3 3^2$ **b.** $\log_7 7^{-1}$ **c.** $5^{\log_5 3}$

    d. $2^{\log_2 6}$

18. Solve $\log_2 x + \log_2 (x - 1) = 1$ for x.

19. The amount of weight, in pounds, a puppy gains in each month of its first year is modeled by a sequence whose general term is $a_n = n + 4$, where n is the number of the month. Write the first five terms of the sequence, and find how much weight the puppy should gain in its fifth month.

20. Use the binomial theorem to expand $(x + 2y)^5$.

APPENDIX A

Table of Squares and Square Roots

n	n^2	$\sqrt{n}$	n	n^2	$\sqrt{n}$
1	1	1.000	51	2,601	7.141
2	4	1.414	52	2,704	7.211
3	9	1.732	53	2,809	7.280
4	16	2.000	54	2,916	7.348
5	25	2.236	55	3,025	7.416
6	36	2.449	56	3,136	7.483
7	49	2.646	57	3,249	7.550
8	64	2.828	58	3,364	7.616
9	81	3.000	59	3,481	7.681
10	100	3.162	60	3,600	7.746
11	121	3.317	61	3,721	7.810
12	144	3.464	62	3,844	7.874
13	169	3.606	63	3,969	7.937
14	196	3.742	64	4,096	8.000
15	225	3.873	65	4,225	8.062
16	256	4.000	66	4,356	8.124
17	289	4.123	67	4,489	8.185
18	324	4.243	68	4,624	8.246
19	361	4.359	69	4,761	8.307
20	400	4.472	70	4,900	8.367
21	441	4.583	71	5,041	8.426
22	484	4.690	72	5,184	8.485
23	529	4.796	73	5,329	8.544
24	576	4.899	74	5,476	8.602
25	625	5.000	75	5,625	8.660
26	676	5.099	76	5,776	8.718
27	729	5.196	77	5,929	8.775
28	784	5.292	78	6,084	8.832
29	841	5.385	79	6,241	8.888
30	900	5.477	80	6,400	8.944
31	961	5.568	81	6,561	9.000
32	1,024	5.657	82	6,724	9.055
33	1,089	5.745	83	6,889	9.110
34	1,156	5.831	84	7,056	9.165
35	1,225	5.916	85	7,225	9.220
36	1,296	6.000	86	7,396	9.274
37	1,369	6.083	87	7,569	9.327
38	1,444	6.164	88	7,744	9.381
39	1,521	6.245	89	7,921	9.434
40	1,600	6.325	90	8,100	9.487
41	1,681	6.403	91	8,281	9.539
42	1,764	6.481	92	8,464	9.592
43	1,849	6.557	93	8,649	9.644
44	1,936	6.633	94	8,836	9.695
45	2,025	6.785	95	9,025	9.747
46	2,116	6.782	96	9,216	9.798
47	2,209	6.856	97	9,409	9.849
48	2,304	6.928	98	9,604	9.899
49	2,401	7.000	99	9,801	9.950
50	2,500	7.071	100	10,000	10.000

Review of Angles, Lines, and Special Triangles

The word **geometry** is formed from the Greek words, **geo,** meaning earth, and **metron,** meaning measure. Geometry literally means to measure the earth.

This section contains a review of some basic geometric ideas. It will be assumed that fundamental ideas of geometry such as point, line, ray, and angle are known. In this appendix, the notation $\angle 1$ is read "angle 1" and the notation $m\angle 1$ is read "the measure of angle 1."

We first review types of angles.

Angles

A **right angle** is an angle whose measure is 90°. A right angle can be indicated by a square drawn at the vertex of the angle, as shown below. An angle whose measure is more than 0° but less than 90° is called an **acute angle.**

An angle whose measure is greater than 90° but less than 180° is called an **obtuse angle.**

An angle whose measure is 180° is called a **straight angle.**

Two angles are said to be **complementary** if the sum of their measures is 90°. Each angle is called the **complement** of the other.

Two angles are said to be **supplementary** if the sum of their measures is 180°. Each angle is called the **supplement** of the other.

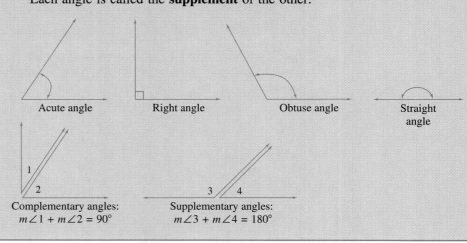

Acute angle   Right angle   Obtuse angle   Straight angle

Complementary angles:
$m\angle 1 + m\angle 2 = 90°$

Supplementary angles:
$m\angle 3 + m\angle 4 = 180°$

EXAMPLE 1 If an angle measures 28°, find its complement.

Solution: Two angles are complementary if the sum of their measures is 90°. The complement of a 28° angle is an angle whose measure is $90° - 28° = 62°$. To check, notice that $28° + 62° = 90°$. ∎

Plane is an undefined term that we will describe. A plane can be thought of as a flat surface with infinite length and width, but no thickness. A plane is two dimensional. The arrows in the following diagram indicate that a plane extends indefinitely and has no boundaries.

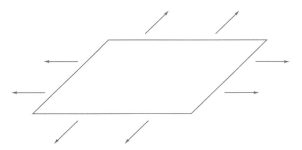

Figures that lie on a plane are called **plane figures.** (See the description of common plane figures in Appendix C.) Lines that lie in the same plane are called **coplanar.**

Lines

Two lines are **parallel** if they lie in the same plane but never meet.

Intersecting lines meet or cross in one point.

Two lines that form right angles when they intersect are said to be **perpendicular.**

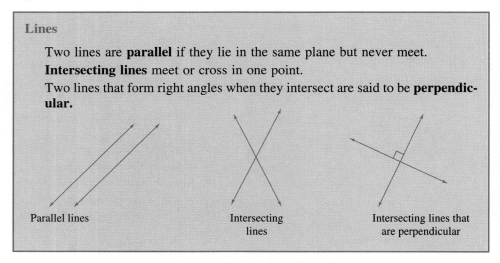

Parallel lines | Intersecting lines | Intersecting lines that are perpendicular

Two intersecting lines form **vertical angles.** Angles 1 and 3 are vertical angles. Also angles 2 and 4 are vertical angles. It can be shown that **vertical angles have equal measures.**

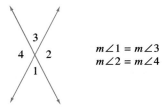

$$m\angle 1 = m\angle 3$$
$$m\angle 2 = m\angle 4$$

Adjacent angles have the same vertex and share a side. Angles 1 and 2 are adjacent angles. Other pairs of adjacent angles are angles 2 and 4, angles 3 and 4, and angles 3 and 1.

A **transversal** is a line that intersects two or more lines in the same plane. Line l is a transversal that intersects lines m and n. The eight angles formed are numbered and certain pairs of these angles are given special names.

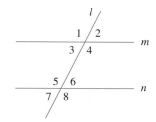

Corresponding angles: $\angle1$ and $\angle5$, $\angle3$ and $\angle7$, $\angle2$ and $\angle6$, and $\angle4$ and $\angle8$.
Exterior angles: $\angle1$, $\angle2$, $\angle7$, and $\angle8$.
Interior angles: $\angle3$, $\angle4$, $\angle5$, and $\angle6$.
Alternate interior angles: $\angle3$ and $\angle6$, $\angle4$ and $\angle5$.

These angles and parallel lines are related in the following manner.

Parallel Lines Cut by a Transversal

1. If two parallel lines are cut by a transversal, then
   a. **corresponding angles are equal** and
   b. **alternate interior angles are equal.**
2. If corresponding angles formed by two lines and a transversal are equal, then the lines are parallel.
3. If alternate interior angles formed by two lines and a transversal are equal, then the lines are parallel.

EXAMPLE 2   Given that lines m and n are parallel and that the measure of angle 1 is 100°, find the measures of angles 2, 3, and 4.

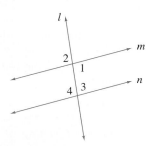

Solution: $m\angle2 = 100°$,   since angles 1 and 2 are vertical angles.

$m\angle4 = 100°$,   since angles 1 and 4 are alternate interior angles.

$m\angle3 = 180° - 100° = 80°$,   since angles 4 and 3 are supplementary angles.   ∎

A **polygon** is the union of three or more coplanar line segments that intersect each other only at each end point, with each end point shared by exactly two segments.

A **triangle** is a polygon with three sides. The sum of the measures of the three angles of a triangle is 180°. In the following figure, $m\angle 1 + m\angle 2 + m\angle 3 = 180°$.

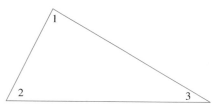

EXAMPLE 3   Find the measure of the third angle of the triangle shown.

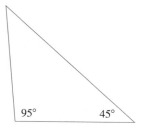

Solution: The sum of the measures of the angles of a triangle is 180°. Since one angle measures 45° and the other angle measures 95°, the third angle measures $180° - 45° - 95° = 40°$.   ∎

Two triangles are **congruent** if they have the same size and the same shape. In congruent triangles, the measures of corresponding angles are equal and the lengths of corresponding sides are equal. The following triangles are congruent.

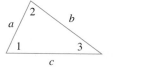
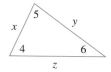

Corresponding angles are equal: m∠1 = m∠4, m∠2 = m∠5, and m∠3 = m∠6. Also, lengths of corresponding sides are equal: $a = x$, $b = y$, and $c = z$.

Any one of the following may be used to determine whether two triangles are congruent.

Congruent Triangles

1. If the measures of two angles of a triangle equal the measures of two angles of another triangle and the lengths of the sides between each pair of angles are equal, the triangles are congruent.

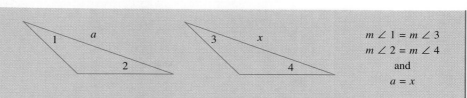

2. If the lengths of the three sides of a triangle equal the lengths of corresponding sides of another triangle, the triangles are congruent.

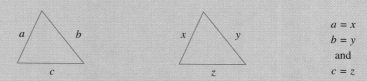

3. If the lengths of two sides of a triangle equal the lengths of corresponding sides of another triangle, and the measures of the angles between each pair of sides are equal, the triangles are congruent.

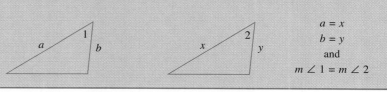

Two triangles are similar if they have the same shape, but not necessarily the same size. In similar triangles, the measures of corresponding angles are equal and corresponding sides are in proportion. The following triangles are similar. (All similar triangles drawn in this appendix will be oriented the same.)

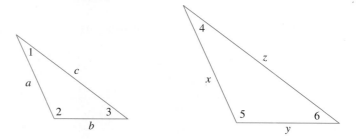

Corresponding angles are equal: $m \angle 1 = m \angle 4$, $m \angle 2 = m \angle 5$, and $m \angle 3 = m \angle 6$. Also, corresponding sides are proportional: $\dfrac{a}{x} = \dfrac{b}{y} = \dfrac{c}{z}$.

Any one of the following may be used to determine whether two triangles are similar.

Similar Triangles

1. If the measures of two angles of a triangle equal the measures of two angles of another triangle, the triangles are similar.

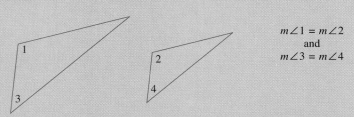

$$m\angle 1 = m\angle 2$$
and
$$m\angle 3 = m\angle 4$$

2. If three sides of one triangle are proportional to three sides of another triangle, the triangles are similar.

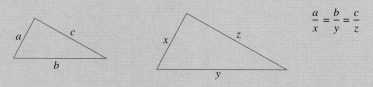

$$\frac{a}{x} = \frac{b}{y} = \frac{c}{z}$$

3. If two sides of a triangle are proportional to two sides of another triangle and the measures of the included angles are equal, the triangles are similar.

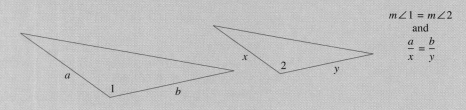

$$m\angle 1 = m\angle 2$$
and
$$\frac{a}{x} = \frac{b}{y}$$

EXAMPLE 4 Given that the following triangles are similar, find the missing length x.

Solution: Since the triangles are similar, corresponding sides are in proportion. Thus, $\frac{2}{3} = \frac{10}{x}$.
To solve this equation for x, we multiply both sides by the LCD $3x$.

$$3x\left(\frac{2}{3}\right) = 3x\left(\frac{10}{x}\right)$$

$$2x = 30$$

$$x = 15$$

The missing length is 15 units. ∎

A **right triangle** contains a right angle. The side opposite the right angle is called the **hypotenuse,** and the other two sides are called the **legs.** The **Pythagorean theorem** gives a formula that relates the lengths of the three sides of a right triangle.

The Pythagorean Theorem

If a and b are the lengths of the legs of a right triangle, and c is the length of the hypotenuse, then $a^2 + b^2 = c^2$.

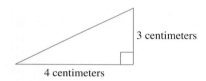

EXAMPLE 5   Find the length of the hypotenuse of a right triangle whose legs have lengths of 3 centimeters and 4 centimeters.

3 centimeters

4 centimeters

Solution: Because we have a right triangle, we use the Pythagorean theorem. The legs are 3 centimeters and 4 centimeters, so let $a = 3$ and $b = 4$ in the formula.

$$a^2 + b^2 = c^2$$

$$3^2 + 4^2 = c^2$$

$$9 + 16 = c^2$$

$$25 = c^2$$

Since c represents a length, we assume that c is positive. Thus, if c^2 is 25, c must be 5. The hypotenuse has a length of 5 centimeters.   ■

APPENDIX B EXERCISE SET

Find the complement of each angle. See Example 1.

1. $19°$

2. $65°$

3. $70.8°$

4. $45\frac{2}{3}°$

5. $11\frac{1}{4}°$

6. $19.6°$

Find the supplement of each angle.

7. $150°$

8. $90°$

9. $30.2°$

10. $81.9°$

11. $79\frac{1}{2}°$

12. $165\frac{8}{9}°$

13. If lines m and n are parallel, find the measures of angles 1 through 7. See Example 2.

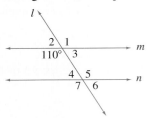

14. If lines m and n are parallel, find the measures of angles 1 through 5. See Example 2.

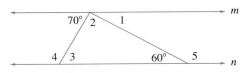

In each of the following, the measures of two angles of a triangle are given. Find the measure of the third angle. See Example 3.

15. 11°, 79°

16. 8°, 102°

17. 25°, 65°

18. 44°, 19°

19. 30°, 60°

20. 67°, 23°

In each of the following, the measure of one angle of a right triangle is given. Find the measures of the other two angles.

21. 45°

22. 60°

23. 17°

24. 30°

25. $39\frac{3}{4}°$

26. 72.6°

Given that each of the following pairs of triangles is similar, find the missing lengths. See Example 4.

27.

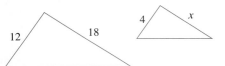

28.

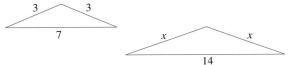

29.

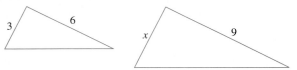

30.

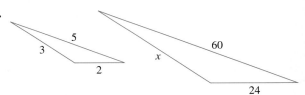

Use the Pythagorean Theorem to find the missing lengths in the right triangles. See Example 5.

31.

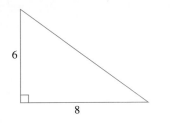

32.

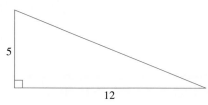

33.

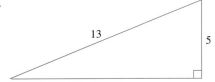

34.

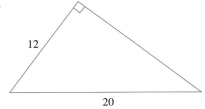

Review of Geometric Figures

Plane figures have length and width but no thickness or depth.		
Name	**Description**	**Figure**
Polygon	Union of three or more coplanar line segments that intersect each other only at each end point, with each end point shared by two segments.	
Triangle	Polygon with three sides (sum of measures of three angles is 180°).	
Scalene triangle	Triangle with no sides of equal length.	
Isosceles triangle	Triangle with two sides of equal length.	
Equilateral triangle	Triangle with all sides of equal length.	
Right triangle	Triangle that contains a right angle.	leg, hypotenuse, leg
Quadrilateral	Polygon with four sides (sum of measures of four angles is 360°).	

Plane figures have length and width but no thickness or depth.

Name	Description	Figure
Trapezoid	Quadrilateral with exactly one pair of opposite sides parallel.	
Isosceles trapezoid	Trapezoid with legs of equal length.	
Parallelogram	Quadrilateral with both pairs of opposite sides parallel and equal in length.	
Rhombus	Parallelogram with all sides of equal length.	
Rectangle	Parallelogram with four right angles.	
Square	Rectangle with all sides of equal length.	
Circle	All points in a plane the same distance from a fixed point called the **center**.	

Solid figures have length, width, and height or depth.		
Name	**Description**	**Figure**
Rectangular solid	A solid with six sides, all of which are rectangles.	
Cube	A rectangular solid whose six sides are squares.	
Sphere	All points the same distance from a fixed point, called the **center**.	radius center
Right circular cylinder	A cylinder consisting of two circular bases that are perpendicular to its altitude.	
Right circular cone	A cone with a circular base that is perpendicular to its altitude.	

Answers
to Selected Exercises

CHAPTER 1
Review of Real Numbers

Exercise Set 1.1
1. $\in$ **3.** $\in$ **5.** $\notin$ **7.** $\subseteq$ **9.** $\not\subseteq$ **11.** $\subseteq$ **13.** true **15.** false **17.** true **19.** false **21.** false
23. $\{7, 8, 9, \ldots\}$ **25.** $\{1, 3, 5, \ldots\}$ **27.** $\{\ \}$ **29.** $\{1, 3, 5, 7\}$ **31.** false **33.** true **35.** false

37. ![number line with points at 1, 3, 5, 7] **39.** ![number line with points at −10, −6, −2] **41.** ![number line marked 0, 1/4, 1/3, 1] **43.** $\{3, 0, \sqrt{36}, -134\}$

45. $\left\{3, 0, \sqrt{36}, \dfrac{2}{5}, -134\right\}$ **47.** $\left\{3, 0, \sqrt{7}, \sqrt{36}, \dfrac{2}{5}, -134\right\}$ **49.** $\notin$ **51.** $\in$ **53.** $\notin$ **55.** $\notin$ **57.** $\subseteq$
59. $\not\subseteq$ **61.** $\subseteq$ **63.** $\subseteq$

Exercise Set 1.2
1. $4c = 7$ **3.** $3(x + 1) = 7$ **5.** $\dfrac{n}{5} = 4$ **7.** $>$ **9.** $=$ **11.** $>$ **13.** $7x \leq -21$ **15.** $-2 + x \neq 10$

17. $2(x - 6) > \dfrac{1}{11}$ **19.** multiplicative identity **21.** commutative property for addition

23. commutative property for addition **25.** distributive property **27.** inverse property for multiplication

29. $\dfrac{1}{5}$ **31.** -4 **33.** does not exist **35.** $-\dfrac{8}{7}$ **37.** 6 **39.** $-\dfrac{4}{7}$ **41.** $\dfrac{2}{3}$ **43.** 0 **45.** $y - 7 = 6$

47. $3y < -17$ **49.** $2(x - 6) = -27$ **51.** $x - 4 \geq 3x$ **53.** $2y - 6 = \dfrac{1}{8}$ **55.** $\dfrac{n + 5}{2} > 2n$

57. $3x + 15$ **59.** $8a + 8b$ **61.** $xa - xb$ **63.** $3x + 3y - 24$ **65.** $6 + 3x$ **67.** 0 **69.** 7
71. 0, since no real number a satisfies $a \cdot 0 = 0 \cdot a = 1$

Exercise Set 1.3
1. -2 **3.** 4 **5.** 0 **7.** -3 **9.** 9 **11.** 1 **13.** 8 **15.** $\dfrac{4}{9}$ **17.** 7 **19.** $\dfrac{1}{3}$ **21.** 4 **23.** 3 **25.** 11
27. 50 **29.** 42 **31.** 9 **33.** 9 **35.** 25 **37.** 0 **39.** 29 **41.** 1 **43.** 2 **45.** 0 **47.** 4 **49.** 26
51. 64 **53.** $\dfrac{3}{2}$ **55.** 28 **57.** 1 **59.** $-\dfrac{2}{5}$ **61.** 6 **63.** 24 **65.** 8 **67.** 16 **69.** 17 **71.** 0

Exercise Set 1.4
1. 5 **3.** -24 **5.** -10 **7.** -4 **9.** -23 **11.** -2 **13.** -60 **15.** 80 **17.** 3 **19.** 0 **21.** -8

23. $-\dfrac{3}{7}$ **25.** -49 **27.** 36 **29.** -8 **31.** 48 **33.** -1 **35.** -3 **37.** -2 **39.** 9 **41.** 1 **43.** -8

45. 3 **47.** -12 **49.** 15 **51.** -12 **53.** 0 **55.** -1 **57.** -21 **59.** $-\dfrac{1}{2}$ **61.** -6 **63.** -17

65. -1 **67.** 14 **69.** 48 **71.** $\dfrac{1}{3}$ **73.** $-\dfrac{1}{3}$ **75.** 17 **77.** 25 **79.** $-\dfrac{5}{14}$ **81.** 38 **83.** -1 **85.** -67

Chapter 1 Review
1. $\{-1, 1, 3\}$ **3.** $\{\ \}$ **5.** $\{6, 7, 8, \ldots\}$ **7.** true   **9.** true   **11.** false   **13.** false   **15.** true   **17.** false

19. true   **21.** true   **23.** false   **25.** $\left\{5, \dfrac{8}{2}, \sqrt{9}\right\}$ **27.** $\left\{5, -\dfrac{2}{3}, \dfrac{8}{2}, \sqrt{9}, 0.3, 1\dfrac{5}{8}, -1\right\}$

29. $\left\{5, -\dfrac{2}{3}, \dfrac{8}{2}, \sqrt{9}, 0.3, \sqrt{7}, 1\dfrac{5}{8}, -1, \pi\right\}$ **31.** $12 = -4x$ **33.** $4(y + 3) = -1$ **35.** $z - 7 = 6$

37. $x - 5 \geq 12$ **39.** $\dfrac{2}{3} \neq 2\left(n + \dfrac{1}{4}\right)$ **41.** associative property for addition

43. additive inverse property   **45.** associative and commutative properties of multiplication

47. multiplication property of zero   **49.** additive identity property   **51.** $\dfrac{3}{4}$ **53.** 0 **55.** $-\dfrac{4}{3}$ **57.** does not exist

59. $5(x - 3z)$ **61.** $2 + (-2)$, for example   **63.** $(3.4)[(0.7)5]$ **65.** $>$ **67.** $<$ **69.** $<$ **71.** 8 **73.** 0

75. 34 **77.** $\dfrac{1}{5}$ **79.** $\dfrac{2}{3}$ **81.** -2 **83.** 8 **85.** 0 **87.** undefined   **89.** 4 **91.** 9 **93.** $-\dfrac{2}{15}$ **95.** 3

97. $-\dfrac{32}{135}$ **99.** $-\dfrac{5}{4}$ **101.** $\dfrac{5}{8}$ **103.** -1 **105.** 1 **107.** -4 **109.** $\dfrac{5}{7}$

Chapter 1 Test
1. true   **2.** false   **3.** false   **4.** false   **5.** true   **6.** false   **7.** -3 **8.** 52 **9.** 43 **10.** -225 **11.** 4

12. -2 **13.** 1 **14.** $\dfrac{7}{5}$ **15.** $2|x + 5| = 30$ **16.** $\dfrac{(6 - y)^2}{7} < -2$ **17.** $\dfrac{9z}{|-12|} \neq 10$ **18.** $3\left(\dfrac{n}{5}\right) = -n$

19. $20 = 2x - 6$ **20.** $-2 = \dfrac{x}{x + 5}$ **21.** distributive property   **22.** associative property for addition

23. additive inverse property   **24.** multiplication property of zero

CHAPTER 2
Solving Equations and Inequalities

Mental Math, Sec. 2.1
1. $8x + 21$ **3.** $6n - 7$ **5.** $-4x - 1$

Exercise Set 2.1
1. $-15x + 18$ **3.** $2k + 10$ **5.** $-3x + 5$ **7.** $4x + 9$ **9.** $\{6\}$ **11.** $\{-1\}$ **13.** $\{-5\}$ **15.** $\{0\}$ **17.** $\{2\}$

19. $\{-9\}$ **21.** $\left\{-\dfrac{10}{7}\right\}$ **23.** $\left\{\dfrac{1}{6}\right\}$ **25.** $\{4\}$ **27.** $\{1\}$ **29.** $\left\{\dfrac{40}{3}\right\}$ **31.** $\{n \mid n$ is a real number$\}$

33. $\{x \mid x$ is a real number$\}$ **35.** $25 - 8x$ centimeters   **37.** $x + (x + 2) + (x + 4)$ or $3x + 6$.   **39.** $4n - 8$ **41.** -24

43. $2x + 10$ **45.** $\{4\}$ **47.** $\{2\}$ **49.** $\{-2\}$ **51.** $\{0\}$ **53.** $\{5\}$ **55.** $\{\ \}$ **57.** $\left\{\dfrac{1}{8}\right\}$ **59.** $\{0\}$ **61.** $\{29\}$

63. $\{-8\}$ **65.** $\left\{\dfrac{20}{11}\right\}$ **67.** $\{\ \}$ **69.** $\{4\}$ **71.** $\{8\}$ **73.** $\{x \mid x$ is a real number$\}$ **75.** $\left\{-\dfrac{15}{47}\right\}$ **77.** $\left\{\dfrac{3}{5}\right\}$

79. $\{17\}$ **81.** $\left\{\dfrac{4}{5}\right\}$ **83.** $(805x - 25)\cancel{c}.$ **85.** $(30 - x)$ days   **87.** $\{1\}$ **89.** $\{3\}$ **93.** 6 **95.** 208 **97.** -55

Mental Math, Sec. 2.2

1. $y = 5 - 2x$ **3.** $a = 5b + 8$ **5.** $k = h - 5j + 6$

Exercise Set 2.2

1. $800 **3.** Yes, he will have $1053.95. **5.** 3.6 hr or 3 hr and 36 min **7.** 3000 Frequent Flier miles

9. 8640 cubic inches **11.** approximately 25,120 miles **13.** $W = \dfrac{P - 2L}{2}$ **15.** $A = \dfrac{J + 3}{C}$ **17.** $r = \dfrac{E}{I} - R$

19. $L = \dfrac{2s}{n} - a$ **21.** 171 packages **23.** 40°C **25.** $1700 **27.** 2.25 hr or 2 hr and 15 min **29.** 75 goldfish

31. 2 gal **33.** approximately 22°C **35.** the 16-in. cheese pizza **37.** 11.7 bags **39.** $g = \dfrac{W}{h}$ **41.** $L = \dfrac{V}{WH}$

43. $y = \dfrac{17 - 2x}{3}$ **45.** $P = \dfrac{A}{rt + 1}$ **47.** $b = \dfrac{A - 5HB}{5H}$ **49.** $h = \dfrac{S - 2\pi r^2}{2\pi r}$ **51.** $R = \dfrac{r_1 r_2}{r_1 + r_2}$ **53.** $d = \dfrac{L - a}{n - 1}$

55. $y_1 = y_2 - m(x_2 - x_1)$ **57.** $v = \dfrac{T + 4ws}{3s + 5w}$ **59.** $3(x + 4)$ **61.** $(2x)[3(-x)]$ **63.** 3 **65.** 6

Exercise Set 2.3

1. 18, 20 and 22 **3.** 10, 11, and 12 **5.** 13 in. **7.** width, 4 ft; length, 10 ft. **9.** 55 mph and 35 mph
11. 300 miles **13.** 31 nickels and 44 dimes **15.** 200 adult tickets and 600 student tickets
17. 4 oz of 20%, 2 oz of 50% **19.** 40 gal **21.** $14,000 at 8%, $10,000 at 9%

23. $34,000 at 18% profit, $16,000 at 11% loss **25.** 27, 28, 29 **27.** $2\dfrac{1}{2}$ hr **29.** 80 lb **31.** 25°, 50°, 105°

33. $13,500 at 9%, $27,000 at 9% **35.** 105 mi. **37.** 2 gal must be replaced **39.** three $10 bills, seven $20 bills.
41. 38°, 38°, 104° **43.** 6 miles **45.** cut a square of side 5 feet **47.** 200 lb of ground sirloin, 300 lb of hamburger
49. 25 skateboards **51.** 800 books **53.** If fewer products are sold a loss results; if more are sold a profit results.

55. $\dfrac{1}{2}(x - 1) = 37$ **57.** $\dfrac{3(x + 2)}{5} = 0$ **59.** $40 = 8(2x + 1)$ **61.** $20 - 8x = -4$

Mental Math, Sec. 2.4

1. 7 **3.** −5 **5.** −6 **7.** 12

Exercise Set 2.4

1. $\{7, -7\}$ **3.** $\{4, -4\}$ **5.** $\{7, -2\}$ **7.** $\{8, 4\}$ **9.** $\{5, -5\}$ **11.** $\{3, -3\}$ **13.** $\{0\}$ **15.** $\{\ \}$

17. $x = \dfrac{1}{5}$ **19.** $\left\{9, -\dfrac{1}{2}\right\}$ **21.** $\left\{-\dfrac{5}{2}\right\}$ **23.** $\{4, -4\}$ **25.** $\{0\}$ **27.** $\{\ \}$ **29.** $\left\{0, \dfrac{14}{3}\right\}$ **31.** $\{2, -2\}$

33. $\{\ \}$ **35.** $\{7, -1\}$ **37.** $\{\ \}$ **39.** $\{\ \}$ **41.** $\left\{-\dfrac{1}{8}\right\}$ **43.** $\left\{\dfrac{1}{2}, -\dfrac{5}{6}\right\}$ **45.** $\left\{2, -\dfrac{12}{5}\right\}$ **47.** $\{3, -2\}$

49. $\left\{-8, \dfrac{2}{3}\right\}$ **51.** $\{\ \}$ **53.** $\{4\}$ **55.** $\{13, -8\}$ **57.** $\{3, -3\}$ **59.** $\{8, -7\}$ **61.** $\{2, 3\}$

63. $\left\{2, -\dfrac{10}{3}\right\}$ **65.** $\left\{\dfrac{3}{2}\right\}$ **67.** $\{\ \}$ **69.** −5 **71.** 8 **73.** $-\dfrac{1}{3}$ **75.** $\dfrac{9}{5}$

Exercise Set 2.5

1. $(-\infty, -3)$ **3.** $[0.3, \infty)$ **5.** $(5, \infty)$

7. $(-2, 5)$ **9.** $(-1, 5)$ **11.** $x > -1$

13. $x \le 2$ **15.** $x < -4$ **17.** $x \le -8$

19. $x \le 11$ **21.** $x \ge \dfrac{8}{3}$ **23.** $x > -13$

25. $x \le 7$ **27.** all reals **29.** $\{\ \}$

31. 1040 lb of luggage and cargo

33. 18 oz

35.  $x > 2$

37. $x \geq -8$

39. $x \leq -1$

41. $x > 0$

43. $x > -2$

45. $x \geq -\dfrac{3}{5}$

47. $x \geq -9$

49. $x > 38$

51. $x \geq 0$

53. $x \leq -5$

55. $x < \dfrac{1}{4}$

57. $x \leq -1$

59. $x \geq -\dfrac{79}{3}$

61. $x < -15$

63. $x \geq 3$

65. $x \geq -\dfrac{37}{3}$

67. $x < 5$

69. The minimum score is 30.   **71.** 200 calls.

75. 5   **77.** 2   **79.** 2   **81.** -2

Exercise Set 2.6

1. $-2 < x < 5$ **3.** $x \geq 6$ **5.** $x \leq -3$

7. $11 < x < 17$ **9.** $1 \leq x \leq 4$

11. $-3 \leq x \leq \dfrac{3}{2}$ **13.** $-21 \leq x \leq -9$

15. $x < -1$ or $x > 0$ **17.** $x \geq 2$

19. all reals   **21.** $-1 < x < 2$

23. all reals   **25.** $x \geq -1$ **27.** $x \geq -5$

29. $\dfrac{3}{2} \leq x \leq 6$ **31.**   **33.** $\{\ \}$

35. $x > -7$ **37.** $-5 < x < \dfrac{5}{2}$

39. $0 < x \leq \dfrac{14}{3}$ **41.** $x \leq -3$

43. $x \leq 1$ or $x > \dfrac{29}{7}$ **45.** $\{\ \}$

47. $-\dfrac{1}{2} \leq x < \dfrac{3}{2}$ **49.** $-\dfrac{4}{3} < x < \dfrac{7}{3}$

51. $6 < x < 12$ **53.** The temperature ranges from 14° to 64.4°F.   **55.** $(6, \infty)$ **57.** $[3, 7]$

59. $(-\infty, -1)$ **61.** -15 **63.** -5 **65.** -3 **67.** 10

Exercise Set 2.7

1. $[-4, 4]$

3. $(1, 5)$

5. $(-5, -1)$

7. $[-10, 3]$

9. $[-5, 5]$

11. $\{\ \}$

13. $[0, 12]$

15. $(-\infty, -3) \cup (3, \infty)$

17. $(-\infty, -24] \cup [4, \infty)$

19. $(-\infty, -4) \cup (4, \infty)$

21. $(-\infty, \infty)$

23. $\left(-\infty, \frac{2}{3}\right) \cup (2, \infty)$

25. $\{0\}$

27. $\left(-\infty, -\frac{3}{8}\right) \cup \left(-\frac{3}{8}, \infty\right)$

29. $[-2, 2]$

31. $(-\infty, -1) \cup (1, \infty)$

33. $(-5, 11)$

35. $\left(-\infty, \frac{2}{3}\right) \cup (2, \infty)$

37. $\{\ \}$

39. $(-\infty, \infty)$

41. $[-2, 9]$

43. $(-\infty, -11] \cup [1, \infty)$

45. $(-\infty, 0) \cup (0, \infty)$

47. $(-\infty, \infty)$

49. $\left[-\frac{1}{2}, 1\right]$

51. $(-\infty, -3) \cup (0, \infty)$

53. $\{\ \}$

55. $(-\infty, \infty)$

57. $\left(-\frac{2}{3}, 0\right)$

59. $(-\infty, \infty)$

61. $(-\infty, -1) \cup (1, \infty)$

63. $(-\infty, -12) \cup (0, \infty)$

65. $(-\infty, -6) \cup (0, \infty)$

67. $\left(-\frac{31}{5}, \frac{11}{5}\right)$

69. $[-1, 8]$

71. $\left[-\frac{23}{8}, \frac{17}{8}\right]$

73. $(-2, 5)$ **75.** $\{5, -2\}$ **77.** $(-\infty, -7] \cup [17, \infty)$ **79.** $\left\{-\frac{9}{4}\right\}$ **81.** $(-2, 1)$ **83.** $\left\{2, \frac{4}{3}\right\}$ **85.** $\{\ \}$

87. $\left\{\frac{19}{2}, -\frac{17}{2}\right\}$ **89.** $\left(-\infty, -\frac{25}{3}\right) \cup \left(\frac{35}{3}, \infty\right)$ **91.** $\{1, 3, 5, 7, 9\}$ **93.** $\{\ \}$ **95.** 34 **97.** 1

Chapter 2 Review

1. $\{3\}$ **3.** $\left\{-\frac{45}{14}\right\}$ **5.** $\{0\}$ **7.** $\{6\}$ **9.** $\{x \mid x \text{ is a real number}\}$ **11.** $\{\ \}$ **13.** $\{-3\}$ **15.** $\left\{\frac{96}{5}\right\}$ **17.** $\{32\}$

19. $\{8\}$ **21.** $\{\ \}$ **23.** $\{2\}$ **25.** $W = \dfrac{V}{LH}$ **27.** $y = \dfrac{5x + 12}{4}$ **29.** $m = \dfrac{y - y_1}{x - x_1}$ **31.** $r = \dfrac{E - IR}{I}$

33. $g = \dfrac{T}{r + vt}$ **35.** $B = \dfrac{2A - hb}{h}$ **37.** $r_1 = 2R - r_2$ **39.** $b = \dfrac{ac}{a - c}$ **41.** $R_2 = \dfrac{RR_1}{R_1 - R}$ **43.** $y = \dfrac{28}{3}x$

45. \$4500 must be invested. **47.** The cylinder holds more ice cream. **49.** 258 miles **51.** 10, 11, 12, and 13

53. $11,500 at 8%, $13,500 at 9%   **55.** $1\frac{2}{3}$ hr or 1 hr 40 min   **57.** level road at 55 mph   **59.** 60 seats in the balcony

61. 50 cL at 15%, 100 cL at 45%   **63.** $66\frac{2}{3}$ gal of water   **65.** width, 40 meters; length, 75 meters

67. 250 calculators must be sold to break even.   **69.** $\{16, -2\}$ **71.** $\{0, -9\}$ **73.** $\left\{2, -\frac{2}{3}\right\}$ **75.** $\{\ \}$ **77.** $\{3, -3\}$

79. $\left\{5, -\frac{1}{3}\right\}$ **81.** $\left\{7, -\frac{8}{5}\right\}$ **83.** $(3, \infty)$ **85.** $(-4, \infty)$ **87.** $(-\infty, 7]$ **89.** $\left(\frac{1}{2}, \infty\right)$ **91.** $[-19, \infty)$

93. It is more economical to use the housekeeper for more than 35 pounds per week.

95. The last judge must give her at least a 9.6 for her to win the silver medal.   **97.** $\left[2, \frac{5}{2}\right]$ **99.** $\left(\frac{1}{8}, 2\right)$

101. $\left(\frac{7}{8}, \frac{27}{20}\right]$ **103.** $(-5, 2]$ **105.** $\left(\frac{11}{3}, \infty\right)$ **107.** $\left(-\frac{8}{5}, 2\right)$

109. $(-\infty, -3) \cup (3, \infty)$ **111.** $\{\ \}$

113. $\left(-\infty, -\frac{22}{15}\right] \cup \left[\frac{6}{5}, \infty\right)$ **115.** $(-\infty, -27) \cup (-9, \infty)$

Chapter 2 Test

1. $\{10\}$ **2.** $\{1\}$ **3.** $\{\ \}$ **4.** $\{n \mid n$ is a real number$\}$ **5.** $\{12\}$ **6.** $\left\{-\frac{80}{29}\right\}$ **7.** $\left\{1, \frac{2}{3}\right\}$ **8.** $\{\ \}$

9. $y = \dfrac{3x - 8}{4}$ **10.** $n = \dfrac{9}{7}m$ **11.** $g = \dfrac{S}{t^2 + vt}$ **12.** $C = \dfrac{5}{9}(F - 32)$ **13.** $(5, \infty)$ **14.** $[2, \infty)$

15. $\left(\frac{3}{2}, 5\right]$ **16.** $(-\infty, -2) \cup \left(\frac{4}{3}, \infty\right)$ **17.** $[5, \infty)$ **18.** $[4, \infty)$ **19.** $[-3, -1)$ **20.** $(-\infty, \infty)$

21. Approximately 8 hunting dogs could be safely kept in the pen.
22. They would need to sell more than 850 sunglasses to make a profit.   **23.** $8500 at 10%, $17,000 at 12%

24. $2\frac{1}{12}$ hr or 2 hr and 5 min   **25.** 11 twenties, 8 tens   **26.** 6.25 liters

Chapter 2 Cumulative Review

1. *(Sec. 1.1, Ex. 3)* a, b, c, and d are true; e and f are false   **2.** *(Sec. 1.2, Ex. 2)* (a) $>$; (b) $=$; (c) $<$

3. *(Sec. 1.3, Ex. 1)* (a) 3; (b) 5; (c) -2; (d) -8; (e) 0   **4.** *(Sec. 1.3, Ex. 3)* (a) 3; (b) 5; (c) $\frac{1}{2}$

5. *(Sec. 1.3, Ex. 6)* (a) $\frac{2}{5}$; (b) 3; (c) 173   **6.** *(Sec. 1.4, Ex. 2)* (a) -6; (b) -7; (c) -16; (d) 19; (e) $\frac{1}{6}$; (f) 0.94

7. *(Sec. 1.4, Ex. 7)* (a) 18; (b) 1; (c) undefined; (d) $\frac{12}{5}$ **8.** *(Sec. 2.1, Ex. 2)* (a) $2x - 2$; (b) $2x + 23$; (c) $3x$

9. *(Sec. 2.1, Ex. 5)* $\{1\}$ **10.** *(Sec. 2.1, Ex. 6)* $\{2\}$ **11.** *(Sec. 2.2, Ex. 3)* Five 1-gal containers of sealer should be bought

12. *(Sec. 2.3, Ex. 4)* 3 nickels; 17 dimes   **13.** *(Sec. 2.4, Ex. 2)* $\left\{\frac{4}{5}, -2\right\}$ **14.** *(Sec. 2.4, Ex. 7)* $\{\ \}$

15. *(Sec. 2.4, Ex. 9)* $\{4\}$ **16.** *(Sec. 2.5, Ex. 3)* (a) $(-\infty, -3)$; (b) $(-3, \infty)$ **17.** *(Sec. 2.5, Ex. 6)* $(-\infty, \infty)$
18. *(Sec. 2.6, Ex. 1)* $(-\infty, 4)$ **19.** *(Sec. 2.7, Ex. 2)* $(4, 8)$ **20.** *(Sec. 2.7, Ex. 7)* $(-\infty, -3] \cup [9, \infty)$

CHAPTER 3
Graphing Linear Equations and Inequalities

Mental Math, Sec. 3.1
1. $(5, 2)$ **3.** $(3, -1)$ **5.** $(-5, -2)$ **7.** $(-1, 0)$

Exercise Set 3.1

1. quadrant I

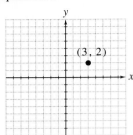

3. quadrant II

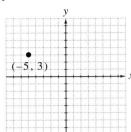

5. quadrant IV

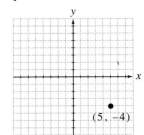

7. y-axis

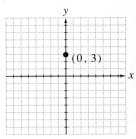

9. quadrant III

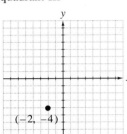

11. 5 units   **13.** 25 units   **15.** $\sqrt{41}$ units   **17.** $\sqrt{34}$ units   **19.** $(4, -2)$

21. $(2, 4)$ **23.** $\left(-5, \dfrac{5}{2}\right)$ **25.** $\left(6, -\dfrac{13}{2}\right)$ **27.** $\sqrt{17}$ units   **29.** 10 units

31. $\sqrt{10}$ units   **33.** $\sqrt{13}$ units   **35.** $\sqrt{41}$ units   **37.** 4 units   **39.** $(5, 7)$

41. $(-2, 1)$ **43.** $\left(-\dfrac{3}{2}, \dfrac{11}{2}\right)$ **45.** $\left(\dfrac{3}{4}, \dfrac{9}{2}\right)$ **47.** quadrant IV

49. quadrants I and IV   **51.** 20 meters   **53.** $\{-5\}$ **55.** $\left\{-\dfrac{1}{10}\right\}$

57. $\{x \mid x \le -5\}$ **59.** $\{x \mid x < -4\}$

Exercise Set 3.2

1. $(0, 5)$ **3.** $(3, 1)$ **5.** $\left(\dfrac{12}{5}, 0\right)$ **7.** $(-2, -11)$ **9.** $\left(0, -\dfrac{3}{2}\right)$ **11.** $\left(\dfrac{15}{2}, 3\right)$

13. x-intercept, 5;
y-intercept, 5

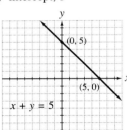

15. x-intercept, 2;
y-intercept, 4

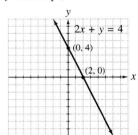

17. x-intercept, -4;
y-intercept, 2

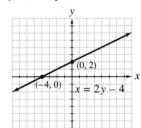

19. x-intercept, -2;
y-intercept, 6

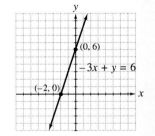

21.

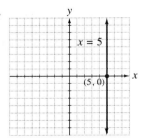

23.

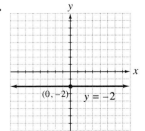

25.

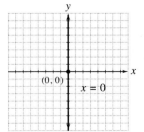

27. x-intercept, 8;
y-intercept, 4

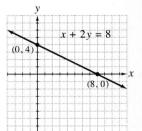

29. x-intercept, -5;

   y-intercept, -2

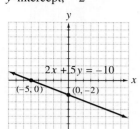

31. no x-intercept;

   y-intercept, -3

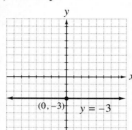

33. x-intercept, $\dfrac{3}{2}$;

   y-intercept, $\dfrac{9}{4}$

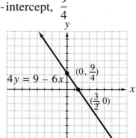

35. x-intercept, 0;

   y-intercept, 0

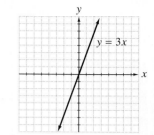

37. x-intercept, 10;

   no y-intercept

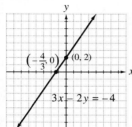

39. x-intercept, $-\dfrac{4}{3}$;

   y-intercept, 2

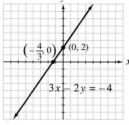

41. x-intercept, 7;

   y-intercept, 2

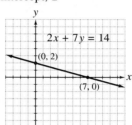

43. x-intercept, 5;

   y-intercept, -5

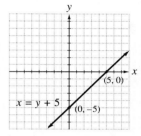

45. x-intercept, 0;
   y-intercept, 0

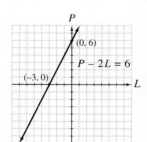

47. (a) ; (b) 14 in.

49. (a) $(0, 500)$, 500 chairs; (b) $(750, 0)$, 750 tables;
   (c) 466 chairs

51.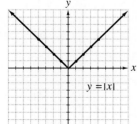

53.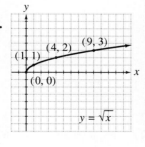

55.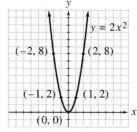

59. $\{9, -3\}$
61. $(-\infty, -4) \cup (-1, \infty)$
63. $\left[\dfrac{2}{3}, 2\right]$

Mental Math, Sec. 3.3
1. positive   **3.** 0   **5.** negative   **7.** negative   **9.** undefined

Exercise Set 3.3
1. $\dfrac{9}{5}$ **3.** $-\dfrac{7}{2}$ **5.** $-\dfrac{5}{6}$ **7.** $\dfrac{1}{3}$ **9.** $-\dfrac{4}{3}$ **11.** 0   **13.** undefined

15. 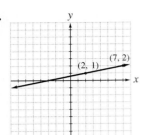   **17.** 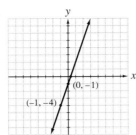   **19.** 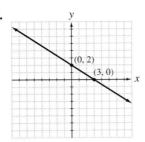   **21.**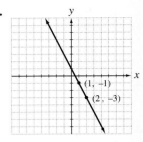

23. parallel   **25.** perpendicular   **27.** perpendicular   **29.** neither   **31.** $\dfrac{1}{4}$ **33.** -1 **35.** undefined

37. $\dfrac{7}{2}$ **39.** -3 **41.** 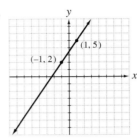   **43.** 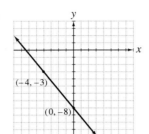   **45.**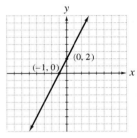

47. 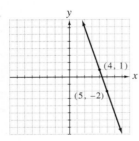   **49.** parallel   **51.** neither   **53.** neither   **55.** perpendicular

57. The slope of the roof is $\dfrac{2}{3}$.   **59.** The slope of his descent is approximately -0.12.

65. $y = \dfrac{-3x + 22}{5}$ **67.** $y = \dfrac{-6x + 9}{2}$ **69.** 6   **71.** -1

Mental Math, Sec. 3.4

1. $m = 6, b = 2$ **3.** $m = -2, b = 5$ **5.** $m = \dfrac{4}{3}, b = -3$ **7.** $m = -\dfrac{3}{5}, b = -\dfrac{7}{2}$

Exercise Set 3.4

1. $3x - y = 1$ **3.** $2x + y = -1$ **5.** $x - 2y = -10$ **7.** $9x + 10y = -27$ **9.** $3x - y = 6$ **11.** $2x + y = 1$

13. $x + 2y = -10$ **15.** $x - 3y = 21$ **17.** $m = -3, b = 8$ **19.** $m = 2, b = -3$ **21.** $m = \dfrac{2}{3}, b = -3$

23. $2x - y = -5$ **25.** $4x - 5y = 15$ **27.** $x + 8y = -12$ **29.** $x = 2$ **31.** $y = 1$ **33.** $x = 0$

35. parallel   **37.** perpendicular   **39.** neither   **41.** $4x - y = 4$ **43.** $3x + y = 1$ **45.** $3x + 2y = -12$

47. $y = 6$ **49.** $x = 0$ **51.** $x = 3$ **53.** $2x - y = -7$ **55.** $x + y = 7$ **57.** $x + 2y = 22$

59. $2x + 7y = -42$ **61.** $4x + 3y = -20$ **63.** $x = -2$ **65.** $x + 2y = 2$ **67.** $y = 12$ **69.** $8x - y = 47$

71. $x = 5$ **73.** $3x + 8y = -58$ **75.** $F = \dfrac{9}{5}C + 32$

77. (a) $S = -250P + 3500$; (b) daily sales will be approximately 1625   **79.** $-4x + y = 4$ **81.** $2x + y = -23$

83. $3x - 2y = -13$ **87.** $-\dfrac{1}{8}$ **89.** 12   **91.** $-\dfrac{1}{8}$ **93.** $-\dfrac{7}{8}$

Exercise Set 3.5

1. domain $\{-1, 0, -2, 5\}$, range $\{7, 6, 2\}$; is a function   **3.** domain $\{-2, 6, -7\}$, range $\{4, -3, -8\}$; not a function

5. domain $\{1\}$, range $\{1, 2, 3, 4\}$; not a function   **7.** domain $\left\{\dfrac{3}{2}, 0\right\}$, range $\left\{\dfrac{1}{2}, -7, \dfrac{4}{5}\right\}$; not a function

9. domain $\{-3, 0, 3\}$, range $\{-3, 0, 3\}$; is a function   **11.** domain $\{-1, 1, 2, 3\}$, range $\{1, 2\}$; is a function

13. domain $\{x \,|\, x \geq 0\}$, range $\{y \,|\, y$ is a real number$\}$; not a function

15. domain $\{x \,|\, -1 \leq x \leq 1\}$, range $\{y \,|\, y$ is a real number$\}$; not a function

17. domain $\{x \,|\, x \geq 2$ or $x \leq -2\}$, range $\{y \,|\, y$ is a real number$\}$; not a function

19. domain $\{x \,|\, x$ is a real number$\}$; is a function   **21.** domain $\{x \,|\, x \geq 0\}$; not a function

23. domain $\{x \,|\, x$ is a real number$\}$; is a function   **25.** domain $\{x \,|\, x \neq -4\}$; is a function

27. domain $\{4, 7, -4\}$, range $\{7, 4, 0\}$; is a function   **29.** domain $\{5, -6, 4, 2\}$, range $\{1, -8\}$; is a function

31. domain $\{0, 2, 5\}$, range $\{10, 9\}$; is a function   **33.** domain $\{x \,|\, x$ is a real number$\}$, range $\{y \,|\, y \leq 0\}$; is a function

35. domain $\{x \,|\, x$ is a real number$\}$, range $\{y \,|\, y$ is a real number$\}$; is a function

37. domain $\{x \,|\, x \geq -1\}$, range $\{y \,|\, y$ is a real number$\}$; not a function

39. domain $\{x \,|\, x$ is a real number$\}$, range $\{y \,|\, y$ is a real number$\}$; is a function

41. domain $\{3\}$, range $\{y \,|\, y$ is a real number$\}$; not a function

43. domain $\{x \,|\, x$ is a real number$\}$, range $\{y \,|\, y$ is a real number$\}$; is a function

45. domain $\{x \,|\, x \geq 0\}$; is a function   **47.** domain $\{x \,|\, -4 \leq x \leq 4\}$; not a function

49. domain $\{x \,|\, x$ is a real number$\}$; is a function   **51.** domain $\{x \,|\, x$ is a real number$\}$; is a function

53. $y = 12x$; is a function   **55.** $y = 29x$; is a function   **57.** $y = 0.08x$; is a function

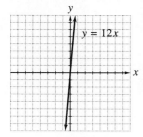

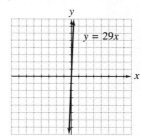

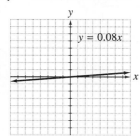

61. $\{x \,|\, -1 < x < 4\}$ **63.** $\{x \,|\, x < 5\}$ **65.** The angles have measures of $45°$ and $135°$.

Exercise Set 3.6

1.

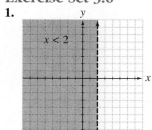

3.

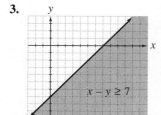

5.

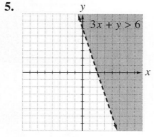

7.

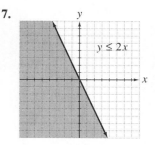

9.

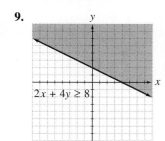

11.

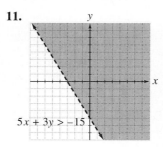

13.

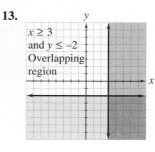

15.

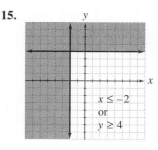

17.
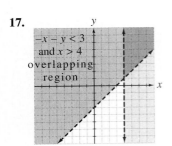
$-x - y < 3$ and $x > 4$ overlapping region

19.
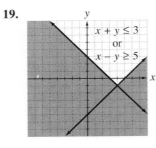
$x + y \leq 3$ or $x - y \geq 5$

21.
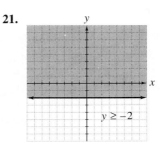
$y \geq -2$

23.
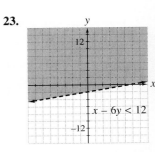
$x - 6y < 12$

25.
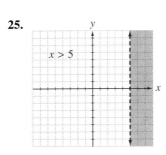
$x > 5$

27.
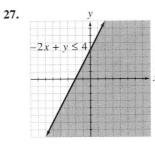
$-2x + y \leq 4$

29.
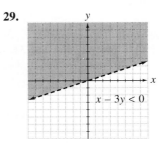
$x - 3y < 0$

31.
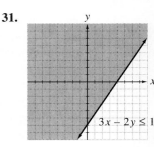
$3x - 2y \leq 1$

33.
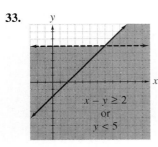
$x - y \geq 2$ or $y < 5$

35.
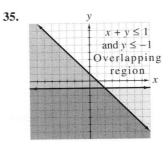
$x + y \leq 1$ and $y \leq -1$ Overlapping region

37.
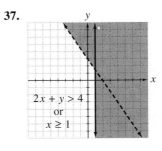
$2x + y > 4$ or $x \geq 1$

39.
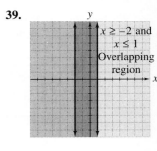
$x \geq -2$ and $x \leq 1$ Overlapping region

41.
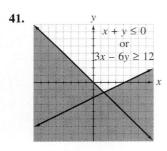
$x + y \leq 0$ or $3x - 6y \geq 12$

43.
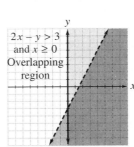
$2x - y > 3$ and $x \geq 0$ Overlapping region

45. $x \geq 2$ **47.** $y \leq -3$ **49.** $y > 4$ **51.** $x < 1$

53.
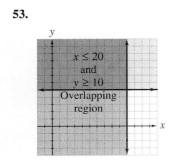
$x \leq 20$ and $y \geq 10$ Overlapping region

55.
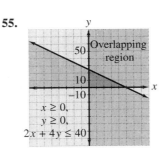
Overlapping region $x \geq 0$, $y \geq 0$, $2x + 4y \leq 40$

59. 8 **61.** -25 **63.** 16 **65.** $\dfrac{27}{125}$

Chapter 3 Review

1.

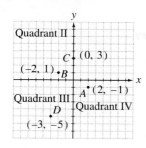

3. $\sqrt{197}$ units **5.** $\sqrt{130}$ units **7.** $(-5, 5)$ **9.** $\left(-\dfrac{15}{2}, 1\right)$

11. (a) yes; (b) no; (c) yes

13.

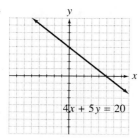

15.

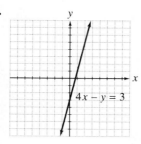

17.

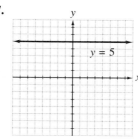

19. -3 **21.** $\dfrac{5}{2}$

23. neither **25.** $y = -1$ **27.** $x = -4$ **29.** $m = \dfrac{2}{5}, b = -\dfrac{4}{3}$ **31.** $3x - y = -14$ **33.** $2x + 3y = 12$

35. $2x + y = -2$ **37.** $3x - 4y = -14$ **39.** $x + 2y = -8$ **41.** $y = 3$

43. domain $\left\{-\dfrac{1}{2}, 6, 0, 25\right\}$, range $\left\{\dfrac{3}{4}, -12, 25\right\}$; is a function

45. domain $\{x \mid x$ is a real number$\}$, range $\{y \mid |y| \geq 1\}$; not a function

47. domain $\{x \mid x$ is a real number$\}$, range $\{4\}$; is a function

49.

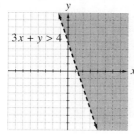

51.

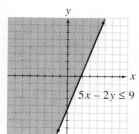

53.

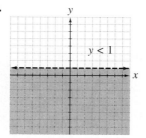

55.

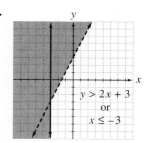

Chapter 3 Test

1. x-intercept, 3; y-intercept, 2 **2.** $(-6, -3)$ **3.**

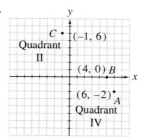

4.

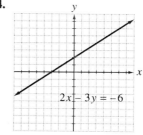

5.
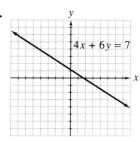

6.
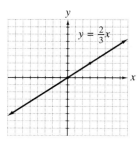

7.
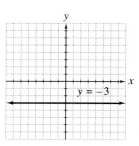

8. $2\sqrt{26}$ units   **9.** $\left(-4, \dfrac{7}{2}\right)$ **10.** $-\dfrac{3}{2}$ **11.** $m = -\dfrac{1}{4}, b = \dfrac{2}{3}$ **12.** $y = -8$ **13.** $x = -4$ **14.** $y = -2$

15. $3x + y = 11$ **16.** $5x - y = 2$ **17.** $x + 2y = 0$ **18.** $3x - y = 4$ **19.** $x + 2y = -1$ **20.** neither

21.
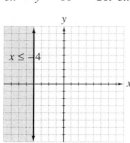

22.
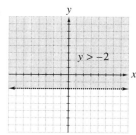

23.
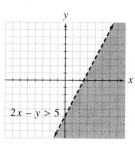

24.
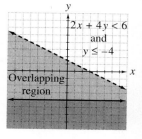

25. The equation represents a function.   **26.** The equation represents a function.   **27.** The graph represents a function.
28. The graph does not represent a function.   **29.** The graph represents a function.   **30.** The graph represents a function.

Chapter 3 Cumulative Review

1. *(Sec. 1.2, Ex. 1)* (a) $x + 5 = 20$; (b) $2(3 + y) = 4$; (c) $x - 8 = 2x$; (d) $\dfrac{z}{9} = 3(z - 5)$

2. *(Sec. 1.2, Ex. 5)* (a) $\dfrac{1}{5}$; (b) $-\dfrac{1}{2}$; (c) $\dfrac{7}{4}$ **3.** *(Sec. 1.3, Ex. 2)* (a) 16; (b) $\dfrac{1}{25}$; (c) $\dfrac{8}{27}$

4. *(Sec. 1.4, Ex. 1)* (a) -14; (b) -4; (c) 5; (d) -9; (e) $\dfrac{1}{4}$; (f) $-\dfrac{5}{21}$

5. *(Sec. 1.4, Ex. 5)* (a) 9; (b) $\dfrac{1}{16}$; (c) -25; (d) 25; (e) -125; (f) -125 **6.** *(Sec. 1.4, Ex. 8)* (a) -2; (b) 9; (c) -1

7. *(Sec. 2.1, Ex. 4)* $\{-4\}$ **8.** *(Sec. 2.1, Ex. 9)* $\{x \mid x$ is a real number$\}$ **9.** *(Sec. 2.2, Ex. 2)* 4.8 hr or 4 hr and 48 min

10. *(Sec. 2.4, Ex. 4)* $\{-1, 1\}$ **11.** *(Sec. 2.5, Ex. 2)* $[-10, \infty)$ **12.** *(Sec. 2.5, Ex. 5)* $\left(-\infty, -\dfrac{7}{3}\right]$

13. *(Sec. 2.6, Ex. 3)* $\left[-9, -\dfrac{9}{2}\right]$ **14.** *(Sec. 2.7, Ex. 5)* $(-\infty, -4) \cup (10, \infty)$

15. *(Sec. 3.1, Ex. 2)* (a) 5 units; (b) $\sqrt{2}$ units   **16.** *(Sec. 3.2, Ex. 3)*   **17.** *(Sec. 3.2, Ex. 5)*

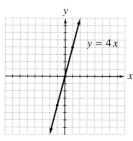

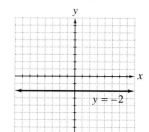

18. *(Sec. 3.3, Ex. 8)* Yes, they are perpendicular.   **19.** *(Sec. 3.4, Ex. 6)* $x = 2$
20. *(Sec. 3.5, Ex. 3)* (a) function; (b) not a function; (c) function; (d) not a function   **21.** *(Sec. 3.6, Ex. 1)*

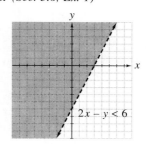

CHAPTER 4
Exponents and Polynomials

Mental Math, Sec. 4.1
1. exponent, 5; base, 3   **3.** exponent, 5; base, 4   **5.** exponent, 2; base, x **7.** exponent, 0; base, $2y + z$

Exercise Set 4.1

1. 16   **3.** -81 **5.** $\dfrac{9}{16}$ **7.** 36   **9.** 4^5 **11.** x^8 **13.** $-20x^2y$ **15.** $-16x^6y^3p^2$ **17.** -1 **19.** 1

21. a^3 **23.** x **25.** $-13z^4$ **27.** $\dfrac{1}{16}$ **29.** $\dfrac{1}{x^8}$ **31.** $\dfrac{5}{a^4}$ **33.** $\dfrac{1}{x^7}$ **35.** $4r^8$ **37.** 1   **39.** x^{7a+5}

41. x^{2t-1} **43.** 3.125×10^7 **45.** 1.6×10^{-2} **47.** 0.0000000036   **49.** 93,000,000   **51.** -36 **53.** $\dfrac{1}{64}$

55. $\dfrac{13}{36}$ **57.** 9   **59.** 64   **61.** x^{15} **63.** $\dfrac{1}{z^3}$ **65.** y^4 **67.** $\dfrac{3}{x}$ **69.** -2 **71.** r^8 **73.** $\dfrac{1}{x^9}$ **75.** $\dfrac{b^7}{9a^7}$

77. $\dfrac{6x^{16}}{5}$ **79.** x^{4a+7} **81.** z^{6x-7} **83.** x^{6t-1} **85.** 6.7413×10^4 **87.** 1.25×10^{-2} **89.** 5.3×10^{-5}

91. 1,278,000   **93.** 7,350,000,000,000   **95.** 0.000000403   **97.** 9.18×10^8 **99.** 5×10^7

105. -2 **107.** $\dfrac{3}{5}$ **109.** function

Mental Math, Sec. 4.2
1. x^{20} **3.** x^9 **5.** y^{42} **7.** z^{20} **9.** z^{18}

Exercise Set 4.2

1. $\dfrac{1}{9}$ **3.** $\dfrac{1}{x^{36}}$ **5.** $\dfrac{1}{y^5}$ **7.** $9x^4y^6$ **9.** $16x^{20}y^{12}$ **11.** $\dfrac{c^{18}}{a^{12}b^6}$ **13.** $\dfrac{y^{15}}{x^{35}z^{20}}$ **15.** $\dfrac{1}{a^2}$ **17.** 4   **19.** $\dfrac{x^4}{4z^2}$

21. x^{9a+18} **23.** x^{12a+2} **25.** 0.3   **27.** 20   **29.** $\dfrac{1}{125}$ **31.** $\dfrac{1}{x^{63}}$ **33.** $\dfrac{343}{512}$ **35.** $16x^4$ **37.** $-\dfrac{y^3}{64}$ **39.** $4^8x^2y^6$

41. $\dfrac{36}{p^{12}}$ **43.** $-\dfrac{a^6}{512x^3y^9}$ **45.** $\dfrac{x^{14}y^{14}}{a^{21}}$ **47.** $\dfrac{x^4}{16}$ **49.** 64   **51.** $\dfrac{1}{y^{15}}$ **53.** $\dfrac{2}{p^2}$ **55.** $\dfrac{3}{8x^8y^7}$ **57.** $\dfrac{1}{x^{30}b^6c^6}$

59. $\dfrac{25}{8x^5y^4}$ **61.** $\dfrac{2}{x^4y^{10}}$ **63.** b^{10x^2-4x} **65.** y^{15a+3} **67.** $16x^{4t+4}$ **69.** 10   **71.** 0.00008   **73.** 11,000,000

75. 0.002 sec   **77.** $\dfrac{8}{x^6y^3}$ cubic meters   **79.** $\dfrac{3x^{18}}{2y^{10}}$ **81.** $27x^{15}y^3z^5$ **83.** $\dfrac{27y^3}{2x^3}$ **85.** -56 **87.** $5x - y = 20$

Exercise Set 4.3

1. 0   **3.** 2   **5.** 3   **7.** binomial of degree 1   **9.** trinomial of degree 2   **11.** monomial of degree 3
13. degree 3; none of these   **15.** $6y$ **17.** $11x - 3$ **19.** $9x^2y - 2xy$ **21.** $xy + 2x - 1$ **23.** $18y^2 - 17$
25. $3x^2 - 3xy + 6y^2$ **27.** $x^2 - 4x + 8$ **29.** $y^2 + 3$ **31.** $-2x^2 + 5x$ **33.** $-2x^2 - 4x + 15$ **35.** 49
37. -51 **39.** 0   **41.** $4x - 13$ **43.** $x^2 + 2$ **45.** $12x^3 + 8x + 8$ **47.** $7x^3 + 4x^2 + 8x - 10$
49. $-18y^2 + 11y + 14$ **51.** $-x^3 + 8a - 12$ **53.** $5x^2 - 9x - 3$ **55.** $-3x^2 + 3$ **57.** $2x^3 + 3x^2 + 8xy - 3$
59. $7y^2 - 3$ **61.** $5x^2 + 22x + 16$ **63.** $-q^4 + q^2 - 3q + 5$ **65.** $15x^2 + 8x - 6$ **67.** $x^4 - 7x^2 + 5$
69. 15   **71.** 38   **73.** 7   **75.** 3   **77.** $(4x^2 - 2x + 1)$ cm   **79.** \$80,000
81. $P(a) = 2a - 3, P(-x) = -2x - 3, P(x + h) = 2x + 2h - 3$
83. $P(a) = 3a^2 + 4a, P(-x) = 3x^2 - 4x$
85. $P(a) = 4a - 1, P(-x) = -4x - 1$
89. $3x^3y$ **91.** $-16x^3y^3$ **93.**   **95.**

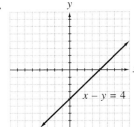 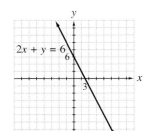

Exercise Set 4.4

1. $-12x^5$ **3.** $9x^2y^3z^3$ **5.** $12x^2 + 21x$ **7.** $-24x^2y - 6xy^2$ **9.** $-4a^3bx - 4a^3by + 12ab$ **11.** $2x^2 - 2x - 12$
13. $2x^4 + 3x^3 - 2x^2 + x + 6$ **15.** $x^4 + 4x^3 + 2x^2 - 4x + 1$ **17.** $3x^2 + 14x + 8$ **19.** $2x^3 - 14x^2 + 22x - 10$
21. $a^4 - 6a^3 + 26a + 15$ **23.** $6a^4 + a^3b + 4a^2b^2 + b^4$ **25.** $x^2 + x - 12$ **27.** $4x^2 - 24x + 32$
29. $3x^2 + 8x - 3$ **31.** $9x^2 - \dfrac{1}{4}$ **33.** $x^2 + 8x + 16$ **35.** $36y^2 - 1$ **37.** $9x^2 - 6xy + y^2$
39. $9b^2 - 36y^2$ **41.** $16b^2 + 32b + 16$ **43.** $4s^2 - 12s + 8$ **45.** $x^2y^2 - 4xy + 4$ **47.** $-12ab^2$
49. $4x^2y + 4y^2 + 4yz$ **51.** $9x^2 + 18x + 5$ **53.** $10x^5 + 8x^4 + 2x^3 + 25x^2 + 20x + 5$ **55.** $49x^2 - 9$
57. $9x^3 + 30x^2 + 12x - 24$ **59.** $16x^2 - \dfrac{2}{3}x - \dfrac{1}{6}$ **61.** $36x^2 + 12x + 1$ **63.** $x^4 - 4y^2$
65. $-30a^4b^4 + 36a^3b^2 + 36a^2b^3$ **67.** $2a^2 - 12a + 16$ **69.** $49a^2b^2 - 9c^2$ **71.** $-9x^2 - 6x + 15$
73. $m^2 - 8m + 16$ **75.** $4y^2 - 4yc + c^2 + 12y - 6c + 9$ **77.** $25x^2 - 20xy + 4y^2 - 16$ **79.** $9x^2 + 6x + 1$
81. $y^2 - 7y + 12$ **83.** $-4x^2 - 4xy - y^2 + 16$ **85.** $2x^3 + 2x^2y + x^2 + xy - x - y$
87. $9x^4 + 12x^3 - 2x^2 - 4x + 1$ **89.** $12x^3 - 2x^2 + 13x + 5$ **91.** $(6x^3 + 36x^2 - 6x)$ cubic ft
93. $10x^{-1} + 15x^{-2} + 5x^{-3}$ **95.** $-4y^2 + x^{-4}$ **97.** $3a^2 + 3ay^{-2} + ay^{-4} + y^{-6}$ **99.** $18a^{n+5} - 36a^4$
101. $9x^{2y} + 42x^y + 49$ **105.** $\dfrac{x^2y^2}{4}$ **107.** $-9x^3y^4$ **109.** $\{14, -4\}$

Exercise Set 4.5

1. $2b^5$ **3.** $\dfrac{2y^3z^2}{x^4b}$ **5.** $-\dfrac{1}{3y^2}$ **7.** $2a + 4$ **9.** $3ab + 4$ **11.** $2x^2 + 3x - 2$ **13.** $x^2 + 2x + 1$
15. $x + 1 + \dfrac{1}{x + 2}$ **17.** $2x - 8$ **19.** $x - \dfrac{1}{2}$ **21.** $2x^2 - \dfrac{1}{2}x + 5$ **23.** $\dfrac{5b^5}{2a^3}$ **25.** $x^3y^3 - 1$ **27.** $a + 3$

29. $2x + 5$ **31.** $4y - 6y^2$ **33.** $2x + 23 + \dfrac{130}{x - 5}$ **35.** $10x + 3y - 6x^2y^2$ **37.** $2x + 4$ **39.** $y + 5$

41. $2x + 3$ **43.** $2x^2 - 8x + 38 - \dfrac{156}{x + 4}$ **45.** $3x + 3 - \dfrac{1}{x - 1}$ **47.** $-2x^3 + 3x^2 - x + 4$

49. $3x^3 + 5x + 4 - \dfrac{2x}{x^2 - 2}$ **51.** $x - \dfrac{5}{3x^2}$ **53.** $(3x - 7)$ in.   **55.** $(x - 5)$ cm   **57.** $2x^2 + \dfrac{1}{2}x - 5$

59. $2x^3 + \dfrac{9}{2}x^2 + 10x + 21 + \dfrac{42}{x - 2}$ **61.** $3x^4 - 2x$ **65.** $=$ **67.** $=$ **69.** $-7 \le x \le 9$

71. $x < -3$ or $x > 2$

Exercise Set 4.6

1. $x + 8$ **3.** $x - 1$ **5.** $x^2 - 5x - 23 - \dfrac{41}{x - 2}$ **7.** $4x + 8 + \dfrac{7}{x - 2}$ **9.** 3 **11.** 73

13. -8 **15.** $x^2 + \dfrac{2}{x - 3}$ **17.** $6x + 7 + \dfrac{1}{x + 1}$ **19.** $2x^3 - 3x^2 + x - 4$ **21.** $3x - 9 + \dfrac{12}{x + 3}$

23. $3x^2 - \dfrac{9}{2}x + \dfrac{7}{4} + \dfrac{47}{8x - 4}$ **25.** $3x^2 + 3x - 3$ **27.** $3x^2 + 4x - 8 + \dfrac{20}{x + 1}$ **29.** $x^2 + x + 1$

31. $x - 6$ **33.** 1 **35.** -133 **37.** 3 **39.** $\dfrac{-187}{81}$ **41.** $\dfrac{95}{32}$ **45.** -5 **47.** 29 **49.** 3

51. 0 **53.** -1

Exercise Set 4.7

1. 15 **3.** 3 **5.** 4 **7.** $\dfrac{1}{3}$ **9.**

11.

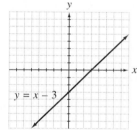

13.

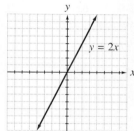

15. $(f + g)(x) = x^2 + 5x + 1$ **17.** $(f - g)(x) = x^2 - 5x + 1$ **19.** $\left(\dfrac{g}{f}\right)(x) = \dfrac{5x}{x^2 + 1}$

21. $(f \circ g)(x) = 25x^2 + 1$ **23.** 5 **25.** 51 **27.** 0 **29.** $(f + g)(x) = x^2 - 2x + 2$ **31.** $\left(\dfrac{f}{g}\right)(x) = -\dfrac{2x}{x^2 + 2}$

33. $(g - h)(x) = x^2 - 4x - 1$ **35.** 19 **37.** $(f \circ g)(x) = -2x^2 - 4$ **39.** 27 **41.** $(h + f)(x) = 2x + 3$

43. $(f \circ f)(x) = 4x$ **45.** $f(a + b) = -2a - 2b$ **47.** $\left(\dfrac{f}{h}\right)(x) = -\dfrac{2x}{4x + 3}$ **49.** $(g \circ h)(x) = 16x^2 + 24x + 11$

51.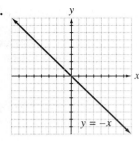

53.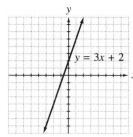

55.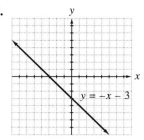

57. $(f \circ g)(x) = x$, $(g \circ f)(x) = x$ **59.** $P(x) = R(x) - C(x)$ **61.** $A(10) = 100\pi$ sq cm

63. (a) $W = 2H$; (b) $W = 130$ lb   **65.** $63x^2 - 7x$ **67.** $81x^2y - 36xy$ **69.** $\{0\}$ **71.** $\left\{\dfrac{17}{6}\right\}$

Chapter 4 Review

1. 4   **3.** -4 **5.** 1   **7.** $-\dfrac{1}{16}$ **9.** $-x^2y^7z$ **11.** $\dfrac{1}{a^9}$ **13.** $\dfrac{1}{x^{11}}$ **15.** $\dfrac{1}{y^5}$ **17.** -3.62×10^{-4}

19. 410,000   **21.** $\dfrac{a^2}{16}$ **23.** $\dfrac{1}{16x^2}$ **25.** $\dfrac{1}{8^{18}}$ **27.** $-\dfrac{1}{8x^9}$ **29.** $-\dfrac{27y^6}{x^6}$ **31.** $\dfrac{xz}{4}$ **33.** $\dfrac{2}{27z^3}$

35. $2y^{x-7}$ **37.** -2.21×10^{-11} **39.** $\dfrac{x^3y^{10}}{3z^{12}}$ **41.** 4   **43.** 1   **45.** 1   **47.** $-4xy^3 - 3x^3y$

49. $-4x^2 + 10y^2$ **51.** $-4x^3 + 4x^2 + 16xy - 9x + 18$ **53.** $x^2 - 6x + 3$ **55.** 290   **57.** 110

59. $9x^2 + 18xh + 9h^2 - 7x - 7h + 8$ **61.** $-24x^2y^4 + 36x^2y^3 - 6xy^3$ **63.** $2x^2 + x - 36$

65. $9x^2a^2 - 24xab + 16b^2$ **67.** $15x^2 + 18x - 81$ **69.** $9x^2 - 6xy + y^2$ **71.** $x^2 - 9y^2$

73. $-9a^2 + 6ab - b^2 + 16$ **75.** $y^4 - 18y^3 + 87y^2 - 54y + 9$ **77.** $16y^6 + 24y^3x^2 + 9x^4$

79. $16x^2y^{2z} - 8xy^zb + b^2$ **81.** $\dfrac{x^4b^9}{3y^6}$ **83.** $1 + \dfrac{x}{2y} - \dfrac{9}{4xy}$ **85.** $3x^3 + 9x^2 + 2x + 6 - \dfrac{2}{x-3}$

87. $2x^3 + 2x - 2$ **89.** $3x^2 + 2x - 1$ **91.** $3x^2 + 6x + 24 + \dfrac{44}{x-2}$ **93.** $x^4 - x^3 + x^2 - x + 1 - \dfrac{2}{x+1}$

95. $3x^3 + 13x^2 + 51x + 204 + \dfrac{814}{x-4}$ **97.** 3043   **99.** $\dfrac{113}{81}$ **101.** $(f + g)(x) = x^2 + x - 1$

103. $\left(\dfrac{h}{g}\right)(x) = \dfrac{x^3 - x^2}{x+1}$ **105.** $(f \circ g)(x) = x^2 + 2x - 1$ **107.** 18   **109.** -2 **111.**

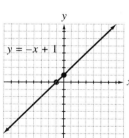

Chapter 4 Test

1. -8 **2.** $\dfrac{1}{36}$ **3.** $-12x^2z$ **4.** $\dfrac{1}{81x^2}$ **5.** $-\dfrac{y^{40}}{z^5}$ **6.** $\dfrac{12x^5z^3}{y^7}$ **7.** 6.3×10^8 **8.** 1.2×10^{-2}

9. 0.000005   **10.** 0.0009   **11.** $-5x^3 - 11x - 9$ **12.** $14x^6 - 3xy^2 + 6y^2 + 15$ **13.** 95

14. $-12x^2y - 3xy^2$ **15.** $12x^2 - 5x - 28$ **16.** $81x^4 + 72x^2y + 36x^2 + 16y^2 + 16y + 4$

17. $-16x^2 + 8xy - y^2 - 24x + 6y + 9$ **18.** $\dfrac{4xy}{3z} + \dfrac{3}{z} + \dfrac{1}{3x}$

19. $x^5 + 5x^4 + 8x^3 + 16x^2 + 33x + 63 + \dfrac{128}{x-2}$ **20.** $4x^3 - \dfrac{1}{3}x^2 + \dfrac{16}{9}x + \dfrac{5}{27} - \dfrac{71}{81\left(x - \dfrac{2}{3}\right)}$

21. 91 **22.** $(h - g)(x) = x^2 - 7x + 12$ **23.** $(h \cdot f)(x) = x^3 - 6x^2 + 5x$ **24.** $(g \circ f)(x) = x - 7$

25. $(g \circ h)(x) = x^2 - 6x - 2$ **26.**

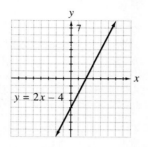

$y = 2x - 4$

Chapter 4 Cumulative Review

1. (a) *(Sec. 1.2, Ex. 3)* $5 + y \geq 7$; (b) $11 \neq z$; (c) $20 < 5 - x$ **2.** *(Sec. 1.2, Ex. 6)* (a) -8; (b) $-\dfrac{1}{5}$; (c) 9

3. *(Sec. 1.4, Ex. 6)* (a) -5; (b) 3; (c) $-\dfrac{1}{8}$; (d) -4; (e) $\dfrac{1}{4}$ **4.** *(Sec. 2.1, Ex. 3)* $\{2\}$

5. *(Sec. 2.1, Ex. 10)* (a) $10 - x$ ft; (b) $(100 - 3x)°$; (c) $2x + 1$ **6.** *(Sec. 2.3, Ex. 1)* 9, 11, 13

7. *(Sec. 2.4, Ex. 3)* $\{24, -20\}$

8. *(Sec. 3.1, Ex. 1)*
(a) fourth quadrant;
(b) on y-axis;
(c) second quadrant;
(d) on x-axis;
(e) third quadrant

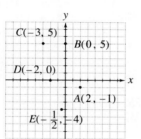

9. *(Sec. 3.3, Ex. 1)* 1 **10.** *(Sec. 3.3, Ex. 5)*

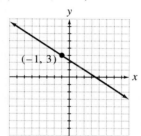

11. *(Sec. 3.4, Ex. 1)* $3x + y = -2$ **12.** *(Sec. 3.5, Ex. 2)* (a), (c), (d), (e) are functions

13. *(Sec. 3.6, Ex. 3)*

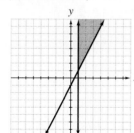

14. *(Sec. 4.1, Ex. 3)* (a) $15x^7$; (b) $-8x^4 p^{12}$

15. *(Sec. 4.1, Ex. 6)* (a) $\dfrac{1}{25}$; (b) $\dfrac{2}{x^3}$; (c) $\dfrac{1}{3x}$; (d) $\dfrac{1}{m^{10}}$; (e) $\dfrac{1}{27}$; (f) $\dfrac{11}{18}$; (g) t^5

16. *(Sec. 4.3, Ex. 5)* $12x^3 - 12x^2 - 9x + 2$

17. *(Sec. 4.4, Ex. 3)* (a) $2x^2 + 11x + 15$; (b) $10x^5 - 12x^4 + 14x^3 - 15x^2 + 18x - 21$

18. *(Sec. 4.5, Ex. 2)* $2x^2 - x + 4$ **19.** *(Sec. 4.5, Ex. 6)* $2x - 3 - \dfrac{3}{3x - 5}$

20. *(Sec. 4.6, Ex. 1)* $2x^2 + 5x + 2 + \dfrac{7}{x - 3}$

21. *(Sec. 4.7, Ex. 4)* (a) $x^2 + 2x + 1$; (b) 5

CHAPTER 5
Factoring Polynomials

Mental Math, Sec. 5.1
1. 6   **3.** 5   **5.** x **7.** $7x$

Exercise Set 5.1
1. 6   **3.** 28   **5.** 4   **7.** 6   **9.** a^3 **11.** y^2z^2 **13.** $3x^2y$ **15.** $5xz^3$ **17.** $6(3x - 2)$ **19.** $4y^2(1 - 4xy)$
21. $2x^3(3x^2 - 4x + 1)$ **23.** $4ab(2a^2b^2 - ab + 1 + 4b)$ **25.** $(x + 3)(6 + 5a)$ **27.** $(z + 7)(2x + 1)$
29. $(x^2 + 5)(3x - 2)$ **31.** $(a + 2)(b + 3)$ **33.** $(a - 2)(c + 4)$ **35.** $(x - 2)(2y - 3)$ **37.** $(4x - 1)(3y - 2)$
39. $3(2x^3 + 3)$ **41.** $x^2(x + 3)$ **43.** $4a(2a^2 - 1)$ **45.** $-4xy(5x - 4y^2)$ **47.** $5ab^2(2ab + 1 - 3b)$
49. $3b(3ac^2 + 2a^2c - 2a + c)$ **51.** $(y - 2)(4x - 3)$ **53.** $(n - 8)(2m - 1)$ **55.** $3x^2y^2(5x - 6)$
57. $(2x + 3)(3y + 5)$ **59.** $(x + 3)(y - 5)$ **61.** $(2a - 3)(3b - 1)$ **63.** $(6x + 1)(2y + 3)$ **65.** $(2x + 3y)(x + 2)$
67. $(5x - 3)(x + y)$ **69.** $(x^2 + 4)(x + 3)$ **71.** $(x^2 - 2)(x - 1)$ **73.** $2\pi r(r + h)$ **75.** $P(1 + RT)$
77. $x^n(x^{2n} - 2x^n + 5)$ **79.** $2x^{3a}(3x^{5a} - x^{2a} - 2)$ **81.** $x^{-2}(3 + 8x)$ **83.** $xy^{-3}(3x + 2y^2)$
85. $2x^{-2}y^{-4}(3y^3 - 1 + 4x^3y^2)$ **91.** $-14y^4$ **93.** $16y^{12}$ **95.** $-16xy + 5$

Mental Math, Sec. 5.2
1. 5 and 2   **3.** 8 and 3

Exercise Set 5.2
1. $(x + 3)(x + 6)$ **3.** $(x - 8)(x - 4)$ **5.** $(x + 12)(x - 2)$ **7.** $(x - 6)(x + 4)$ **9.** $3(x - 2)(x - 4)$
11. $4z(x + 2)(x + 5)$ **13.** $2(x + 18)(x - 3)$ **15.** $(5x + 1)(x + 3)$ **17.** $(2x - 3)(x - 4)$ **19.** prime polynomial
21. $(2x - 3)^2$ **23.** $2(3x - 5)(2x + 5)$ **25.** $y^2(3y + 5)(y - 2)$ **27.** $2x(3x^2 + 4x + 12)$ **29.** $(x + 7z)(x + z)$
31. $(2x + y)(x - 3y)$ **33.** $(x^2 + 3)(x^2 - 2)$ **35.** $(5x + 8)(5x + 2)$ **37.** $(x^3 - 4)(x^3 - 3)$ **39.** $(a - 3)(a + 8)$
41. $(x - 4)(x + 3)$ **43.** $2(7y + 2)(2y + 1)$ **45.** $(2x - 3)(x + 9)$ **47.** $(x - 27)(x + 3)$ **49.** $(x - 18)(x + 3)$
51. $3(x - 1)^2$ **53.** $(3x + 1)(x - 2)$ **55.** $(4x - 3)(2x - 5)$ **57.** $3x^2(2x + 1)(3x + 2)$ **59.** $3(a + 2b)^2$
61. prime polynomial   **63.** $(2x + 13)(x + 3)$ **65.** $(3x - 2)(2x - 15)$ **67.** $(x^2 - 6)(x^2 + 1)$
69. $x(3x + 1)(2x - 1)$ **71.** $(4a - 3b)(3a - 5b)$ **73.** $(3x + 5)^2$ **75.** $y(3x - 8)(x - 1)$ **77.** $2(x + 3)(x - 2)$
79. $(x + 2)(x - 7)$ **81.** $(2x^3 - 3)(x^3 + 3)$ **83.** $2x(6y^2 - z)^2$ **85.** $h(3h + 4)(h - 2)$ **87.** $(x^n + 8)(x^n + 2)$
89. $(x^n - 6)(x^n + 3)$ **91.** $(2x^n + 1)(x^n + 5)$ **93.** $(2x^n - 3)^2$ **95.** $x^3 - 8$ **97.** -9 **99.** -8

Exercise Set 5.3
1. $(x + 3)^2$ **3.** $(2x - 3)^2$ **5.** $3(x - 4)^2$ **7.** $x^2(3y + 2)^2$ **9.** $(x + 5)(x - 5)$ **11.** $(3 + 2z)(3 - 2z)$
13. $(y + 9)(y - 5)$ **15.** $4(4x + 5)(4x - 5)$ **17.** $(x + 3)(x^2 - 3x + 9)$ **19.** $(m + n)(m^2 - mn + n^2)$
21. $b(a + 2b)(a^2 - 2ab + 4b^2)$ **23.** $(z - 1)(z^2 + z + 1)$ **25.** $y^2(x - 3)(x^2 + 3x + 9)$
27. $(5y - 2x)(25y^2 + 10yx + 4x^2)$ **29.** $(x + 3 + y)(x + 3 - y)$ **31.** $(x - 5 + y)(x - 5 - y)$
33. $(2x + 1 + z)(2x + 1 - z)$ **35.** $(x + 4)(x - 4)$ **37.** $(3x + 7)(3x - 7)$ **39.** $(x^2 + 9)(x + 3)(x - 3)$
41. $(x + 4 + 2y)(x + 4 - 2y)$ **43.** $(x + 2y + 3)(x + 2y - 3)$ **45.** $(x - 1)(x^2 + x + 1)$
47. $(x + 5)(x^2 - 5x + 25)$ **49.** prime polynomial   **51.** $(2a + 9b)(2a - 9b)$ **53.** $2y(3x + 1)(3x - 1)$
55. $(4x + 9)(4x - 11)$ **57.** $(x + 4)(x^2 - 4x + 16)$ **59.** $(x^2 - y)(x^4 + x^2y + y^2)$
61. $(x + 8 + x^2)(x + 8 - x^2)$ **63.** $(y + x + 3)(y - x - 3)$ **65.** $6(x + 2)(x - 2)$ **67.** $4(x^2 + 4)$
69. $(x^4 + 1)(x^2 + 1)(x + 1)(x - 1)$ **71.** $3y^2(x^2 + 3)(x^4 - 3x^2 + 9)$ **73.** $3(3x + 2)(x - 2)$
75. $(x + y + 5)(x^2 + 2xy + y^2 - 5x - 5y + 25)$ **77.** $(x - y + 1)(x^2 - 2xy + y^2 - x + y + 1)$
79. $(2x - y + 3)(4x^2 - 4xy + y^2 - 6x + 3y + 9)$ **81.** $(2x - 1)(4x^2 + 20x + 37)$
83. $\pi R^2 - \pi r^2 = \pi(R + r)(R - r)$ **85.** $(x^n + 5)(x^n - 5)$ **87.** $(x^n + 3)(x^n - 3)$ **89.** $(6x^n + 7)(6x^n - 7)$
91. $(x^{2n} + 4)(x^n + 2)(x^n - 2)$ **95.** $x^4 + 2x^3 + 4x^2 + 8x + 16$ **97.** $\dfrac{1}{27}$ **99.** y^5

Exercise Set 5.4
1. $(x + 3)(x - 3)$ **3.** $(x - 9)(x + 1)$ **5.** prime polynomial   **7.** $(x - 4 + y)(x - 4 - y)$ **9.** $x(x - 1)(x^2 + x + 1)$
11. $2xy(7x - 1)$ **13.** $2ab(4a - 3b)$ **15.** $x^3(x - 1)$ **17.** $x^2(x + 2)(x - 2)$ **19.** $4(x + 2)(x - 2)$
21. $x(1 + 3x)(1 - 3x)$ **23.** $(3x - 11)(x + 1)$ **25.** $4(x + 3)(x - 1)$ **27.** $(4x + 3)(x + 5)$ **29.** $(3x^2 - 4)(2 + y)$

31. $2(x + 3)(y + 4)$ **33.** $(x + 2)(x - 2)(x + 3)$ **35.** $(2x + 9)^2$ **37.** $(3x - 5)^2$ **39.** prime polynomial
41. $3(4x - 1)^2$ **43.** prime polynomial   **45.** $2(x^2 + 1)(x + 1)(x - 1)$ **47.** $(a - 2b)(a^2 + 2ab + 4b^2)$
49. $(5 - x)(25 + 5x + x^2)$ **51.** $(2x + 3y)(4x^2 - 6xy + 9y^2)$ **53.** $2(a + 4b)(a^2 - 4ab + 16b^2)$
55. $6a^2(b^2 + 2a)(b^2 - 2a)$ **57.** $2y^3(2xy + 1)(2xy - 1)$ **59.** $(x + 3 + y)(x + 3 - y)$
61. $2(x - 5 + y)(x - 5 - y)$ **63.** $3(a - 1 + b)(a - 1 - b)$ **65.** $x(3x - y)(9x^2 + 3xy + y^2)$
67. $2y(2xy + 3)(4x^2y^2 - 6xy + 9)$ **69.** $8x^2(2y - 1)(4y^2 + 2y + 1)$ **71.** $3(x + 1)(3x + 2)$
73. $(x^2 + 7)(x + 1)(x - 1)$ **75.** $(x^3 + 3)(x + 1)(x^2 - x + 1)$ **77.** $(x + y + 3)(x + y - 3)$
79. $(a - 3)(a^2 - 9a + 21)$ **81.** $(x + 5 + y)(x^2 + 10x - xy - 5y + y^2 + 25)$ **83.** $2(x + 2)(4x^2 - 2x + 7)$
85. $-x(x^2 + 9x + 27)$ **87.** $12(x + 3)$ sq in.   **89.** $(x^n + 5)(x^n - 5)$ **91.** $(x^{2n} + 4)(x^n + 2)(x^n - 2)$

93. $(y^n - 3)(y^{2n} + 3y^n + 9)$ **95.** $\{5\}$ **97.** $\left\{-\dfrac{1}{3}\right\}$ **99.** 24   **101.** 3

Mental Math, Sec. 5.5
1. $\{3, -5\}$ **3.** $\{3, -7\}$ **5.** $\{0, 9\}$

Exercise Set 5.5
1. $\left\{-3, \dfrac{4}{3}\right\}$ **3.** $\left\{\dfrac{5}{2}, -\dfrac{3}{4}\right\}$ **5.** $\{-3, -8\}$ **7.** $\left\{\dfrac{1}{4}, -\dfrac{2}{3}\right\}$ **9.** $\{1, 9\}$ **11.** $\left\{\dfrac{3}{5}, -1\right\}$ **13.** $\{0\}$ **15.** $\{6, -3\}$

17. $\left\{\dfrac{2}{5}, -\dfrac{1}{2}\right\}$ **19.** $\left\{\dfrac{3}{4}, -\dfrac{1}{2}\right\}$ **21.** $\left\{-2, 7, \dfrac{8}{3}\right\}$ **23.** $\{0, 3, -3\}$ **25.** $\{2, 1, -1\}$ **27.** length, 12

29. The integers are $-4, -3,$ and -2 or 2, 3, and 4.   **31.** $\left\{-\dfrac{7}{2}, 10\right\}$ **33.** $\{0, 5\}$ **35.** $\{5, -3\}$ **37.** $\left\{\dfrac{1}{3}, -\dfrac{1}{2}\right\}$

39. $\{9, -4\}$ **41.** $\left\{\dfrac{4}{5}\right\}$ **43.** $\{0, -5, 2\}$ **45.** $\left\{0, \dfrac{4}{5}, -3\right\}$ **47.** $\{\ \}$ **49.** $\{-7, 4\}$ **51.** $\{4, 6\}$ **53.** $\left\{-\dfrac{1}{2}\right\}$

55. $\{-4, -3, 3\}$ **57.** $\{0, -5, 5\}$ **59.** $\{-6, 5\}$ **61.** $\left\{0, -\dfrac{1}{3}, 1\right\}$ **63.** $\left\{0, -\dfrac{1}{3}\right\}$ **65.** $\left\{-\dfrac{7}{8}\right\}$

67. $\left\{\dfrac{31}{4}\right\}$ **69.** $\{1\}$ **71.** The numbers are -11 and -6 or 6 and 11.

73. 75 ft   **75.** 12 cm and 9 cm   **77.** 2 in   **79.** 4 sec   **85.** $\dfrac{4x}{y}$ **87.** $\dfrac{3a^3}{b^2}$ **89.** function   **91.** function

Mental Math, Sec. 5.6
1. upward   **3.** downward

Exercise Set 5.6
1.
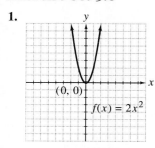

3.
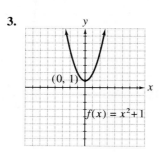

5.
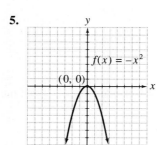

7. $(-4, -9)$

9. $(-1, 1)$ **11.** $(5, 30)$

13.
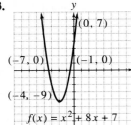
$f(x) = x^2 + 8x + 7$

15.
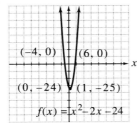
$f(x) = x^2 - 2x - 24$

17.
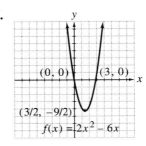
$f(x) = 2x^2 - 6x$

19.
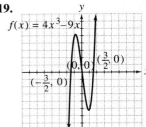
$f(x) = 4x^3 - 9x$

21.
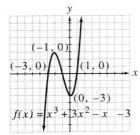
$f(x) = x^3 + 3x^2 - x - 3$

23.
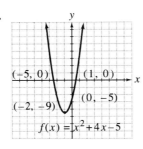
$f(x) = x^2 + 4x - 5$

25.
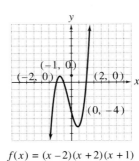
$f(x) = (x-2)(x+2)(x+1)$

27.
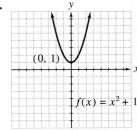
$f(x) = x^2 + 1$

29.
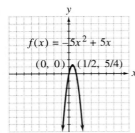

31.
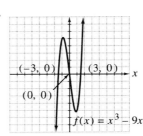

33.
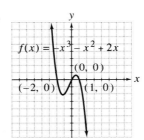

35.
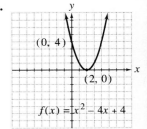

37.
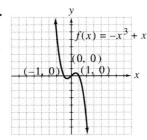

39.
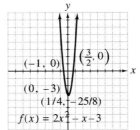

41.
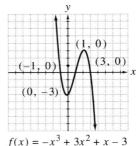
$f(x) = -x^3 + 3x^2 + x - 3$

43.
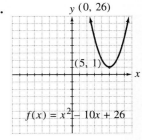
$f(x) = x^2 - 10x + 26$

45.
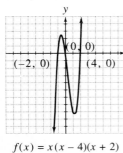
$f(x) = x(x-4)(x+2)$

49. domain, $\{x \mid x$ is a real number$\}$; range, $\{y \mid y \geq -3\}$

51. domain, $\{x \mid -4 \leq x \leq 4\}$; range, $\{y \mid -2 \leq y \leq 2\}$

Chapter 5 Review

1. 12   **3.** x^2 **5.** $2x$ **7.** $8x^2(2x - 3)$ **9.** $3xy^2z(5x^2y^2 - z + 2x)$ **11.** $2ab(3b + 4 - 2ab)$
13. $(a + 3b)(6a - 5)$ **15.** $(x - 6)(y + 3)$ **17.** $(p - 5)(q - 3)$ **19.** $(x^2 - 2)(x - 1)$ **21.** $(x - 18)(x + 4)$
23. $2(x - 2)(x - 7)$ **25.** $6xy(x + 2)(x - 2)$ **27.** $(2x - 9)(x + 1)$ **29.** $(6x + 5)(x + 2)$ **31.** $(2x - 3)^2$
33. $(5x + 6y)^2$ **35.** $2(2x - 3)(x + 2)$ **37.** $(x + 4)(x + 3)$ **39.** $(x^2 - 8)(x^2 + 2)$ **41.** $(x + 10)(x - 10)$
43. $2(x + 4)(x - 4)$ **45.** $(9 + x^2)(3 + x)(3 - x)$ **47.** $(y + 7)(y - 3)$ **49.** $(y - 2)(y^2 + 2y + 4)$
51. $(x + 6)(x^2 - 6x + 36)$ **53.** $(2 - 3y)(4 + 6y + 9y^2)$ **55.** $6xy(x + 2)(x^2 - 2x + 4)$
57. $(x - 1 + y)(x - 1 - y)$ **59.** $(2x + 1 + 3y)(2x + 1 - 3y)$ **61.** $3xy(2y - 1)$ **63.** $25(x + 2)(x - 2)$
65. $(3x - 4)(x - 2)$ **67.** $(x + 2)(x - 2)(y + 3)$ **69.** prime polynomial   **71.** $(2x + 5)(2x - 5)$
73. $(2x - y)(4x^2 + 2xy + y^2)$ **75.** $2x^3y(x + 3y)(x^2 - 3xy + 9y^2)$ **77.** $4(a - 3 + b)(a - 3 - b)$

79. $2x^3y^5(y - 4)(y^2 + 4y + 16)$ **81.** $\left\{\dfrac{1}{3}, -7\right\}$ **83.** $\left\{0, 4, \dfrac{9}{2}\right\}$ **85.** $\{0, 6\}$ **87.** $\left\{-\dfrac{1}{3}, 2\right\}$ **89.** $\{-4, 1\}$

91. $\{0, 6, -3\}$ **93.** $\{0, -2, 1\}$ **95.**

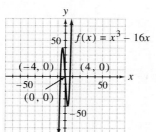

97.

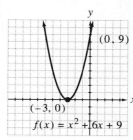

$f(x) = (x - 1)(x^2 - 2x - 3)$

99.

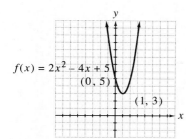

101.

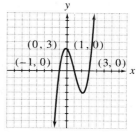

Chapter 5 Test

1. $4x^2y(4x - 3y^3)$ **2.** $5ab^2(4a^2 - 7b)$ **3.** $(x + 3)(y - 5)$ **4.** $(x^2 + 3)(x + 2)$ **5.** $(x - 3)(x - 8)$
6. $(x - 15)(x + 2)$ **7.** $(2x - 1)(x + 9)$ **8.** $(2y + 5)^2$ **9.** $3(2x + 1)(x - 3)$ **10.** $x(x^2 + 1)(x^2 + 2)$
11. $(2x + 5)(2x - 5)$ **12.** $3(x + 1)(3x - 1)$ **13.** $(x + 4)(x^2 - 4x + 16)$ **14.** $(2x - 5)(4x^2 + 10x + 25)$
15. $3y(x + 3y)(x - 3y)$ **16.** $(4x + 3)(2x - 3)$ **17.** $6(x^2 + 4)$ **18.** $4y^2(x^2 - 2y)(x^4 + 2x^2y + 4y^2)$

19. $(x + 3)(x - 3)(y - 3)$ **20.** $4(2x - 1)(x - 5)$ **21.** $\left\{4, -\dfrac{8}{7}\right\}$ **22.** $\{2, 3\}$ **23.** $\{0, -2, 2\}$

24. $\left\{-\dfrac{5}{2}, -2, 2\right\}$ **25.** $\left\{\dfrac{5}{3}, -1\right\}$ **26.**

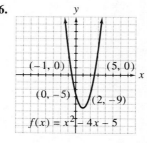

27.

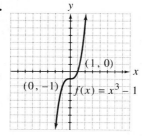

Chapter 5 Cumulative Review

1. *(Sec. 1.1, Ex. 4)* (a) true; (b) false; (c) false   **2.** *(Sec. 1.3, Ex. 5)* (a) 8; (b) 72; (c) 4; (d) $\dfrac{8}{5}$

3. *(Sec. 1.4, Ex. 4)* (a) 8; (b) $-\dfrac{1}{3}$; (c) -9; (d) 0; (e) $-\dfrac{2}{11}$; (f) 42; (g) 0   **4.** *(Sec. 2.1, Ex. 7)* $\left\{\dfrac{21}{11}\right\}$ **5.** *(Sec. 2.3, Ex. 2)* 24 cm

6. *(Sec. 2.5, Ex. 1)*

(a) $[2, \infty)$;   (b) $(-\infty, -1)$;   (c) $(0.5, 3]$

7. *(Sec. 2.7, Ex. 1)* $[-3, 3]$ **8.** *(Sec. 3.1, Ex. 3)* $\left(-1, \dfrac{3}{2}\right)$ **9.** *(Sec. 3.4, Ex. 2)* $5x - 8y = 20$

10. *(Sec. 3.4, Ex. 10)* $x = 1$

11. *(Sec. 3.6, Ex. 2)*

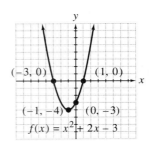

$3x \geq y$

12. *(Sec. 4.1, Ex. 1)* (a) 81; (b) 8; (c) 7; (d) $\dfrac{8}{27}$; (e) -16; (f) 16; (g) 32

13. *(Sec. 4.1, Ex. 9)* (a) 7.3×10^5; (b) 1.04×10^{-6}

14. *(Sec. 4.3, Ex. 7)* $12z^5 + 3z^4 - 13z^3 - 11z$ **15.** *(Sec. 4.4, Ex. 7)* $6x^2 - 29x + 28$

16. *(Sec. 4.5, Ex. 7)* $3x^2 + 2x + 3 + \dfrac{-6x + 9}{x^2 - 1}$ **17.** *(Sec. 5.1, Ex. 8)* $(a + 2)(b - 6)$

18. *(Sec. 5.2, Ex. 3)* $5x(x + 1)(x - 7)$ **19.** *(Sec. 5.3, Ex. 2)* $3x(a - 2b)^2$

20. *(Sec. 5.3, Ex. 6)* $(p + 3q)(p^2 - 3pq + 9q^2)$ **21.** *(Sec. 5.5, Ex. 4)* $\left\{-\dfrac{1}{6}, 3\right\}$

22. *(Sec. 5.6, Ex. 2)*
vertex, $(-1, -4)$;
x-intercept, -3 and 1;
y-intercept, -3

$(-3, 0)$ $(1, 0)$
$(-1, -4)$ $(0, -3)$
$f(x) = x^2 + 2x - 3$

CHAPTER 6
Rational Expressions

Exercise Set 6.1

1. none   **3.** $x = 0$ **5.** $x = 2$ **7.** $x = 2$ or $x = -2$ **9.** $\dfrac{5x^2}{9}$ **11.** $\dfrac{x^4}{2y^2}$ **13.** $\dfrac{1}{2(q - 1)}$ **15.** 1

17. $-\dfrac{1}{x + 1}$ **19.** $\dfrac{1}{7 - x}$ **21.** $\dfrac{1}{x^2 - 3x + 9}$ **23.** $\dfrac{2x^2 + 4x + 8}{3}$ **25.** $\dfrac{y + 2}{x - 3}$ **27.** $\dfrac{10y^2z}{4y^3z}$ **29.** $\dfrac{3x^2 + 15x}{2x^2 + 9x - 5}$

31. $\dfrac{x^2 - 4}{x + 2}$ **33.** $x = -11$ **35.** $x = 2$ **37.** none   **39.** $x = 0$ or $x = 2$ **41.** $x = 0, x = -2,$ or $x = 1$ **43.** $\dfrac{2}{9}$

45. $2b$ **47.** -1 **49.** $\dfrac{4}{3}$ **51.** -2 **53.** $\dfrac{x + 1}{x - 3}$ **55.** $\dfrac{2x + 6}{x - 3}$ **57.** $\dfrac{3}{x}$ **59.** $\dfrac{1}{x - 2}$ **61.** $\dfrac{x + 1}{x^2 + 1}$

63. $x^2 - 4$ **65.** $x - 4$ **67.** $\dfrac{2x + 1}{x - 1}$ **69.** $-x^2 - 5x - 25$ **71.** $\dfrac{4x^2 + 6x + 9}{2}$ **73.** $\dfrac{x + 5}{x^2 + 5}$ **75.** $\dfrac{30m^2}{6m^3}$

77. $\dfrac{35}{5m - 10}$ **79.** $\dfrac{y^2 + 8y + 16}{y^2 - 16}$ **81.** $\dfrac{12x^2 + 24x}{x^2 + 4x + 4}$ **83.** $\dfrac{x^2 - 2x + 4}{x^3 + 8}$ **85.** $\dfrac{ab - 3a}{ab - 3a + 2b - 6}$ **87.** -1

89. $\dfrac{1}{x^n - 4}$ **91.** $\dfrac{1}{x^k + 4}$ **93.** $\dfrac{1}{y^3}$ **95.** $\dfrac{1}{a^3}$ **99.** $\dfrac{1}{18}$ **101.** 3 **103.**

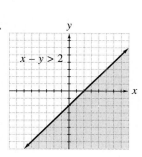

Exercise Set 6.2

1. $\dfrac{x}{2}$ **3.** $\dfrac{c}{4}$ **5.** $\dfrac{x}{3}$ **7.** $-\dfrac{4}{5}$ **9.** $-\dfrac{6a}{2a + 1}$ **11.** $\dfrac{1}{6}$ **13.** $-\dfrac{8}{3}$ **15.** $\dfrac{x}{2}$ **17.** $\dfrac{4}{c}$ **19.** $\dfrac{x}{3}$ **21.** $\dfrac{5}{3}$

23. $\dfrac{4}{(x + 2)(x + 3)}$ **25.** $\dfrac{1}{2}$ **27.** $\dfrac{49y^7}{2x^2}$ **29.** $-\dfrac{2}{3x^3y^2}$ **31.** $\dfrac{4}{ab^6}$ **33.** $\dfrac{1}{4a(a - b)}$ **35.** $\dfrac{x^2 + 5x + 6}{4}$

37. $\dfrac{3}{2(x - 1)}$ **39.** $\dfrac{4a^2}{a - b}$ **41.** $\dfrac{2x^2 - 18}{5(x^2 - 8x - 15)}$ **43.** $\dfrac{x + 2}{x + 3}$ **45.** $\dfrac{3b}{a - b}$ **47.** $\dfrac{3a}{a - b}$ **49.** $\dfrac{1}{4}$ **51.** -1

53. $\dfrac{8}{3}$ **55.** $\dfrac{8a - 16}{3(a + 2)}$ **57.** $\dfrac{8}{x^2y}$ **59.** $\dfrac{(y + 5)(2x - 1)}{(y + 2)(5x + 1)}$ **61.** $\dfrac{15a + 10}{a}$ **63.** $\dfrac{5x^2 - 2}{(x - 1)^2}$ **65.** $\dfrac{5}{x - 2}$ square meters

67. $\dfrac{x^3 - 3x + 2}{x^5}$ ft **69.** $2x^2(x^n + 2)$ **71.** $\dfrac{1}{10y(y^n + 3)}$ **73.** $\dfrac{y^n + 1}{2(y^n - 1)}$ **75.** $\dfrac{7}{5}$ **77.** $\dfrac{1}{12}$

79. $x^2 - 5x - 2 - \dfrac{6}{x - 1}$

Exercise Set 6.3

1. $-\dfrac{3}{x}$ **3.** $\dfrac{x + 2}{x - 2}$ **5.** $x - 2$ **7.** $\dfrac{1}{2 - x}$ **9.** $35x$ **11.** $x(x + 1)$ **13.** $(x + 7)(x - 7)$ **15.** $6(x + 2)(x - 2)$

17. $(a + b)(a - b)^2$ **19.** $\dfrac{17}{6x}$ **21.** $\dfrac{35 - 4y}{14y^2}$ **23.** $\dfrac{-13x + 4}{x^2 - 16}$ **25.** $\dfrac{2x + 4}{(x - 5)(x + 4)}$ **27.** 0 **29.** $\dfrac{x^2}{x - 1}$

31. $\dfrac{-x + 1}{x - 2}$ **33.** $\dfrac{y^2 + 2y + 10}{(y + 4)(y - 4)(y - 2)}$ **35.** $\dfrac{5x^2 + 5x - 20}{(3x + 2)(x + 3)(2x - 5)}$ **37.** $\dfrac{x^2 + 5x + 21}{(x - 2)(x + 1)(x + 3)}$

39. $\dfrac{-2x + 5}{2(x + 1)}$ **41.** $\dfrac{2x^2 + 2x - 42}{(x + 3)^2(x - 3)}$ **43.** $\dfrac{3}{x^2y^3}$ **45.** $-\dfrac{5}{x}$ **47.** $\dfrac{25}{6(x + 5)}$ **49.** $\dfrac{-2x - 1}{x^2(x - 3)}$ **51.** $\dfrac{2ab - b^2}{(a + b)(a - b)}$

53. $\dfrac{2x + 16}{(x + 2)^2(x - 2)}$ **55.** $\dfrac{5a + 1}{(a + 1)^2(a - 1)}$ **57.** $\dfrac{2x^2 + 9x - 18}{6x^2}$ **59.** $\dfrac{4}{3}$ **61.** $\dfrac{4a^2}{9(a - 1)}$ **63.** 4

65. $\dfrac{6x}{(x + 3)(x - 3)^2}$ **67.** $-\dfrac{4}{x - 1}$ **69.** $-\dfrac{32}{x(x + 2)(x - 2)}$ **71.** $\dfrac{4x}{x + 5}$ ft **73.** $\dfrac{3}{2x}$ **75.** $\dfrac{4 - 3x}{x^2}$

77. $\dfrac{-3x + 1}{x^3}$ **79.** 10 **81.** $4 + x^2$ **83.** 10

Exercise Set 6.4

1. $\dfrac{5}{6}$ **3.** $\dfrac{8}{5}$ **5.** 4 **7.** $\dfrac{7}{13}$ **9.** $\dfrac{4}{x}$ **11.** $\dfrac{9x - 18}{9x^2 - 4}$ **13.** $\dfrac{1 - x}{1 + x}$ **15.** $\dfrac{xy^2}{x^2 + y^2}$ **17.** $\dfrac{2b^2 + 3a}{b^2 - ab}$ **19.** $\dfrac{x}{x^2 - 1}$

21. $\dfrac{x + 1}{x + 2}$ **23.** $\dfrac{10}{69}$ **25.** $\dfrac{2x + 2}{2x - 1}$ **27.** $\dfrac{x^2 + x}{6}$ **29.** $\dfrac{x}{2 - 3x}$ **31.** $-\dfrac{y}{x + y}$ **33.** $-\dfrac{2x^3}{xy - y^2}$ **35.** $\dfrac{2x + 1}{y}$

37. $\dfrac{x - 3}{9}$ **39.** $\dfrac{1}{x + 2}$ **41.** $\dfrac{x}{5x - 10}$ **43.** $\dfrac{x - 2}{2x - 1}$ **45.** $-\dfrac{x^2 + 4}{4x}$ **47.** $\dfrac{x - 3y}{x + 3y}$ **49.** $\dfrac{1 + a}{1 - a}$ **51.** $\dfrac{x^2 + 6xy}{2y}$

53. $\dfrac{5a}{2a + 4}$ **55.** $5xy^2 + 2x^2y$ **57.** $\dfrac{xy}{2x + 5y}$ **59.** $\dfrac{xy}{x + y}$ **61.** $x^2 + x$ **63.** $\dfrac{x - 1}{x}$ **65.** $2x$

67. $3a^2 + 4a + 4$ **69.** $\left\{-\dfrac{5}{6}\right\}$ **71.** $\{2\}$ **73.** $\{54\}$

Exercise Set 6.5

1. $\{72\}$ **3.** $\{2\}$ **5.** $\{6\}$ **7.** $\{2\}$ **9.** $\{3\}$ **11.** $\{\ \}$ **13.** $\{15\}$ **15.** $\{4\}$ **17.** $\{\ \}$ **19.** $\{1\}$ **21.** $\{1\}$

23. $\{-3\}$ **25.** $\left\{\dfrac{5}{3}\right\}$ **27.** $\{10, 2\}$ **29.** $\{2\}$ **31.** $\{3\}$ **33.** $\{\ \}$ **35.** $\{\ \}$ **37.** $\{-1\}$ **39.** $\{9\}$ **41.** $\{1, 7\}$

43. $\left\{\dfrac{1}{10}\right\}$ **45.** $\left\{-\dfrac{2}{3}\right\}$ **47.** $\dfrac{5}{2x}$ **49.** $-\dfrac{y}{x}$ **51.** $\dfrac{-a^2 + 31a + 10}{5(a-6)(a+1)}$ **53.** $\left\{-\dfrac{3}{13}\right\}$ **55.** $\dfrac{-a-8}{4a(a-2)}$

57. $\dfrac{x^2 - 3x + 10}{2(x+3)(x-3)}$ **59.** $\{x \mid x \neq 2 \text{ and } x \neq -1\}$ **61.** $\dfrac{22z - 45}{z(3z-9)}$ **63.** $\left\{\dfrac{1}{9}, -\dfrac{1}{4}\right\}$ **65.** $\{3, 2\}$ **67.** $\{1, 2\}$

69. $\left\{-3, -\dfrac{3}{4}\right\}$ **71.** 73, 74   **73.** $2, \dfrac{1}{2}$ **75.** -9 **77.** -9

Exercise Set 6.6

1. $C = \dfrac{5}{9}(F - 32)$ **3.** $R = \dfrac{R_1 R_2}{R_1 + R_2}$ **5.** $n = \dfrac{2S}{a+L}$ **7.** 1 and 5   **9.** 5   **11.** 6 ohms

13. $\dfrac{1}{r} = \dfrac{1}{r_1} + \dfrac{1}{r_2} + \dfrac{1}{r_3}$; $r = \dfrac{15}{13}$ ohms   **15.** 15.6 hr   **17.** 10 min   **19.** 200 mph   **21.** 15 mph   **23.** $h = \dfrac{2A}{a+b}$

25. $T_2 = T_1 - \dfrac{LH}{kA}$ **27.** $r = \dfrac{E}{I} - R$ **29.** $a_1 = S(1-r) + a_n r$ **31.** $M = -\dfrac{Fr^2}{Gm}$ **33.** -8 and -7 **35.** 36 min

37. 45 mph and 60 mph   **39.** $\dfrac{7}{8}$ **41.** 3 hours   **43.** 22,500 miles from earth   **45.** $\{-5\}$ **47.** $\{2\}$

49. $x(2x+3)(x-6)$ **51.** $(x+2)(x^2 - 2x + 4)$

Exercise Set 6.7

1. $\{27\}$ **3.** $\{-6\}$ **5.** $\{2\}$ **7.** $\{5\}$ **9.** $A = kB$ **11.** $X = \dfrac{k}{Z}$ **13.** $N = kP^2$ **15.** $T = \dfrac{k}{R}$ **17.** $P = kR$

19. 45   **21.** 24 m^3 **23.** 10   **25.** 72 amperes   **27.** $x = kyz$ **29.** $r = kst^3$ **31.** $\{-10\}$

33. $\left\{-\dfrac{3}{4}\right\}$ **35.** $\left\{\dfrac{-23}{2}\right\}$ **37.** $\left\{-\dfrac{8}{15}\right\}$ **39.** 7   **41.** $\dfrac{48}{5}$ **43.** 2   **45.** 2.7   **47.** 4.05 lb   **49.** \$0.80

51. 108 cars   **53.** 16 lb   **55.** $\dfrac{2025}{256}$ lb   **57.** The area is multiplied by 9.   **59.** The intensity is divided by 4.

63. $C = 8\pi$ in.; $A = 16\pi$ sq in.   **65.** $C = 18\pi$ in.; $A = 81\pi$ sq in.

Chapter 6 Review

1. none   **3.** $x = 4$ **5.** $x = 0, x = 3$ or $x = -2$ **7.** 1   **9.** -1 **11.** 2   **13.** $\dfrac{1}{x-7}$ **15.** $\dfrac{y-3}{x+2}$ **17.** $\dfrac{2x^3}{z^3}$

19. $\dfrac{2}{5}$ **21.** $\dfrac{1}{6}$ **23.** $\dfrac{3x}{16}$ **25.** $\dfrac{3c^2}{14a^2 b}$ **27.** $\dfrac{x^2 + 9x + 20}{3}$ **29.** $\dfrac{7x - 28}{2(x-2)}$ **31.** $-\dfrac{1}{x}$ **33.** $\dfrac{8}{9a^2}$ **35.** $\dfrac{6}{a}$

37. $60x^2 y^5$ **39.** $5x(x-5)$ **41.** $\dfrac{2}{5}$ **43.** $\dfrac{2}{x^2}$ **45.** $\dfrac{1}{x-2}$ **47.** $\dfrac{5x^2 - 3y^2}{15x^4 y^3}$ **49.** $\dfrac{-x+5}{(x+1)(x-1)}$

51. $\dfrac{2x^2 - 5x - 4}{x-3}$ **53.** $\dfrac{3x^2 - 7x - 4}{(3x-4)(9x^2 + 12x + 16)}$ **55.** $-\dfrac{12}{x(x+1)(x-3)}$ **57.** $\dfrac{60 + 4x - x^2}{15x^2}$ **59.** $\dfrac{-10x - 25}{(x+5)^2}$

61. $\dfrac{2}{3}$ **63.** $\dfrac{2}{15 - 2x}$ **65.** $\dfrac{y}{2}$ **67.** $\dfrac{20x - 15}{10x^2 - 4}$ **69.** $\dfrac{5xy + x}{3y}$ **71.** $\dfrac{1+x}{1-x}$ **73.** $\dfrac{x-1}{3x-1}$ **75.** $-\dfrac{x^2 + 9}{6x}$

77. $\{6\}$ **79.** $\{2\}$ **81.** $\left\{\dfrac{3}{2}\right\}$ **83.** $\left\{\dfrac{5}{3}\right\}$ **85.** $\left\{-\dfrac{1}{3}, 2\right\}$ **87.** $a = \dfrac{2A}{h} - b$ **89.** $R = \dfrac{E}{I} - r$

91. $A = \dfrac{HL}{k(T_1 - T_2)}$ **93.** 7   **95.** $\dfrac{23}{25}$ **97.** -10 and -8 **99.** 12 hr   **101.** 490 mph   **103.** 8 mph

105. 4 mph   **107.** $\{-10\}$ **109.** 9   **111.** 3.125 ft^3

Chapter 6 Test

1. none   **2.** $x = -\dfrac{3}{2}$ or $x = 5$ **3.** $\dfrac{5x^3}{3}$ **4.** $-\dfrac{7}{8}$ **5.** $\dfrac{2}{x-3}$ **6.** $\dfrac{x}{x+9}$ **7.** $\dfrac{x+2}{5}$ **8.** $\dfrac{3x+3}{10}$ **9.** $\dfrac{5}{3x}$

10. $\dfrac{4a^3b^4}{c^6}$ **11.** $\dfrac{x+2}{2(x+3)}$ **12.** $\dfrac{-8x-36}{5}$ **13.** $75x^3$ **14.** $\dfrac{3}{x^3}$ **15.** -1 **16.** $\dfrac{5x-2}{(x-3)(x+2)(x-2)}$

17. $\dfrac{x-1}{4x^2-2x+1}$ **18.** $\dfrac{-x+30}{6(x-7)}$ **19.** $\dfrac{3}{2}$ **20.** $\dfrac{5x+3y}{xy}$ **21.** $\dfrac{2}{x+y}$ **22.** $\dfrac{1}{5}$ **23.** $\dfrac{64}{3}$ **24.** $\dfrac{7y^2+4y}{6}$

25. $\dfrac{x^2-6x+9}{x-2}$ **26.** $\{7\}$ **27.** $\{2, -2\}$ **28.** $\{8\}$ **29.** $x = \dfrac{7a^2+b^2}{4a-b}$ **30.** 5   **31.** $\dfrac{6}{7}$ hr   **32.** 16
33. 9   **34.** 256 ft

Chapter 6 Cumulative Review

1. *(Sec. 1.1, Ex. 1)* (a) true; (b) true; (c) false   **2.** *(Sec. 2.1, Ex. 1)* (a) $-2x$; (b) $8y$; (c) $4z + 6$

3. *(Sec. 2.2, Ex. 4)* $y = \dfrac{2x+7}{3}$ **4.** *(Sec. 3.2, Ex. 1)* (a) (4, 0); (b) (1, −9); (c) (0, −12)   **5.** *(Sec. 3.4, Ex. 8)* $2x + 3y = 20$

6. *(Sec. 4.1, Ex. 5)* (a) x^3; (b) 5^6; (c) $5x$; (d) $\dfrac{6y^2}{7}$ **7.** *(Sec. 4.3, Ex. 10)* (a) 5; (b) 1; (c) 35; (d) −2

8. *(Sec. 4.4, Ex. 5)* $2x^2 + 11x + 15$

9. *(Sec. 4.7, Ex. 2)*

10. *(Sec. 5.1, Ex. 3)* (a) $4(2x + 1)$; (b) prime; (c) $3x^2(2 - x)$
11. *(Sec. 5.2, Ex. 1)* $(x + 8)(x + 2)$ **12.** *(Sec. 5.2, Ex. 9)* $(2a - 1)(a + 4)$
13. *(Sec. 5.3, Ex. 1)* $(m + 5)^2$ **14.** *(Sec. 5.3, Ex. 5)* $(x + 2)(x^2 - 2x + 4)$
15. *(Sec. 5.5, Ex. 2)* $\left\{\dfrac{1}{2}, -5\right\}$ **16.** *(Sec. 5.5, Ex. 7)* width, 2 meters; length, 8 meters
17. *(Sec. 6.1, Ex. 2)* (a) $\dfrac{3y^4}{x}$; (b) $\dfrac{1}{5x-1}$ **18.** *(Sec. 6.2, Ex. 1)* (a) $\dfrac{y}{18}$; (b) $\dfrac{n-2}{n(n-1)}$
19. *(Sec. 6.3, Ex. 1)* (a) $\dfrac{5+x}{7}$; (b) $\dfrac{3x}{2}$; (c) $x - 7$; (d) $-\dfrac{1}{3y^2}$ **20.** *(Sec. 6.5, Ex. 2)* $\{-2\}$
21. *(Sec. 6.7, Ex. 1)* $\{-5\}$

CHAPTER 7
Rational Exponents, Radicals, and Complex Numbers

Exercise Set 7.1

1. 7   **3.** 3   **5.** $\dfrac{1}{2}$ **7.** 13   **9.** −3   **11.** not a real number   **13.** −8   **15.** −1   **17.** 8   **19.** 16

21. not a real number   **23.** $\dfrac{1}{16}$ **25.** $\dfrac{1}{16}$ **27.** not a real number   **29.** $a^{7/3}$ **31.** $\dfrac{8u^3}{v^9}$ **33.** $-b$ **35.** $y - y^{7/6}$

37. $2x^{5/3} - 2x^{2/3}$ **39.** $4x^{2/3} - 9$ **41.** $x^{8/3}(1 + x^{2/3})$ **43.** $x^{1/5}(x^{1/5} - 3)$ **45.** $x^{-1/3}(5 + x)$ **47.** 5   **49.** 4

51. −3   **53.** not a real number   **55.** $\dfrac{1}{7}$ **57.** 8   **59.** $\dfrac{1}{4}$ **61.** not a real number   **63.** $\dfrac{1}{25}$

65. not a real number   **67.** $\dfrac{4}{3}$ **69.** $\dfrac{343}{125}$ **71.** $\dfrac{1}{x^{7/6}}$ **73.** $w^{1/2}$ **75.** $\dfrac{y^{1/4}}{x^{3/5}}$ **77.** $\dfrac{1}{a^{4/3}b^6}$ **79.** $4x^2y$ **81.** $\dfrac{1}{y^{5/2}}$

83. $\dfrac{a^{3/8}}{b^{3/4}}$ **85.** $24x^{19/6} - 4x^{2/3}y^{2/3}$ **87.** $x - x^{1/2} - 6$ **89.** $x - 4x^{3/2} + 4x^2$ **91.** $x^{3/4}(9 - x)$

93. $3x^{5/3}(3 + 5x^{2/3})$ **95.** $y^{-1/5}(3 - 5y^{7/5})$ **97.** $x^{8a/15}$ **99.** $\dfrac{1}{y^{a/6}}$ **101.** $\dfrac{1}{x^{3b}y}$ **105.** $x(x - 2)(x - 4)$

107. $(x + 3y)(x^2 - 3xy + 9y^2)$ **109.** $-x - 3$

Exercise Set 7.2

1. 6   **3.** 3   **5.** $7x^3$ **7.** $-5x$ **9.** $3^{1/2}$ **11.** $y^{5/3}$ **13.** $4^{1/5}y^{7/5}$ **15.** $(y + 1)^{3/2}$ **17.** $2x^{1/2} - 3y^{1/2}$

19. $\sqrt[7]{a^3}$ **21.** $2\sqrt[3]{x^5}$ **23.** $\dfrac{1}{\sqrt[6]{(4t)^5}}$ **25.** $\sqrt[5]{(4x - 1)^3}$ **27.** 4   **29.** x **31.** $|a|$ **33.** $|x - 4|$ **35.** $4\sqrt{2}$

37. $4\sqrt[3]{3}$ **39.** $25\sqrt{3}$ **41.** $10x^2\sqrt{x}$ **43.** $2y^2\sqrt[3]{2y}$ **45.** $a^2b\sqrt[4]{b^3}$ **47.** $\dfrac{\sqrt{3}}{5}$ **49.** $\dfrac{7}{2x}$ **51.** $\dfrac{y^2\sqrt[3]{y}}{2x^2}$ **53.** 11

55. $2x$ **57.** $y^2\sqrt{y}$ **59.** $2\sqrt{5}$ **61.** $5ab\sqrt{b}$ **63.** $-3x^3$ **65.** a^4b **67.** $x^4\sqrt[3]{50x^2}$ **69.** $-2x^2\sqrt[5]{y}$

71. $-4a^4b^3\sqrt{2b}$ **73.** $\dfrac{\sqrt{6}}{7}$ **75.** $\dfrac{\sqrt{5x}}{2y}$ **77.** $-\dfrac{z^2\sqrt[3]{z}}{3x}$ **79.** $\dfrac{x\sqrt[4]{x^3}}{2}$ **83.** $\dfrac{1}{5}$ **85.** $-\dfrac{1}{8}$ **87.** $\dfrac{1}{x^5}$

Mental Math, Sec. 7.3

1. $6\sqrt{3}$ **3.** $3\sqrt[3]{x}$ **5.** $12\sqrt[3]{x}$

Exercise Set 7.3

1. $-2\sqrt{2}$ **3.** $10x\sqrt{2x}$ **5.** $17\sqrt{2} - 15\sqrt{5}$ **7.** $-\sqrt[3]{2x}$ **9.** $5b\sqrt{b}$ **11.** $\dfrac{31\sqrt{2}}{15}$ **13.** $\dfrac{\sqrt[3]{11}}{3}$

15. $\dfrac{5\sqrt{5x}}{9}$ **17.** $14 + \sqrt{3}$ **19.** $7 - 3y$ **21.** $6\sqrt{3} - 6\sqrt{2}$ **23.** $-23\sqrt[3]{5}$ **25.** $2b\sqrt{b}$ **27.** $20y\sqrt{2y}$

29. $2y\sqrt[3]{2x}$ **31.** $6\sqrt[3]{11} - 4\sqrt{11}$ **33.** $4x\sqrt[4]{x^3}$ **35.** $\dfrac{2\sqrt{3}}{3}$ **37.** $\dfrac{5x\sqrt[3]{x}}{7}$ **39.** $\dfrac{5\sqrt{7}}{2x}$ **41.** $\dfrac{\sqrt[3]{2}}{6}$

43. $\dfrac{14x\sqrt[3]{2x}}{9}$ **45.** $15\sqrt{3}$ in.   **47.** $22\sqrt{5}$ ft   **49.** $x^2 - 16$ **51.** $4x^2 + 20x + 25$ **53.** { }

Exercise Set 7.4

1. $\sqrt{35} + \sqrt{21}$ **3.** $7 - 2\sqrt{10}$ **5.** $3\sqrt{x} - x\sqrt{3}$ **7.** $6x - 13\sqrt{x} - 5$ **9.** $\sqrt[3]{a^2} + \sqrt[3]{a} - 20$ **11.** $\dfrac{\sqrt{3}}{3}$

13. $\dfrac{\sqrt{5}}{5}$ **15.** $\dfrac{4\sqrt[3]{9}}{3}$ **17.** $\dfrac{3\sqrt{2x}}{4x}$ **19.** $\dfrac{3\sqrt[3]{2x}}{2x}$ **21.** $-\dfrac{5(2 + \sqrt{7})}{3}$ **23.** $\dfrac{7(3 + \sqrt{x})}{9 - x}$ **25.** $-5 + 2\sqrt{6}$

27. $\dfrac{2a + 2\sqrt{a} + \sqrt{ab} + \sqrt{b}}{4a - b}$ **29.** $6\sqrt{2} - 12$ **31.** $2 + 2x\sqrt{3}$ **33.** $-16 - \sqrt{35}$ **35.** $x - y^2$

37. $3 + 2x\sqrt{3} + x^2$ **39.** $5x - 3\sqrt{10x} - 3\sqrt{15x} + 9\sqrt{6}$ **41.** $2\sqrt[3]{2} - \sqrt[3]{4}$

43. $-4\sqrt[6]{x^5} + \sqrt[3]{x^2} + 8\sqrt[3]{x} - 4\sqrt{x} + 7$ **45.** $\dfrac{\sqrt{10}}{5}$ **47.** $\dfrac{3\sqrt{3a}}{a}$ **49.** $\dfrac{3\sqrt[3]{4}}{2}$ **51.** $\dfrac{2\sqrt{21}}{7}$ **53.** $-\dfrac{8(1 - \sqrt{10})}{9}$

55. $-\dfrac{2\sqrt{3x}}{3x}$ **57.** $\dfrac{a\sqrt[3]{4x}}{4}$ **59.** $\dfrac{x^2y^3\sqrt{6z}}{z}$ **61.** $3 - 2\sqrt{2}$ **63.** $\dfrac{x + 2\sqrt{x} + 1}{x - 1}$ **65.** $\dfrac{4x - 4\sqrt{xy} - y\sqrt{x} + y\sqrt{y}}{x - y}$

67. $\dfrac{x - \sqrt{xy}}{x - y}$ **69.** $\dfrac{y\sqrt{y} + 4y + 3\sqrt{y}}{y^2 - y}$ **71.** 5   **73.** $\dfrac{4(\sqrt{x} + 2\sqrt{y})}{x - 4y}$ **75.** $\dfrac{4\sqrt{7} + 4\sqrt{x} - \sqrt{21x} - x\sqrt{3}}{7 - x}$

77. $(2\sqrt{3} + \sqrt{30})$ cubic cm   **79.** $\dfrac{2}{5\sqrt{2}}$ **81.** $\dfrac{1}{\sqrt{10} - 2\sqrt{2}}$ **83.** $\dfrac{x - 9}{x - 3\sqrt{x}}$ **87.** $\left\{-2, -\dfrac{4}{5}\right\}$ **89.** $\{0, 1, -1\}$

91. $-\dfrac{20 + 16y}{3y}$

Exercise Set 7.5

1. $\{8\}$ **3.** $\{7\}$ **5.** $\{\ \}$ **7.** $\{7\}$ **9.** $\{6\}$ **11.** $\left\{-\dfrac{9}{2}\right\}$ **13.** $\{29\}$ **15.** $\{4\}$ **17.** $\{-4\}$ **19.** $\{\ \}$

21. $\{7\}$ **23.** $3\sqrt{5}$ ft **25.** $2\sqrt{10}$ meters **27.** $\{9\}$ **29.** $\{50\}$ **31.** $\{\ \}$ **33.** $\left\{\dfrac{15}{4}\right\}$ **35.** $\{7\}$ **37.** $\{5\}$

39. $\{-12\}$ **41.** $\{9\}$ **43.** $\{-3\}$ **45.** $\{1\}$ **47.** $\{1\}$ **49.** $\left\{\dfrac{1}{2}\right\}$ **51.** $2\sqrt{10}$ m **53.** $7\sqrt{2}$ mm

55. 17 ft **57.** $45\sqrt{2}$ in. **59.** 2744 deliveries per day **61.** $50\sqrt{145}$ lb **63.** $\{-1, 2\}$ **65.** $\{-8, -6, 0, 2\}$

67. $\dfrac{7}{9}$ **69.** $m = -1$ **71.** not a function

Mental Math, Sec. 7.6

1. $9i$ **3.** $i\sqrt{7}$ **5.** -4 **7.** $8i$

Exercise Set 7.6

1. $2i\sqrt{6}$ **3.** $-6i$ **5.** $24i\sqrt{7}$ **7.** $-3\sqrt{6}$ **9.** $6 - 4i$ **11.** $-2 + 6i$ **13.** $-2 - 4i$ **15.** $18 + 12i$

17. 7 **19.** $12 - 16i$ **21.** $-4i$ **23.** $\dfrac{28}{25} - \dfrac{21}{25}i$ **25.** $4 + i$ **27.** $-\sqrt{14}$ **29.** $-5\sqrt{2}$ **31.** $4i$

33. $i\sqrt{3}$ **35.** $2\sqrt{2}$ **37.** 1 **39.** i **41.** $-i$ **43.** -1 **45.** 63 **47.** $2 - i$ **49.** 20 **51.** $6 - 3i\sqrt{3}$

53. 2 **55.** $-5 + \dfrac{16}{3}i$ **57.** $17 + 144i$ **59.** $\dfrac{3}{5} - \dfrac{1}{5}i$ **61.** $5 - 10i$ **63.** $\dfrac{1}{5} - \dfrac{8}{5}i$ **65.** $8 - i$ **69.** $40°$

71. $83°$ **73.** $[-6, 2]$

Chapter 7 Review

1. $\dfrac{1}{3}$ **3.** $-\dfrac{1}{3}$ **5.** -27 **7.** not a real number **9.** $\dfrac{9}{4}$ **11.** $a^{13/6}$ **13.** a **15.** $\dfrac{1}{a^{9/2}}$ **17.** $\dfrac{y^2}{x}$ **19.** $\dfrac{1}{x^{11/12}}$

21. $\dfrac{y^2}{x^3}$ **23.** $x^2 + x$ **25.** 3 **27.** not a real number **29.** x^{32} **31.** 4 **33.** $5^{1/5}x^{2/5}y^{3/5}$ **35.** $5\sqrt[3]{xy^2z^5}$

37. $|x^2 - 4|$ **39.** -5 **41.** Not a real number, for $x \neq 0$ and is 0 when $x = 0$. **43.** $-y$ **45.** $-a^2b^3$

47. $2ab^2$ **49.** $2x\sqrt{3xy}$ **51.** $9a^2b^2\sqrt{ab}$ **53.** $16ab\sqrt[4]{ab}$ **55.** $17\sqrt{2} - 15\sqrt{5}$ **57.** $-4ab\sqrt[4]{2b}$

59. 6 **61.** $x - 6\sqrt{x} + 9$ **63.** $-10 + \sqrt{2}$ **65.** $4x - 9y$ **67.** $a - 5\sqrt{a} + 6$ **69.** $\sqrt[3]{a^2} + 4\sqrt[3]{a} + 4$

71. $6x - \sqrt{xy} - y$ **73.** $\dfrac{\sqrt{5x}}{5}$ **75.** $\dfrac{\sqrt{2xy}}{2y}$ **77.** $\dfrac{x^2y^2\sqrt[3]{15yz}}{z}$ **79.** $-\dfrac{10 + 5\sqrt{7}}{3}$ **81.** $-5 + 2\sqrt{6}$

83. $\{-2\}$ **85.** $\{29\}$ **87.** $\{\ \}$ **89.** $\{21\}$ **91.** $\{\ \}$ **93.** $3\sqrt{2}$ **95.** $\sqrt{65}$ **97.** $2i\sqrt{2}$ **99.** $6i$

101. $15 - 4i$ **103.** $\sqrt{3} + 4\sqrt{2} - 2i\sqrt{2}$ **105.** $10 + 4i$ **107.** $1 + 5i$ **109.** $\dfrac{3}{2} - i$

Chapter 7 Test

1. -6 **2.** $-x^{16}$ **3.** $\dfrac{1}{5}$ **4.** 5 **5.** $\dfrac{4x^2}{9}$ **6.** $-a^6b^3$ **7.** $\dfrac{8a^{1/3}c^{2/3}}{b^{5/12}}$ **8.** $a^{7/12} - a^{7/3}$ **9.** $|4xy|$ **10.** -27

11. $\dfrac{3\sqrt{y}}{y}$ **12.** $\dfrac{\sqrt{2xy}}{2y}$ **13.** $\dfrac{8 - 6\sqrt{x} + x}{8 - 2x}$ **14.** $\dfrac{\sqrt[3]{b^2}}{b}$ **15.** $-x\sqrt{5x}$ **16.** $4a\sqrt[4]{2ab^2}$ **17.** $4\sqrt{3} - \sqrt{6}$

18. $x + 2\sqrt{x} + 1$ **19.** $\sqrt{6} - 4\sqrt{3} + \sqrt{2} - 4$ **20.** -20 **21.** $\{2, 3\}$ **22.** $\{\ \}$ **23.** $\{6\}$ **24.** $i\sqrt{2}$

25. $-2i\sqrt{2}$ **26.** $-3i$ **27.** 40 **28.** $7 + 24i$ **29.** $-\dfrac{3}{2} + \dfrac{5}{2}i$ **30.** $x = \dfrac{5\sqrt{2}}{2}$ **31.** $y = 3\sqrt{7}$

Chapter 7 Cumulative Review

1. *(Sec. 2.1, Ex. 8)* { }   **2.** *(Sec. 2.4, Ex. 8)* $\left\{\dfrac{3}{4}, 5\right\}$ **3.** *(Sec. 2.7, Ex. 3)* $\left[-2, \dfrac{8}{5}\right]$

4. *(Sec. 3.3, Ex. 6)* L_1 and L_2 are parallel   **5.** *(Sec. 3.6, Ex. 4)*   **6.** *(Sec. 4.1, Ex. 2)* (a) 128; (b) x^{10}; (c) y^7

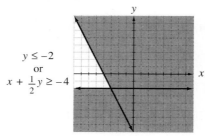

$y \le -2$
or
$x + \dfrac{1}{2}y \ge -4$

7. *(Sec. 4.2, Ex. 2)* (a) $125x^6$; (b) $\dfrac{8}{27}$; (c) $\dfrac{9p^8}{q^{10}}$; (d) $64y^2$; (e) $\dfrac{y^{14}}{x^{35}z^7}$ **8.** *(Sec. 4.2, Ex. 5)* (a) $4x^b$; (b) y^{5a+6}

9. *(Sec. 4.4, Ex. 2)* (a) $10x^2 - 8x$; (b) $-12x^4 + 18x^3 - 3x^2$; (c) $-7x^3y^2 - 3x^2y^2 + 11xy$

10. *(Sec. 4.5, Ex. 4)* $\dfrac{x^3}{2y} - \dfrac{5x}{2y^2} - \dfrac{1}{xy^3}$ **11.** *(Sec. 5.1, Ex. 2)* $5x^2$ **12.** *(Sec. 5.2, Ex. 5)* $(2x + 5)(x + 3)$

13. *(Sec. 5.3, Ex. 7)* $(y - 4)(y^2 + 4y + 16)$ **14.** *(Sec. 6.1, Ex. 3)* (a) 1; (b) -1; (c) $-\dfrac{2(3 + x)}{x + 1}$

15. *(Sec. 6.2, Ex. 3)* (a) $\dfrac{x^3}{15y^2}$; (b) $-\dfrac{m^2}{15(m + 2)}$ **16.** *(Sec. 6.4, Ex. 3)* $\dfrac{xy + 2x^3}{y - 1}$ **17.** *(Sec. 6.5, Ex. 5)* $\{-1, -6\}$

18. *(Sec. 7.1, Ex. 3)* (a) 8; (b) -8; (c) not a real number; (d) 9; (e) $\dfrac{1}{27}$

19. *(Sec. 7.2, Ex. 3)* (a) $\sqrt[5]{x}$; (b) $5\sqrt[3]{x^2}$; (c) $3\sqrt{p + q}$

20. *(Sec. 7.3, Ex. 1)* (a) $8\sqrt{5}$; (b) $-6\sqrt[3]{2}$; (c) $3\sqrt{3x} - 6\sqrt{x} + 6\sqrt{2x}$; (d) $\sqrt[3]{98} + 7\sqrt{2}$; (e) $3y\sqrt[3]{6y}$

21. *(Sec. 7.5, Ex. 1)* $x = 42$ **22.** *(Sec. 7.6, Ex. 1)* (a) $6i$; (b) $i\sqrt{5}$ **23.** *(Sec. 7.6, Ex. 6)* (a) $-i$; (b) 1; (c) -1; (d) 1

CHAPTER 8
Quadratic Equations and Inequalities and Conic Sections

Exercise Set 8.1

1. $\{4, -4\}$ **3.** $\{\sqrt{7}, -\sqrt{7}\}$ **5.** $\{3\sqrt{2}, -3\sqrt{2}\}$ **7.** $\{\sqrt{10}, -\sqrt{10}\}$ **9.** $\{3i, -3i\}$

11. $\{\sqrt{6}, -\sqrt{6}\}$ **13.** $\{2i\sqrt{2}, -2i\sqrt{2}\}$ **15.** $\{-8, -2\}$ **17.** $\{6 - 3\sqrt{2}, 6 + 3\sqrt{2}\}$ **19.** $\left\{\dfrac{3 - 2\sqrt{2}}{2}, \dfrac{3 + 2\sqrt{2}}{2}\right\}$

21. $\{1 - 4i, 1 + 4i\}$ **23.** $\{-7 - \sqrt{5}, -7 + \sqrt{5}\}$ **25.** $\{-3 - 2i\sqrt{2}, -3 + 2i\sqrt{2}\}$

27. $x^2 + 16x + 64 = (x + 8)^2$ **29.** $z^2 - 12z + 36 = (z - 6)^2$ **31.** $p^2 + 9p + \dfrac{81}{4} = \left(p + \dfrac{9}{2}\right)^2$

33. $x^2 + x + \dfrac{1}{4} = \left(x + \dfrac{1}{2}\right)^2$ **35.** $\{-5, -3\}$ **37.** $\{-3 - \sqrt{7}, -3 + \sqrt{7}\}$ **39.** $\left\{\dfrac{-1 - \sqrt{5}}{2}, \dfrac{-1 + \sqrt{5}}{2}\right\}$

41. $\{-1 - i, -1 + i\}$ **43.** $\{3 + \sqrt{6}, 3 - \sqrt{6}\}$ **45.** $\left\{\dfrac{-1 - i\sqrt{3}}{2}, \dfrac{-1 + i\sqrt{3}}{2}\right\}$ **47.** $\left\{\dfrac{6 - \sqrt{30}}{3}, \dfrac{6 + \sqrt{30}}{3}\right\}$

49. $\left\{\dfrac{3 - \sqrt{11}}{2}, \dfrac{3 + \sqrt{11}}{2}\right\}$ **51.** $\left\{\dfrac{1}{2}, -4\right\}$ **53.** $\{-2 - i\sqrt{2}, -2 + i\sqrt{2}\}$ **55.** $\left\{\dfrac{-15 - 7\sqrt{5}}{10}, \dfrac{-15 + 7\sqrt{5}}{10}\right\}$

57. $\left\{\dfrac{1 - i\sqrt{47}}{4}, \dfrac{1 + i\sqrt{47}}{4}\right\}$ **59.** $\{5, -1\}$ **61.** $\{-4 - \sqrt{15}, -4 + \sqrt{15}\}$ **63.** $\left\{\dfrac{-3 - \sqrt{21}}{3}, \dfrac{-3 + \sqrt{21}}{3}\right\}$

65. $\left\{\dfrac{5}{2}, -1\right\}$ **67.** $\{-5 - i\sqrt{3}, -5 + i\sqrt{3}\}$ **69.** $\{-4, 1\}$ **71.** $\left\{\dfrac{2 - i\sqrt{2}}{2}, \dfrac{2 + i\sqrt{2}}{2}\right\}$

73. $\left\{\dfrac{-3 - \sqrt{69}}{6}, \dfrac{-3 + \sqrt{69}}{6}\right\}$ **75.** We want to use the rule $ab = 0 \Leftrightarrow a = 0$ or $b = 0$.

77. 34   **79.** 27   **81.** $\dfrac{1}{y^{1/4}}$

Exercise Set 8.2

1. $\{-6, 1\}$ **3.** $\left\{-\dfrac{3}{5}, 1\right\}$ **5.** $\{3\}$ **7.** $\left\{\dfrac{-7 - \sqrt{33}}{2}, \dfrac{-7 + \sqrt{33}}{2}\right]$ **9.** $\left\{\dfrac{1 - \sqrt{57}}{8}, \dfrac{1 + \sqrt{57}}{8}\right\}$

11. $\left\{\dfrac{7 - \sqrt{85}}{6}, \dfrac{7 + \sqrt{85}}{6}\right\}$ **13.** $\{1 - \sqrt{3}, 1 + \sqrt{3}\}$ **15.** $\left\{-\dfrac{3}{2}, 1\right\}$ **17.** $\left\{\dfrac{3 - \sqrt{11}}{2}, \dfrac{3 + \sqrt{11}}{2}\right\}$

19. $\left\{\dfrac{3 - i\sqrt{87}}{8}, \dfrac{3 + i\sqrt{87}}{8}\right\}$ **21.** $\{-2 - \sqrt{11}, -2 + \sqrt{11}\}$ **23.** $\left\{\dfrac{-5 - i\sqrt{5}}{10}, \dfrac{-5 + i\sqrt{5}}{10}\right\}$

25. two real solutions   **27.** two complex solutions   **29.** two complex solutions   **31.** one real solution

33. $\left\{\dfrac{-5 - \sqrt{17}}{2}, \dfrac{-5 + \sqrt{17}}{2}\right\}$ **35.** $\left\{\dfrac{5}{2}, 1\right\}$ **37.** $\left\{\dfrac{3 - \sqrt{29}}{2}, \dfrac{3 + \sqrt{29}}{2}\right\}$ **39.** $\left\{\dfrac{-1 - \sqrt{19}}{6}, \dfrac{-1 + \sqrt{19}}{6}\right\}$

41. $\{-3 - 2i, -3 + 2i\}$ **43.** $\left\{\dfrac{-1 - i\sqrt{23}}{4}, \dfrac{-1 + i\sqrt{23}}{4}\right\}$ **45.** $\{1\}$ **47.** $\left\{\dfrac{19 - \sqrt{345}}{2}, \dfrac{19 + \sqrt{345}}{2}\right\}$

49. 12 or -8 **51.** 8.53 hours   **53.** $\left\{\dfrac{\sqrt{3}}{3}\right\}$ **55.** $\left\{\dfrac{-\sqrt{2} - i\sqrt{2}}{2}, \dfrac{-\sqrt{2} + i\sqrt{2}}{2}\right\}$

57. $\left\{\dfrac{\sqrt{3} - \sqrt{11}}{4}, \dfrac{\sqrt{3} + \sqrt{11}}{4}\right\}$ **59.** $\left\{\dfrac{11}{5}\right\}$ **61.** $-2\sqrt{5}$ **63.** $5\sqrt{3x}$

Exercise Set 8.3

1. $\{3 - \sqrt{7}, 3 + \sqrt{7}\}$ **3.** $\left\{\dfrac{3 - \sqrt{57}}{4}, \dfrac{3 + \sqrt{57}}{4}\right\}$ **5.** $\left\{\dfrac{1 - \sqrt{29}}{2}, \dfrac{1 + \sqrt{29}}{2}\right\}$ **7.** $\left\{1, \dfrac{-1 - i\sqrt{3}}{2}, \dfrac{-1 + i\sqrt{3}}{2}\right\}$

9. $\left\{0, -3, \dfrac{3 - 3i\sqrt{3}}{2}, \dfrac{3 + 3i\sqrt{3}}{2}\right\}$ **11.** $\{4, -2 - 2i\sqrt{3}, -2 + 2i\sqrt{3}\}$ **13.** $\{-2, 2, -2i, 2i\}$

15. $\left\{-\dfrac{1}{2}, \dfrac{1}{2}, -i\sqrt{3}, i\sqrt{3}\right\}$ **17.** $\{-3, 3, -2, 2\}$ **19.** $\{125, -8\}$ **21.** $\left\{-\dfrac{1}{8}, 27\right\}$ **23.** $\left\{-\dfrac{1}{125}, \dfrac{1}{8}\right\}$

25. width, $4\sqrt{10}$ ft; length, $10\sqrt{10}$ ft   **27.** $\{-\sqrt{2}, \sqrt{2}, -\sqrt{3}, \sqrt{3}\}$ **29.** $\left\{\dfrac{-9 - \sqrt{201}}{6}, \dfrac{-9 + \sqrt{201}}{6}\right\}$

31. $\{-4, 2 - 2i\sqrt{3}, 2 + 2i\sqrt{3}\}$ **33.** $\{27, 125\}$ **35.** $\{1, -3i, 3i\}$ **37.** $\left\{\dfrac{1}{8}, -8\right\}$

39. $\left\{5, \dfrac{-5 + 5i\sqrt{3}}{2}, \dfrac{-5 - 5i\sqrt{3}}{2}\right\}$ **41.** $\{-3\}$ **43.** $\{-\sqrt{5}, \sqrt{5}, -2i, 2i\}$ **45.** $\left\{-3, \dfrac{3 - 3i\sqrt{3}}{2}, \dfrac{3 + 3i\sqrt{3}}{2}\right\}$

47. $\left\{-\dfrac{1}{3}, \dfrac{1}{3}, -\dfrac{i\sqrt{6}}{3}, \dfrac{i\sqrt{6}}{3}\right\}$ **49.** base, $2\sqrt{42}$ cm; height, $\sqrt{42}$ cm

51. width, $20\sqrt{2}$ in.; height, $30\sqrt{2}$ in.   **53.** (a) width, $5\sqrt{10}$ cm; length, $15\sqrt{10}$ cm; (b) perimeter, $40\sqrt{10}$ cm

55. $\left[\dfrac{11}{4}, \infty\right)$ **57.** $(-\infty, 5)$ **59.** 4   **61.** $2\sqrt[3]{4}$

Exercise Set 8.4

$(-\infty, -5) \cup (-1, \infty)$

1. ←———|————|———→
       −5   −1

$[-4, 3]$

3. ←——|███████|——→
    −4   3

$[2, 5]$

5. ←———|███|———→
    2   5

$\left(-5, -\frac{1}{3}\right)$

7. ←——|███|——→
   −5 $-\frac{1}{3}$

$(2, 4) \cup (6, \infty)$

9. ←——|██|—|███→
   2 4 6

$(-\infty, -4] \cup [0, 1]$

11. ←███|———|██|→
   −4   0 1

$(-\infty, -3) \cup (-2, 2) \cup (3, \infty)$

13. ←██|—|███|—|██→
  −3 −2   2 3

$(-7, 2)$

15. ←|███████████|→
  −7   2

$(-1, \infty)$

17. ←———|███████→
    −1

$(-\infty, -1] \cup (4, \infty)$

19. ←███|———|███→
   −1   4

$\left(-\infty, 2\right) \cup \left(\frac{11}{4}, \infty\right)$

21. ←███|—|█████→
    2 $\frac{11}{4}$

$(0, 2] \cup [3, \infty)$

23. ←——|██|—|███→
   0 2 3

$(-\infty, -7) \cup (8, \infty)$

25. ←███|———|███→
   −7 8

$\left[-\frac{5}{4}, \frac{3}{2}\right]$

27. ←————|███|————→
    $-\frac{5}{4}$ $\frac{3}{2}$

$(-\infty, 0) \cup (1, \infty)$

29. ←███|———|███→
    0 1

$(-\infty, -4] \cup [4, 6]$

31. ←███|————|██|→
   −4   4 6

$\left(-\infty, -\frac{2}{3}\right] \cup \left[\frac{3}{2}, \infty\right)$

33. ←███|——|███→
    $-\frac{2}{3}$ $\frac{3}{2}$

$(-\infty, -5] \cup [-1, 1] \cup [5, \infty)$

35. ←██|——|██|——|██→
   −5 −1 1 5

$\left(-\infty, -\frac{5}{3}\right) \cup \left(\frac{7}{2}, \infty\right)$

37. ←███|———|███→
    $-\frac{5}{3}$ $\frac{7}{2}$

$(0, 10)$

39. ←————|███|————→
    0   10

$(-\infty, -4) \cup [5, \infty)$

41. ←███|——————|███→
   −4   5

$(-\infty, -6] \cup (-1, 0] \cup (7, \infty)$

43. ←██|———|██|——|██→
  −6 −1 0   7

$(-\infty, 1) \cup (2, \infty)$

45. ←███|———|███→
    1   2

$(-\infty, -8] \cup (-4, \infty)$

47. ←██|————|███→
   −8 −4

$(-\infty, 0) \cup \left(5, \frac{11}{2}\right]$

49. ←███|——|██|→
    0 5 $\frac{11}{2}$

$(0, \infty)$

51. ←————|███████→
      0

53. The number is any number less than -1 or between 0 and 1.

57. $\{-6, 2\}$ **59.** $\{-5 - \sqrt{26}, -5 + \sqrt{26}\}$

Exercise Set 8.5

1. $(-4, -9)$ **3.** $(5, 30)$ **5.** $(1, -2)$ **7.** $\left(\frac{1}{2}, \frac{5}{4}\right)$

9.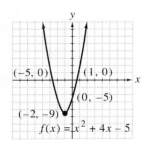

11.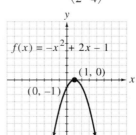

13.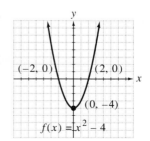

15.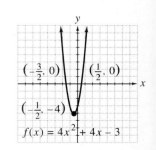

17.

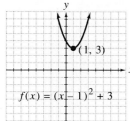

(1, 3)

$f(x) = (x - 1)^2 + 3$

19.

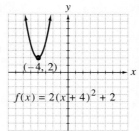

(−4, 2)

$f(x) = 2(x + 4)^2 + 2$

21.

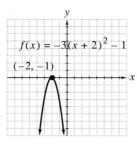

$f(x) = -3(x + 2)^2 - 1$

(−2, −1)

23.

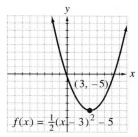

(3, −5)

$f(x) = \frac{1}{2}(x - 3)^2 - 5$

25.

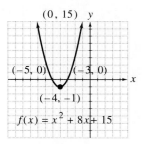

(0, 15)

(−5, 0)   (−3, 0)

(−4, −1)

$f(x) = x^2 + 8x + 15$

27.

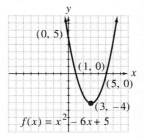

(0, 5)

(1, 0)

(5, 0)

(3, −4)

$f(x) = x^2 - 6x + 5$

29.

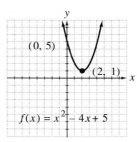

(0, 5)

(2, 1)

$f(x) = x^2 - 4x + 5$

31.

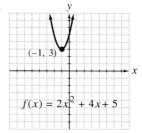

(−1, 3)

$f(x) = 2x^2 + 4x + 5$

33.

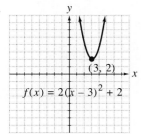

$f(x) = -2x^2 + 12x$

(3, 18)

35. 30 and 30

37.

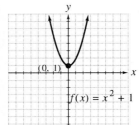

(0, 2)

$f(x) = x^2 + 2$

39.

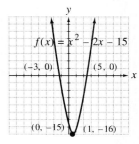

(−1, 4)

$f(x) = (x + 1)^2 + 4$

41.

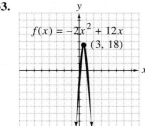

(3, 2)

$f(x) = 2(x - 3)^2 + 2$

43.

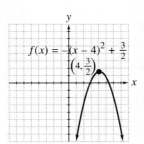

$f(x) = -(x - 4)^2 + \frac{3}{2}$

$\left(4, \frac{3}{2}\right)$

45.

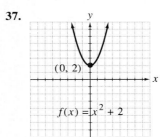

(0, 1)

$f(x) = x^2 + 1$

47.

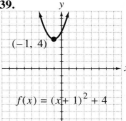

$f(x) = x^2 - 2x - 15$

(−3, 0)   (5, 0)

(0, −15) (1, −16)

49.

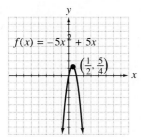

$f(x) = -5x^2 + 5x$

$\left(\frac{1}{2}, \frac{5}{4}\right)$

51.

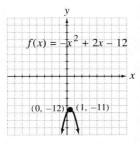

$f(x) = -x^2 + 2x - 12$

(0, −12) (1, −11)

53.

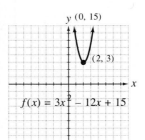

(0, 15)

(2, 3)

$f(x) = 3x^2 - 12x + 15$

55.

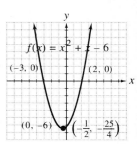

$f(x) = x^2 + x - 6$

(−3, 0)   (2, 0)

(0, −6) $\left(-\frac{1}{2}, -\frac{25}{4}\right)$

57.

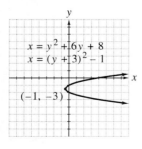

$f(x) = -2x^2 - 3x + 35$

59. 144 ft   **61.** (a) 200 bicycles; (b) \$12,000   **63.**

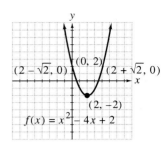

65.

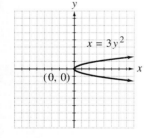

67.

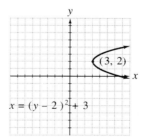

71. 10   **73.** 7   **75.** not a function

Mental Math, Sec. 8.6
1. upward   **3.** to the left   **5.** downward

Exercise Set 8.6

1.

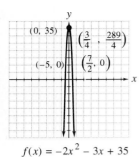

3.

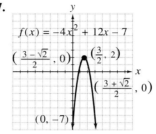

5.

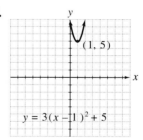

7.

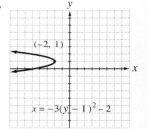

9.

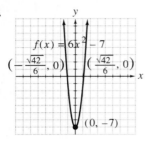

11.

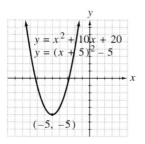

13.

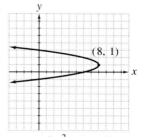

$x = -2y^2 + 4y + 6$
$x = -2(y - 1)^2 + 8$

15.

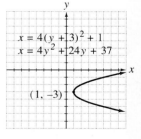

17.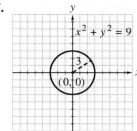
$x^2 + y^2 = 9$
3
(0, 0)

19.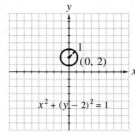
1
(0, 2)
$x^2 + (y - 2)^2 = 1$

21.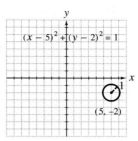
$(x - 5)^2 + (y - 2)^2 = 1$
1
(5, -2)

23. $(x - 2)^2 + (y - 3)^2 = 36$ **25.** $x^2 + y^2 = 4$ **27.** $(x + 5)^2 + (y - 4)^2 = 5$

29.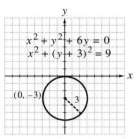
$x^2 + y^2 + 6y = 0$
$x^2 + (y + 3)^2 = 9$
(0, -3)
3

31.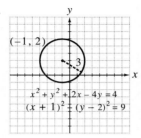
(-1, 2)
3
$x^2 + y^2 + 2x - 4y = 4$
$(x + 1)^2 + (y - 2)^2 = 9$

33.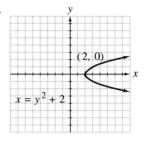
(2, 0)
$x = y^2 + 2$

35.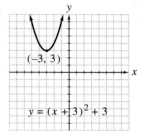
(-3, 3)
$y = (x + 3)^2 + 3$

37.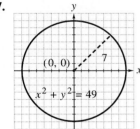
(0, 0)
7
$x^2 + y^2 = 49$

39.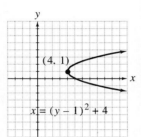
(4, 1)
$x = (y - 1)^2 + 4$

41.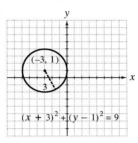
(-3, 1)
3
$(x + 3)^2 + (y - 1)^2 = 9$

43.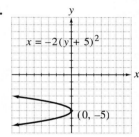
$x = -2(y + 5)^2$
(0, -5)

45.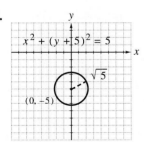
$x^2 + (y + 5)^2 = 5$
$\sqrt{5}$
(0, -5)

47.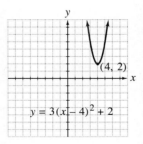
(4, 2)
$y = 3(x - 4)^2 + 2$

49.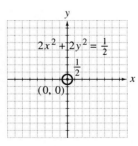
$2x^2 + 2y^2 = \frac{1}{2}$
$\frac{1}{2}$
(0, 0)

51.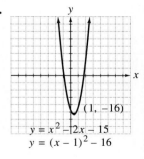
(1, -16)
$y = x^2 - 2x - 15$
$y = (x - 1)^2 - 16$

53.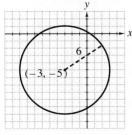
6
(-3, -5)
$x^2 + y^2 + 6x + 10y - 2 = 0$
$(x + 3)^2 + (y + 5)^2 = 36$

55.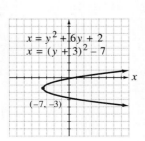
$x = y^2 + 6y + 2$
$x = (y + 3)^2 - 7$
(-7, -3)

57.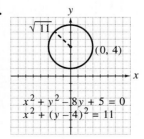
$\sqrt{11}$
(0, 4)
$x^2 + y^2 - 8y + 5 = 0$
$x^2 + (y - 4)^2 = 11$

59.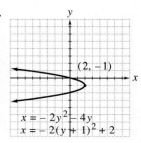
(2, -1)
$x = -2y^2 - 4y$
$x = -2(y + 1)^2 + 2$

61.

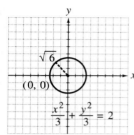

$$\frac{x^2}{3} + \frac{y^2}{3} = 2$$

63.

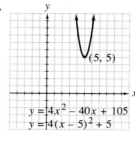

$$y = 4x^2 - 40x + 105$$
$$y = 4(x - 5)^2 + 5$$

65.

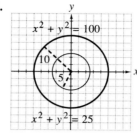

$$x^2 + y^2 = 100$$
$$x^2 + y^2 = 25$$

67.

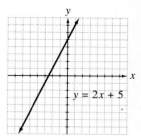

$$y = 2x + 5$$

69.

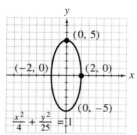

$$y = 3$$

71. $\dfrac{\sqrt{3}}{3}$ **73.** $\dfrac{2\sqrt{42}}{3}$

Exercise Set 8.7

1.

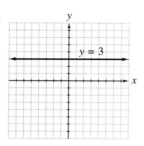

$$\frac{x^2}{4} + \frac{y^2}{25} = 1$$

3.

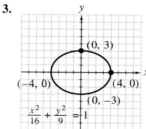

$$\frac{x^2}{16} + \frac{y^2}{9} = 1$$

5.

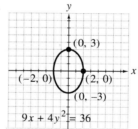

$$9x + 4y^2 = 36$$

7.

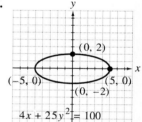

$$4x + 25y^2 = 100$$

9.

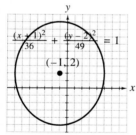

$$\frac{(x+1)^2}{36} + \frac{(y-2)^2}{49} = 1$$

11.

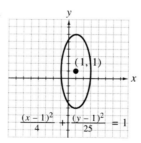

$$\frac{(x-1)^2}{4} + \frac{(y-1)^2}{25} = 1$$

13.

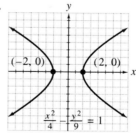

$$\frac{x^2}{4} - \frac{y^2}{9} = 1$$

15.

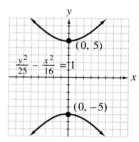

$$\frac{y^2}{25} - \frac{x^2}{16} = 1$$

17.

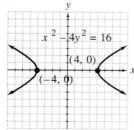

$$x^2 - 4y^2 = 16$$

19.

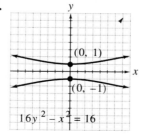

$$16y^2 - x^2 = 16$$

21. circle;

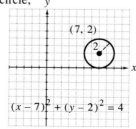

$$(x - 7)^2 + (y - 2)^2 = 4$$

23. parabola;

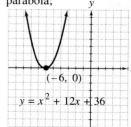

$$y = x^2 + 12x + 36$$

25. hyperbola;

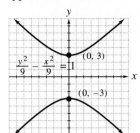

$$\frac{y^2}{9} - \frac{x^2}{9} = 1$$

(0, 3)
(0, −3)

27. ellipse;

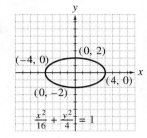

(−4, 0) (0, 2) (4, 0) (0, −2)

$$\frac{x^2}{16} + \frac{y^2}{4} = 1$$

29. parabola;

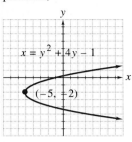

$$x = y^2 + 4y - 1$$

(−5, −2)

31. hyperbola;

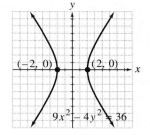

(−2, 0) (2, 0)

$$9x^2 - 4y^2 = 36$$

33. ellipse;

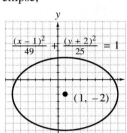

$$\frac{(x-1)^2}{49} + \frac{(y+2)^2}{25} = 1$$

(1, −2)

35. circle;

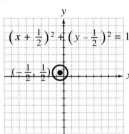

$$\left(x + \tfrac{1}{2}\right)^2 + \left(y - \tfrac{1}{2}\right)^2 = 1$$

$$\left(-\tfrac{1}{2}, \tfrac{1}{2}\right)$$

37. $\dfrac{x^2}{1.69 \times 10^{16}} + \dfrac{y^2}{1.5625 \times 10^{16}} = 1$

39.

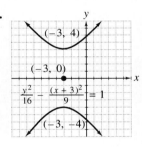

$$\frac{(x-1)^2}{4} + \frac{(y+1)^2}{25} = 1$$

(3, −1)
(−1, −1) (1, −1)

41.

(−3, 4)
(−3, 0)
(−3, −4)

$$\frac{y^2}{16} - \frac{(x+3)^2}{9} = 1$$

43.

$$\frac{(x+5)^2}{16} + \frac{(y+2)^2}{25} = 1$$

(−9, −2) (−1, −2)
(−5, −2)

45. $(-\infty, 1)$

47. $[2, \infty)$

49. $-8x^5$

51. $-4x^2$

Chapter 8 Review

1. $\{14, -1\}$ **3.** $\left\{\dfrac{4}{5}, -\dfrac{1}{2}\right\}$ **5.** $\{-7, 7\}$ **7.** $\left\{-\dfrac{4}{9}, \dfrac{2}{9}\right\}$ **9.** $\left\{\dfrac{-3 - \sqrt{5}}{2}, \dfrac{-3 + \sqrt{5}}{2}\right\}$

11. $\left\{\dfrac{-3 - i\sqrt{7}}{8}, \dfrac{-3 + i\sqrt{7}}{8}\right\}$ **13.** two complex solutions   **15.** two real solutions   **17.** $\{8\}$

19. $\{-i\sqrt{11}, i\sqrt{11}\}$ **21.** $\left\{\dfrac{5 - i\sqrt{143}}{12}, \dfrac{5 + i\sqrt{143}}{12}\right\}$ **23.** $\left\{\dfrac{21 - \sqrt{41}}{50}, \dfrac{21 + \sqrt{41}}{50}\right\}$

25. $\left\{3, \dfrac{-3 + 3i\sqrt{3}}{2}, \dfrac{-3 - 3i\sqrt{3}}{2}\right\}$ **27.** $\left\{\dfrac{2}{3}, 5\right\}$ **29.** $\{-5, 5, -2i, 2i\}$

31. $\{1, 125\}$ **33.** $\{-1, 1, -i, i\}$ **35.** 6 and 8   **37.** (a) 20 ft; (b) $\dfrac{15 + \sqrt{321}}{16}$ seconds

39. The integers are 20, 22, and 24.   **41.** The number is -5.   **43.**

(−5, 5)
−5   5

45. $(-\infty, -\frac{5}{4}] \cup [\frac{3}{2}, \infty)$

$-\frac{5}{4}$ $\frac{3}{2}$

47. $(5, \ 6)$

5   6

49. $(-\infty, -6) \cup (-\frac{3}{4}, 0) \cup (5, \infty)$

-6 $-\frac{3}{4}$ 0   5

51. $[-4, \frac{1}{2})$

-4 $\frac{1}{2}$

53. $\{x \mid x \neq -5 \text{ and } x \neq 3\}$

-5 3

55.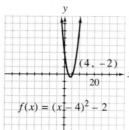

$(4, -2)$

$f(x) = (x - 4)^2 - 2$

57.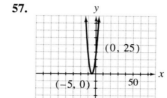

$(0, 25)$

$(-5, 0)$

$f(x) = x^2 + 10x + 25$

59.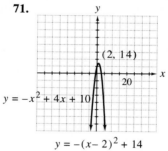

$\left(-\frac{1}{2}, 0\right)$ $\left(\frac{1}{2}, 0\right)$

$(0, -1)$

$f(x) = 4x^2 - 1$

61. The numbers are both 210.   **63.** $(x + 4)^2 + (y - 4)^2 = 9$ **65.** $(x + 7)^2 + (y + 9)^2 = 11$

67.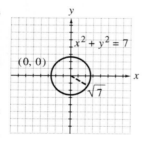

$x^2 + y^2 = 7$

$(0, 0)$

$\sqrt{7}$

69.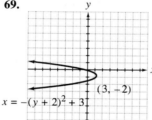

$(3, -2)$

$x = -(y + 2)^2 + 3$

71.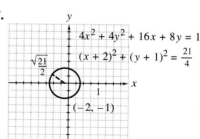

$(2, 14)$

$y = -x^2 + 4x + 10$

$y = -(x - 2)^2 + 14$

73.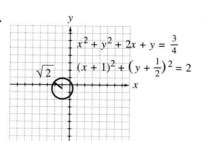

$x = \frac{1}{2} y^2 + 2y + 1$

$x = \frac{1}{2} (y + 2)^2 - 1$

$(-1, -2)$

75.

$x^2 + y^2 + 2x + y = \frac{3}{4}$

$\sqrt{2}$ $(x + 1)^2 + \left(y + \frac{1}{2}\right)^2 = 2$

77.

$4x^2 + 4y^2 + 16x + 8y = 1$

$\frac{\sqrt{21}}{2}$ $(x + 2)^2 + (y + 1)^2 = \frac{21}{4}$

$(-2, -1)$

79.
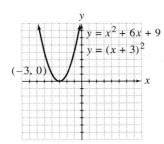
$y = x^2 + 6x + 9$
$y = (x + 3)^2$
$(-3, 0)$

81. $(x - 5.6)^2 + (y + 2.4)^2 = 9.61$

83.
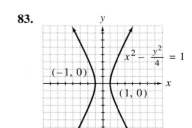
$x^2 - \dfrac{y^2}{4} = 1$
$(-1, 0)$
$(1, 0)$

85.
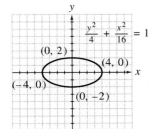
$\dfrac{y^2}{4} + \dfrac{x^2}{16} = 1$
$(0, 2)$
$(4, 0)$
$(-4, 0)$
$(0, -2)$

87.
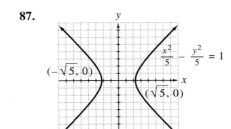
$\dfrac{x^2}{5} - \dfrac{y^2}{5} = 1$
$(-\sqrt{5}, 0)$
$(\sqrt{5}, 0)$

89.
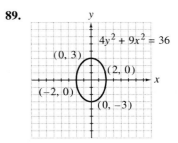
$4y^2 + 9x^2 = 36$
$(0, 3)$
$(2, 0)$
$(-2, 0)$
$(0, -3)$

91.
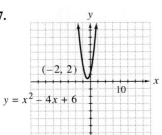
$\dfrac{(x + 3)^2}{9} + \dfrac{(y - 4)^2}{25} = 1$
$(-3, 4)$
10

93.
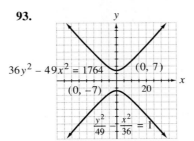
$36y^2 - 49x^2 = 1764$
$(0, 7)$
$(0, -7)$
20
$\dfrac{y^2}{49} - \dfrac{x^2}{36} = 1$

95.
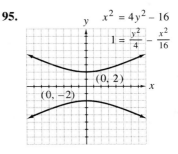
$x^2 = 4y^2 - 16$
$1 = \dfrac{y^2}{4} - \dfrac{x^2}{16}$
$(0, 2)$
$(0, -2)$

97.
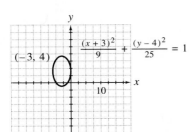
$(-2, 2)$
10
$y = x^2 - 4x + 6$

99.
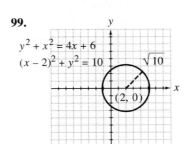
$y^2 + x^2 = 4x + 6$
$(x - 2)^2 + y^2 = 10$
$\sqrt{10}$
$(2, 0)$

101.
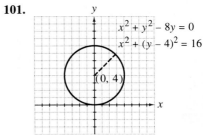
$x^2 + y^2 - 8y = 0$
$x^2 + (y - 4)^2 = 16$
$(0, 4)$

103.
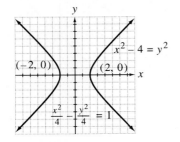
$x^2 - 4 = y^2$
$(-2, 0)$
$(2, 0)$
$\dfrac{x^2}{4} - \dfrac{y^2}{4} = 1$

105.
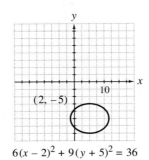
10
$(2, -5)$
$6(x - 2)^2 + 9(y + 5)^2 = 36$

107.
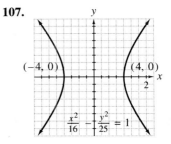
$(-4, 0)$
$(4, 0)$
2
$\dfrac{x^2}{16} - \dfrac{y^2}{25} = 1$

Chapter 8 Test

1. $\left\{\dfrac{7}{5}, -1\right\}$ **2.** $\left\{-2, 1 - i\sqrt{3}, 1 + i\sqrt{3}\right\}$ **3.** $\left\{\dfrac{1 + i\sqrt{31}}{2}, \dfrac{1 - i\sqrt{31}}{2}\right\}$ **4.** $\{3 - \sqrt{7}, 3 + \sqrt{7}\}$

5. $\left\{-\dfrac{1}{7}, -1\right\}$ **6.** $\left\{\dfrac{3 + \sqrt{29}}{2}, \dfrac{3 - \sqrt{29}}{2}\right\}$ **7.** $\{-2 - \sqrt{11}, -2 + \sqrt{11}\}$ **8.** $\{-3, 3, -i, i\}$

9. $\{-1, 1, -i, i\}$ **10.** $\{6, 7\}$ **11.** $\{3 - \sqrt{7}, 3 + \sqrt{7}\}$ **12.** $\left\{\dfrac{2 - i\sqrt{6}}{2}, \dfrac{2 + i\sqrt{6}}{2}\right\}$

13. $\left[-\dfrac{7}{6}, \dfrac{12}{7}\right]$

14. $(-\infty, -5) \cup (-4, 4) \cup (5, \infty)$

15. $[-\infty, -4) \cup \left[\dfrac{11}{2}, \infty\right)$

16. $(-\infty, -3) \cup [2, 3)$

17.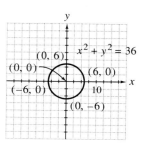

18.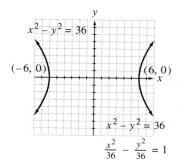

19.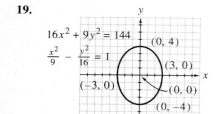

20.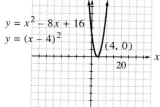

21.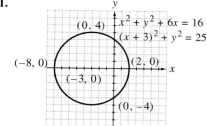

22.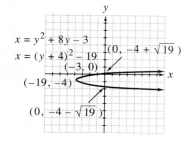

23.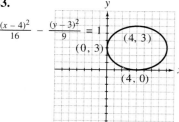

24.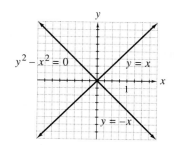

25. (a) 256 ft; (b) 4 seconds   **26.** 41, 43, and 45

Chapter 8 Cumulative Review

1. *(Sec. 2.4, Ex. 1)* $\{2, -2\}$ **2.** *(Sec. 2.5, Ex. 4)* $\left[\dfrac{5}{2}, \infty\right)$ **3.** *(Sec. 3.3, Ex. 7)* $m = -5$

4. *(Sec. 4.1, Ex. 4)* (a) 1; (b) -1; (c) 1; (d) 2   **5.** *(Sec. 4.1, Ex. 7)* (a) $\dfrac{1}{x^{11}}$; (b) p^7; (c) $\dfrac{1}{4}$; (d) $\dfrac{y^7}{5x^8}$; (e) $\dfrac{3}{x^7}$

6. *(Sec. 4.3, Ex. 8)* $6x^3y^2 - 4x^2y^2 - 2y^2$ **7.** *(Sec. 4.4, Ex. 10)* $4a^2 + 4ab + b^2 + 12a + 6b + 9$

8. *(Sec. 4.7, Ex. 3)* (a) $3x - 4$; (b) $-x + 2$; (c) $2x^2 - 5x + 3$; (d) $\dfrac{x - 1}{2x - 3}$ **9.** *(Sec. 5.1, Ex. 4)* $17x^3y^2(1 - 2x)$

10. *(Sec. 5.1, Ex. 9)* $(m^2 - 2)(n^2 + 1)$ **11.** *(Sec. 5.2, Ex. 10)* $(5x^2 - 6)(x^2 + 7)$

12. *(Sec. 5.3, Ex. 3)* (a) $(x + 3)(x - 3)$; (b) $(4y + 3)(4y - 3)$; (c) $2(5 - 2y)(5 + 2y)$

13. *(Sec. 5.5, Ex. 5)* $\{0, 2, -2\}$ **14.** *(Sec. 5.6, Ex. 4)* x-intercepts, $-2, 0, 2$

15. *(Sec. 6.1, Ex. 4)* (a) $x^2 - 2x + 4$; (b) $\dfrac{2}{y - 5}$

16. *(Sec. 6.2, Ex. 4)* $\dfrac{2(4x^2 - 10x + 25)}{x^2 + 1}$

17. *(Sec. 6.3, Ex. 3)* (a) $\dfrac{6x + 5}{3x^3y}$; (b) $\dfrac{5z^2 - 2z}{(z + 2)(z - 2)}$; (c) $\dfrac{5k^2 - 7k + 4}{(k + 2)(k - 2)(k - 1)}$

18. *(Sec. 6.5, Ex. 3)* $\{\ \}$

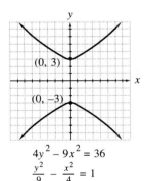

19. *(Sec. 7.2, Ex. 1)* (a) 5; (b) -2; (c) 0; (d) not a real number; (e) $\dfrac{1}{6}$; (f) $-\dfrac{1}{2}$; (g) x^3; (h) x^5

20. *(Sec. 7.4, Ex. 3)* (a) $\dfrac{10 - 6\sqrt{2}}{7}$; (b) $\dfrac{\sqrt{30} + 3\sqrt{2} + 2\sqrt{5} + 2\sqrt{3}}{2}$; (c) $\dfrac{7\sqrt{3xy}}{6x}$; (d) $\dfrac{6\sqrt{xm} - 2m}{9x - m}$

21. *(Sec. 8.2, Ex. 1)* $\left\{-\dfrac{1}{3}, -5\right\}$ **22.** *(Sec. 8.4, Ex. 1)* $(-\infty, -3] \cup [3, \infty)$

23. *(Sec. 8.6, Ex. 5)*   **24.** *(Sec. 8.7, Ex. 5)*

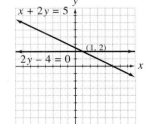

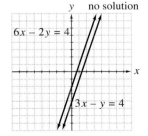

CHAPTER 9
Systems of Equations

Exercise Set 9.1

1.

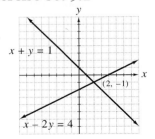

3.

5.

7. $\{(2, 8)\}$ **9.** $\{(0, -9)\}$ **11.** $\{(1, -1)\}$ **13.** $\{(-5, 3)\}$ **15.** $\left\{\left(\dfrac{5}{2}, \dfrac{5}{4}\right)\right\}$ **17.** $\{(1, -2)\}$ **19.** $\{(9, 9)\}$

21. $\{(7, 2)\}$ **23.** $\{\ \}$ **25.** $\{(x, y)\,|\,3x + y = 1\}$ **27.** $\left\{\left(\dfrac{3}{2}, 1\right)\right\}$ **29.** $\{(2, -1)\}$ **31.** $\{(-5, 3)\}$

33. $\{(x, y)\,|\,3x + 9y = 12\}$ **35.** $\{\ \}$ **37.** $\left\{\left(\dfrac{1}{2}, \dfrac{1}{5}\right)\right\}$ **39.** $\{(8, 2)\}$ **41.** $\{(x, y)\,|\,x = 3y + 2\}$ **43.** $\left\{\left(-\dfrac{1}{4}, \dfrac{1}{2}\right)\right\}$

45. $\{(3, 2)\}$ **47.** $\{(7, -3)\}$ **49.** $\{\ \}$ **51.** $\{(3, 4)\}$ **53.** $\{(-2, 1)\}$ **55.** 15 and 30   **57.** $\left\{\left(\dfrac{1}{4}, 8\right)\right\}$

59. $\left\{\left(\dfrac{1}{3}, \dfrac{1}{2}\right)\right\}$ **61.** $\left\{\left(\dfrac{1}{4}, -\dfrac{1}{3}\right)\right\}$ **63.** $\{\ \}$ **65.**

67.

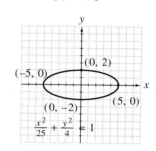

69. $2i$ **71.** $2i\sqrt{5}$

Exercise Set 9.2
1. $\{(-2, 5, 1)\}$ **3.** $\{(-2, 3, -1)\}$ **5.** $\{(x, y, z)\,|\,x - 2y + z = -5\}$ **7.** $\{\ \}$ **9.** $\{(0, 0, 0)\}$
11. $\{(-3, -35, -7)\}$ **13.** $\{(6, 22, -20)\}$ **15.** $\{\ \}$ **17.** $\{(3, 2, 2)\}$ **19.** $\{(x, y, z)\,|\,x + 2y - 3z = 4\}$
21. $\{(-3, -4, -5)\}$ **23.** $\{(12, 6, 4)\}$ **25.** $\{(1, 1, 0, 2)\}$ **27.** $\{(1, -1, 2, 3)\}$ **29.** $\{-2 - \sqrt{3}, -2 + \sqrt{3}\}$
31. $\left\{\dfrac{1 - i\sqrt{23}}{6}, \dfrac{1 + i\sqrt{23}}{6}\right\}$ **33.** $\dfrac{x}{5 - 3x}$

Exercise Set 9.3
1. 10 and 8   **3.** plane, 520 mph; wind, 40 mph   **5.** 20 quarts of 4%; 40 quarts of 1%   **7.** length, 52 ft; width, 26 ft
9. 9 large frames; 13 small frames   **11.** -10 and -8 **13.** tablets, \$0.80; pens, \$0.20
15. plane, 630 mph; wind, 90 mph   **17.** 5 in., 7 in., 7 in., and 10 in.   **19.** 24 nickels; 15 dimes   **21.** 18, 13, and 9
23. \$2000 in sales   **25.** 22 \$10 bills; 63 \$20 bills   **27.** 26 dimes, 13 nickels, and 17 pennies
29. Two units of mix A, 3 units of mix B, and 1 unit of mix C   **31.** $(x + 3)(x + y)$ **33.** $x^2 + 10x + 25$
35. $4x^2 - 4xy + y^2$

Exercise Set 9.4
1. 26   **3.** -19 **5.** 0   **7.** $\{(1, 2)\}$ **9.** $\{(x, y)\,|\,3x + y = 1\}$ **11.** $\{(9, 9)\}$ **13.** 8   **15.** 0   **17.** 54

19. $\{(-2, 0, 5)\}$ **21.** $\{(6, -2, 4)\}$ **23.** 16   **25.** 15   **27.** $\dfrac{13}{6}$ **29.** 0   **31.** 56   **33.** $\{(-3, -2)\}$

35. $\{\ \}$ **37.** $\{(-2, 3, -1)\}$ **39.** $\{(3, 4)\}$ **41.** $\{(-2, 1)\}$ **43.** $\{(x, y, z)\,|\,x - 2y + z = -3\}$
45. $\{(0, 2, -1)\}$ **47.** -125 **49.** 24   **53.** $3y + 8z = 18$ **55.** $9 + 5i$ **57.** $-77 + 36i$

Exercise Set 9.5
1. $\{(2, -1)\}$ **3.** $\{(-4, 2)\}$ **5.** $\{(-2, 5, -2)\}$ **7.** $\{(1, -2, 3)\}$ **9.** $\{\ \}$ **11.** $\{(x, y)\,|\,x - y = 3\}$
13. $\{(4, -3)\}$ **15.** $\{(2, 1, -1)\}$ **17.** $\{(9, 9)\}$ **19.** $\{\ \}$ **21.** $\{\ \}$ **23.** $\{(1, -4, 3)\}$ **25.** 6   **27.** 4
29. width, 5 km; length, 9 km

Exercise Set 9.6
1. $\{(3, -4), (-3, 4)\}$ **3.** $\{(\sqrt{2}, \sqrt{2}), (-\sqrt{2}, -\sqrt{2})\}$ **5.** $\{(4, 0), (0, -2)\}$

7. $\{(-\sqrt{5}, -2), (-\sqrt{5}, 2), (\sqrt{5}, -2), (\sqrt{5}, 2)\}$ **9.** $\{\ \}$ **11.** $\{(1, -2), (3, 6)\}$

13. $\{(2, 4), (-5, 25)\}$ **15.** $\{\ \}$ **17.** $\{(1, -3)\}$ **19.** $\{(-1, -2), (-1, 2), (1, -2), (1, 2)\}$

21. $\{(0, -1)\}$ **23.** $\{(-1, 3), (1, 3)\}$ **25.** $\{(-\sqrt{3}, 0), (\sqrt{3}, 0)\}$ **27.** $\{\ \}$ **29.** $\{(-6, 0), (6, 0), (0, -6)\}$

31. 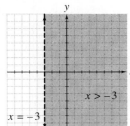 **33.** 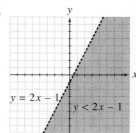 **35.** $(8x - 25)$ in. **37.** $(4x^2 + 6x + 2)$ meters

Exercise Set 9.7

1. 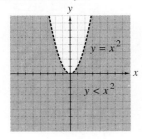 **3.** 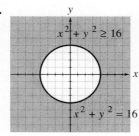 **5.** 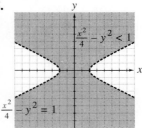 **7.**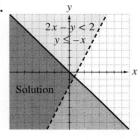

9. 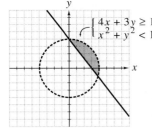 **11.** 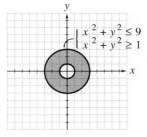 **13.** 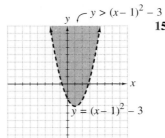 **15.**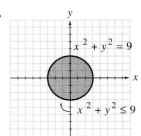

17. 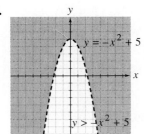 **19.** 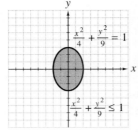 **21.** 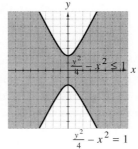 **23.**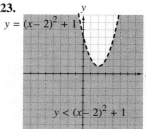

25.

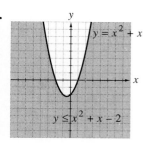

$y = x^2 + x - 2$

$y \le x^2 + x - 2$

27.

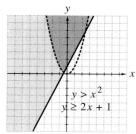

$y > x^2$

$y \ge 2x + 1$

29.

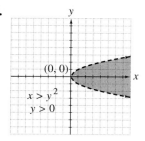

$(0, 0)$

$x > y^2$

$y > 0$

31.

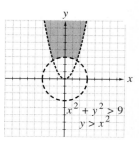

$x^2 + y^2 \ge 9$

$y > x^2$

33.

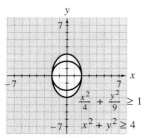

$\frac{x^2}{4} + \frac{y^2}{9} \ge 1$

$x^2 + y^2 \ge 4$

35.

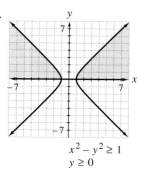

$x^2 - y^2 \ge 1$

$y \ge 0$

37.

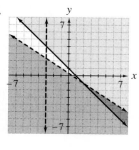

39.

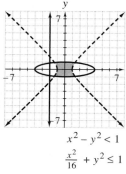

$x^2 - y^2 < 1$

$\frac{x^2}{16} + y^2 \le 1$

$x \ge -2$

41. $y = \dfrac{4x}{3}$ **43.** $y = \dfrac{a - c}{2}$ **45.** $(-\infty, -2) \cup (1, \infty)$ **47.** $(0, 3)$

Chapter 9 Review

1. $\{(-3, 1)\}$;

3. $\{\ \}$;

5. $\left\{3, \dfrac{8}{3}\right\}$;

7. $\{(2, 0, -3)\}$

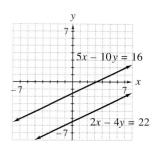

$3x + 10y = 1$

$x + 2y = -1$

$5x - 10y = 16$

$2x - 4y = 22$

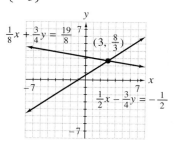

$\frac{1}{8}x + \frac{3}{4}y = \frac{19}{8}$

$\left(3, \dfrac{8}{3}\right)$

$\frac{1}{2}x - \frac{3}{4}y = -\frac{1}{2}$

9. $\{(-1, 2, 0)\}$ **11.** $\{(x, y, z) \mid x + 2y + 3z = 11\}$ **13.** $\{(3, 1, 1)\}$ **15.** 24 dimes, 23 nickels, and 48 pennies
17. Sue is 17 years old and Pat is 1 year old.   **19.** width, 37 ft; length, 111 ft
21. 30 lbs of creme-filled; 5 lbs of chocolate-covered nuts; 10 lbs of chocolate-covered raisins.
23. larger investment, 9.5%; smaller investment, 7.5%   **25.** 120, 115, and 60   **27.** 17   **29.** -72
31. $\left\{\left(\dfrac{1}{3}, \dfrac{7}{6}\right)\right\}$ **33.** $\left\{\left(0, \dfrac{2}{3}\right)\right\}$ **35.** $\{(x, y) \mid x - 2y = 4\}$ **37.** $\{(2, 0, -3)\}$ **39.** $\left\{\left(\dfrac{3}{7}, -2, -\dfrac{1}{7}\right)\right\}$
41. $\{(-1, 2, 0)\}$ **43.** $\{(x, y) \mid x - 2y = 4\}$ **45.** $\left\{\left(\dfrac{1}{3}, \dfrac{7}{6}\right)\right\}$ **47.** $\{(-7, -15)\}$ **49.** $\{(2, 1)\}$ **51.** $\{(2, 0, -3)\}$
53. $\{(-1, 2, 0)\}$ **55.** $\{\ \}$ **57.** $\{\ \}$ **59.** $\{(5, 1), (-1, 7)\}$ **61.** $\{(0, 2), (0, -2)\}$ **63.** $\left\{\left(2, \dfrac{5}{2}\right), (-7, -20)\right\}$

65. $\{(-2, -1), (-2, 1), (2, -1), (2, 1)\}$ **67.**

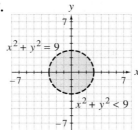

69.

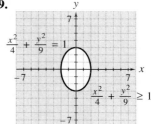

71.

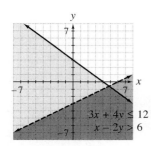

73.

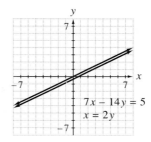

75.

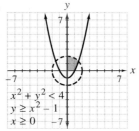

Chapter 9 Test

1. 34   **2.** -6 **3.** $\{(1, 3)\}$ **4.** $\{\ \}$

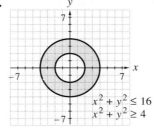

5. $\{(2, -3)\}$ **6.** $\{(x, y)\,|\,10x + 4y = 10\}$ **7.** $\{(-1, -2, 4)\}$ **8.** $\{\ \}$ **9.** $\{(2, -1)\}$ **10.** $\{3, 6\}$
11. $\{(3, -1, 2)\}$ **12.** $\{(5, 0, -4)\}$ **13.** $\{(x, y)\,|\,x - y = -2\}$ **14.** $\{(5, -3)\}$ **15.** $\{(-1, -1, 0)\}$ **16.** $\{\ \}$
17. $\{(-12, 5), (12, -5)\}$ **18.** $\{(-5, -1), (-5, 1), (5, -1), (5, 1)\}$ **19.** $\{(6, 12), (1, 2)\}$ **20.** $\{(1, 1), (-1, -1)\}$

21.

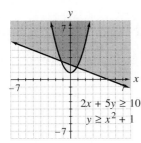

22.

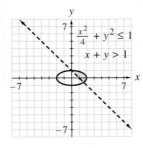

23.

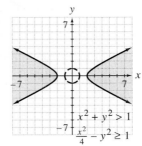

24.

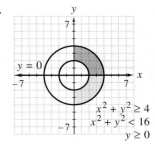

25. Dean jogs 7 mph, Tom bicycles 17 mph, and Dean jogged 7/6 miles.   **26.** 58

Chapter 9 Cumulative Review

1. *(Sec. 2.6, Ex. 2)* $(-3, 2)$ **2.** *(Sec. 3.3, Ex. 2)* $-\dfrac{13}{8}$ **3.** *(Sec. 3.4, Ex. 9)* $y = -2x - 5$

4. *(Sec. 4.1, Ex. 10)* (a) 770,000,000; (b) 0.001025   **5.** *(Sec. 4.2, Ex. 4)*(a) $\dfrac{z^2}{9x^4y^{20}}$; (b) $\dfrac{27a^4x^6}{2}$

6. *(Sec. 4.4, Ex. 1)* (a) $10x^9$; (b) $-7xy^{15}z^9$ **7.** *(Sec. 4.5, Ex. 3)* $\dfrac{7a}{2b} - 1$

8. *(Sec. 5.2, Ex. 7)* $2xy\,(3x - 4)(2x - 1)$ **9.** *(Sec. 5.5, Ex. 6)* $\{-1, 1, -5\}$ **10.** *(Sec. 5.5, Ex. 8)* -7 and -9

11. *(Sec. 6.3, Ex. 4)* (a) $\dfrac{-4x + 15}{x - 3}$; (b) $\dfrac{4}{x - y}$ **12.** *(Sec. 6.5, Ex. 4)* $\left\{\dfrac{8}{15}\right\}$

13. *(Sec. 7.1, Ex. 6)* (a) $z - z^{17/3}$; (b) $x^{2/3} - 3x^{1/3} - 10$ **14.** *(Sec. 7.4, Ex. 2)* (a) $\dfrac{3\sqrt{15}}{5}$; (b) $\dfrac{8\sqrt{x}}{3x}$; (c) $\dfrac{\sqrt[3]{4}}{2}$

15. *(Sec. 8.1, Ex. 5)* $\left\{\dfrac{7 - \sqrt{53}}{2}, \dfrac{7 + \sqrt{53}}{2}\right\}$ **16.** *(Sec. 8.3, Ex. 1)* $\left\{\dfrac{-1 + \sqrt{33}}{4}, \dfrac{-1 - \sqrt{33}}{4}\right\}$

17. *(Sec. 8.4, Ex. 3)* $(-\infty, -2] \cup [1, 5]$ **18.** *(Sec. 8.6, Ex. 8)* $x^2 + y^2 + 4x - 8y = 16$

19. *(Sec. 9.2, Ex. 1)* $\{(-4, 2, -1)\}$
20. *(Sec. 9.4, Ex. 3)* (a) -2; (b) -2
21. *(Sec. 9.5, Ex. 3)* $\{\ \}$
22. *(Sec. 9.6, Ex. 3)*
$\{(-2, \sqrt{3}), (-2, -\sqrt{3}), (2, \sqrt{3})(2, -\sqrt{3})\}$

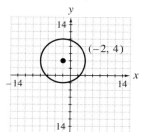

CHAPTER 10
Exponential and Logarithmic Functions

Exercise Set 10.1
1. one-to-one   **3.** one-to-one   **5.** one-to-one   **7.** one-to-one   **9.** not one-to-one   **11.** one-to-one
13. not one-to-one

15. $f^{-1}(x) = x - 4$ **17.** $f^{-1}(x) = \dfrac{x + 3}{2}$ **19.** $f^{-1}(x) = \dfrac{3x + 4}{12}$

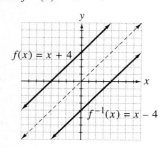

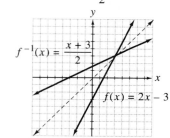

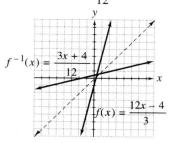

21. not one-to-one   **23.** not one-to-one   **25.** one-to-one; $f^{-1} = \{(-8, -4), (-12, -6), (-16, -8), (-18, -9)\}$

27. not a function   **29.** one-to-one   **31.** not a function   **33.** $f^{-1}(x) = \dfrac{x-1}{3}$

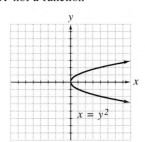

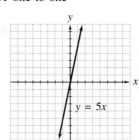

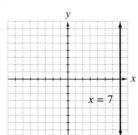

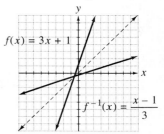

35. $f^{-1}(x) = 5x + 2$ **37.** $g^{-1}(x) = 2x + 8$ **41.** 1   **43.** 3   **45.** $\sqrt{41}$ miles

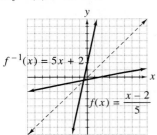

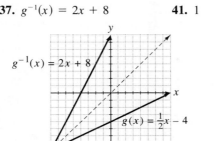

Exercise Set 10.2

1. 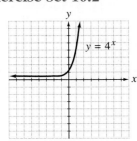   **3.** 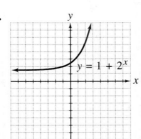   **5.** 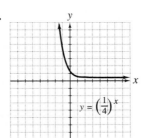   **7.**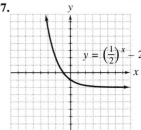

9. 3   **11.** $\dfrac{3}{4}$ **13.** $\dfrac{8}{5}$ **15.** $-\dfrac{2}{3}$ **17.** $\sim$ 24.6 pounds   **19.** $\sim$\$7621.42

21. 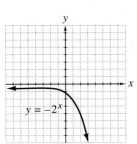   **23.** 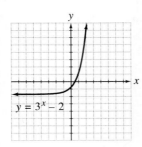   **25.** 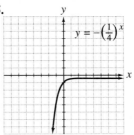   **27.**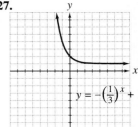

29. 4   **31.** $\dfrac{3}{2}$ **33.** $-\dfrac{1}{3}$ **35.** -2 **37.** $y = 3^x$ **39.** $y = \left(\dfrac{1}{2}\right)^x$ **41.** $\sim$519 rats   **43.** $\sim$22.5%

45. $\sim$ 9,060,000   **47.** $\sim$ \$4065.59   **51.** {4}   **53.** { }   **55.** {2, 3}

Exercise Set 10.3

1. 3   **3.** -2 **5.** $\dfrac{1}{2}$ **7.** -1 **9.** {2}   **11.** {81}   **13.** {7}   **15.** {-3}   **17.** 3   **19.** 3   **21.** 1

23. 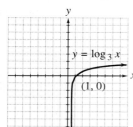   **25.** 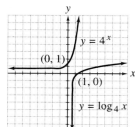   **27.** 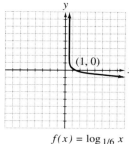   **29.**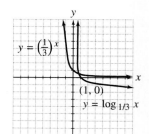

31. 0   **33.** 4   **35.** 2   **37.** 5   **39.** 4   **41.** -3 **43.** {-3}   **45.** {2}   **47.** {2}   **49.** $\left\{\dfrac{27}{64}\right\}$

51. {10}   **53.** 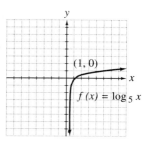   **55.** 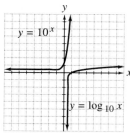   **57.** 0.0827   **61.** -1 **63.** $x-5$

65. 3   **67.** $\dfrac{y-9}{y^2-1}$

Exercise Set 10.4

1. $\log_5 14$ **3.** $\log_4 (9x)$ **5.** $\log_{10} (10x^2 + 20)$ **7.** $\log_5 3$ **9.** $\log_2\left(\dfrac{x}{y}\right)$ **11.** 1   **13.** $\log_2 25$

15. $\log_5 (x^3 z^6)$ **17.** $\log_{10}\left(\dfrac{x^3 - 2x}{x+1}\right)$ **19.** $\log_3 4 + \log_3 y - \log_3 5$ **21.** $3\log_2 x - \log_2 y$

23. $\dfrac{1}{2}\log_b 7 + \dfrac{1}{2}\log_b x$ **25.** 0.2   **27.** 1.2   **29.** 0.23   **31.** $\log_4 35$ **33.** $\log_3 4$ **35.** $\log_7\left(\dfrac{9}{2}\right)$ **37.** $\log_4 48$

39. $\log_2\left[\dfrac{x^{7/2}}{(x+1)^2}\right]$ **41.** $\log_8 (x^{16/3})$ **43.** $\log_7 5 + \log_7 x - \log_7 4$ **45.** $3\log_5 x + \log_5 (x+1)$

47. $2\log_6 x - \log_6 (x+3)$ **49.** 1.29   **51.** -0.68 **53.** -0.125 **57.**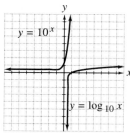
59. {14}   **61.** { }

Exercise Set 10.5

1. 0.9031   **3.** 0.3636   **5.** 2   **7.** -3 **9.** $10^{1.3} \approx 19.9526$ **11.** $\dfrac{10^{1.1}}{2} \approx 6.2946$ **13.** 0.6931   **15.** -2.6367

17. 2   **19.** $\dfrac{1}{4}$ **21.** $e^{1.4} \approx 4.0552$ **23.** $\dfrac{4 + e^{2.3}}{3} \approx 4.6581$ **25.** 4.2   **27.** \$2542.50   **29.** 1.5850

31. -2.3219 **33.** 1.1004   **35.** 1.6094   **37.** 1.6180   **39.** 3   **41.** 2   **43.** -4 **45.** $\dfrac{1}{2}$

47. $10^{2.3} \approx 199.5262$ **49.** $e^{-2.3} \approx 0.1003$ **51.** $\dfrac{10^{-0.5} - 1}{2} \approx -0.3419$ **53.** $\dfrac{e^{0.18}}{4} \approx 0.2993$ **55.** 1.5850

57. -1.6309 **59.** 0.8617   **61.** 5.3   **63.** \$3656.38   **65.** 13 weeks   **67.** 21 weeks   **71.** $\left\{\dfrac{4}{7}\right\}$

73. $x = \dfrac{3y}{4}$ **75.** $\{(2, -3)\}$

Exercise Set 10.6

1. $\left\{\dfrac{\log 6}{\log 3}\right\}$; $\{1.6309\}$ **3.** $\left\{\dfrac{\log 3.8}{2\log 3}\right\}$; $\{0.6076\}$ **5.** $\left\{3 + \dfrac{\log 5}{\log 2}\right\}$; $\{5.3219\}$ **7.** $\{11\}$ **9.** $\{9, -9\}$

11. $\left\{\dfrac{1}{2}\right\}$ **13.** $\left\{\dfrac{3}{4}\right\}$ **15.** $\{2\}$ **17.** $\left\{\dfrac{1}{8}\right\}$ **19.** 103 wolves   **21.** 8.8 years   **23.** $\left\{\dfrac{\log 5}{\log 9}\right\}$; $\{0.7325\}$

25. $\left\{\dfrac{\log 3}{\log 4} - 7\right\}$; $\{-6.2075\}$ **27.** $\left\{\dfrac{1}{3}\left(4 + \dfrac{\log 11}{\log 7}\right)\right\}$; $\{1.7441\}$ **29.** $\left\{\dfrac{\ln 5}{6}\right\}$; $\{0.2682\}$ **31.** $\{11\}$ **33.** $\{4, -1\}$

35. $\left\{\dfrac{9}{5}\right\}$ **37.** $\{64\}$ **39.** $\left\{\dfrac{-5 + \sqrt{33}}{2}\right\}$ **41.** $\left\{\dfrac{192}{127}\right\}$ **43.** $\left\{\dfrac{2}{3}\right\}$ **45.** 8,850,000 inhabitants.   **47.** 10 years

49. $1\dfrac{3}{4}$ years   **51.** 56 inches   **53.** 11.9 lb/in.²   **55.** 3.2 miles   **57.** $-\dfrac{5}{3}$ **59.** $\dfrac{17}{4}$ **61.** $f^{-1}(x) = \dfrac{x - 2}{5}$

Chapter 10 Review

1. one-to-one; $h^{-1} = \{(14, -9), (8, 6), (12, -11), (15, 15)\}$ **3.** not one-to-one   **5.** not one-to-one

7. $f^{-1}(x) = \dfrac{x - 11}{6}$ **9.** $q^{-1}(x) = \dfrac{x - b}{m}$ **11.** $r^{-1}(x) = \dfrac{2(x + 4)}{13}$

13. 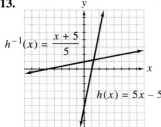   **15.** $\{-2\}$ **17.** $\left\{\dfrac{3}{2}\right\}$ **19.** $\left\{\dfrac{8}{9}\right\}$ **21.** 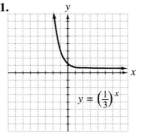   **23.**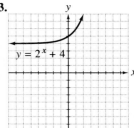

25. \$1131.82   **27.** $\log_2\left(\dfrac{1}{16}\right) = -4$ **29.** $0.4^3 = 0.064$ **31.** $\{9\}$ **33.** $\{3\}$ **35.** $\{3\}$ **37.** $\{-2\}$ **39.** $\{9\}$

41. $\{2\}$ **43.** $\{-8, 1\}$ **45.** 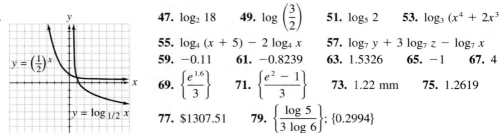   **47.** $\log_2 18$ **49.** $\log\left(\dfrac{3}{2}\right)$ **51.** $\log_5 2$ **53.** $\log_3 (x^4 + 2x^3)$

55. $\log_4 (x + 5) - 2\log_4 x$ **57.** $\log_7 y + 3\log_7 z - \log_7 x$

59. -0.11 **61.** -0.8239 **63.** 1.5326   **65.** -1 **67.** 4

69. $\left\{\dfrac{e^{1.6}}{3}\right\}$ **71.** $\left\{\dfrac{e^2 - 1}{3}\right\}$ **73.** 1.22 mm   **75.** 1.2619

77. \$1307.51   **79.** $\left\{\dfrac{\log 5}{3 \log 6}\right\}$; $\{0.2994\}$

81. $\left\{\dfrac{1}{3}\left(\dfrac{\log 9}{\log 4} - 2\right)\right\}$; $\{-0.1383\}$ **83.** $\left\{\dfrac{1}{4}\left(\dfrac{\log 3}{\log 8} + 2\right)\right\}$; $\{0.6321\}$ **85.** $\left\{\dfrac{\log \frac{2}{3}}{\log 4} - 5\right\}$; $\{-5.2925\}$ **87.** $\left\{\dfrac{9}{10}\right\}$

89. $\left\{\dfrac{3e^2}{e^2 - 3}\right\}$ **91.** $\{\ \}$ **93.** 224,310,000   **95.** 21 years   **97.** 33 years   **99.** 8.6 years

Chapter 10 Test

1.

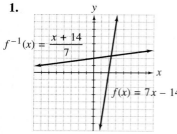

$f^{-1}(x) = \dfrac{x + 14}{7}$

$f(x) = 7x - 14$

2. one-to-one   **3.** not one-to-one   **4.** $f^{-1}(x) = \dfrac{-x + 6}{2}$

5. $f^{-1} = \{(0, 0), (3, 2), (5, -1\}$ **6.** $\log_3 24$ **7.** $\log_5\left(\dfrac{x^4}{x + 1}\right)$

8. $\log_6 2 + \log_6 x - 3\log_6 y$ **9.** -1.53 **10.** 1.0686 **11.** $\{-1\}$

12. $\left\{\dfrac{1}{2}\left(\dfrac{\log 4}{\log 3} - 5\right)\right\}$; $\{-1.8691\}$ **13.** $\left\{\dfrac{1}{9}\right\}$ **14.** $\left\{\dfrac{1}{2}\right\}$ **15.** $\{22\}$ **16.** $\left\{\dfrac{25}{3}\right\}$

17. $\left\{\dfrac{43}{21}\right\}$ **18.** $\{-1.0979\}$ **19.**

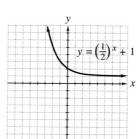

$y = \left(\dfrac{1}{2}\right)^x + 1$

20.

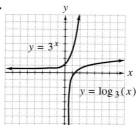

$y = 3^x$

$y = \log_3(x)$

21. \$5234.58   **22.** 6 years

23. 64,913 prairie dogs

24. 15 years   **25.** 1.2%

Chapter 10 Cumulative Review

1. *(Sec. 1.4, Ex. 3)* (a) 6; (b) -7; (c) -11; (d) 5; (e) -3 **2.** *(Sec. 2.6, Ex. 4)* $\left(\infty, \dfrac{13}{5}\right] \cup [4, \infty)$

3. *(Sec. 3.2, Ex. 2)*

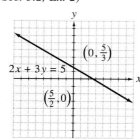

$\left(0, \dfrac{5}{3}\right)$

$2x + 3y = 5$

$\left(\dfrac{5}{2}, 0\right)$

4. *(Sec. 3.3, Ex. 4)* 0   **5.** *(Sec. 4.2, Ex. 1)* (a) x^{35}; (b) 64; (c) $\dfrac{1}{25}$; (d) y^{12}

6. *(Sec. 4.3, Ex. 6)* $12x^3 - 12x^2 - 9x + 2$

7. *(Sec. 4.4, Ex. 8)* (a) $x^2 + 10x + 25$; (b) $x^2 - 18x + 81$; (c) $9x^2 + 12xz + 4z^2$; (d) $16m^4 - 24m^2n + 9n^2$

8. *(Sec. 5.2, Ex. 2)* $(x - 5)(x - 7)$ **9.** *(Sec. 5.3, Ex. 8)* $q^2(5 - n)(25 + 5n + n^2)$

10. *(Sec. 6.2, Ex. 2)* (a) $-\dfrac{4x^2}{2x + 1}$; (b) $-(a + b)^2$

11. *(Sec. 6.4, Ex. 2)* (a) $\dfrac{x(x - 2)}{2(x + 2)}$; (b) $\dfrac{x}{y}$

12. *(Sec. 7.1, Ex. 4)* (a) $\dfrac{1}{8}$; (b) $\dfrac{1}{9}$

13. *(Sec. 7.2, Ex. 7)* (a) $\dfrac{\sqrt{5}}{7}$; (b) $-\dfrac{2}{3}$; (c) $\dfrac{6x^2}{5y}$; (d) $\dfrac{2x\sqrt[3]{2}}{z^2}$ **14.** *(Sec. 7.7, Ex. 3)* (a) $13 - 18i$; (b) $3 - 4i$; (c) 58

15. *(Sec. 8.2, Ex. 4)* $\left\{\dfrac{-1 - i\sqrt{35}}{6}, \dfrac{-1 + i\sqrt{35}}{6}i\right\}$

16. *(Sec. 8.5, Ex. 2)*   **17.** *(Sec. 8.6, Ex. 6)*   **18.** *(Sec. 9.2, Ex. 2)* { }

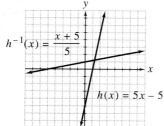

$h^{-1}(x) = \dfrac{x + 5}{5}$

$h(x) = 5x - 5$

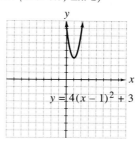

$y = 4(x - 1)^2 + 3$

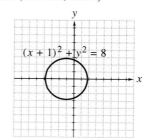

$(x + 1)^2 + y^2 = 8$

19. *(Sec. 9.7, Ex. 2)*

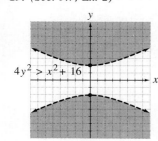

$4y^2 > x^2 + 16$

20. *(Sec. 10.1, Ex. 3)*

$$f^{-1}(x) = \frac{x + 5}{3}$$

21. *(Sec. 10.4, Ex. 3)* (a) $\log_5 72$;

(b) $\log_9 \left[x^2 (x + 1) \right]$

22. *(Sec. 10.6, Ex. 1)* $\{1.7712\}$

CHAPTER 11
Sequences, Series, and the Binomial Theorem

Exercise Set 11.1

1. 5, 6, 7, 8, 9   **3.** $-1, 1, -1, 1, -1$ **5.** $\dfrac{1}{4}, \dfrac{1}{5}, \dfrac{1}{6}, \dfrac{1}{7}, \dfrac{1}{8}$ **7.** 75   **9.** 118   **11.** $\dfrac{6}{5}$ **13.** $a_n = 4n - 1$

15. $a_n = -2^n$ **17.** $a_n = \dfrac{1}{3^n}$ **19.** 48 ft, 80 ft, and 112 ft   **21.** $a_n = 0.10(2)^{n-1}$, \$819.20   **23.** 2, 4, 6, 8, 10

25. $-1, -4, -9, -16, -25$ **27.** 2, 4, 8, 16, 32   **29.** 7, 9, 11, 13, 15   **31.** $-1, 4, -9, 16, -25$

33. 729   **35.** $\dfrac{4}{7}$ **37.** $\dfrac{1}{8}$ **39.** -95 **41.** $-\dfrac{1}{25}$ **43.** 2400 cases; 75 cases originally   **45.** 50; extinct in 1996

Exercise Set 11.2
1. 4, 6, 8, 10, 12   **3.** 6, 4, 2, 0, -2 **5.** 1, 3, 9, 27, 81   **7.** 48, 24, 12, 6, 3   **9.** 33   **11.** -875 **13.** -60

15. 96   **17.** -28 **19.** 1250   **21.** 31   **23.** 20   **25.** $a_1 = \dfrac{2}{3}$ and $r = -2$ **27.** $a_n = 4n + 50$; 130 seats

29. $a_n = 6(3)^{n-1} = 2(3)^n$ **31.** $a_1 = 2; d = 2$ **33.** $a_1 = 5; r = 2$ **35.** $a_1 = \dfrac{1}{2}; r = \dfrac{1}{5}$ **37.** $a_1 = x; r = 5$

39. $a_1 = p; d = 4$ **41.** 19   **43.** $-\dfrac{8}{9}$ **45.** $\dfrac{17}{2}$ **47.** $\dfrac{8}{81}$ **49.** -19

51. 486, 162, 54, 18, 6, 2; $a_n = \dfrac{486}{3^{n-1}}$; seven bounces   **53.** $a_n = 125n + 875$; \$2250   **55.** 25 grams

Exercise Set 11.3

1. -2 **3.** 60   **5.** 20   **7.** $\dfrac{73}{168}$ **9.** $\displaystyle\sum_{i=1}^{5} (2i - 1)$ **11.** $\displaystyle\sum_{i=1}^{4} 4(3)^{i-1}$ **13.** $\displaystyle\sum_{i=1}^{6} (-3i + 15)$ **15.** -20

17. 30   **19.** 1, 2, 3, $\cdots$, 10; 55 trees   **21.** $a_n = 6(2)^{n-1}$; 96 units   **23.** $\dfrac{11}{36}$ **25.** 60   **27.** 74   **29.** 62

31. $\dfrac{241}{35}$ **33.** $\displaystyle\sum_{i=1}^{4} \dfrac{4}{3^{i-2}}$ **35.** $\displaystyle\sum_{i=1}^{7} i^2$ **37.** -24 **39.** 0   **41.** 82

43. $a_n = 50(2)^n$; n represents the number of 12 hr periods; 800 bacteria

45. 30 opossums; 68 opossums   **47.** 6.25 lb; 93.75 lb   **49.** 16.4 in.; 134.5 in.

Exercise Set 11.4
1. 36   **3.** 63   **5.** 55   **7.** 16   **9.** 4000, 3950, 3900, 3850, 3800; 3450 cars; 44,700 cars

11. Firm A (Firm A, $265,000; Firm B, $254,000)   **13.** 484   **15.** 2.496   **17.** $39,930; $139,230

19. 20 min; 123 min   **21.** 24   **23.** $\frac{1}{9}$ **25.** 180 ft   **27.** 185   **29.** $\frac{381}{64}$ or 5.95   **31.** $-\frac{33}{4}$ or -8.25

33. $-\frac{75}{2}$ **35.** $\frac{56}{9}$ **37.** -20 **39.** $\frac{16}{9}$ **41.** $\frac{4}{9}$ **43.** player A, 45 points; player B, 75 points

45. $3050   **47.** $10,737,418.23   **49.** $\frac{8}{10} + \frac{8}{100} + \frac{8}{1000} + \cdots ; \frac{8}{9}$

Exercise Set 11.5

1. $m^3 + 3m^2n + 3mn^2 + n^3$ **3.** $c^5 + 5c^4d + 10c^3d^2 + 10c^2d^3 + 5cd^4 + d^5$
5. $y^5 - 5y^4x + 10y^3x^2 - 10y^2x^3 + 5yx^4 - x^5$ **7.** 8   **9.** 42   **11.** 360   **13.** 56
15. $a^7 + 7a^6b + 21a^5b^2 + 35a^4b^3 + 35a^3b^4 + 21a^2b^5 + 7ab^6 + b^7$
17. $a^5 + 10a^4b + 40a^3b^2 + 80a^2b^3 + 80ab^4 + 32b^5$ **19.** $32a^5 - 80a^4b + 80a^3b^2 - 40a^2b^3 + 10ab^4 - b^5$
21. $c^6 - 12c^5d + 60c^4d^2 - 160c^3d^3 + 240c^2d^4 - 192cd^5 + 64d^6$ **23.** $6x^2y^2$ **25.** $30a^9b$
27. $q^9 + 9q^8r + 36q^7r^2 + 84q^6r^3 + 126q^5r^4 + 126q^4r^5 + 84q^3r^6 + 36q^2r^7 + 9qr^8 + r^9$
29. $1024a^5 + 1280a^4b + 640a^3b^2 + 160a^2b^3 + 20ab^4 + b^5$ **31.** $625a^4 - 1000a^3b + 600a^2b^2 - 160ab^3 + 16b^4$
33. $8a^3 + 36a^2b + 54ab^2 + 27b^3$ **35.** $x^5 + 10x^4 + 40x^3 + 80x^2 + 80x + 32$ **37.** $5cd^4$ **39.** d^7
41. $-40r^2s^3$

Chapter 11 Review

1. $-3, -12, -27, -48, -75$ **3.** $\frac{1}{100}$ **5.** $a_n = \frac{1}{6n}$ **7.** 144 ft, 176 ft, 208 ft

9. 450, 1350, 4050, 12,150, 36,450; 36,450 infected people in 1992   **11.** $-2, -\frac{4}{3}, -\frac{8}{9}, -\frac{16}{27}, -\frac{32}{81}$

13. 111   **15.** -83 **17.** $a_1 = 3; d = 5$ **19.** $a_n = \frac{3}{10^n}$ **21.** $a_1 = \frac{8}{3}, r = \frac{3}{2}$ **23.** $a_1 = 7x, r = -2$
25. 8, 6, 4.5, 3.4, 2.5, 1.9; good   **27.** $a_n = 2^{n-1}$, 512, $536,870,912$ **29.** $a_n = 150n + 750$; $1650/month
31. $1 + 3 + 5 + 7 + 9 = 25$ **33.** $\frac{1}{4} - \frac{1}{6} + \frac{1}{8} = \frac{5}{24}$ **35.** -4 **37.** -10 **39.** $\sum_{i=1}^{6} 3^{i-1}$ **41.** $\sum_{i=1}^{4} \frac{1}{4^i}$
43. $a_n = 20(2)^n$; n represents the number of 8 hr periods; 1280 yeast   **45.** Job A, $23,900; Job B, $23,800
47. 150   **49.** 900   **51.** -410 **53.** 936   **55.** 10   **57.** -25 **59.** $30,418; $99,868   **61.** $58; $553
63. 2696 mosquitoes   **65.** $\frac{5}{9}$ **67.** $x^5 + 5x^4z + 10x^3z^2 + 10x^2z^3 + 5xz^4 + z^5$
69. $16x^4 + 32x^3y + 24x^2y^2 + 8xy^3 + y^4$
71. $b^8 + 8b^7c + 28b^6c^2 + 56b^5c^3 + 70b^4c^4 + 56b^3c^5 + 28b^2c^6 + 8bc^7 + c^8$
73. $256m^4 - 256m^3n + 96m^2n^2 - 16mn^3 + n^4$ **75.** $35a^4b^3$

Chapter 11 Test

1. $-\frac{1}{5}, \frac{1}{6}, -\frac{1}{7}, \frac{1}{8}, -\frac{1}{9}$ **2.** $-3, 3, -3, 3, -3$ **3.** 247   **4.** 39,999   **5.** $a_n = \frac{2}{5}\left(\frac{1}{5}\right)^{n-1}$

6. $a_n = (-1)^n 9n$ **7.** 155   **8.** -330 **9.** $\frac{144}{5}$ **10.** 1   **11.** 10   **12.** -60

13. $a^6 - 6a^5b + 15a^4b^2 - 20a^3b^3 + 15a^2b^4 - 6ab^5 + b^6$ **14.** $32x^5 + 80x^4y + 80x^3y^2 + 40x^2y^3 + 10xy^4 + y^5$
15. $y^8 + 8y^7z + 28y^6z^2 + 56y^5z^3 + 70y^4z^4 + 56y^3z^5 + 28y^2z^6 + 8yz^7 + z^8$
16. $128p^7 + 448p^6r + 672p^5r^2 + 560p^4r^3 + 280p^3r^4 + 84p^2r^5 + 14pr^6 + r^7$ **17.** 925 people; 250 people initially
18. $1 + 3 + 5 + 7 + 9 + 11 + 13 + 15$; 64 shrubs   **19.** 33.75 cm; 218.75 cm   **20.** 320 cm   **21.** 304 ft; 1600 ft
22. $\frac{14}{33}$

Chapter 11 Cumulative Review

1. *(Sec. 2.2, Ex. 5)* $b = \dfrac{2A - bh}{h}$ **2.** *(Sec. 2.7, Ex. 6)* $(-\infty, \infty)$ **3.** *(Sec. 3.4, Ex. 4)* $3x - 12y = 8$

4. *(Sec. 4.2, Ex. 3)* (a) $\dfrac{y^6}{4}$; (b) x^9; (c) $\dfrac{49}{4}$; (d) $\dfrac{y^{16}}{25x^5}$ **5.** *(Sec. 4.4, Ex. 9)* (a) $x^2 - 9$; (b) $16y^2 - 1$; (c) $x^4 - 4y^2$

6. *(Sec. 5.1, Ex. 7)* $(7x - 1)(x^2 + 5y)$ **7.** *(Sec. 5.3, Ex. 4)* (a) $(p^2 + 4)(p + 2)(p - 2)$; (b) $(x + 9)(x - 3)$

8. *(Sec. 6.3, Ex. 5)* $\dfrac{7x^2 - 9x - 13}{(2x + 1)(x - 5)(3x - 2)}$ **9.** *(Sec. 7.1, Ex. 1)* (a) 2; (b) 4; (c) 3; (d) 0; (e) -3

10. *(Sec. 7.4, Ex. 1)* (a) $5\sqrt{3} + 3\sqrt{10}$; (b) $\sqrt{35} + \sqrt{5} - \sqrt{42} - \sqrt{6}$; (c) $21x - 7\sqrt{5x} + 15\sqrt{x} - 5\sqrt{5}$; (d) $49 - 8\sqrt{3}$;
(e) $2x - 25$

11. *(Sec. 7.5, Ex. 5)* $\left\{\dfrac{2}{9}\right\}$ **12.** *(Sec. 8.4, Ex. 2)* $(-\infty, 0] \cup [4, \infty)$

13. *(Sec. 8.5, Ex. 4)*

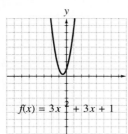

$f(x) = 3x^2 + 3x + 1$

14. *(Sec. 8.7, Ex. 3)*

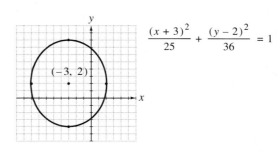

$\dfrac{(x + 3)^2}{25} + \dfrac{(y - 2)^2}{36} = 1$

$(-3, 2)$

15. *(Sec. 9.2, Ex. 3)* $\left\{\left(\dfrac{1}{2}, 0, \dfrac{3}{4}\right)\right\}$ **16.** *(Sec. 9.6, Ex. 2)* $\{(2, \sqrt{2})\}$ **17.** *(Sec. 10.3, Ex. 3)* (a) 2; (b) -1; (c) 3; (d) 6

18. *(Sec. 10.6, Ex. 3)* $\{2\}$ **19.** *(Sec. 11.1, Ex. 4)* 5, 6, 7, 8, 9; the puppy should gain 9 lb in its fifth month.
20. *(Sec. 11.5, Ex. 4)* $x^5 + 10x^4y + 40x^3y^2 + 80x^2y^3 + 80xy^4 + 32y^5$

Index